From Biomass to Biobased Products

Edited by Eduardo Jacob-Lopes,
Leila Queiroz Zepka
and Rosangela Rodrigues Dias

Published in London, United Kingdom

From Biomass to Biobased Products
http://dx.doi.org/10.5772/intechopen.107581
Edited by Eduardo Jacob-Lopes, Leila Queiroz Zepka and Rosangela Rodrigues Dias

Contributors
A. Venkata Ramayya, Abdulbari A. Ahmad, Abdulraqeeb Alwahbi, Adewale Elijah Fadeyi, Aleksandra M. Rozhkova, Ana Cristna de Oliveira Macedo, Antonio Donizeti de Oliveira, Arkady P. Sinitsyn, Aseeya Wahid, Asfaw Gezae Daful, Benito Buendía Quispe, Carlos Yesid Varela Bernal, Celia Kirsch, Daniela Eusébio, Danyang Ying, Derese T. T. Nega, Eduardo Jacob-Lopes, Ekaterina A. Rubtsova, Fabian Giebner, Hani Dammag, Henok Mekonnen Mossissa, Hisashi Miyafuji, Ishka Bless, Jaqueline Torres de Souza, Jordan Pennells, Jose Antonio Mayoral Chavando, Joseph Akpan, João Graciano Mendonca Filho, João Sousa Cardoso, Julio Javier Armijo Carranza, Laila A. Al Khateeb, Lalita Pal, Leila Queiroz Zepka, Luiz Guilherme Costa dos Santos, Luís A. C. Tarelho, Marie Loridon, Mattaparthi Lakshmi Durga, Meegalla R. Chandraratne, Million M. Afessa, Milton Cézar da Silva, Moses Jeremiah Barasa Kabeyi, Muhammad Ahsan Amjed, Natascha Eggers, Noelia del Valle Franco Rondón, Olga G. Korotkova, Oludolapo Akanni Olanrewaju, Pablo Juliano, Raymundo Erazo Erazo, Roberto Robles Calderón, Rosangela Rodrigues Dias, Saheed Olatunbosun Akiode, Sammy N. Aso, Tais Freitas, Takashi Hosoya, Thiago dos Santos Barbosa, Torsten Birth, Valter Silva, Yuki Hirano

First published in London, United Kingdom, 2024 by IntechOpen
IntechOpen is the global imprint of INTECHOPEN LIMITED, registered in England and Wales, registration number: 11086078, 167-169 Great Portland Street, London, W1W 5PF, United Kingdom

British Library Cataloguing-in-Publication Data
A catalogue record for this book is available from the British Library

Additional hard and PDF copies can be obtained from orders@intechopen.com

From Biomass to Biobased Products
Edited by Eduardo Jacob-Lopes, Leila Queiroz Zepka and Rosangela Rodrigues Dias
p. cm.
Print ISBN 978-1-83769-002-2
Online ISBN 978-1-83769-003-9
eBook (PDF) ISBN 978-1-83769-004-6

For EU product safety concerns:
IN TECH d.o.o., Prolaz Marije Krucifikse Kozulić 3, 51000 Rijeka, Croatia,
info@intechopen.com or visit our website at intechopen.com.

Meet the editor

Eduardo Jacob-Lopes is currently an associate professor in the Department of Food Technology and Science, Federal University of Santa Maria (UFSM), Brazil. He has more than 15 years of teaching and research experience. He has coordinated/coordinates more than fifty research projects and/or technological development financed by public funding agencies and private initiatives. He has published more than 650 scientific publications/communications, including 20 books, 70 book chapters, 150 original research papers, and 400 research communications at national and international conferences. He has also registered twelve patents. He is a member of the editorial board of ten journals and acts as a reviewer for several national and international journals. His research interest includes bioprocess engineering with an emphasis on microalgal biotechnology.

Leila Queiroz Zepka is currently an associate professor in the Department of Food Technology and Science, Federal University of Santa Maria (UFSM), Brazil. She has more than 15 years of teaching and research experience. She has published more than 450 scientific publications/communications, including 18 books, 61 book chapters, 120 original research papers, 256 research communications in national and international conferences, and 12 patents. She is a member of the editorial board of five journals and acts as a reviewer for several national and international journals. Her research interest includes microalgal biotechnology with an emphasis on microalgae-based products.

Rosangela Rodrigues Dias is a researcher at the Department of Food Technology and Science, Federal University of Santa Maria (UFSM), Brazil. Currently, she holds a junior scientist position at the Bioprocess Intensification Group (UFSM). She has published more than sixty scientific publications/communications, including one book, eighteen book chapters, fifteen original research papers, and twenty-seven research communications at national and international conferences. She acts as a reviewer for international journals. Her research interest includes microalgae biotechnology with an emphasis on microalgae-based processes and products.

Contents

Preface

Realizing that our current economic system could lead to the collapse of civilization as we know it, the emphasis is now on an economy that is positive for people and the planet. The aim is to turn the tide and bring nature back to the center of our economy. The change from the linear economic system to circular models with bio-resources as its true engine is the pivot to build a resilient and sustainable future. Therefore, in this book, we explore the use of biomass as a source of a multitude of sustainable biological products. This book consists of 18 chapters written by renowned researchers from every part of the world. Each chapter provides an overview of basic information on the subject in question. We hope that the chapters in this work will benefit students, researchers, and professionals engaged in biomass-based technology and inspire research ideas and future directions.

Eduardo Jacob-Lopes, Leila Queiroz Zepka and Rosangela Rodrigues Dias
Food Science and Technology Department,
Federal University of Santa Maria, UFSM,
Santa Maria, RS, Brazil

Chapter 1

Introductory Chapter: From Biomass to Bio-Based Products

Rosangela Rodrigues Dias, Leila Queiroz Zepka and Eduardo Jacob-Lopes

1. Introduction

Fossil resources have served the global economy well, creating wealth and human well-being unprecedented in recent decades. However, the issue is that there is no future for an economy powered by the consumption of nonrenewable fossils. It is time for change to build an economy that works for people and the planet [1, 2].

Leaders in all disciplines are tracing a course to meet the needs of an ever-increasing population and stem the uncontrolled climate change that threatens to lead to the collapse of civilization as we know it [3]. That threat hangs like a sword over our heads. Therefore, every option is important, every choice is important, and every action matters. Moving toward the transition to a bio-based economy is part of this puzzle. Furthermore, for various reasons, the transition to a bio-economy can bring the planetary boundaries back into safe operating space [4].

In this vision, with nature back at the heart of the economy, the prognosis is for equitable and resilient growth. In general, bio-based products are more efficient in the use of resources, are by definition renewable and less polluting, and are more likely to heed the appeal to circularity. Industries that produce renewable bio-based products from biomass are creating durable green jobs that pave the way for long-term sustainable development [5].

Furthermore, the bio-based products market is predicted to be a leading market. Demand for renewable and bio-based products will increase in response to market pressure driven by investors, media, regulators, and other stakeholders moving toward sustainability [6]. In particular, the support of governments, introducing regulations and promoting their adoption in different sectors, has been the backbone for incorporating demand. Reducing barriers to introducing biological products in the market and identifying them as one of the priority lines to focus investments and innovation is a sensible approach to meet the diverse demands of today's world [7].

2. Conclusion

The linear economy and overly reliant on fossil fuels is about to reach its physical limitations. The transition to a biocircular economy seems to have gone from an option to an urgent imperative. Thus, against this background, in this book, the chapters presented intend to explore the use of biomass for energy, chemicals and materials, food and feed, pharmaceuticals and personal care, and bioremediation and address key issues including market, legal issues, technological nodes, and disruptive technologies capable of unlocking the black box of biomass-based technology. This book brings together the latest advances for those already working in the field and, simultaneously, provides an overview for those attentive to the promises of biomass as a source of a myriad of specialty products.

Author details

Rosangela Rodrigues Dias, Leila Queiroz Zepka and Eduardo Jacob-Lopes*
Food Science and Technology Department, Federal University of Santa Maria, UFSM, Santa Maria, RS, Brazil

*Address all correspondence to: ejacoblopes@gmail.com

References

[1] Suchek N, Fernandes CI, Kraus S, Filser M, Sjögrén H. Innovation and the circular economy: A systematic literature review. Business Strategy and the Environment. 2021;**30**(8):3686-3702

[2] King LC, Savin I, Drews S. Shades of green growth scepticism among climate policy researchers. Nature Sustainability. 2023;**2023**:1-5

[3] Kocubovski M, Ristovska G, Petrova A, Bukovetz J, Disho K, Tosev TH, et al. Review of policies and actions on tackling global climate change-COP26-is this the last chance? Archives of Public Health. 2023;**15**(1):5-15

[4] Elustondo D, Stocchero A, Gaunt D. A Case for Circular Bio-Based Cities. Available at SSRN 4497088. 2023

[5] Vanhamaki S, Medkova K, Malamakis A, Kontogianni S, Marisova E, Dellago DH, et al. Bio-based circular economy in European national and regional strategies. International Journal of Sustainable Development and Planning. 2019;**14**(1):31-43

[6] Zwicker MV, Brick C, Gruter GJM, van Harreveld F. Consumer attitudes and willingness to pay for novel bio-based products using hypothetical bottle choice. Sustainable Production and Consumption. 2023;**35**:173-183

[7] Salvador R, Barros MV, Donner M, Brito P, Halog A, Antonio C. How to advance regional circular bioeconomy systems? Identifying barriers, challenges, drivers, and opportunities. Sustainable Production and Consumption. 2022;**32**:248-269

Chapter 2

Analysis and Characterization of Lignocellulosic Biomass Extracted from Selected Agricultural Wastes

Adewale Elijah Fadeyi and Saheed Olatunbosun Akiode

Abstract

Agricultural wastes are a viable source of lignocellulosic biomass for biofuel production. Precise biomass characterization is required to assess the new potential lignocellulose biosource. This study assesses the characterization and compositional analysis of three agricultural wastes (melon husk, moringa pod, and mango endocarp) obtained from Sheda Village, Federal Capital Territory of Abuja, Nigeria. Gravimetric method was used to assess the samples' lignocellulose biomass composition and they were further characterized using FTIR. The findings indicated that hemicellulose content ranged from 19.38 to 27.74%, with melon husk having the highest concentration. The highest concentration of cellulose was found in the mango endocarp (45.84%). Melon husk possesses the highest lignin content (29.87%). FTIR spectroscopic examination revealed a broad spectrum around 3422.85 cm^{-1}, which represented bonded -OH groups. A peak observed at around 1637 cm^{-1} is attributed to the stretching of C=C bonds in conjugated carboxylic acids. Peaks, obtained at 1205.72, 1204.50, and 1206.24 cm^{-1}, reveal the vibrations of the aliphatic chains, $-CH_2$- and $-CH_3$, which constitute the fundamental structure of cellulose material. The findings demonstrate that the samples contain a sizable amount of lignocellulosic biomass. Therefore, wastes generated from agricultural wastes may be useful in the production of biofuel.

Keywords: agricultural wastes, bioethanol, biomass, biofuel, lignocelluloses

1. Introduction

The European Union must reduce its overall greenhouse gas emissions by 8% from 1990 levels between 2008 and 2012 [1]. This can be made possible by substituting common fossil fuels with low-carbon fuels for energy generation. Wastes of various origins, such as industrial or agricultural wastes, are usually disposed *via* incineration. This act can be replaced by utilizing them as sources of energy. The cost of landfill, environmental pollution/degradation, and the diminishing space available for landfilling trash, particularly in densely populated places, has become a source of concern lately. This in turn has become major driver for waste-to-energy and it is receiving more and more attention. Due to emerging issues from lack of land for landfills, CO_2 and CH_4 emissions, potential groundwater pollution, and the need for more recycling and reuse of trash, policy and regulation of waste disposal have been a front-burner issue in many nations. As stated in the EU legislation that is now being prepared, landfilling of garbage that can serve as a source of renewable energy will be prohibited. Organic materials derived from plants are commonly referred to biomass. The chemical energy in biomass can be extracted when it undergoes combustion and it is transformed into heat or electricity. Green biomass refers to biomass resources that are sustainably managed as they have become renewable energy sources and do not contribute to global warming. Sustainable management of plants has helped in reabsorbing carbon dioxide released during the combustion of other biomass thus leaving no net carbon dioxide emissions in the atmosphere. Both valuable compounds and energy (biofuel) can be obtained from biomass [2]. Fossil fuels, in addition to increasing the amount of carbon, sulfur, nitrogen, and other air pollutants by thousands of tons, also cause environmental hazards. These adverse effects serve as a pointer to the need to find creative and affordable alternative fuels. A reliable alternative consists of renewable energy sources. Many researchers across the globe have been conducting extensive studies on these sources to examine their uses [3, 4].

Renewable fuels are processed from biomass, which consists of organic waste products obtained from sources such as industry, forestry, and agriculture. Agro-based biomass is mostly biomass obtained from considerable organic wastes that are left in fields after crops have been harvested. Although some of these wastes are used as animal feed, the majority lie on the fields and are usually burnt or left as a nuisance in the environment. Crop wastes, such as wheat and paddy straw, have been utilized as raw materials in biomass-based power plants for the production of electricity. Biomass such as melon seed husk, mango endocarp, and moringa pod are just a few of the abundant agricultural by-products obtainable in an agrarian nation like Nigeria. Worthy of note is that Nigeria ranks as Number 10 among the world's top producers of mangoes [5]. About 35 to 60% of the mango fruit are being wasted after processing [6]. Also, more than a million tons of mango seeds are produced yearly as waste and are currently not adding value to the nation's economy [7]. A typical moringa tree (15–20 feet tall) can produce thousands of seed pods, yielding an incalculable number of moringa seeds [8]. The majority of agricultural wastes lack commercially viable conversion technology as they are in most cases set ablaze on the field. A significance of agricultural wastes is that they serve as one of the alternatives to grains for the generation of ethanol or gasoline without compromising the security of the food supply. Owing to their quantity and renewability, agricultural wastes are a major raw material in the commercial manufacture of bioethanol [9]. The lignocellulose content of agricultural wastes has been acknowledged as a significant and promising source of biofuels and other products of immense value [10]. The effect of global warming can be significantly alleviated by reducing greenhouse gas emissions through the manufacture of biofuels using lignocellulosic materials [11]. The majority of the global biomass supply comes from lignocelluloses. Lignocellulose can be obtained in significant quantities from municipal solid wastes, crop leftovers, forest residues, and specific energy crops [12, 13]. The components of plants known as extractives are those found outside of the cell wall; they are made up of low or medium molecular weight materials and may be readily extracted using particular solvents such as acetone, toluene, alcohol, and water [14–16]. Studies have shown that lignocelluloses are converted to biofuel at a greater rate when cellulose's crystallinity is reduced significantly [17]. Four main steps are involved in the synthesis of biofuel (e.g., ethanol) from lignocellulose biomass: These are pretreatment, hydrolysis, fermentation, and product purification. Monomeric sugars are produced when carbohydrate polymers undergo hydrolysis, and they (monomeric sugars) are then fermented to produce ethanol. To make it simpler for the enzyme to convert carbohydrate polymers into fermentable sugars, the pretreatment step lyses the lignin seal and disrupts the crystalline portion of the cellulose [12]. The amount of ethanol output (L/mg) depends on the precise assessment of the lignocellulose content in the biomass [18]. A major factor that determines how effectively the processes can convert lignocellulosic biomass into ethanol is the composition of the lignocellulose. To the best of the authors' knowledge, little has been done to determine the lignocellulose biomass content of mango endocarp, moringa pods, and melon husks. This study is therefore aimed at characterizing and analyzing the lignocellulose biomass of these selected agricultural wastes in order to determine their morphological, chemical, and compositional characteristics.

2. Materials and method

The materials, mango endocarp, melon husk, and moringa seed pod were sourced for at Kwali Area Council, an agrarian community of the Federal Capital Territory of Nigeria. Each of the biomass materials was subjected to compositional analysis using the gravimetric method reported [19].

2.1 Extractives

About 2.5 g of each of the dried samples was extracted with 150 mL acetone using a Soxhlet extractor setup. After extraction, the samples were air-dried at room temperature. The extracted samples were put in a convection oven at 105°C to achieve a constant weight. The %(w/w) of the extractives content was evaluated as the difference in weight between the raw extractive-laden biomass and extractive-free biomass [20–22].

2.2 Hemicellulose

About 150 mL of 0.5 M NaOH was added to 1 g of extractives-free dried biomass in a 250-ml Erlenmeyer flask. The mixture was boiled for 3.5 h, allowed to cool, filtered, and washed with distilled water until a neutral pH was obtained. The residue was dried to a constant weight at 105°C in a convection oven. The difference between the sample weight before and after this treatment is the hemicellulose content (%w/w) of dry biomass [13, 20–22].

2.3 Lignin

About 3 mL of 72% H_2SO_4 was added to 0.3 g of dried extractive-free samples. The mixture was kept at room temperature for 2 h while carefully shaking at intervals of 30 minutes to allow for initial hydrolysis. The second step of hydrolysis was carried out by adding 84 mL of distilled water and the resulting mixture was placed in an autoclave for 1 h at 121°C. The slurry was then cooled at room temperature. Hydrolysates were filtered through a vacuum using a filtering crucible. The residue is dried at 105°C and then incinerated in a muffle furnace at 575°C. The lignin content was calculated as the summation of acid-insoluble lignin (AIL) and acid-soluble lignin (ASL). The acid-soluble lignin was quantified by UV-visible spectroscopy at 320 nm, and the acid-insoluble lignin was determined by gravimetric analysis [23].

$$\text{Lignin} = ASL + AIL \qquad (1)$$

2.4 Cellulose

The cellulose content (%w/w) was calculated by difference.

2.5 Fourier transform infrared (FT-IR) spectroscopy

Nicolet IS 5 Thermo Fisher Scientific, USA FTIR spectrophotometer was used to analyze the surface chemistry of the agricultural waste samples. The dried waste samples were analyzed by scanning between wavelengths 400 and 4000 cm^{-1}. The distinctive functional groups on the surface of the melon seed shell, moringa seed pod, and mango seed endocarp can be observed from the samples' FTIR spectra.

3. Result and discussion

3.1 Composition of lignocellulose biomass

Figure 1 displays the lignocellulose biomass content of the selected agricultural wastes. Among them, mango endocarp was shown to possess the highest amount of cellulose (45.84%) while hemicellulose quantity was the highest in moringa pod. Generally, the amounts of lignin in the samples do not differ noticeably. As indicated in **Figure 2**, the main components of lignocellulosic biomass feedstock are cellulose, hemicelluloses, lignin, extractives, and ash. The results of the investigation align with findings from previous works, as indicated in **Table 1**.

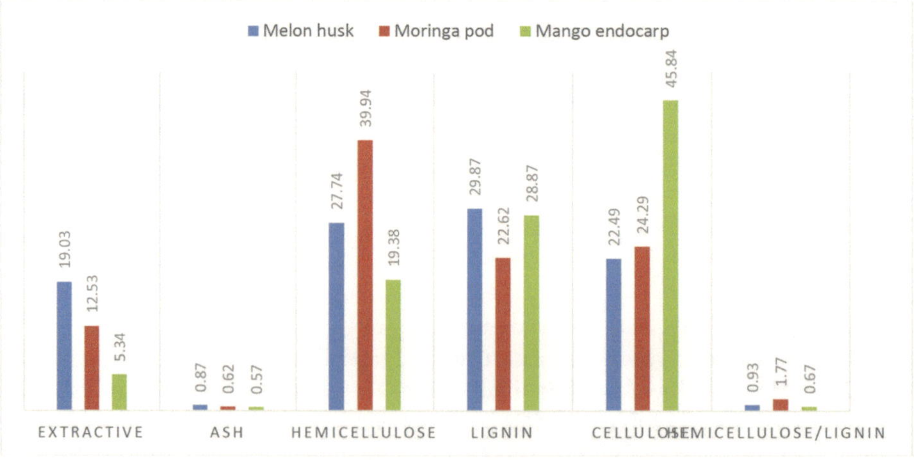

Figure 1.
Analysis of the raw lignocellulose content of the samples (%).

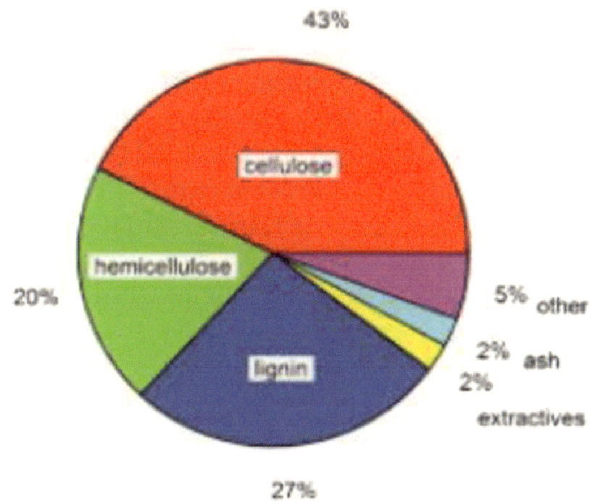

Figure 2.
Composition of lignocellulosic biomass [26].

	Cellulose	Hemicellulose	Lignin	Ash	Extractives	References
Sugarcane bagasse	35–49	16–33	11–25	4–9	2	[19]
Siam weed	4.8 ± 0.9	40.2 ± 2.3	29.9 ± 0.7	23.2 ± 5.3	0.9 ± 3.1	[19]
Rice straw	32–47	19–27	5–24	12.4		[24]

Table 1.
Reported compositional analysis of raw lignocelluloses of sugarcane bagasse, Siam weed, and Rice straw (%w/w).

The availability of various lignocellulose components of biomass and the ease of their usage in the production of biofuel is mostly determined by their quantity in the source. Among, the lignocellulose biomass, the cellulose and hemicellulose components are the most abundant renewable organic resource [25, 26]. The high cellulose content found in the mango seed endocarp and the considerable hemicellulose content of the moringa seed pod from the present investigation are an indication of their potential to be useful in the production of biofuel. In the cell walls of plants, the bonds holding cellulose, hemicellulose, and lignin together differ. Molecules of cellulose and hemicellulose, or lignin are generally held together by hydrogen bonds. In addition to the hydrogen bond, lignin and hemicellulose also chemically bind together, causing the lignin to always contain a tiny amount of carbohydrates when it is separated from natural lignocelluloses [25]. This interaction influences the release of simple sugars during hydrolysis. In this study, the proportion of hemicellulose to lignin ratio of the mango endocarp sample was the lowest (0.67), followed by that of the melon husk (0.93) and the moringa pod (1.77). Therefore, the degree of hydrolysis of these agricultural wastes may be in order: melon seed husk, mango endocarp, and moringa pod.

3.2 Fourier transform infrared (FT-IR) spectroscopy

Functional groups in complicated chemical mixtures are usually identified and compared using Fourier transform infrared spectroscopy. **Figures 3–5** show the

Fourier transform infrared spectra of the agricultural waste samples components. The characteristic prominent peaks for samples are provided in **Table 2**.

The FTIR spectroscopy examination revealed spectra at 3422.99, 3422.66, and 3422.85 cm^{-1} for mango seed endocarp, melon shell, and moringa seed pod, respectively. These spectra signify the presence of bonded —OH groups, as illustrated in the peaks in **Figures 3–5** and **Table 2**. Conjugated carboxylic acids undergo a C=C stretching at the peak of about 1637 cm^{-1}. At 1205.72, 1204.50, and 1206.24 cm^{-1}, the

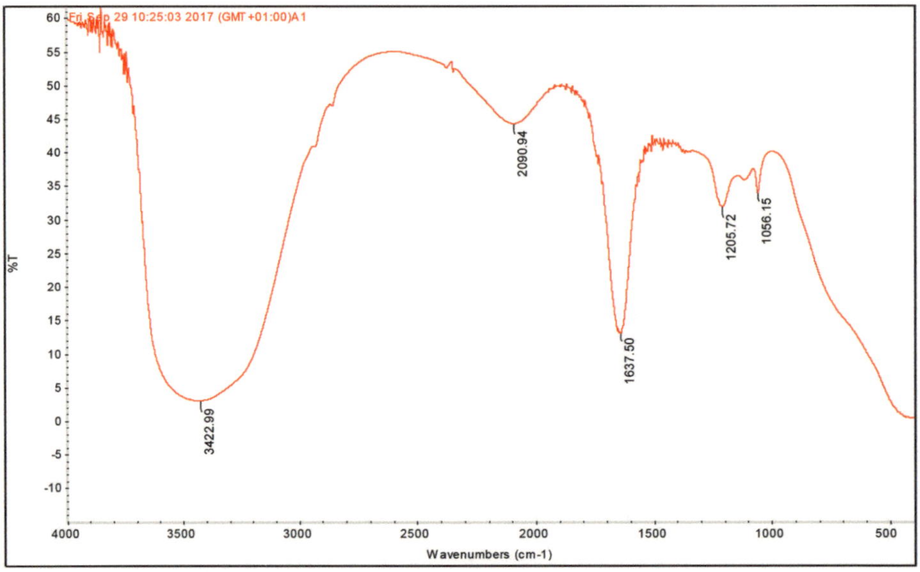

Figure 3.
FT-IR spectrum for mango endocarp.

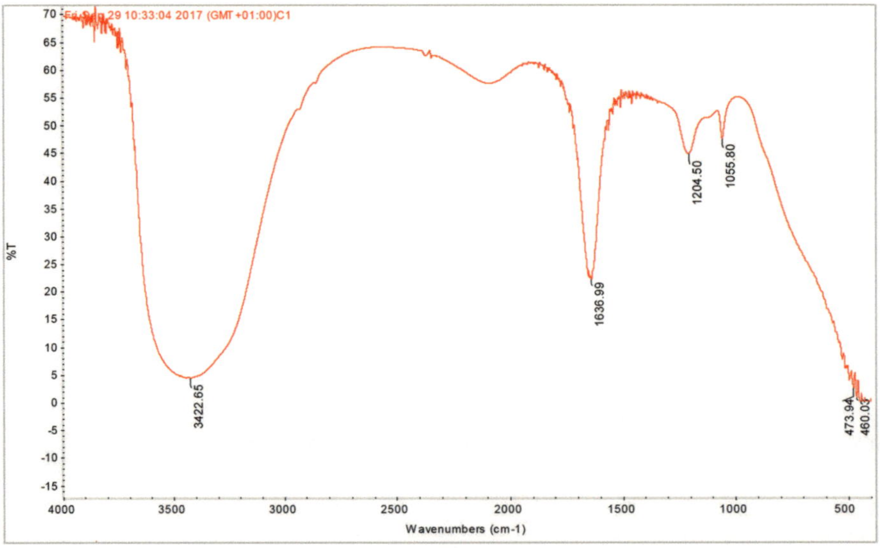

Figure 4.
Melon husk FT-IR Spectrum.

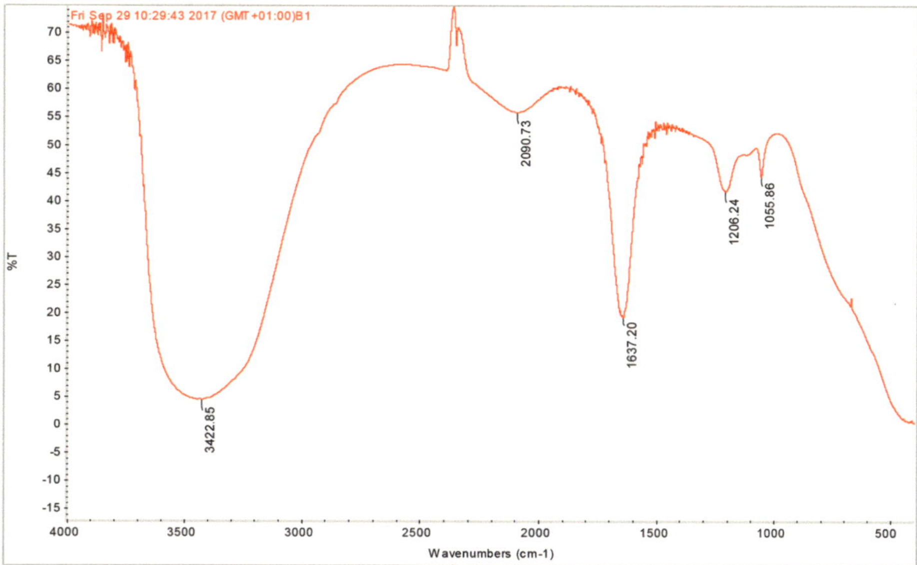

Figure 5.
FT-IR spectrum for Moringa pod.

Functional groups	Mango endocarp	Melon husk	Moringa pod
O−H stretch	3422.99 cm^{-1}	3422.66 cm^{-1}	3422.85 cm^{-1}
C=C stretch	1637.50 cm^{-1}	1636.99 cm^{-1}	1637.20 cm^{-1}
$-CH_2$- and $-CH_3$- vibration	1205.72 cm^{-1}	1204.50 cm^{-1}	1206.24 cm^{-1}
C−O−R (ester)	1056.15 cm^{-1}	1035.80 cm^{-1}	1055.86 cm^{-1}

Table 2.
FT-IR of acid hydrolysis of extractive-free mango endocarp, melon husk, and moringa seed pod.

vibrations of the aliphatic chains, $-CH_2$- and $-CH_3$-, which constitute the fundamental structure of cellulose material, are observed. Peaks at 1056.15, 1035.80, and 1055.86 cm–1 may be attributed to the vibration of C−O−R or C−O−R (alcohols or esters) [27, 28]. The observation in this study is in agreement with that reported by [29, 30]. These bands are therefore common to those observed in cellulose, hemicelluloses, and lignin FT-IR spectra [31, 32].

4. Conclusion

The amount and quality of lignocellulosic biomass, particularly cellulose and hemicellulose, obtained from the selected agro wastes from this study is an indication that such and even more agricultural and food residues could be useful raw materials in the production of biofuel. Thus, rather than constituting an environmental nuisance, agricultural wastes (such as melon husk, mango seed endocarp, and moringa seed pod) may serve as alternate and affordable natural source of raw material in biofuel production.

Acknowledgements

The authors wish to acknowledge Sheda Science and Technology Complex (SHESTCO) for the use of the laboratory facilities at their Chemistry Advanced Research Centre (CARC).

Conflict of interest

The authors here declare that there is no known conflict of interest.

Author details

Adewale Elijah Fadeyi[1*] and Saheed Olatunbosun Akiode[2]

1 Chemistry Advanced Research Centre, Sheda Science and Technology Complex, Abuja, Nigeria

2 Biotechnology Advanced Research Centre, Sheda Science and Technology Complex, Abuja, Nigeria

*Address all correspondence to: wale.fade@gmail.com

References

[1] Gavrilescu D. Energy from biomass in pulp and paper mills. Environmental Engineering and Management Journal. 2008;**7**(5):537-546

[2] Popa VI, Volf I. Contribution to the complex processing of biomass. In: Proceedings of the 2008 Nordic Wood Biorefinery Conference. Vol. 229. Sweden: Stockholm; 2008

[3] Asthana AK. Biomass as Fuel in Small Boilers. APO 2009 Report; 2009. ISBN: 92-833-7078-3

[4] Balat M, Ayar G. Biomass energy in the world, use of biomass and potential trends. Energy Sources. 2005;**27**:931-940

[5] Priyadharshini R. A study on the export performance of fresh mangoes from India. International Journal of Interdisciplinary and Multidisciplinary Studies (IJIMS). 2015;**2**(6):134-140

[6] O'Shea N, Arendt E, Gallagher E. Dietary fibre and phytochemical characteristics of fruit and vegetable by-products and their recent applications as novel ingredients in food products. Innovative Food Science & Emerging Technologies. 2012;**16**:1-10

[7] Leanpolchareanchai J, Padois K, Falson F, Bavovada R, Pithayanukul P. Microemulsion system for topical delivery of Thai mango seed kernel extract: Development, physicochemical characterization, and ex-vivo skin permeation studies. Molecules. 2014;**19**(11):17107-17129

[8] Ayotunde EO, Fagbenro OA, Adebanyo OT. Toxicity of aqueous extract of Moringa olifera seed powder to Nile tilapia (Oreochromisniloticu) fingerlings. International Research Journal of Agricultural Sciences. 2011;**1**:142-150

[9] Mtui YS. Recent advances in pretreatment of lignocellulosic wastes and production of value-added products. African Journal of Biotechnology. 2009;**8**(8):1398-1415

[10] Elumalai S, Agarwal B, Runge TM, Sangwan RS. Advances in transformation of lignocellulosic biomass to carbohydrate-derived fuel precursors. In: Kumar S, Sani R, editors. Biorefining of Biomass to Biofuels. Biofuel and Biorefinery Technologies. Vol. 4. Cham: Springer; 2018. DOI: 10.1007/978-3-319-67678-4_4

[11] Jeswani HK, Chilvers A, Azapagic A. Environmental sustainability of biofuels: A review. Proceedings of the Royal Society A. 2020;**476**:20200351. DOI: 10.1098/rspa.2020.0351

[12] Kusch S, Morar MV. Integration of lignocellulosic biomass into renewable energy generation concepts. Pro Environment. 2009;**2**:32-37

[13] Ayeni AO, Hymore FKS, Mudliar N, Deskmukh SC, Satpute DB, Omoleye JA, et al. Hydrogen peroxide and lime based oxidative pretreatment of wood waste to enhance enzymatic hydrolysis for a biorefinery: Process parameters optimization using response surface methodology, Fuel. 2013;**106**:187-194

[14] Silvério FO, Barbosa LCA, Gomide JL, Reis FP, Piló-Veloso D. Methodology of extraction and determination of extractive contents in eucalypt woods. Revistaárvore, Viçosa. 2006;**30**(6):1009-1016

[15] Coelho C, Michelin M, Cerqueira MA, Gonçalves C, Tonon RV, Pastrana LM, et al. Cellulose nanocrystals from grape pomace: Production, properties and cytotoxicity assessment. Carbohydrate Polymers. 2018;**192**:327-336

[16] Reddy JP, Rhim J-W. Extraction and characterization of cellulose microfibers from agricultural wastes of onion and garlic. Journal of Natural Fibers. 2018;**15**:465-473

[17] Hall M, Bansal P, Lee JH, Realff MJ, Bommarius AS. Cellulose crystallinity –a key predictor of the enzymatic hydrolysis rate. The FEBS Journal. 2010;**277**:1571-1582

[18] Sluiter JB, Ruiz RO, Scarlata CJ, Sluiter AD, Templeton DW. Compositional analysis of Lignocellulosic Feedstocks. Review and description of methods. Journal of Agricultural and Food Chemistry. 2010;**58**(16):9043-9053

[19] Ayeni AO, Adeeyo OA, Oresegun OM, Oladimeji TE. Compositional analysis of lignocellulosic materials: Evaluation of an economically viable method suitable for woody and non-woody biomass. American Journal of Engineering Research (AJER). 2015;**4**(4):14-19

[20] Blasi CD, Signorelli G, Russo CD, Rea G. Product distribution from pyrolysis of wood and agricultural residues. Industrial and Engineering Chemistry Research. 1999;**38**:2216-2224

[21] Li S, Xu S, Liu S, Yang C, Lu Q. Fast pyrolysis of biomass in free-fall reactor for hydrogen-rich gas. Fuel Processing Technology. 2004;**85**:1201-1211

[22] Lin L, Yan R, Liu Y, Jiang W. In-depth investigation of enzymatic hydrolysis of biomass wastes based on three major components: Cellulose, hemicelluloses, and lignin. Bioresource Technology. 2010;**101**:8217-8223

[23] Sluiter A, Hames B, Ruiz R, Scarlata C, Sluiter J, Templeton D, et al. Determination of Structural Carbohydrates and Lignin in Biomass: Laboratory Analytical Procedure (LAP). Golden, CO: National Renewable Energy Laboratory; 2008

[24] Sarkar N, Ghosh SK, Bannerjee S, Aikat K. Bioethanol production from agricultural wastes: An overview. Renewable Energy. 2012;**37**(1):19-27

[25] Chen H. Chemical composition and structure of natural lignocellulose. In: Biotechnology of Lignocellulose: Theory and Practice. Springer; 2014. pp. 25-71

[26] Ali EN, Jamaludin MZ. Possibility of producing ethanol from Moringa Oleifera pod husk. Journal of Advanced Research Design. 2015;**5**(1):1-9

[27] Champagne P, Li C. Enzymatic hydrolysis of cellulosic municipal wastewater treatment process residuals as feedstock for the recovery of simple sugars. Bioresource Technology. 2009;**100**(23):5700-5706

[28] Zapata B, Balmaseda J, Fregoso-Israel E, Torres-Garcia E. Thermokinetics study of orange peel in air. Journal of Thermal Analysis and Calorimetry. 2009;**98**(1):309-315

[29] Sanchez RO, Balderras PH, Roa GM, Urena FN, Orozco JV, Lugo VL, et al. Fruit residue to ethanol. Bio Resources. 2014;**9**(2):1873-1885

[30] Bello OS, Adegoke KA, Akinyunni OO. Preparation and characterization of a novel adsorbent from Moringa oleifera leaf. Applied Water Science. 2017;**2017**(7):1295-1305

[31] Bodirlau R, Teaca CA, Spiridon I. Chemical modification of beech wood: Effect on thermal stability. BioResources. 2008;**3**(3):789-800

[32] Xu F, Yu J, Tesso T, Dowell F, Wang D. Qualitative and quantitative analysis of lignocellulose biomass using infrared techniques: A mini-review. Applied Energy. 2013;**104**(1):801-809

Chapter 3

Moisture and Solid Mass as Indicators of the Ideal Transformation Technology

Carlos Yesid Varela Bernal

Abstract

The waste generated from human activities consists of molecules composed of hydrogen, carbon, minerals, metals, and water, representing the moisture content of the biomass. This moisture content plays a crucial role in determining the most suitable technology for extracting its potential. This chapter proposes a technological route based on the humidity content, offering optimal alternatives for biomass transformation. Additionally, a quantitative method is presented to aid in the selection of the ideal transformation technology.

Keywords: hydrocarbons, biofuels, fermentation, thermochemistry, residual biomass

1. Introduction

Efforts must be made to recover and repurpose biodegradable organic waste, minimizing its quantity and promoting a circular economy where nothing is discarded. With the growing reality of climate change, it becomes imperative to establish a self-sustaining and environmentally friendly economy to protect soils and ecosystems from the impact of industrial and urban waste.

Biomass naturally degrades and changes its chemical composition, providing an opportunity to direct this process toward obtaining a valuable end product. By defining the guidelines and characteristics of this chemical transformation, a new resource or raw material can be obtained with practical applications.

2. Characterization and outline of the technological route

Grande [1], fully characterizing the chemical composition of waste and having specific data on production volume, physical state, stability, and storage conditions are essential when considering waste as a potential raw material. When deciding on waste transformation, options include biodigestion, fermentation, mechanical methods, and thermal methods to extract desired molecules or by-products. It is crucial to ensure that all substances involved in the process do not pose a threat to human health and the environment [2].

Considering the workplace's needs and circumstances, various technologies can be employed to achieve the desired final product. A selection process becomes necessary to evaluate and qualify the technologies and conditions for biomass transformation (**Figure 1**).

Given the variations in waste production percentages among countries, a strategy based on biomass humidity becomes even more crucial. This technological route can be extended by considering specific technologies suitable for treating particular biomass based on its moisture content.

2.1 Humidity greater than 75%

2.1.1 Wet and dry biodigestion

When dealing with high humidity content, wet and dry biodigestion are the recommended options. Wet biodigestion requires less thermal and mechanical intervention and is suitable for urban waste, while dry biodigestion is more convenient for residual biomass from agro-industries. Mathematical expressions can provide a better understanding of the transformation process and the production of by-products, according to [4].

$$Vol\ CH_4Nm^3 = \frac{(Methane\ molar\ mass\ \frac{g}{mol})(dry\ weight\ mass\ kg)}{(molar\ mass\ dry\ weight\ \frac{g}{mol})\left(Methane\ density\ \frac{kg}{m^3}\right)} \quad (1)$$

Where:

Nm^3 = Normal cubic meter.

Methane molar mass = 234 $\frac{g}{mol.}$

Molar mass dry weight = 521 $\frac{g}{mol}$

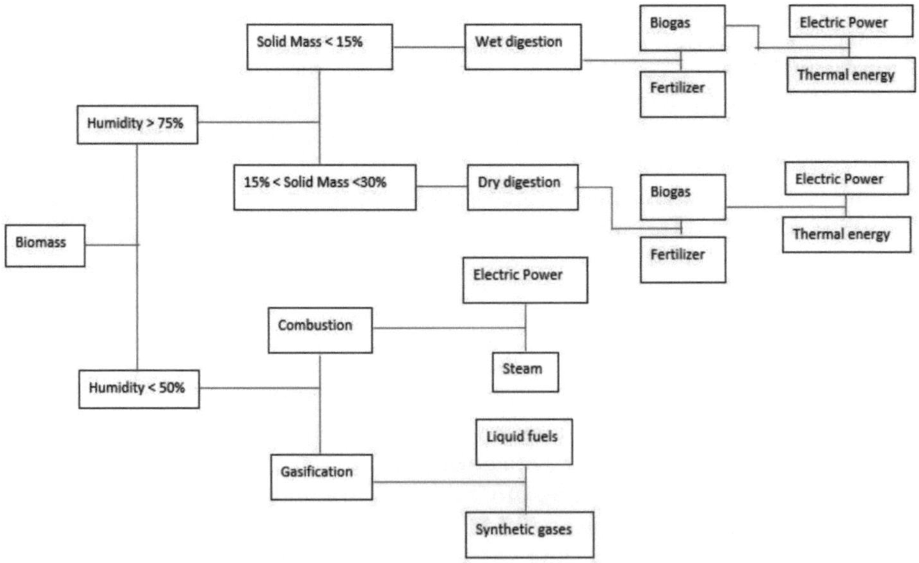

Figure 1.
Technological routes for the transformation of biomass into bioenergy according to their moisture content.
Source: *[3]*.

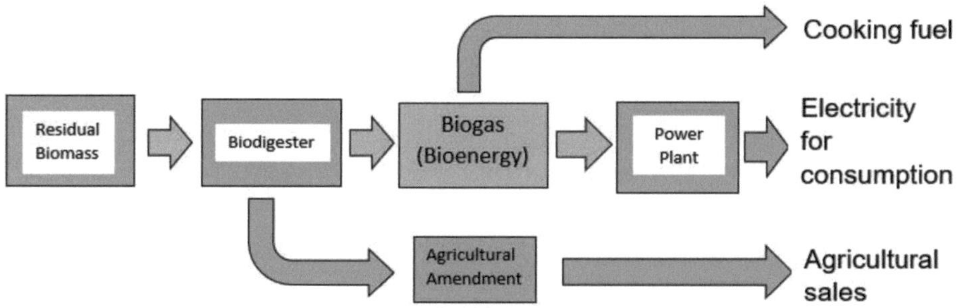

Figure 2.
Scheme for the use of residual biomass to produce three by-products (3).

Methane density = 0.717 $\frac{kg}{m^3}$
Additionally, carbon dioxide is produced during this process.

$$Vol\ CO_2Nm^3 = \frac{\left(CO_2\ molar\ mass\ \frac{g}{mol}\right)(dry\ weight\ mass\ kg)}{\left(molar\ mass\ dry\ weight\ \frac{g}{mol}\right)\left(CO_2\ density\ \frac{kg}{m^3}\right)} \tag{2}$$

Where:
Nm^3 = Normal cubic meter
Carbon dioxide molar mass = 589 $\frac{g}{mol}$
Molar mass dry weight = 521 $\frac{g}{mol}$
Carbon dioxide density = 1.978 $\frac{kg}{m^3}$

A continuous model for receiving high-moisture residual biomass and producing biogas at a corresponding rate for electricity generation is proposed (**Figure 2**).

While the general scheme is illustrated, variations may be necessary depending on the specific case, such as rural or urban environments (**Figure 3**).

2.2 Humidity less than 50%

2.2.1 Combustion

For lower humidity levels, other technologies are considered for biomass transformation. These technologies are more suitable for biomass containing carbohydrates, starches, sugars, fats, and lignocellulosic fibers, indicating forest residues (**Figure 4**).

If the biomass moisture content exceeds 50%, a drying process is necessary prior to incineration. The energy required to reduce the humidity can be estimated by subtracting the current content from 50% and multiplying it by the water vaporization value (2250 kJ/kg).

Pelletizing the biomass for improved transport and presentation may be an option, although it incurs additional costs. The energy delivered can be estimated using the lower calorific power (LCP) formula.

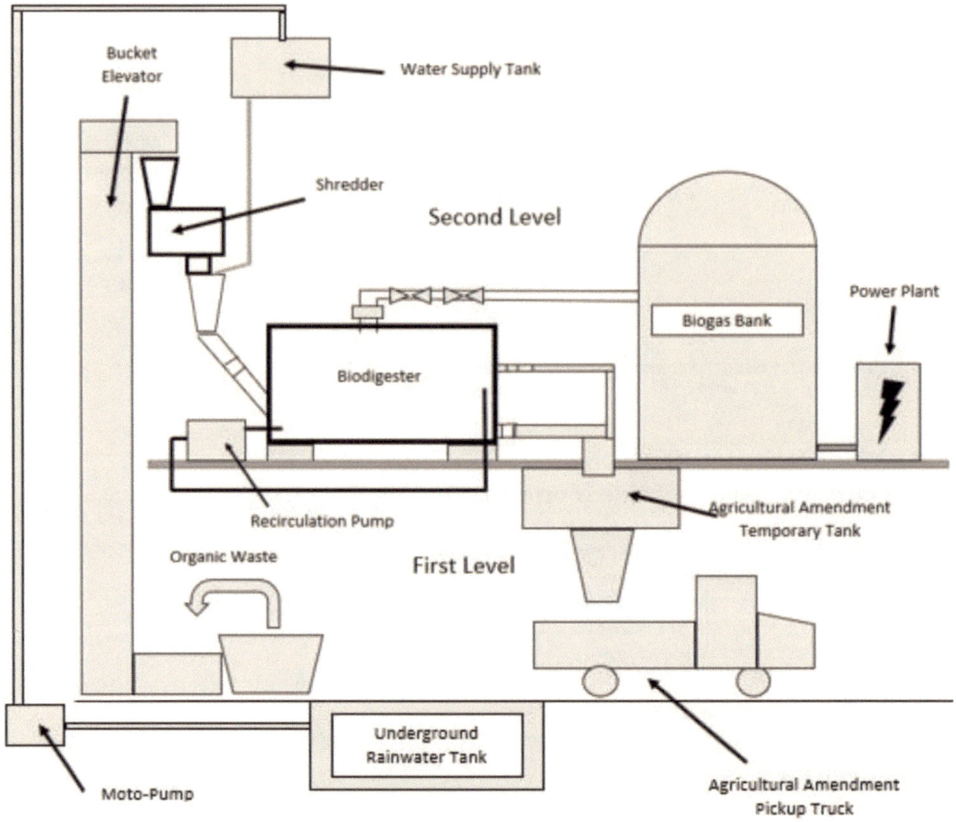

Figure 3.
Model of the biomass use system to produce energy, heat, and agricultural amendment.

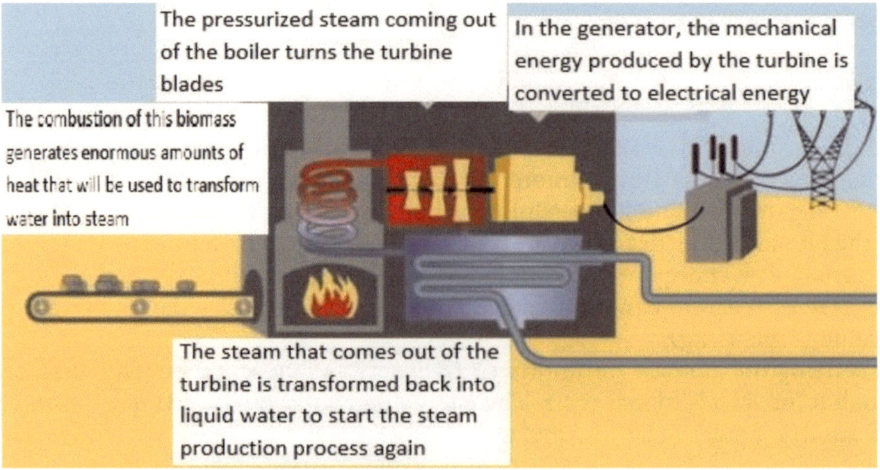

Figure 4.
Biomass combustion process for electricity generation.

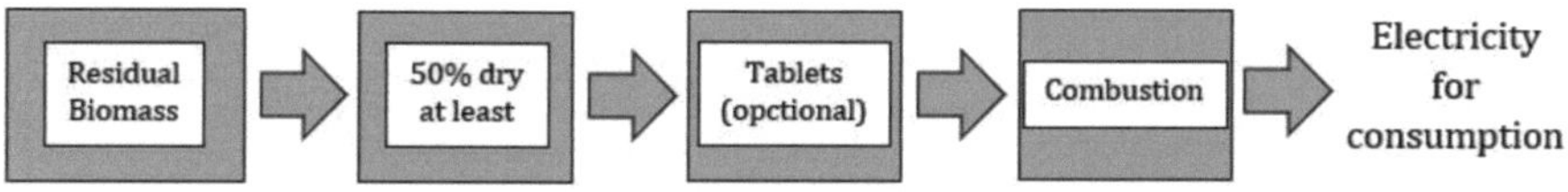

Figure 5.
Scheme of operation of the combustion as a technology of energetic transformation of the residual biomass.

$$PCI = PCS * \frac{1-w}{100} - 2.447 * \frac{w}{100} - \left(\frac{h}{100} * 2\right) * 18.02 * 2.447 * \left(\frac{1-w}{100}\right) [\![MJ]\!] \tag{3}$$

Where:
$LCP =$ Lower calorific power $[\![65{,}460\ MJ]\!]$
$HCP =$ Higher calorific power $[\![MJ]\!]$
$w =$Moisture (%)
$h =$Hydrogen content (% weight)
Subtracting the costs of humidity reduction and pellet production from the obtained value provides the useful energy for electricity generation through a turbine (**Figure 4**). The biomass combustion scheme is presented in **Figure 5**.

The choice of technology depends on the specific circumstances, such as geographical location and temperature conditions. Cold areas have different combustion efficiencies compared to tropical or desert areas due to heat dissipation and bacterial activity during biodigestion.

2.2.2 Gasification

According to (Probiomasa, FAO [5]), the following are the stages of gasification within an industrial plant.

Conditioning of the biomass: The gasifiers can run powered by different types of biomasses.

a. Fine biomass: peel of rice, peanuts, pomace, etc.

b. Biomass derived from wood: chips, pellets, briquettes.

Gasification: It takes place in reactors or gasifiers, where they carry out a complex process with thermochemicals.

Adequacy of synthesis gas (syngas): For its transformation into electric power, gasification gas must be cleaned to remove tars, and cool down. If one has to use internal combustion motors (ICM), they must be treated in ultracool systems.

Energy generation, thermal or electric: On generators or turbo-steam cogenerators, internal combustion engines or cycle engines must be combined.

Water treatment process: Current gasification systems operate in a closedcycle pattern, pointing to the maximum utilization of the resources.

The gasification process yields 2.5 to 3 normal m^3. of gas per kilogram of dry biomass, with a calorific value ranging from 1000 to 1300 normal kcal/m^3.

Gasification offers an economical option for many, producing synthesis gas (syngas) with specific chemical compositions, which are presented in the following **Table 1**.

N_2	**44–55%**
CO	15–20%
H_2	15–20%
CO_2	8–12%
CH_4	1–4%

Table 1.
Chemical composition of synthesis gas (syngas).

2.2.3 Other energy and non-energy products

Fermentation, a part of bioethanol production, yields by-products and useful raw materials depending on the biomass type and composition. Major bioethanol producers like Brazil and the United States primarily, close to 60% [1], utilize sugar cane and cereals (such as corn and wheat), respectively.

Fermentation of biomass results in the production of by-products, such as processing of cassava, and this must go through a hydrolysate with enzyme α-amylase, which is a pretreatment before fermentation. After this, dextrins and maltodextrins are produced, useful for the production of antibiotics such as penicillin, cephalosporin, and streptomycin, and other organic acids (citric, lactic, gluconic, and itaconic), some amino acids, xanthan gum, glucans, dietary fibers, flavorings, dietary supplements, aroma compounds, acetone, enzymes [1], etc.

2.3 Selection of the appropriate technology as the case may be

With a wide range of technologies available, the chemical composition of biomass becomes a crucial factor in the transformation process. Theoretical estimates and mathematical expressions that represent biomass characteristics and energy potential help determine the ideal technology. Selection methods, such as the analytical hierarchical process developed by Professor Thomas Satty, aid in quantitative decision-making by considering independent criteria or variables that may conflict. Comparative matrices and pairwise comparison tables assist in evaluating and prioritizing alternative technologies.

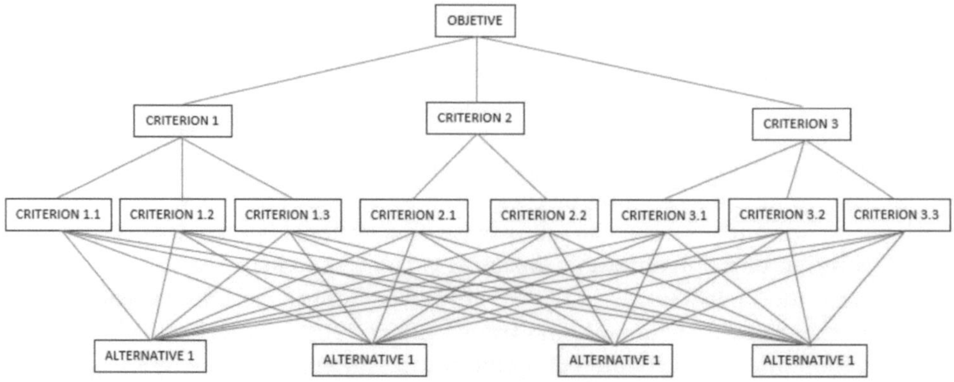

Figure 6.
Hierarchy of importance of selection criteria with respect to the alternatives.

Worth	Definition	Comments
1	Equal importance	Criterion A is just as important as Criterion B
3	Moderate importance	Experience and judgment slightly favor Criterion A over B
5	Great importance	Experience and judgment strongly favor Criterion A over B
7	Very great importance	Criterion A is much more important than criterion B
9	Extreme importance	The greater importance of Criterion A over B is beyond any doubt
2, 4, 6, and 8	Intermediate values in previous ones, when it is necessary to clarify	

Table 2.
Matrix of qualifiers between pairs to be compared of the analytical hierarchical process (AHP).

This selection method is very useful in decision-making, having both quantitative and qualitative aspects at hand, which makes it a multi-attribute method, being widely used in the field of business, economics, or operations research [6], like in this particular case of choosing the most appropriate technology to transform biomass into electricity.

Select an alternative among several proposals based on a series of criteria or variables, normally hierarchical, see **Figure 6** [6].

It is important that criteria are independent of each other, which makes it easier for the comparative matrices to yield more reliable results, with the support of the comparison **Table 2** between pairs, see **Figure 2**, which is essential when qualifying the selection priorities of the alternatives.

3. Conclusions

It is the same biomass with its physical and chemical characteristics, which gives indications of the best transformation and recycling treatments, whether in energy conversion or in new organic products useful in the food, cosmetics, and medicine industries, among others.

The energetic potential of the residual biomass resides in its solid mass, where the organic elements based on carbon are found, susceptible to decompose and later become fuel. If the biomass is very humid, this characteristic can help this transformation through biodigestion, water being the environment that allows bacteria to perform this task.

The alternative to transform the residual biomass into electricity with less humidity and the greatest amount of solid mass to be seen is gasification, which offers a synthesis gas (syngas), being an accessible option of economic production for many people.

Another method that takes advantage of bacterial activity in energy transformation is the production of bioethanol, with fermentation, the latter is also crucial in the production of other products, also biologically based, useful as inputs for other industries.

Acknowledgements

I express my sincere gratitude to the district university and its academic community for teaching me what I know about engineering and having the power to provide solutions to humanity's challenges.

Conflict of interest

The authors declare no conflict of interest.

Author details

Carlos Yesid Varela Bernal
District University, Bogota, Colombia

*Address all correspondence to: cyvarelab@correo.udistrital.edu.co

References

[1] Grande CD. Residuos Agroindustriales Biocombustibles. Cali: Bonaventura; 2016

[2] Saval S. Aprovechamiento de residuos agroindustriales: pasado, presente y futuro. Biotecnología. 2012;**16**(2):14-37

[3] German, Lopez CB. The livestock residual biomass as an energy resource in Colombia. Visión Electrónica. 2018;**12**: 180-188

[4] Varela C, López G. Bioenergía a partir de biomasa residual en plazas de mercado de Bogotá. Vol. 11. Bogota, Colombia: Letras Conciencia Tecnológica; 2022

[5] Probiomasa, FAO. Gasificacion de la Biomasa. n.d. Retrieved from: http://www.probiomasa.gob.ar/_pdf/04Gasificacion_hojaTecnica.pdf

[6] Piqueras VY. Proceso Analítico Jerárquico (Analytic Hierarchy Process, AHP). 2023. Retrieved from: https://victoryepes.blogs.upv.es/2018/11/27/proceso-analitico-jerarquico-ahp/

Chapter 4

Biomass-Based Activated Carbon

Abdulbari A. Ahmad, Abdulraqeeb Alwahbi, Laila A. Al Khatib and Hani Dammag

Abstract

Biomass is a renewable and eco-friendly energy source, which is easily regenerated, pollution-free, and widely available. It is also naturally carbonaceous and has low disposal costs. Biomass activated carbon (BAC) is a highly effective adsorbent that can remove a wide range of organic and inorganic pollutants, as well as polar and nonpolar compounds in aqueous or gaseous environments. Additionally, it is also utilized for energy storage purposes. Converting biomass into activated carbon for carbon dioxide (CO_2) adsorption is a practical solution for managing solid waste and reducing anthropogenic greenhouse gas emissions. Activated carbon is a microporous form of carbon that possesses a well-developed high internal surface area, pore volume, pore structure, and surface chemistry. The production of biomass-derived activated carbons is dependent on pyrolysis temperatures and physical and chemical activation conditions, which can alter their surface characteristics and adsorption behavior. Literature indicates that biomass-derived activated carbons possess a high surface and adsorption capacity, making them a suitable option for environmental remediation and energy storage.

Keywords: biomass, biomass activated carbon, pyrolysis temperatures, activation, environmental remediation

1. Introduction

Biomass has the potential to be transformed into solid, liquid, and gaseous biofuels, as well as certain chemicals to generate bioenergy. This is due to the renewability of biomass, which enables CO_2-neutral conversion, making biofuel combustion widely accepted as not contributing to the greenhouse effect. The increasing focus on bioenergy as a *viable* alternative to fossil energy has been driven by concerns over global warming, which largely stems from the combustion of fossil fuels [1]. In recent years, there has been a significant global effort to shift toward biomass as an alternative to fossil fuels for energy production [2]. Biomass is a diverse group of materials that includes wood, woody biomass, herbaceous and agricultural biomass, aquatic biomass, animal and human biomass waste, semi-biomass, and biomass mixtures, all of which can be utilized to produce biofuels and biochemicals [3]. Currently, around 95–97% of the world's bioenergy is generated through direct biomass combustion, and many countries are exploring the potential of large-scale combustion of natural biomass and its co-combustion with semi-biomass and solid fossil fuels, such as coal, peat, and petroleum coke, to promote biofuel adoption [4]. Numerous studies [5, 6] have been conducted worldwide to investigate the advantages and disadvantages of using biomass fuels for various thermochemical processes such as combustion, pyrolysis, gasification, and liquefaction, as well as biochemical processes such as anaerobic digestion, alcoholic fermentation, and aerobic biodegradation. Additionally, research has been conducted on co-combustion, co-pyrolysis, and co-gasification of biomass with other solid fuels. The main technologies for producing biochar and activated carbon from biomass are pyrolysis, carbonization, gasification, and torrefaction [7]. All of these technologies involve thermal treatment of the biomass under oxygen-limited conditions to increase the carbon content [8]. Pyrolysis refers to the process of thermally breaking down organic matter, such as sawdust, tire waste, and sewage sludge, in the absence of oxygen, typically within a temperature range of 400–800°C. During this process, the resulting primary products from biomass are commonly known as condensable (tars) and non-condensable volatiles, as well as char [9]. During pyrolysis, the moisture and light volatiles are initially released, followed by the aromatic components and hydrogen gas. The remaining solid residues form biochar, which possesses well-defined porous structures and abundant carbon content. The specific surface area, porosity, total pore volume, and surface chemical properties of biochar depend on the pyrolysis conditions such as the heating rate, pyrolysis temperature, and residence time, as demonstrated in previous studies [10]. In general, higher heating rates and temperatures result in lower biochar yields but a higher surface area and pore volume [11]. Higher temperatures also increase the ash and fixed carbon content, while decreasing the volatile matter.

Due to the rapid growth of the world's population, increased urbanization, agricultural demands, and industrial development, global biomass has significantly expanded. By 2030, the world's population is projected to reach 8.5 billion, and solid waste production is expected to reach 2.59 billion tons [12]. **Figure 1** illustrates the global distribution of research activities on converting biomass into biochar, as well as the research collaborations among different countries. Notably, China, India, South Korea, Australia, and Malaysia are among the countries that prioritize the efficient use of biomass resources to address environmental challenges and produce biochar as part of their energy development strategies.

Figure 1.
Shows the bibliometric mapping of the conversion of biomass into biochar, as well as the research interrelations between various countries.

The objective of this chapter is to provide a comprehensive summary and analysis of the recent advancements and research conducted on biomass resources, structure of biomass, and biomass-based activated carbons. This review will cover various preparation methods for biomass-based activated carbons and their applications.

2. Biomass resources

Biomass resources encompass a wide range of materials, including wood waste, agricultural crops, industrial waste, municipal solid waste, animal waste, and by-products from food processing. Annually, more than 140 billion metric tons of biomass waste are produced from agriculture alone, equivalent to approximately 50 billion tons of oil [13]. These materials can be utilized to produce renewable energy and fuels, referred to as bioenergy and biofuels. Through thermal and chemical treatment, agricultural waste can generate an array of valuable products, including biosolids, biogases, biooils, and biofuels [14]. The composition of the activated carbon derived from agricultural waste depends on whether it undergoes biochemical or thermochemical conversion processes [13]. Biomass is primarily composed of three natural fiber components: cellulose, hemicellulose, and lignin, with varying compositions depending on the material. Typically, cellulose, hemicellulose, and lignin make up 20–55 wt%, 20–45 wt%, and 15–35 wt% of biomass, respectively [15]. Activated carbon possesses desirable properties such as high porosity, excellent surface textural characteristics, and good adsorption capacity. **Figure 2** shows the chemical structures of cellulose, hemicellulose, and lignin, which are all components of lignocellulosic materials.

Agricultural waste is a plentiful, renewable, and eco-friendly resource that is also available at a low cost [17]. Examples of agricultural waste include bamboo waste [18], rattan sawdust [19], peanut hull [20], sesame seed shell [21], garlic peel [22], wood [23], cotton stalk [24], coffee grounds [25], almond shell [26], and waste tea [27]. Although activated carbon has numerous applications, there are certain limitations

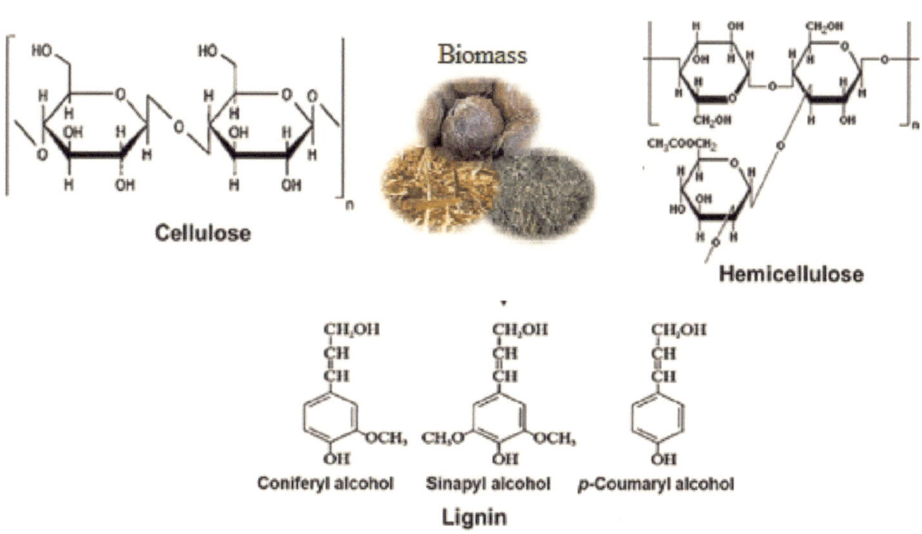

Figure 2.
Lignocellulosic chemical structures of cellulose, hemicellulose, and lignin [16].

that must be considered. In a study by Abbas and Ahmed [28], it was found that leaves are not suitable for activated carbon due to their low carbon content, high volume-to-weight ratio, and ash content. Furthermore, research indicates that the digestibility of amorphous lignin is higher than that of biomass cellulose by the activating agent used in chemical activation. Thus, it is crucial to comprehend the properties of any substance utilized [17]. Activated carbon, which can be a promising CO_2 adsorbent, can be produced from various materials, including carbonaceous raw materials, biomass-based residues, and lignocellulosic agricultural by-products [15].

3. Biomass-based activated carbon

Biomass is an abundant energy source that possesses several desirable characteristics, including high porosity, renewability, ease of handling, storability, low pollution, wide availability, and inherent carbon content. Utilizing biomass energy has gained significant attention as a means of reducing carbon footprint [29]. In recent years, researchers have been actively exploring substitutes for solid adsorbents that are low in cost, easy to regenerate, and have a minimal environmental impact. Biomass-based adsorbents have emerged as a promising alternative. The choice of biomass for BAC production depends on the application area, such as wastewater treatment, supercapacitors, energy generation, and air pollution control [30]. The carbon content, such as cellulose, hemicellulose, and lignin, is a crucial criterion for selecting the appropriate biomass for pollutant removal processes, as these elements contribute to the porous structure of the BAC. Wood-type biomass, which contains wood biomass yields higher carbon content [31]. Cellulose is the most abundant lignocellulosic component in biomass, followed by hemicelluloses and lignin [32]. The carbon content of biomass varies depending on location and geographical conditions. Herbaceous biomass generally has a lower carbon content than wood-type biomass [33]. Lignocellulosic biomasses with unique characteristics are suitable for producing BAC in an easy and cost-effective manner. The energy required for activation is dependent on the properties of the biomass waste, including its structure and the chemical behavior of its constituents [34]. Selecting the appropriate biomass waste is crucial for BAC production while converting agricultural waste into value-added products. These factors impact the quality of the end product of BAC [35]. Thus, the next section will discuss the available carbonization methods for producing biochar from biomass and activation techniques.

4. Activated carbon

Activated carbon is a category of amorphous carbon-based materials that possess a high degree of porosity and a substantial surface area. The properties of activated carbon depend on the starting materials used and the conditions of activation, resulting in a wide range of activated carbons with specific uses. The most common method used to prepare activated carbon involves the carbonization of precursors at high temperatures in an inert atmosphere followed by activation. Activation can be achieved through physical, chemical, or physiochemical. Physical activation involves the treatment of char with oxidizing gases, such as steam or carbon dioxide, at high temperatures to create a porous structure [36]. Chemical activation involves impregnating a chemical agent with precursors and then heat-treating them in an inert

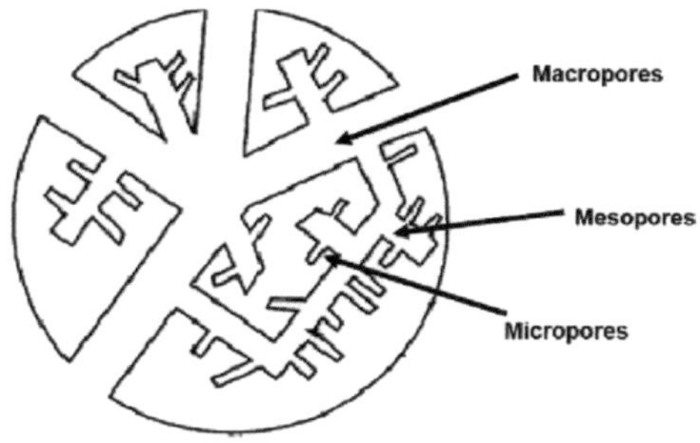

Figure 3.
Structure of activated carbon.

atmosphere to develop pores through dehydration and oxidation reactions. Activated carbons have various applications, including their use as adsorbents, catalysts, or catalyst supports. Activated carbons can be classified into gas-adsorbing and liquid-phase carbons, based on their pore size distribution. **Figure 3** shows the pore structure of activated carbons is categorized into three groups based on pore size: micropores (pore size <2 nm), mesopores (pore size 2–50 nm), and macropores (pore size >50 nm), according to the International Union of Pure and Applied Chemistry classification [37]. Gas-adsorbing carbons have more pore volume in the micropore and macropore ranges, whereas liquid-phase carbons have significant pore volume in the mesopore or transitional pore range, allowing easy access for liquids to the micropore structure [30]. Activated carbons are widely used in industries, such as food processing, pharmaceuticals, chemical, petroleum, nuclear, and automobile, due to their high surface area and extensive internal pore structure. They are also commonly used in environmental processes for removing toxic gases and in wastewater and potable water treatments.

5. Preparing biomass-based activated carbon

The production process of biomass-based activated carbons typically consists of two stages: activation of the carbonized material, as depicted in **Figure 4**. The ultimate physical and chemical characteristics of the activated carbons, including pore volume, pore size distribution, type of surface functional groups, and surface area, are heavily influenced by factors such as the chemical composition of the biomass precursor materials, the techniques used for carbonization and activation, and the respective conditions employed.

5.1 Pretreatment of biomass precursor

Prior to the carbonization of biomass materials, it is crucial to remove any impurities such as soil by washing the materials with hot or cold deionized water, acid, or base. The materials should then be dried at a temperature between 65 and 105°C. To ensure a uniform and efficient carbonization process, the particle size of the dried biomass should be reduced through milling, as discussed in a study by Ahmed et al. [13].

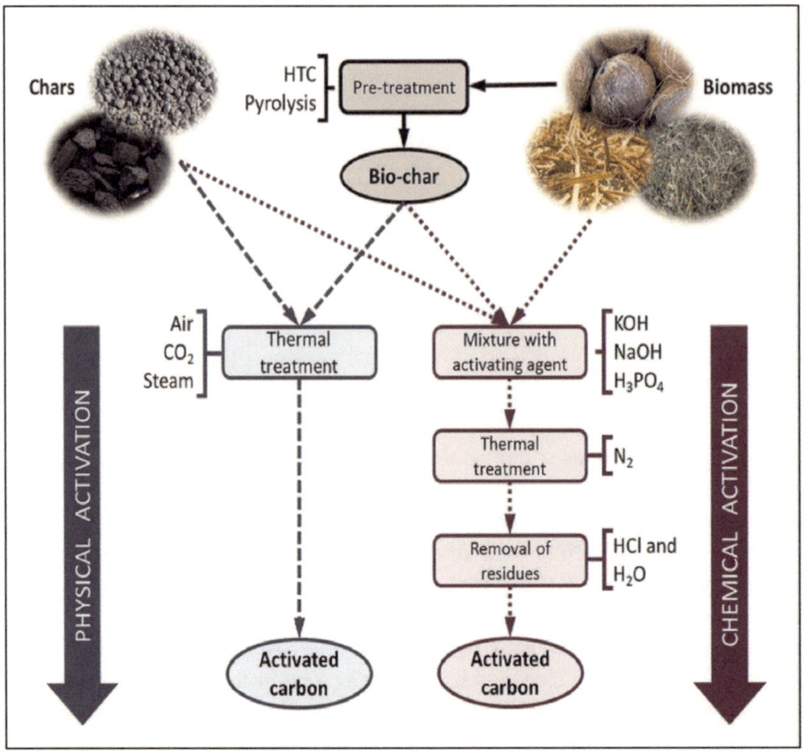

Figure 4.
The schematic production process of activated carbons, from biomass and char, by physical or chemical activation [38].

5.2 Carbonization

The thermal decomposition of biomass material at a temperature range of 600-800°C in an inert atmosphere is known as carbonization, which produces a fixed mass of carbon [39]. The most commonly used gas during carbonization is N_2 because of its cleanliness, ease of handling, and low cost [40]. Thermal decomposition eliminates moisture, low molecular weight substances, and non-carbon elements such as H_2, N_2, O_2, and S from the biomass precursor [41]. The carbonization process involves heating the biomass material in an electric furnace to the carbonization temperature while maintaining an inert gas flow rate, followed by cooling the produced biochar to room temperature before the activation stage. The key factors that influence the carbonization process include carbonization temperature, carbonization time, heating rate, and flow rate of the inert gas [42].

5.3 Activation

The activation process is designed to increase the porosity and surface area of activated carbon derived from biomass (**Figure 5**). The type and degree of the activation process can affect the physical and chemical properties, as well as the yield of biomass-based activated carbons [42]. The activation process involves the opening of

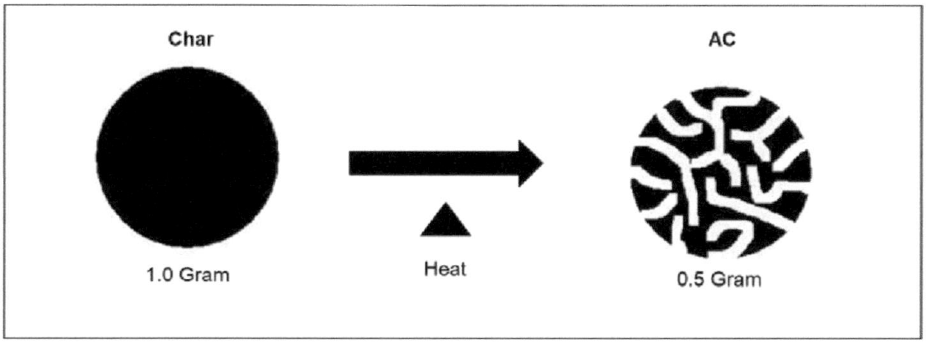

Figure 5.
Activation of the porosity.

inaccessible pores, the development of new pores, and the widening of existing pores, leading to increased porosity. Three methods are used for activating carbonized biomass materials: physical, chemical, and physiochemical activation. Physical activation employs steam, N_2, air, or CO_2 as activating agents, while chemical activation uses different chemical agents such as $ZnCl_2$, KOH, NaOH, HNO_3, H_2SO_4, K_2CO_3, and H_3PO_4, and physiochemical are physical and chemical methods [15]. During carbonization, tar products fill the pore structures of the carbon material, obstructing their openings. To enhance porosity and remove these obstructions, activation is necessary, as depicted in **Figure 5**.

5.3.1 Physical activation

The physical activation method involves heating carbonized biomass material to a predetermined activation temperature, typically ranging from 700 to 1100°C [40]. The activation agents used in this method include air, nitrogen, steam, or CO_2 [40]. The use of activation gas can prevent CO_2 formation and generate a wide range of pore sizes in the activated carbon during the activation process [43]. Carbon dioxide is the most commonly used activation agent in the physical activation process due to its ease of handling [44], which leads to higher carbon yield and the development of uniform pores [45]. However, the steam activation process results in lower carbon yield, which could be attributed to the rapid degradation of biomass materials at high temperatures [46]. Furthermore, physical activation with CO_2 produces microporous carbons, whereas activation with steam at identical conditions results in mesoporous and macroporous carbons [15]. Activation reactions of O_2, CO_2, and steam are given below [15]:

Oxygen	$C + O_2 \rightarrow CO_2$
	$2C + 2O_2 \rightarrow 2CO_2$
Carbon dioxide	$2C + O_2 \rightarrow 2CO$
Steam	$2C + 2H_2O \rightarrow 2H_2 + 2CO$
	$CO + H_2O \rightarrow H_2 + CO_2$
	$C + 2H_2 \rightarrow CH_4$

The main parameters that affect the physical activation process are activation temperature, activation time, heating rate, and activator flow rate.

5.3.2 Chemical activation

The chemical activation process involves the use of a chemical activator to enhance the surface properties of biochar such as porosity, pore size distribution, and surface area. Various chemicals can be used as activation agents, including H_3PO_4, H_2SO_4, KOH, K_2CO_3, and $ZnCl_2$ [47]. It is crucial to ensure complete mixing between the biomass material and the activation agent before the activation step to achieve the desired quality of the produced activated carbon. Chemical activators work by degrading and dehydrating organic carbon molecules of the starting biomass materials to increase the porosity and surface area of the produced BAC [40]. Chemical activation can be carried out in a single step or in two steps. In the single-step method, the biomass material is impregnated with the activator and then carbonized, while in the two-step method, the biomass material is initially carbonized in the first step and then impregnated with the activation agent and subjected to thermal activation in the second step. **Figure 4** shows the steps of the main methods for the preparation of biomass-based activated carbons.

5.3.3 Activation agents

Chemical activation has proven to be a successful method for producing activated carbons using a range of chemical reagents such as KOH, NaOH, $ZnCl_2$, H_2SO_4, and H_3PO_4.

5.3.3.1 H_3PO_4

The use of H_3PO_4 as an activation agent has been shown to improve the porosity and adsorptive properties of activated carbon due to cross-linking reactions with H_3PO_4 [48]. The products of the reactions between H_3PO_4 and organic substances in biomass materials diffuse through the char matrix, which can lead to the enlargement of existing pores or the generation of new pores [49].

5.3.3.2 KOH

According to Ji et al. [50], the primary reaction between KOH and amorphous carbon at the activation temperature is:

$$4KOH + C \rightarrow K_2CO_3 + K_2O + 2H_2. \qquad (1)$$

However, during the KOH activation process, various side reactions occur as described by Raymund et al. [51]:

$$2KOH \rightarrow K_2O + H_2O. \qquad (2)$$

$$C + H_2O^{Steam} \rightarrow H_2 + CO.$$

$$CO + H_2O \rightarrow H_2 + CO_2.$$

$$K_2O + CO_2 \rightarrow 2K_2CO_3.$$

$$K_2O + H_2 \rightarrow 2K + H_2O.$$
$$K_2O + C \rightarrow 2K + CO. \qquad (3)$$
$$K_2CO_3 + 2C \rightarrow 2K + 3CO.$$

The steam produced during KOH activation plays a crucial role in the creation of pores by converting amorphous carbon to carbon monoxide. Moreover, the activation process results in the formation of K_2CO_3, which serves as an activation agent to enhance the porosity and surface area of KOH-treated biomass.

5.3.3.3 $ZnCl_2$

The activation of biomass char using $ZnCl_2$ involves inhibiting the formation of tar and other liquids that can clog the pores of the biomass material [47]. This process allows for the free movement of volatiles through open pores and their release from the carbon surface, which leads to an increase in the porosity and surface area of the activated carbon. Moreover, $ZnCl_2$ activation converts the carbon skeleton of the activated material from aliphatic to aromatic form through cellulose dehydration, as demonstrated by Yorgun et al. [52].

5.3.3.4 H_2SO_4

The activation method involves the diffusion of H_2SO_4 into a char matrix, which prevents the formation of tar substances and promotes the development of oxygen-carbon functional groups on the surface of activated carbon [53]. However, the yield of carbon obtained by this method is typically low due to the carbon burn-off caused by the water vapor generated during H_2SO_4 dehydration [54]. The chemical equation for the reaction between H_2SO_4 and carbon is:

$$2H_2SO_4 + C \rightarrow 2SO_2 + CO_2 + 2H_2O. \qquad (4)$$

5.3.3.5 NaOH

The following reactions may occur during NaOH activation:

$$6NaOH + C \rightarrow 2Na + 3H_2 + 2Na_2CO_3. \qquad (5)$$
$$Na_2CO_3 + C \rightarrow Na_2O + 2CO_2.$$
$$2Na + CO_2 \rightarrow Na_2O + CO.$$

The chemical reactions between alkaline, carbonate metals, CO, CO_2, and H_2 gases have a crucial impact on the development of microporosity and mesoporosity and the stabilization and enlargement of pores in activated carbon. The ratio of NaOH to char is a significant determinant of BET surface area and pore volume, which consequently influences adsorption capacity [55]. An increase in the NaOH to char ratio can enhance adsorption capacity, but an excessive amount of NaOH can damage the carbon structure, decrease the BET surface area and adsorption capacity, and reduce carbon yield due to the gasification reaction of NaOH and the deposition of excessive amounts on the carbon pore wall [56]. The activation process leads to the formation of an irregular pore structure, as shown in **Figure 6**, in activated carbon.

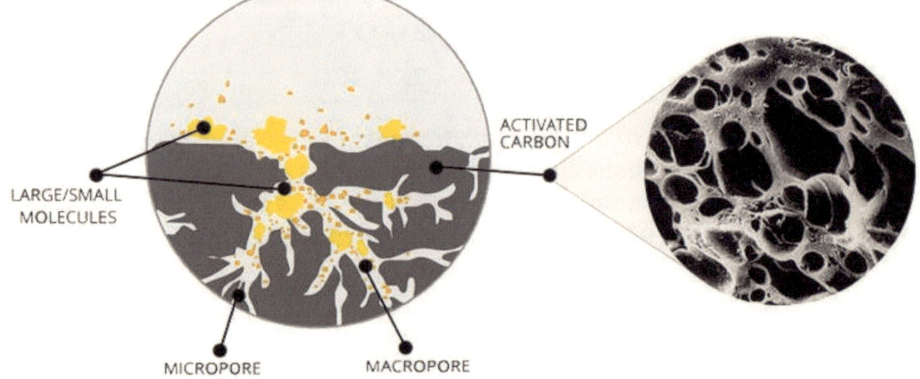

Figure 6.
Irregular pore structure of activated carbon, as a result of its activation treatment [57].

5.3.4 Physiochemical activation

The physiochemical activation method is a combination of both physical and chemical activation methods, and it is considered a recent advancement in the production of activated carbon. In this method, the activated carbon precursor is first activated chemically and then subjected to thermal treatment at a high temperature ranging between 600 and 1000°C. Initially, the char was subjected to chemical impregnation and dried at 85°C. **Figure 6** shows the carbonization process was carried out in a stainless-steel vertical tubular reactor that was placed inside a tube furnace. The reactor was purged with purified nitrogen (99.995%) at a flow rate of 150 cm^3/min, and the temperature was raised to 450°C, a fixed heating rate of 10°C/min, where it was maintained for 4 hours. The resulting mixture was dehydrated by placing it in an oven overnight at 85 C and then activated by gradually raising the temperature to 850°C at the same heating rate. Once the final temperature was reached, the flow of nitrogen gas was replaced by carbon dioxide, and the activation process was allowed to continue more 3 hours. Finally, the activated carbon was cooled to room temperature under a nitrogen flow of 150 cm^3/min [58]. The two-step physiochemical method for producing activated carbon is illustrated in **Figure** 7. This approach leads to the production of activated carbon with unique surface properties such as a very large surface area and high pore volume. However, this method tends to induce more mesopores, which is an important characteristic needed for liquid-phase applications. Therefore, the physiochemical activation method is currently a priority in activated carbon production technology.

5.3.5 Comparison among physical, chemical, and physiochemical activation

The selection of an activation method for deriving activated carbon depends on the precursor's properties and the intended application [59]. Physical activation is often preferred for industrial-scale production due to its ability to optimize the pyrolysis stage and achieve better control over microporosity development [60]. Physical activation also results in activated carbon that does not require chemical neutralization, reducing both the process costs and associated pollution. Moreover, the higher physical strength of physically activated carbon makes them ideal for use in high-pressure

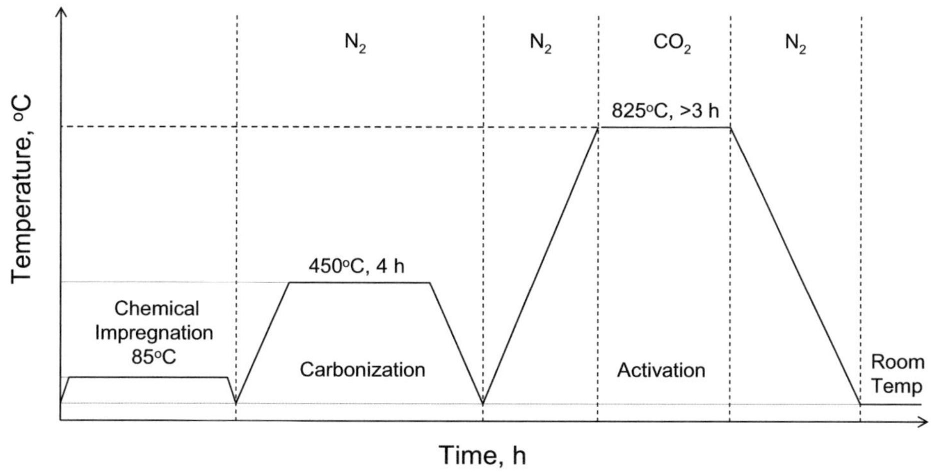

Figure 7.
Thermal treatment scheme of two-step physiochemical method for preparation of activated carbon [58].

columns [61]. As a result, physical activation is commonly used to prepare activated carbon for water treatment.

In contrast, chemical activation offers lower activation temperatures, shorter treatment times, reduced energy requirements, and higher activated carbon yields with a larger surface area and pore volume [62]. However, the process and chemical costs associated with chemical activation are higher than those of physical activation [63]. The lower activation temperature and homogeneously developed internal micropores result in superior physical and chemical properties compared to physical activation [64]. Physical activation results in a nonuniform shape and pore development, leading to higher weight losses and lower yields compared to chemical activation. However, the physiochemical activation method can produce activated carbon with even higher specific surface area, porosity, and pore volume than chemical or physical activation due to improved diffusion mass transfer within the carbon matrix, providing superior adsorption capacity [65]. **Table 1** shows the optimal conditions for preparing biomass-based activated carbon using physical, chemical, or physiochemical activation methods and their corresponding maximum adsorption capacities.

6. Environmental remediation

Activated carbon has a wide range of applications, including purifying water and gas, treating sewage, extracting metals, producing medication, storing energy, controlling air pollution, and facilitating catalytic processes. Water treatment and purification, which involve removing contaminants from liquid-phase drinking water, groundwater, and wastewater, are the most significant areas of demand for activated carbon globally. High-quality activated carbon can effectively eliminate phenols, organic and inorganic toxic compounds from both drinking water and wastewater. In addition, activated carbon materials are effective adsorbents for heavy metals such as Cr, Pb, Cu, Cd, Zn, and Hg and can remove a wide range of contaminants and carcinogenic compounds [78]. The use of activated carbon for gas purification is an effective and environmentally friendly method for removing gaseous pollutants from

Biomass waste	Activation method	Activated carbon properties			Maximum adsorption capacity (mg/g)	References
		Surface area (BET, m^2/g)	Pore volume (cm^3/g)	Pore diameter (nm)		
Coconut shell	Physical	1022.4	0.492	1.516	45.2	[66]
Sugarcane bagasse	Chemical	1145	1.3	4.54	315.00	[67]
Pistachio wood	Chemical	1884	0.994	2.11	190.2	[68]
Date press cake	Chemical	2025.9	0.932	1.839	282.8	[69]
Coffee ground	Chemical	704.23	0.293	2.20	665.9	[70]
Avocado kernel seeds	Physical	299.9	0.172	1.34	19.99	[71]
Prawn carapace	Physical	56.3	0.17	—	218.3	[72]
Corn style		47.3	0.12	—	180.2	[72]
Pumpkin seed hulls	Physiochemical	737.90	0.37	2.261S	260.79	[73]
Palm oil fronds	Physiochemical	1237.13	0.667	2.16	—	[74]
Argania Spinosa tree nutshells	Physical	1159	0.64	—	121.9	[75]
Argania Spinosa tree nutshells	Physical	1159	0.64	—	175.4	[75]
Langsat empty fruit bunch	Chemical	1070.36	0.83	3.671	261.2	[76]
Date stones	Physiochemical	852	0.671	3.15	—	[77]

Table 1.
Preparation of activated carbon from biomass using either physical, chemical, or physiochemical activation method, properties, and maximum adsorption capacities.

industrial and domestic emissions. It is widely used in various industries, including petrochemicals, food and beverage, pharmaceuticals, and semiconductor manufacturing. Activated carbon is also a promising candidate for capturing CO_2, which contributes to climate change and poses a threat to human health. Activated carbon, derived from waste biomass materials, exhibits CO_2 uptake that is comparable to that of commercial adsorbents. Due to its large inner surface area, it is regarded as the most promising adsorbent for CO_2 capture. In addition to CO_2, activated carbon can effectively adsorb various gaseous pollutants such as CH_4, H_2S, NO_2, and H_2, making it an effective adsorbent for air pollution control [79]. Despite its relatively low density, activated carbon can be used for hydrogen storage due to its high surface area, abundant pore volume, acceptable pore size, large microporosity, and controlled pore size distribution. Highly microporous activated carbon is a suitable candidate for hydrogen storage at 77 K [80]. Various biomass-based activated carbons have been synthesized for hydrogen storage applications, providing sustainable and environmentally friendly alternatives to traditional fossil fuel-based materials. Activated carbons with a high surface area of up to 2700 m^2/g have been obtained [81]. Overall, the use of activated carbon for hydrogen storage is an active area of research with great potential for practical applications. As new applications are discovered and more

DOI: http://dx.doi.org/10.5772/intechopen.111852

sources of raw materials are utilized for its production, the use of activated carbon is expected to continue to increase.

7. Conclusions

The utilization of activated carbon derived from biomass in gases and aqueous solutions has grown significantly in various industrial applications. However, due to its high cost and nonrenewable sourcing, alternative sources of these carbons are being sought after. This review highlights biomass as a resource, the structure of biomass, and the production methods for activated carbon using biomass materials, as well as various carbonization and activation techniques. Activated carbon has demonstrated great potential in water treatment, often surpassing commercial carbons, depending on the target contaminant. While activated carbons have been extensively studied, there is a lack of research on how different production methods affect their use in continuous bed columns and regeneration potential. Biomass materials' composition varies depending on origin and agronomic practices, necessitating more detailed studies on the production, optimization, and application of biomass-based activated carbon.

Acknowledgements

The corresponding author gratefully acknowledges the fellowship awarded by the Council for At-Risk Academics (Cara).

Author details

Abdulbari A. Ahmad[1*], Abdulraqeeb Alwahbi[2], Laila A. Al Khatib[3] and Hani Dammag[4]

1 Institute for Infrastructure and Environment, University of Edinburgh, UK

2 Chemical Engineering Department, Hadhramout University, Yemen

3 Environmental Engineering Department, Al-Hussein Bin Talal University, Jordan

4 Chemistry of Science Department, Sana'a University, Yemen

*Address all correspondence to: abdulbari.saeed@ed.ac.uk

References

[1] Vassilev SV, Vassileva CG, Vassilev VS. Advantages and disadvantages of composition and properties of biomass in comparison with coal: An overview. Fuel. 2015;**158**: 330-350

[2] Vassilev SV, Vassileva CG, Baxter D. Trace element concentrations and associations in some biomass ashes. Fuel. 2014;**129**:292-313

[3] Vassilev SV, Baxter D, Andersen LK, Vassileva CG, Morgan TJ. An overview of the organic and inorganic phase composition of biomass. Fuel. 2012;**94**: 1-33

[4] Vassilev SV, Baxter D, Andersen LK, Vassileva CG. An overview of the composition and application of biomass ash. Part 1. Phase mineral and chemical composition and classification. Fuel. 2013;**105**:40-76

[5] Chiang K-Y, Chien K-L, Lu C-H. Characterization and comparison of biomass produced from various sources: Suggestions for selection of pretreatment technologies in biomass-to-energy. Applied Energy. 2012;**100**:164-171

[6] Darvell L, Jones J, Gudka B, Baxter X, Saddawi A, Williams A, et al. Combustion properties of some power station biomass fuels. Fuel. 2010;**89**: 2881-2890

[7] Meyer S, Glaser B, Quicker P. Technical, economical, and climate-related aspects of biochar production technologies: A literature review. Environmental Science & Technology. 2011;**45**:9473-9483

[8] Ouyang J, Zhou L, Liu Z, Heng JY, Chen W. Biomass-derived activated carbons for the removal of pharmaceutical mircopollutants from wastewater: A review. Separation and Purification Technology. 2020;**253**: 117536

[9] Zabara B, Ahmad AA. Biomass waste in Yemen: Management and challenges. Waste Management in MENA Regions. 2020:313-336

[10] Stefanidis SD, Kalogiannis KG, Iliopoulou EF, Michailof CM, Pilavachi PA, Lappas AA. A study of lignocellulosic biomass pyrolysis via the pyrolysis of cellulose, hemicellulose and lignin. Journal of Analytical and Applied Pyrolysis. 2014;**105**:143-150

[11] Tripathi M, Sahu JN, Ganesan P. Effect of process parameters on production of biochar from biomass waste through pyrolysis: A review. Renewable and Sustainable Energy Reviews. 2016;**55**:467-481

[12] Peng X, Jiang Y, Chen Z, Osman AI, Farghali M, Rooney DW, et al. Recycling municipal, agricultural and industrial waste into energy, fertilizers, food and construction materials, and economic feasibility: A review. Environmental Chemistry Letters. 2023;**21**:765-801

[13] Ahmad AA, Al-Raggad M, Shareef N. Production of activated carbon derived from agricultural by-products via microwave-induced chemical activation: A review. Carbon Letters. 2021;**31**: 957-971

[14] Martí Rosselló T, Li J, Lue L. Kinetic models for biomass pyrolysis. Archives of Industrial Biotechnology. 2006;**1**:4-7

[15] Maniarasu R, Rathore SK, Murugan S. Biomass-based activated carbon for CO2 adsorption a review.

Energy & Environment. 2022: 0958305X221093465

[16] Khaire KC, Moholkar VS, Goyal A. Bioconversion of sugarcane tops to bioethanol and other value added products: An overview. Materials Science for Energy Technologies. 2021;**4**: 54-68

[17] Ukanwa KS, Patchigolla K, Sakrabani R, Anthony E, Mandavgane S. A review of chemicals to produce activated carbon from agricultural waste biomass. Sustainability. 2019;**11**:6204

[18] Ahmad A, Hameed B. Reduction of COD and color of dyeing effluent from a cotton textile mill by adsorption onto bamboo-based activated carbon. Journal of Hazardous Materials. 2009;**172**: 1538-1543

[19] Ahmad A, Hameed B, Ahmad A. Removal of disperse dye from aqueous solution using waste-derived activated carbon: Optimization study. Journal of Hazardous Materials. 2009;**170**:612-619

[20] Zhang Z, Qu W, Peng J, Zhang L, Ma X, Zhang Z, et al. Comparison between microwave and conventional thermal reactivations of spent activated carbon generated from vinyl acetate synthesis. Desalination. 2009;**249**: 247-252

[21] Sharif YM, Saka C, Baytar O, Ãm Å. Preparation and characterization of activated carbon from sesame seed shells by microwave and conventional heating with zinc chloride activation. Analytical Letters. 2018;**51**:2733-2746

[22] Hameed B, Ahmad A. Batch adsorption of methylene blue from aqueous solution by garlic peel, an agricultural waste biomass. Journal of Hazardous Materials. 2009;**164**:870-875

[23] Wang T, Tan S, Liang C. Preparation and characterization of activated carbon from wood via microwave-induced $ZnCl_2$ activation. Carbon. 2009;**47**: 1880-1883

[24] Deng H, Li G, Yang H, Tang J, Tang J. Preparation of activated carbons from cotton stalk by microwave assisted KOH and K_2CO_3 activation. Chemical Engineering Journal. 2010;**163**:373-381

[25] Hirata M, Kawasaki N, Nakamura T, Matsumoto K, Kabayama M, Tamura T, et al. Adsorption of dyes onto carbonaceous materials produced from coffee grounds by microwave treatment. Journal of Colloid and Interface Science. 2002;**254**:17-22

[26] Du C, Yang H, Wu Z, Ge X, Cravotto G, Ye B-C, et al. Microwave-assisted preparation of almond shell-based activated carbon for methylene blue adsorption. Green Processing and Synthesis. 2016;**5**:395-406

[27] Yagmur E, Ozmak M, Aktas Z. A novel method for production of activated carbon from waste tea by chemical activation with microwave energy. Fuel. 2008;**87**:3278-3285

[28] Abbas AF, Ahmed MJ. Mesoporous activated carbon from date stones (Phoenix dactylifera L.) by one-step microwave assisted K_2CO_3 pyrolysis. Journal of water. Process Engineering. 2016;**9**:201-207

[29] Bisinella V, Hulgaard T, Riber C, Damgaard A, Christensen TH. Environmental assessment of carbon capture and storage (CCS) as a post-treatment technology in waste incineration. Waste Management. 2021; **128**:99-113

[30] Abuelnoor N, AlHajaj A, Khaleel M, Vega LF, Abu-Zahra MR. Activated

carbons from biomass-based sources for CO_2 capture applications. Chemosphere. 2021;**282**:131111

[31] Yargicoglu EN, Sadasivam BY, Reddy KR, Spokas K. Physical and chemical characterization of waste wood derived biochars. Waste Management. 2015;**36**:256-268

[32] Nowrouzi M, Younesi H, Bahramifar N. High efficient carbon dioxide capture onto as-synthesized activated carbon by chemical activation of Persian Ironwood biomass and the economic pre-feasibility study for scale-up. Journal of Cleaner Production. 2017; **168**:499-509

[33] Kargbo H, Harris JS, Phan AN. Drop-in fuel production from biomass: Critical review on techno-economic feasibility and sustainability. Renewable and Sustainable Energy Reviews. 2021;**135**: 110168

[34] Kantarli IC, Yanik J. Activated carbon from leather shaving wastes and its application in removal of toxic materials. Journal of Hazardous Materials. 2010;**179**:348-356

[35] Gopalan J, Buthiyappan A, Raman AAA. Insight into metal-impregnated biomass based activated carbon for enhanced carbon dioxide adsorption: A review. Journal of Industrial and Engineering Chemistry. 2022;**113**:72-95

[36] Abdullah MO, Tan IAW, Lim LS. Automobile adsorption air-conditioning system using oil palm biomass-based activated carbon: A review. Renewable and Sustainable Energy Reviews. 2011; **15**:2061-2072

[37] Everett D. "IUPAC manual of symbols and terminology", appendix 2, part 1, colloid and surface chemistry. Pure and Applied Chemistry. 1972;**31**: 578-621

[38] Castro-Gutiérrez J, Celzard A, Fierro V. Energy storage in supercapacitors: Focus on tannin-derived carbon electrodes. Frontiers in materials. 2020;7:217

[39] Azmi NZM, Buthiyappan A, Raman AAA, Patah MFA, Sufian S. Recent advances in biomass based activated carbon for carbon dioxide capture-a review. Journal of Industrial and Engineering Chemistry. 2022;**116**: 1-20

[40] Danish M, Ahmad T. A review on utilization of wood biomass as a sustainable precursor for activated carbon production and application. Renewable and Sustainable Energy Reviews. 2018;**87**:1-21

[41] Neme I, Gonfa G, Masi C. Activated carbon from biomass precursors using phosphoric acid: A review. Heliyon. 2022;**8**:11940

[42] Malini K, Selvakumar D, Kumar N. Activated carbon from biomass: Preparation, factors improving basicity and surface properties for enhanced CO_2 capture capacity a review. Journal of CO_2 Utilization. 2023;**67**:102318

[43] Rodriguez-Reinoso F, Molina-Sabio M. Activated carbons from lignocellulosic materials by chemical and/or physical activation: An overview. Carbon. 1992;**30**:1111-1118

[44] Rashidi NA, Yusup S. A review on recent technological advancement in the activated carbon production from oil palm wastes. Chemical Engineering Journal. 2017;**314**:277-290

[45] Mohammad-Khah A, Ansari R. Activated charcoal: Preparation,

characterization and applications: A review article. International Journal of ChemTech Research. 2009;**1**:859-864

[46] Zaini MAA, Zhi LL, Hui TS, Amano Y, Machida M. Effects of physical activation on pore textures and heavy metals removal of fiber-based activated carbons. Materials Today: Proceedings. 2021;**39**:917-921

[47] Yang J, Qiu K. Preparation of activated carbons from walnut shells via vacuum chemical activation and their application for methylene blue removal. Chemical Engineering Journal. 2010;**165**: 209-217

[48] Liu Q-S, Zheng T, Wang P, Guo L. Preparation and characterization of activated carbon from bamboo by microwave-induced phosphoric acid activation. Industrial Crops and Products. 2010;**31**:233-238

[49] Li W, Yang K, Peng J, Zhang L, Guo S, Xia H. Effects of carbonization temperatures on characteristics of porosity in coconut shell chars and activated carbons derived from carbonized coconut shell chars. Industrial Crops and Products. 2008;**28**: 190-198

[50] Ji Y, Li T, Zhu L, Wang X, Lin Q. Preparation of activated carbons by microwave heating KOH activation. Applied Surface Science. 2007;**254**: 506-512

[51] Raymundo-Pinero E, Azaïs P, Cacciaguerra T, Cazorla-Amorós D, Linares-Solano A, Béguin F. KOH and NaOH activation mechanisms of multiwalled carbon nanotubes with different structural organisation. Carbon. 2005;**43**:786-795

[52] Yorgun S, Vural N, Demiral H. Preparation of high-surface area activated carbons from paulownia wood by $ZnCl_2$ activation. Microporous and Mesoporous Materials. 2009;**122**:189-194

[53] Izquierdo M, Rubio B, Mayoral C, Andres J. Modifications to the surface chemistry of low-rank coal-based carbon catalysts to improve flue gas nitric oxide removal. Applied Catalysis B: Environmental. 2001;**33**:315-324

[54] Guo J, Xu WS, Chen YL, Lua AC. Adsorption of NH_3 onto activated carbon prepared from palm shells impregnated with H_2SO_4. Journal of Colloid and Interface Science. 2005;**281**:285-290

[55] Foo K, Hameed B. Coconut husk derived activated carbon via microwave induced activation: Effects of activation agents, preparation parameters and adsorption performance. Chemical Engineering Journal. 2012;**184**:57-65

[56] Lillo-Ródenas MA, Cazorla-Amorós D. Understanding chemical reactions between carbons and NaOH and KOH: An insight into the chemical activation mechanism. Carbon. 2003;**41**: 267-275

[57] De Decker J. Functionalized Metal-Organic Frameworks as Selective Metal Adsorbents. Ghent, Belgium: Ghent University. Faculty of Sciences [Ph.D. Thesis]; 2017.

[58] Hameed B, Salman J, Ahmad A. Adsorption isotherm and kinetic modeling of 2, 4-D pesticide on activated carbon derived from date stones. Journal of Hazardous Materials. 2009;**163**: 121-126

[59] Hesas RH, Arami-Niya A, Daud WMAW, Sahu J. Preparation of granular activated carbon from oil palm shell by microwave-induced chemical activation: Optimisation using surface

response methodology. Chemical Engineering Research and Design. 2013; **91**:2447-2456

[60] Pallarés J, González-Cencerrado A, Arauzo I. Production and characterization of activated carbon from barley straw by physical activation with carbon dioxide and steam. Biomass and Bioenergy. 2018;**115**:64-73

[61] Yang K, Peng J, Xia H, Zhang L, Srinivasakannan C, Guo S. Textural characteristics of activated carbon by single step CO_2 activation from coconut shells. Journal of the Taiwan Institute of Chemical Engineers. 2010;**41**:367-372

[62] Supong A, Bhomick PC, Baruah M, Pongener C, Sinha UB, Sinha D. Adsorptive removal of Bisphenol a by biomass activated carbon and insights into the adsorption mechanism through density functional theory calculations. Sustainable Chemistry and Pharmacy. 2019;**13**:100159

[63] Bergna D, Hu T, Prokkola H, Romar H, Lassi U. Effect of some process parameters on the main properties of activated carbon produced from peat in a lab-scale process. Waste and Biomass Valorization. 2020;**11**:2837-2848

[64] Ideta K, Kim D-W, Kim T, Nakabayashi K, Miyawaki J, Park J-I, et al. 19F ex situ solid-state NMR study on structural differences in pores of activated carbon series derived from chemical and physical activation processes for EDLCs. The Journal of Physical Chemistry C. 2020;**124**: 12457-12465

[65] Aguayo-Villarreal I, Bonilla-Petriciolet A, MuÃiz-Valencia R. Preparation of activated carbons from pecan nutshell and their application in the antagonistic adsorption of heavy metal ions. Journal of Molecular Liquids. 2017;**230**:686-695

[66] Wang Y, Peng C, Padilla-Ortega E, Robledo-Cabrera A, LÃpez-Valdivieso A. Cr (VI) adsorption on activated carbon: Mechanisms, modeling and limitations in water treatment. Journal of environmental. Chemical Engineering. 2020;**8**:104031

[67] El Naga AOA, El Saied M, Shaban SA, El Kady FY. Fast removal of diclofenac sodium from aqueous solution using sugar cane bagasse-derived activated carbon. Journal of Molecular Liquids. 2019;**285**:9-19

[68] Sajjadi S-A, Meknati A, Lima EC, Dotto GL, Mendoza-Castillo DI, Anastopoulos I, et al. A novel route for preparation of chemically activated carbon from pistachio wood for highly efficient Pb (II) sorption. Journal of Environmental Management. 2019;**236**:34-44

[69] Norouzi S, Heidari M, Alipour V, Rahmanian O, Fazlzadeh M, Mohammadi-Moghadam F, et al. Preparation, characterization and Cr (VI) adsorption evaluation of NaOH-activated carbon produced from date press cake; an agro-industrial waste. Bioresource Technology. 2018;**258**:48-56

[70] Jung K-W, Choi BH, Hwang M-J, Jeong T-U, Ahn K-H. Fabrication of granular activated carbons derived from spent coffee grounds by entrapment in calcium alginate beads for adsorption of acid orange 7 and methylene blue. Bioresource Technology. 2016;**219**: 185-195

[71] Salomón-Negrete MÁ, Reynel-Ãvila HE, Mendoza-Castillo DI, Bonilla-Petriciolet AN, Duran-Valle CJ. Water defluoridation with avocado-based adsorbents: Synthesis, physicochemical characterization and thermodynamic

studies. Journal of Molecular Liquids. 2018;**254**:188-197

[72] Das D, Das N. Optimization of parameters for cerium (III) biosorption onto biowaste materials of animal and plant origin using 5-level box-Behnken design: Equilibrium, kinetic, thermodynamic and regeneration studies. Journal of Rare Earths. 2014;**32**:745-758

[73] Njoku V, Foo K, Hameed B. Microwave-assisted preparation of pumpkin seed hull activated carbon and its application for the adsorptive removal of 2, 4-dichlorophenoxyacetic acid. Chemical Engineering Journal. 2013;**215**: 383-388

[74] Salman JM. Optimization of preparation conditions for activated carbon from palm oil fronds using response surface methodology on removal of pesticides from aqueous solution. Arabian Journal of Chemistry. 2014;**7**:101-108

[75] Daou C, Rafqah S, Najjar F, Anane H, Piram A, Hamade A, et al. TiO2 and activated carbon of Argania Spinosa tree nutshells composites for the adsorption photocatalysis removal of pharmaceuticals from aqueous solution. Journal of Photochemistry and Photobiology A: Chemistry. 2020;**388**: 112183

[76] Njoku V, Islam MA, Asif M, Hameed B. Adsorption of 2, 4-dichlorophenoxyacetic acid by mesoporous activated carbon prepared from H_3PO_4-activated langsat empty fruit bunch. Journal of Environmental Management. 2015;**154**:138-144

[77] Darweesh TM, Ahmed MJ. Batch and fixed bed adsorption of levofloxacin on granular activated carbon from date (Phoenix dactylifera L.) stones by KOH chemical activation. Environmental Toxicology and Pharmacology. 2017;**50**: 159-166

[78] Neolaka YA, Riwu AA, Aigbe UO, Ukhurebor KE, Onyancha RB, Darmokoesoemo H, et al. Potential of activated carbon from various sources as a low-cost adsorbent to remove heavy metals and synthetic dyes. Results. Chemistry. 2023;**5**:100711

[79] Reza MS, Yun CS, Afroze S, Radenahmad N, Bakar MSA, Saidur R, et al. Preparation of activated carbon from biomass and itsâ€™ applications in water and gas purification, a review. Arab Journal of Basic and Applied Sciences. 2020;**27**:208-238

[80] Balathanigaimani M, Shim WG, Kim SC. Biomass-based carbon materials for energy storage and environmental applications. Applied Chemistry for Engineering. 2017;**28**:8-16

[81] Sevilla M, Mokaya R. Energy storage applications of activated carbons: Supercapacitors and hydrogen storage. Energy & Environmental Science. 2014; **7**:1250-1280

Chapter 5

Evaluation of Ethanol Production Process by the Fermentation of Blue Cabuya Juice (*Agave americana*), Using Yeast Strains (*Saccharomyces cerevisiae*) as Potential Application for Blue Agave Liquor

Roberto Robles Calderón, Raymundo Erazo Erazo,
Julio Javier Armijo Carranza and Benito Buendía Quispe

Abstract

The blue cabuya (*Agave americana L.*) grows wild on large tracts of land in various regions of Peru. The research is aimed at taking advantage of this renewable plant biomass as a potential substrate in one of its possible applications such as cabuya azul liquor, for which, firstly, the yields of the anaerobic fermentation process of this biomass at temperatures of 26, 28 and 30 C and a pH of 3.5 and 4.0, in a batch bioreactor of 120 liters and nutrient concentrations of diammonium phosphate (DAP) of 1 and 2 g/L and a substrate with reducing sugars (AR) of 60 g/L from the district of La Merced, province of Churcampa, Huancavelica Region. The experimental tests were carried out in the laboratories of the Faculties of Biological Sciences and Chemistry and Chemical Engineering of the Universidad Nacional Mayor de San Marcos. Using *Saccharomyces cerevisiae yeast strains* (D 47, LALVIN-Canada) at 28 C, pH 3.5 and DAP nutrient concentration of 2 g/L during 26 h of fermentation, was obtained total biomass yields $Y(_{x/s})$ =. 0.1606 and ethanol $Y(_{p/s})$ = 0.4739. It is concluded that the production of ethanol for use in its industrialization as blue cabuya liquor is viable and for the transfer of technology environmental aspects and economic studies must be integrated.

Keywords: blue cabuya, substrate, *Saccharomyces cerevisiae*, diammonium phosphate, ethanol production

1. Introduction

Given the particularity of different microclimates in Peru, the blue cabuya (*Agave americana L.*) grows wild mainly in the Regions of Huancavelica, Huánuco, Ancash, and Cajamarca where their inter-Andean valleys have mild weather between 18 C to 26 C.

In addition to this, the favorable conditions of the composition of organic nutrients and compounds of mineral origin of the soils are ideal for the development of *Agave americana L*. Thus, the maturation time of this plant fluctuates between five to ten years less than the seven to nine years required for the maturation of the same variety in Mexico.

In studies carried out in Mexico, the authenticity of parameters was determined in agricultural soils belonging to the region that grants the Denomination of Origin Tequila (DOT), characterizing 26 samples belonging to Jalisco, Michoacán, Nayarit, and Sinaloa. The characterization consisted of determining the texture, apparent density, color, pH, humidity, organic matter, total carbon, and trace metals. The trace presence of the minerals Cr, Ba, and Zr are useful as auxiliary tracers to determine the authenticity of soils and agaves belonging to the DOT region [1].

The preferable soils for growing tequilana agave from Blue Weber variety must have good drainage. The sucker roots of the agave explore the surface soil when water is available and are practically eliminated in dry seasons. For proper functioning, gas exchange is required in the soil. Therefore, flood areas must be avoided. It is essential that the pH of the soil is not very acidic, since at a pH below 5.5 the plants present nutritional imbalances that if not managed properly can cause damage such as the red ring, a disease that causes the appearance of red areas in the leaves, which form a belt around them that constricts them, many times a hardening and delay in the growth of the plant occurs [2].

Recently, new auxiliary analytical techniques have been analyzed in the current verification process for the detection of counterfeits and/or adulterations of tequila. According to the Official Mexican Standard NOM-006-SCFI-2012, tequila is a regional alcoholic beverage obtained by distillation of musts and prepared directly from the extracted material in the factory facilities of an authorized producer. Tequila is made from tequilana Weber Agave from the blue variety, cultivated in specific protected regions of Mexico that make up the geographical territory of Denomination of Origin Tequila (DOT). According to Mexican regulations, Tequila has a classification based on its category (100% agave Tequila and Tequila) and its classes (silver, gold, extra-aged, and ultra-aged). Based on 2021 data from the Tequila Regulatory Council (CRT), the tequila industry produced 352 million liters and exported to 120 countries [3].

The authenticity and traceability of Tequila are established through a permanent inspection carried out by the CRT at the facilities of the tequila producers, verifying the production process in all its stages and according to the Official Mexican Standard NOM-006-SCFI- 2012. However, the consumption of tequila has increased in volume, being marketed in more countries around the world. To achieve this, certain competitive criteria must be met to demonstrate the quality, safety, and authenticity of the Tequila.

The diversity and dynamics of yeasts and bacteria during small-scale spontaneous fermentations of agave juice have been identified and described. Observing a high heterogeneity in the microbial populations and fermentation parameters that shows the bacteria have a greater diversity than the yeast. The central microorganisms identified were *Saccharomyces cerevisiae* and *Lactobacillus fermentum*. Bacterial growth and the concomitant production of lactic acid were associated with low ethanol production. Therefore, bacteria could be defined as contaminants in tequila fermentation and efforts to control them should be implemented [4].

The most important stage in tequila processing is the fermentation of sugar, mainly fructose, which is transformed into ethanol and volatile compounds that give tequila its unique characteristics [5]. Tequila companies often prefer to commercially inoculate *Saccharomyces cerevisiae* strains that provide batch-to-batch homogeneity; however, low yields are still generated as the strains are not well adapted to agave juice, a totally different environment from the origin of the strain [6].

The *Agave tequilana* Weber of blue var. is a semelparous plant, that is, once it blooms, it dies. Its reproduction can be sexual by seeds; or asexual by shoots of rhizomes or bulbils of the inflorescence. For Tequila production, instead of allowing the plant to develop the flower to attract pollinating birds, bats, and moths that feed on the nectar, the flower stem is cut as soon as it begins to bud. Thus, interrupting the process of flowering. The reason is that during the flowering process, the plant consumes its carbohydrate reserve and then dies [7].

The biological cycle of agave cultivation oscillates between six and eight years. During that time of its growth, the pineapple or head can weigh between 35 kg and 120 kg, although some reach up to 150 kg depending on the conditions of the cultivation process. The maturity of the plant is manifested in the appearance of the so-called "quiote", but this must be cut quickly since its permanence in the plant consumes the sugars accumulated for years [8].

Lavado, Robles, and Yenque (2015), in Peru, studied the physicochemical properties of blue cabuya juice obtained in the Huanca Huanca district, Angaraes province, Huancavelica Region [9]. **Table 1** shows the physicochemical properties of blue cabuya juice.

The water-soluble carbohydrates (WSC) contained in the agave heads were: 28,3 ± 0,1% g/100 g (fresh weight) and 86,7 ± 1,3% g/100 g (dry weight) [10]. These high values are consistent with previous reports by other researchers [11, 12].

It has been studied that the variables that influence the alcoholic fermentation process are mainly temperature, degrees Brix (°Brix), consumption of AF fermentable sugars (fructose and glucose), biomass concentration, and nitrogen concentration [13]. On the other hand, It has been studied the effect of pH and airflow on the production of biomass, ethanol, the synthesis of aromatic compounds, the consumption of reducing sugars, the yields, and speeds of production of biomass and ethyl alcohol of two strains of *Saccharomyces cerevisiae* cultivated continuously [14]. Thus, the effect of the dilution speed and the addition of nutrients on the fermentation and synthesis of volatile compounds of two native genera of *Saccharomyces cerevisiae*, S1 and S2, in continuous cultures fed with *Agave tequilana* juice have also been studied [15].

Other researchers studied the effect of temperature and pH variables on the yield of alcoholic fermentation of Agave cocui must. First, they studied the temperature: 27, 31, 33, 35, and 37 C at constant pH. Then, they evaluated the pH at values of 3.0, 4.0, and 5.0 at a constant temperature. The results obtained indicated that the maximum yield of alcoholic fermentation is at a temperature of 33 C and a pH of 4.0 with an ethanol productivity of 1.14 g/(L.h) and a yield of 81.5% [16].

For modern literature standardization purposes, fermentation is defined in microbiology as the type of metabolism of a carbon source in which energy is generated by phosphorylation at the substrate level where organic molecules function as the final acceptor of electrons (or as acceptors of reducing equivalents), generated during the decomposition of carbon-containing compounds or catabolism. As is well known,

Inulin	**°Brix**	**Density**	**Viscosity**
(g/100 g juice)	**(ss/100 mL sol.)**	**(g/mL)**	**(cp)**
56,60	14,00	1,27	1,56
Source [9].			

Table 1.
Physicochemical properties of blue cabuya juice.

the process is called respiration when the final acceptor is an inorganic compound. Respiration is called aerobic if the final acceptor is oxygen and anaerobic when it is some other inorganic compound apart from oxygen, for example, sulfate or nitrate [17].

The musts for alcoholic fermentation are inoculated within the range of 5 x 10^6 cells/ml, equivalent to 1.5 g/L to 6.0 g/L in wet weight and 0.3–1.2 g/L in dry weight. The inoculum is related to the gravity of the wort and should produce as fast fermentation as possible without compromising the quality of the beer or the size of the yeast crop. The pitching rate influences both the fermentation rate and the degree of yeast growth [18].

In this context, the research was oriented to evaluate the production process of ethanol by fermentation of the juice of the blue cabuya (*Agave americana L.*) using yeast strains (*saccharomyces cerevisiae*). First, the temperature and pH of the alcoholic fermentation process must be determined to achieve the best production yield. Then, the concentration of diammonium phosphate, which is a nutrient, will be determined. Finally, the ethanol production yield will be determined by its potential application as a blue cabuya liquor.

2. Methodology, results, and discussion

2.1 Methodology

The development of the research was carried out at the laboratory level and was planned with the purpose of evaluating the yield of ethanol production by fermentation using blue cabuya juice as a substrate and a yeast *saccharomyces cerevisiae*, for which experimentation is used by manipulating and controlling the independent variables and analyzing their effects on the yield of ethanol production (dependent variable).

2.1.1 Type of research

According to the classification proposed by Hernández Sampieri, Fernández, and Baptista [19], this research is descriptive and explanatory with independent variables such as temperature, pH, substrate concentration, and nutrient concentration (diammonium phosphate, DAP).

2.1.2 Research design

The research design is experimental with a 3 x 2 x 2 factorial arrangement with experimental tests carried out with two repetitions. The temperature was considered with three levels: 24 C, 28 C, and 32 C, the pH with two levels: 3.5 and 4.5; and the nutrient concentration with two levels: 1 g/L and 2 g/L. The response variable was the ethanol concentration.

Table 2 shows the experimental design of the first stage of the alcoholic fermentation process of blue cabuya juice at different conditions: temperature, pH, and DAP nutrient concentration, for a substrate concentration of 60 g/L [20].

2.1.3 Obtaining cabuya juice

The development of organoleptic properties such as sweetness, acidity, aromas, and flavorstakes place in the state of complete maturity of the blue cabuya; therefore, it is important to extract the juice when the plant is fully mature. The blue cabuya completes its life cycle after flowering, so the juice is extracted before the floral scape comes out.

Test	Temperature °C	pH	Nutrient (DAP), (g/L)
1	28	4,5	1
2	28	4,5	2
3	32	4,5	1
4	32	4,5	2
5	28	3,5	1
6	32	3,5	2
7	32	3,5	1
8	28	3,5	2
9	24	3,5	1
10	24	3,5	2
11	24	4,5	1
12	24	4,5	2

Table 2.
Experimental design of the first stage of alcoholic fermentation of blue cabuya juice with a substrate concentration of 60 g/L.

Some physical characteristics that indicate that the plant is mature according to the ancestral knowledge of the inhabitants of the La Merced district, Junín region, Peru are described below:

- The leaves of the plant become thicker.
- The outer leaves of the plant droop downward.
- The trunk of the plant is more voluminous.
- The leaves around the upper part of the trunk are thinner and more erect.

Once the mature plant has been selected, the external leaves are cut in such a way that there is free access to the trunk, leaving the rest of it with its respective leaves. Then, a hole is made approximately 30 centimeters (cm) deep and 20 to 25 cm in diameter in the upper part of the pineapple or trunk to accumulate the juice [21].

Depending on the size of the mature plants, the juice is collected between two to three months, with the collection time being longer in larger plants. The collection is carried out two to three times a day, depending on the size of the plant: the first extraction is carried out between 4 and 5 in the morning; the second, at noon; and the third extraction, between 3 and 5 in the afternoon. In each extraction, about three liters of juice obtained. After each extraction, the walls of the hole are scraped so that they do not seal and prevent the juice from accumulating in the hole, while it is cleaned to avoid fermentation of the juice.

2.1.4 Determination of the physicochemical properties of blue cabuya juice

The blue cabuya juice was characterized to determine the following physicochemical parameters: °Brix, density, pH, and viscosity. The following laboratory instruments were used in the measurements: portable refractometer, series X0015RURCN; pycnometer; Milwaukee brand potentiometer, MW 102, pH/Meter, USA, and QUIMIS viscometer, Brazil. The results are shown in **Table 3**.

2.1.5 Alcoholic fermentation process

To carry out the alcoholic fermentation process, the yeast *Saccharomyces cerevisiae* variety ellipsoideus, D 47 - LALVIN - Canada was used. This yeast is lyophilized and comes in 5 g sachets, so it was necessary to reproduce it up to the concentration necessary to inoculate 20 L of cabuya juice.

To obtain the inoculum to 3 L of blue cabuya juice, 6 g of DAP was added as a nutrient. Subsequently, the solution was sterilized at 110 C and transferred to a 5 L flask where the aeration system was installed. The solution was cooled to the incubation temperature of 24 C, the yeast strains were added, and sterile air was supplied at a flow of 1 vvm for 24 h until reaching a concentration of 1,15 g of biomass/L. This biomass concentration is equivalent to 2875 x 10^7 cells/mL, which is within the range required for must fermentation that considers 1,0 g of cells ≅ 1,4 x 10^7 cells [22].

The experimental tests of alcoholic fermentation were carried out in the environmental microbiology and biotechnology laboratories of the Faculty of Biological Sciences and Unitary Operations of the UNMSM. A 120 L capacity bioreactor equipped with automatic temperature control and a 0.5 HP motor was used, with a stirring speed of 90 rpm and a supply of sterile air with a variable volumetric flow rate of 20 L/min. at 60 L/min. For each experimental test, 20 L of cabuya juice (minimum amount of bioreactor operation) was used. The cabuya juice was sterilized at 110 C in the bioreactor before carrying out the experimental tests of the alcoholic fermentation process at different temperatures, pH, and DAP nutrient concentrations.

The tests were carried out under the following conditions: constant substrate concentration, AR of 60 g/L, pH of 3.5 and 4.5 [23], at temperatures of 24 C, 28 C, and 32 C, and DAP nutrient concentration of 1 and 2 g/L. To 20 L of cabuya juice, previously sterilized, the yeast inoculum was supplied at a concentration of 1.15 g/L (dry basis), sterile air at a flow of 1vvm during the first 8 h of the process and the agitation speed was 90 rpm. The process was monitored by taking samples every 2 h for 26 h for the respective analysis of biomass growth, sugar consumption and ethanol production.

2.1.6 Measurement of reducing sugars, ethanol, and microbial biomass

Reducing sugars were determined by Miller's method using 3,5-Dinitrosalicylic acid (DNS), and ethyl alcohol concentration was determined by spectrophotometry using potassium dichromate, at a maximum wavelength of 580 nm.

To measure the concentration of microbial biomass, a 10 ml sample is taken from the broth during fermentation and centrifuged at 4000 rpm for 15 minutes. The sediment is deposited in a crucible and 10 mL of 0,85% w/v alkaline NaCl solution is

Fructans	°Brix	Density	Viscosity
(mg/mL)	ss/100 mL solution	(g/mL)	(cp)
37,34	14,40	1,29	1,58

Table 3.
Physicochemical properties of blue cabuya juice.

added. Then, this solution is dehydrated in the oven at a temperature of 80 C for 6 h until it reaches a constant weight. The amount of biomass is determined by the difference in weight between the crucible with dry biomass and the crucible without sample.

2.2 Results

Table 4 indicates the tests carried out in the investigation at different temperatures, pH, and DAP concentrations for a substrate concentration of 60 g/L of the blue cabuya juice to determine the conditions in which the highest ethanol yield is obtained.

Figure 1 shows the biomass yields of the experimental tests from the blue cabuya juice alcoholic fermentation process. As shown below, test 8 has the highest biomass yield.

Figure 2 shows the graphs of Ln (biomass) Vs time in hours of the alcoholic fermentation process experimental tests. It is observed that the kinetics of microbial growth is greater at a temperature of 28 C, pH 3.5, and DAP concentration of 2 g/L (test 8).

In **Figure 3**, the AR substrate consumption of the experimental tests of the alcoholic fermentation of cabuya azul juice is shown. It is observed that the speed of consumption of AR is greater in the conditions of test 8.

In **Figure 4**, the ethanol production of the experimental tests from the alcoholic fermentation process is observed. The highest yield of ethanol is obtained in test 8.

From the experimental results, it can be appreciated that test 8 at a temperature of 28 C, pH 3.5, Diammonium phosphate (DAP) nutrient concentration of 2 g/L, and substrate concentration (AR) of 60 g/L represents the best conditions in ethanol and biomass yield with $Yp|s = 0.4739$ and $Yx|s = 0.166$, respectively, as shown in **Figure 5**.

2.2.1 Analysis of the results of the consumption of reducing sugars

Table 5 shows the statistical treatment of the experimental tests performed on reducing sugar consumption.

Test	Temperature °C	pH	Nutrient DAP, g/L
1	28	4,5	1
2	28	4,5	2
3	32	4,5	1
4	32	4,5	2
5	28	3,5	1
6	32	3,5	2
7	32	3,5	1
8	28	3,5	2
9	24	3,5	1
10	24	3,5	2
11	24	4,5	1
12	24	4,5	2

Table 4.
Experimental tests performed at different temperatures, pH, and DAP concentrations for a substrate concentration of 60 g/L.

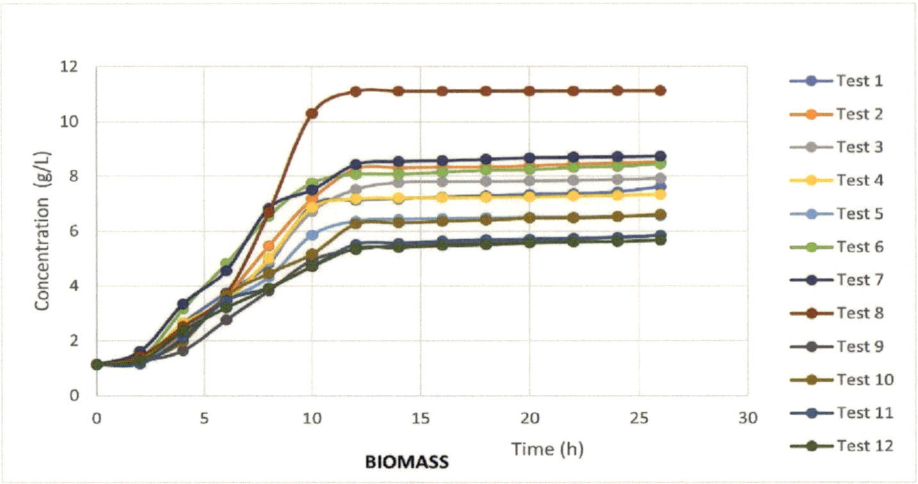

Figure 1.
Biomass production of alcoholic fermentation process in the first stage of experimental tests.

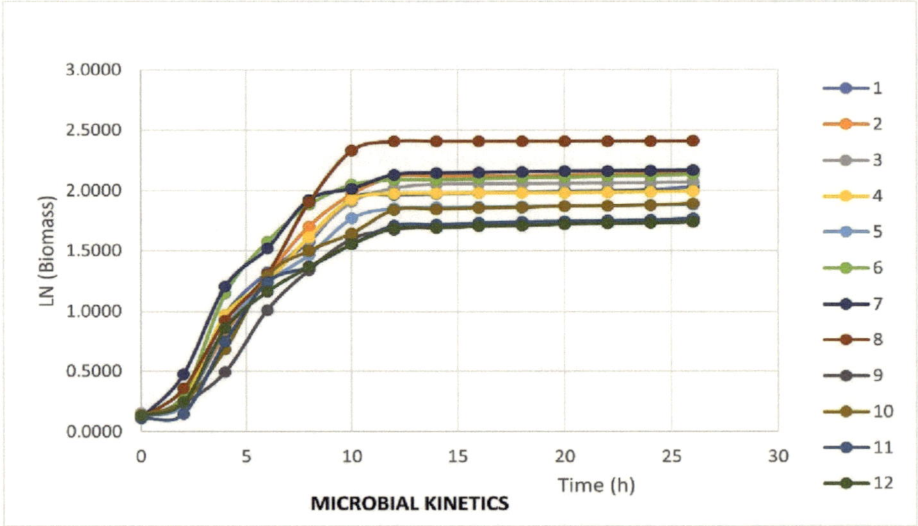

Figure 2.
Microbial kinetics from the 12 experimental tests of the alcoholic fermentation process.

In **Table 5** of inter-subject effects, the significant value of 0.000 ˂ 0.05 is observed. Therefore, Ha is accepted. This value indicates that there is a significant difference between the groups (Test and time).

Figure 6 of estimated marginal means shows that test 8 has the highest consumption of reducing sugars.

2.2.2 Analysis of biomass yield results

Table 6 shows the statistical treatment of the biomass production in the experimental tests performed.

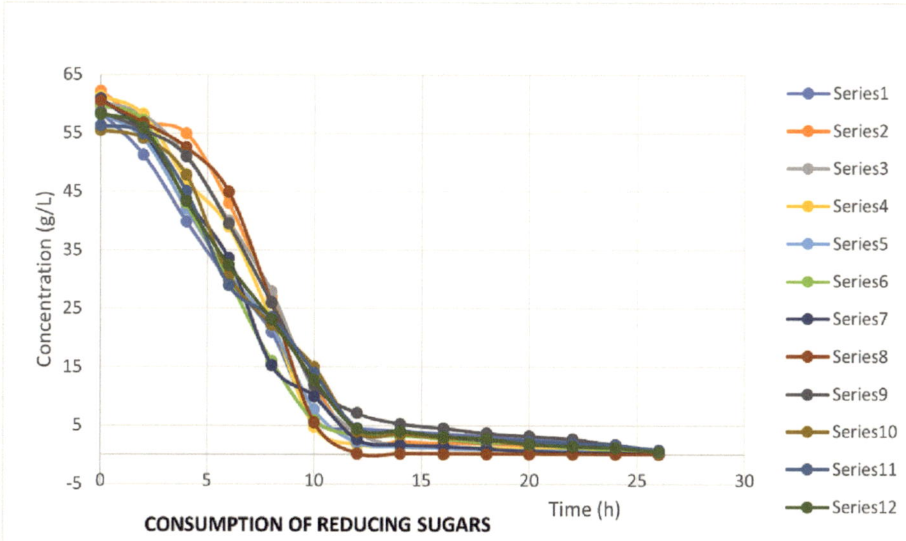

Figure 3.
Consumption of reducing sugars from the 12 experimental tests of the alcoholic fermentation process.

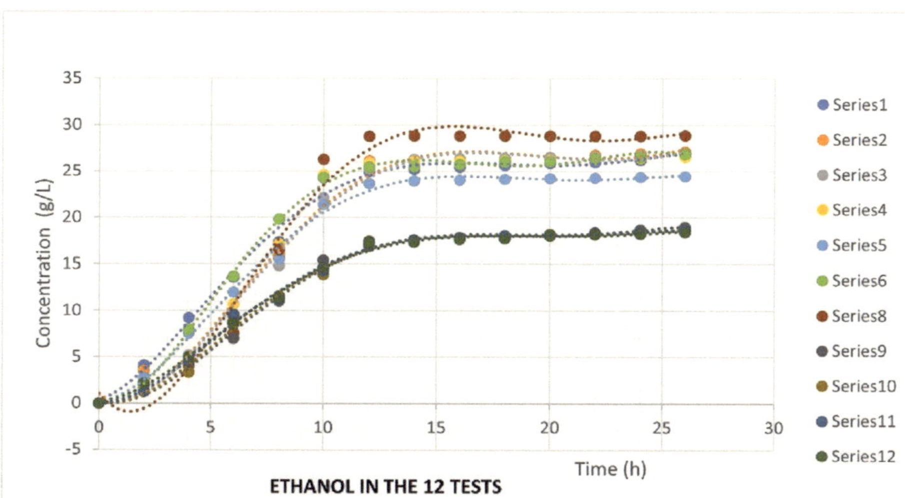

Figure 4.
Ethanol production in the 12 experimental tests of the alcoholic fermentation process.

In **Table 6**, showing the inter-subject effects, a significant value of 0.000 ˂ 0.05 is observed. Therefore, Ha is accepted. This value indicates that there is a significant difference between the groups (Test and time).

In **Figure** 7 of estimated marginal means, it is shown that test 8 has the highest biomass yield.

2.2.3 Ethanol yield results analysis

Table 7 shows the statistical treatment of the ethanol yield from the experimental tests performed.

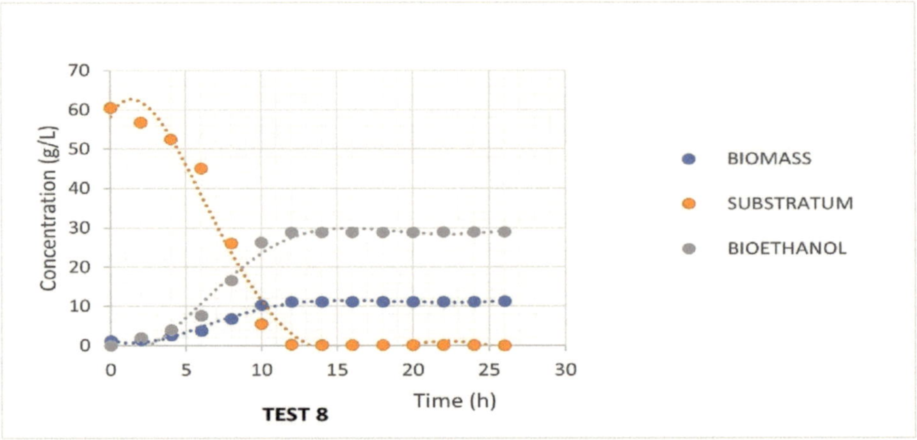

Figure 5.
AR consumption, biomass, and ethanol concentrations during the alcoholic fermentation process of test 8.

between-subjects effects tests					
dependent variable: Reducing sugars					
Origen	**Type III sum of squares**	**gl**	**Root mean square**	**F**	**Sig.**
corrected model	77548,985[a]	24	3231,208	509,808	,000
Intersection	50686,980	1	50686,980	7997,202	,000
Test	201,498	11	18,318	2,890	,002
Time	77347,487	13	5949,807	938,738	,000
Mistake	906,347	143	6,338		
Total	129142,312	168			
Total corrected	78455,332	167			

aR squared =,988 (R adjusted squared =,987).

Table 5.
Statistical treatment of AR consumption of the 12 experimental tests carried out at different temperatures, pH, and DAP concentrations.

In **Table** 7 of inter-subject effects, a significant value of 0.000 ˂ 0.05 is observed. Therefore, Ha is accepted. This value indicates that there is a significant difference between the groups (Test and time).

In **Figure** 8 of estimated marginal means, it is shown that trial 8 has the highest yield of ethanol.

2.3 Discussion

About 20 or more species of maguey are found in all the states of the Republic of Mexico, except in Tabasco and the Yucatan peninsula. A large part of the harvested

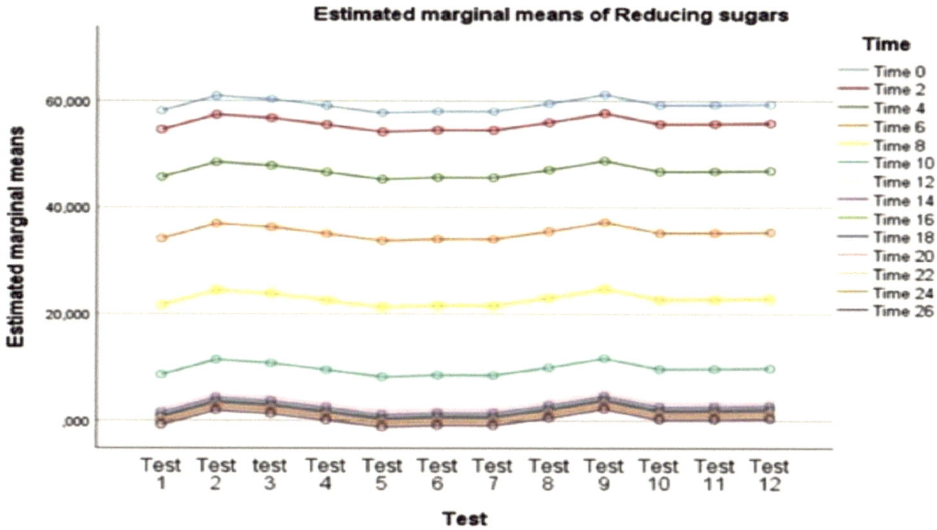

Figure 6.
Estimated marginal means from the 12 experimental tests performed on the consumption of reducing sugars.

between-subjects effects tests					
Dependent variable: Biomass					
Origen	**Type III sum of squares**	**gl**	**root mean square**	**F**	**Sig.**
corrected model	1142,363[a]	24	47,598	95,298	,000
Intersection	5369,762	1	5369,762	10750,927	,000
Test	202,802	11	18,437	36,912	,000
Time	939,561	13	72,274	144,701	,000
Mistake	71,424	143	,499		
Total	6583,549	168			
Total corrected	1213,787	167			

[a]*R squared =,941 (R adjusted squared =,931).*

Table 6.
Statistical treatment of the biomass yield of the 12 experimental tests carried out at different conditions of temperature, pH, and DAP concentration.

area corresponds to the species of the Americana group, mostly used to produce alcoholic beverages, mainly tequila, which has a denomination of origin [24].

Among the few varieties of cabuya existing in Peru, only two grow in large areas of rural land: the blue cabuya (*Agave americana L.*), which is distinguished by the bluish-green color of its leaves, and the other variety is the lineño (green cabuya), which belongs to the *A. angustifolia* Haw complex that is generally used in rural areas for the manufacturing of ropes, while in Mexico it is used to produce a distilled beverage that belongs to the mezcal family [25].

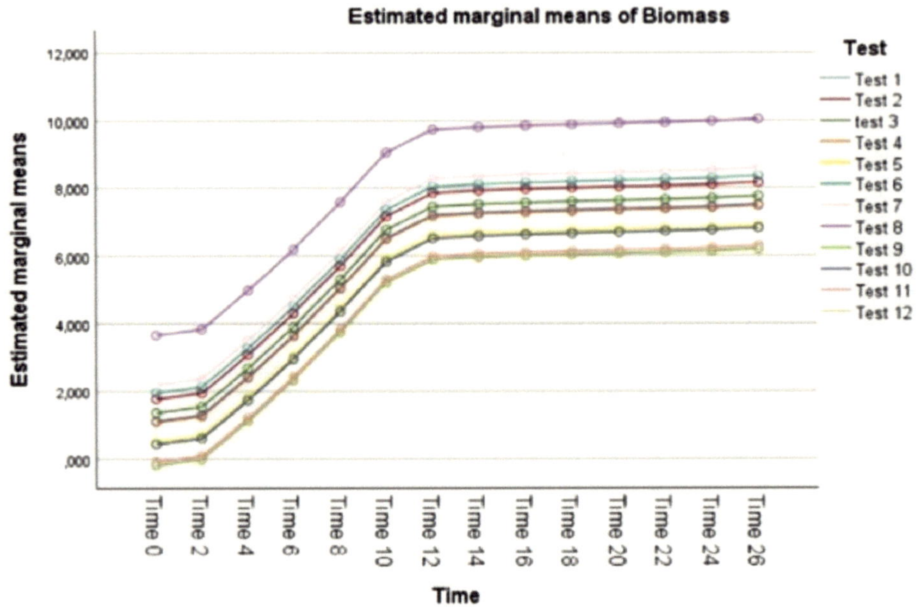

Figure 7.
Estimated marginal means from the 12 experimental tests performed on biomass yield.

between-subjects effects tests					
dependent variable: Ethanol					
Origen	**Type III sum of squares**	**gl**	**Root mean square**	**F**	**Sig.**
corrected model	13992,299[a]	24	583,012	160,230	,000
Intersection	50642,558	1	50642,558	13918,149	,000
Test	1556,292	11	141,481	38,883	,000
Time	12436,007	13	956,616	262,908	,000
mistake	520,320	143	3,639		
Total	65155,177	168			
Total corrected	14512,619	167			

[a]*R squared =,964 (R adjusted squared =,958).*

Table 7.
Statistical treatment of the ethanol yield from the 12 experimental tests performed at different temperatures, pH, and DAP concentration.

Even though blue cabuya grows wild in large tracts of land in the inter-Andean valleys of Peru with temperate climates (Cajamarca and Huánuco Regions), this resource is not used. In the southern regions of the country such as Huancavelica, Ayacucho, and Apurímac, blue cabuya is used for the artisanal production of chancaca (a solid sugar concentrate) and the sale of the juice as a drink. While in Mexico, some 120,000 hectares of blue agave [26] are intensively cultivated with an average of 2500 to 2800 plants per hectare. In 2018, according to the Tequila Regulatory Council (CRT), 1,138,800 tons of tequilana cabuya were used to produce 309.1 million liters of tequila at 40% alcohol volume.

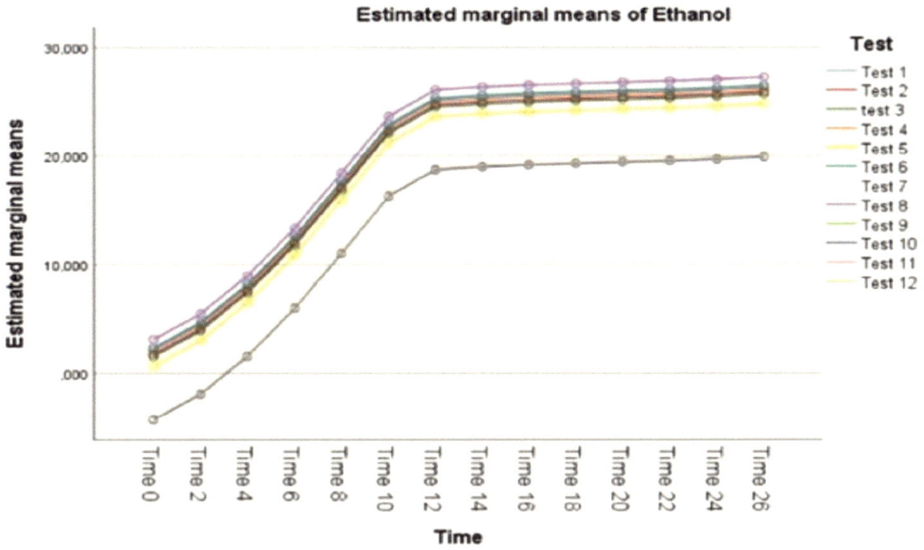

Figure 8.
Estimated marginal means from the 12 experimental tests performed on ethanol yield.

Due to climatic differences, land composition, and geographical location in which the blue cabuya grows in Peru; its physicochemical, physiological, metabolic, and morphological properties differ from the Mexican plants. One of the main differences is the maturation time of the plant. In our country, it takes 5 to 10 years but matures in less time in the inter-Andean valleys with temperate and rainy climates, which are between 1500 to 2200 meters above sea level. In Mexico, the plant matures in an average of 8 years [27]. Therefore, this shorter maturation time is an advantage for agro-industrial exploitation in Peru.

In Mexico, the cabuya tequilana is exploited on a large scale. Therefore, the pineapple plant is generally treated in ovens for the hydrolysis of the polysaccharides present in the juice [28]. Then, add water for the extraction of the juice in the milling at the ratio of 1 to 2.5 L per kg of pineapple pulp to obtain about 120 L of juice per pineapple. However, obtaining blue cabuya juice in Peru is done by hand. The juice is extracted about three times a day in an average of 3 L for each collection for two to three months depending on the plant size. Thus, an average of 270 L of juice is collected per month per plant. This constitutes an advantage for its exploitation on a small scale to produce the liquor-type tequila.

The industries that produce alcoholic beverages are based on the fact that, under proper fermentation conditions, sugar is transformed into alcohol. The organoleptic characteristic of the finished product depends on the raw material that provides the sugar and the conditions for transforming the sugar into alcohol. In the investigation, the factors that affect the alcoholic fermentation process of the blue cabuya juice were examined to obtain the highest ethanol yield, which was: substrate concentration of 60 g/L; temperature 24, 28, and 32 C; pH 3.5 and 4.5; and DAP nutrient concentration 1 and 2 g/L, during 26 h of the process, with an air supply with a flow of 1 vvm, during the first 8 h of the process, these results being very satisfactory. and they agree with the results obtained by other researchers who have worked in similar conditions but with different substrates.

The must was inoculated with *saccharomyces cerevisiae* yeast strains, genus D 47–Lalvin, at a concentration of 1,15 g/L (4,3 x 10^7 cells/mL). This genus of yeast was used for its property of fermenting white and red musts, in addition to having a good performance at low and high fermentation temperatures, which has been demonstrated with the statistical analysis, see **Figure 8** of estimated marginal means that shows that trial 8 has the highest yield of ethanol.

From the experimental results obtained, 12 tests, it is observed that test 8 is the one that gave the highest yield of biomass Y(x/s) = 0.1606 and ethanol Y(p/s) = 0.4739 at the conditions of substrate concentration, AR of 60 g/L; temperature 28 C; pH 3.5; and nutrient concentration DAP 2 g/L. The ethanol yield obtained represents approximately 93% of the theoretical maximum yield of alcoholic fermentation, which is 51.0% [29].

3. Conclusion

The ethanol yield in the alcoholic fermentation process of cabuya azul juice was 47.4%, which represents approximately 93% of the theoretical yield of ethanol production by fermentation, observing that it meets the minimum yield, 45%, that requires a fermentation industry for the production of liquors, therefore, the yield obtained in the investigation is very good.

The experimental results showed that the yield of ethanol in the fermentation process of cabuya azul meets the minimum requirements for its industrialization, it is a product of biology from renewable biomass, which will allow this natural resource to be used in liquor production as part of a sustainable development program.

Acknowledgements

We appreciate the great support received from the authorities of the Faculties of Biology and Chemical Engineering of the Universidad Nacional Mayor de San Marcos (UNMSM) for assisting us with the use of their laboratory facilities during the experimental execution of this research.

Author details

Roberto Robles Calderón[1], Raymundo Erazo Erazo[1*], Julio Javier Armijo Carranza[1] and Benito Buendía Quispe[2]

1 Universidad Nacional Mayor de San Marcos, Lima, Perú

2 Universidad Nacional Daniel Alcides Carrión, Pasco, Perú

*Address all correspondence to: rerazoe@unmsm.edu.pe

References

[1] Espinoza-Angulo C, Warren-Vega WM, Muñoz-Miranda M, Sánchez-Franco SG, Carrasco-Marín F, Miguel-Cruz F, et al. Determination of inorganic tracers in agave plants and agricultural soils as an auxiliary parameter to ensure the authenticity of the region that grants the denomination of origin of tequila. Industrial Crops and Products. 2022;**187**:1-13. [Internet]. DOI: 10.1016/j.indcrop.2022.115401

[2] Ruiz JA, Uvalle JX, Vélez C, Ramírez A, Vicente I, del Real Laborde JI, et al. Conocimiento y prácticas agronómicas para la producción de Agave tequilana Weber en la zona de denominación de origen del tequila. Instituto Nacional de Investigaciones Forestales [Internet]. Pérez JF, del Real Laborde JI, editors. Vol. 4. Tepatitlán de Morelos, Jalisco: Prometeo Editores, S.A. de C.V; 2007. URL: https://books.google.com/books/about/Conocimiento_y_pr%C3%A1cticas_agron%C3%B3micas_p.html?id=P2TdYgEACAAJ

[3] Warren-Vega WM, Fonseca-Aguiñaga R, González-Gutiérrez LV, Romero-Cano LA. A critical review on the assessment of the quality and authenticity of tequila by different analytical techniques: Recent advances and perspectives, food chemistry. Food Chemistry. 2023;**408**:135223 [Internet]. URL: https://www.sciencedirect.com/science/article/pii/S0308814622031855

[4] Aldrete-TapiaJA, Escalante-MinakataP, Martínez-Peniche RA, Tamplin ML, Hernández-IturriagaM. Yeastandbacterial diversity, dynamics and fermentative kinetics during small-scale tequila spontaneous fermentation. Food Microbiology. 2020;**86**:103339. [Internet]. DOI: 10.1016/j.fm.2019.103339

[5] Diaz-Montano DM, Delia ML, Estarron-Espinosa M, Strehaiano P. Fermentative capability and aroma compound production by yeast strains isolated from Agave tequilana weber juice. Enzyme and Micrbial. 2008;**42**(7):608-616. [Internet]. DOI: 10.1016/j.enzmictec.2007.12.007

[6] López-Alvarez A, Díaz-Pérez AL, Sosa-Aguirre C, Macías-Rodríguez L, Campos-García J. Ethanol yield and volatile compound content in fermentation of agave must by Kluyveromyces marxianus UMPe-1 comparing with Saccharomyces cerevisiae baker's yeast used in tequila production. Journal of Bioscience and Bioengineering. 2012;**113**:614-618. [Internet]. DOI: 10.1016/j.jbiosc.2011.12.015

[7] Alemán-Nava GS, Gatti IA, Parra-Saldivar R, Dallemand J-F, Rittmann BE, Iqbal HMN. Biotechnological revalorization of tequila waste and by-product streams for cleaner production – A review from bio-refinery perspective. Journal of Cleaner Production. 2018;**172**:3713-3720. [Internet]. DOI: 10.1016/j.jclepro.2017.07.134

[8] Ibarra B, Botero F, Cortés C. Operaciones unitarias en la producción de tequila. 1a ed., Bogotá: Grupo Ingeniería Institucional. Facultad; 2010

[9] Lavado M, Robles R, Yenque J. Análisis fisicoquímico de la cabuya azul. Industrial Data. 2015;**18**(1):98-100. DOI: 10.15381/idata.v18i1.12071

[10] Waleckx E, Gschaedler A, Colonna-Ceccaldi B, Monsan P.

Hydrolysis of fructans from Agave tequilana Weber var. Azul during the cooking step in a traditional tequila elaboration process. Food Chemistry. 2008;**108**:40-48. [Internet]. DOI: 10.1016/j.foodchem.2007.10.028

[11] Cedeño M, Álvarez-Jacobs J. Production of Tequila from Agave: Historical Influences and Contemporary Processes [Internet]. In: Jaques KA, Lyons TP, Kelsall DR, editors. The alcohol textbook: A reference for the beverage, fuel and industrial alcohol industries. 4th ed. Nottingham University Press; 2003:225-246. Available from: https://www.virtualpro.co/biblioteca/produccion-de-tequila-a-partir-de-agave-influencias-historicas-y-procesos-contemporaneos

[12] Mancilla-Margalli NA, López MG. Water-soluble carbohydrates and Fructan structure patterns from agave and Dasylirion species. Journal of Agricultural and Food Chemistry. 2006;**54**(20):7832-7839. [Internet]. DOI: 10.1021/jf060354v

[13] Godines CI. Caracterización y eficiencia de la fermentación en la elaboración del mezcal potosino [Internet]. San Luis de Potosí: Universidad Autonoma San Luis De Potosí; 2012 URL: https://repositorioinstitucional.uaslp.mx/xmlui/handle/i/3670

[14] Morán-Marroquín GA, Córdova-López J, Estarrón-Espinoza M, Díaz-Montaño DM. Efecto del ph y de la aireación sobre las capacidades fermentativas y aromáticas de dos cepas tequileras de s. cerevisiae cultivadas en continuo [Internet]. Guadalajara Jalisco México. [cited 2023 May 6]. URL: https://smbb.mx/congresos%20smbb/acapulco09/TRABAJOS/AREA_X/CX-20.pdf

[15] Morán-Marroquín GA, Córdova J, Valle-Rodríguez JO, Estarrón-Espinosa M, Díaz-Montaño DM. Effect of dilution rate and nutrients addition on the fermentative capability and synthesis of aromatic compounds of two indigenous strains of Saccharomyces cerevisiae in continuous cultures fed with Agave tequilana juice. International Journal of Food Microbiology. 2011;**151**(1):87-92. DOI: 10.1016/j.ijfoodmicro.2011.08.008

[16] Leal I, Tarantino G, Hernández R, Morán H. Efecto de la temperatura y el pH en la fermentación del mosto de Agave cocui. MULTICIEN. 2014;**14**(4):375-381 [Internet] URL: https://www.redalyc.org/articulo.oa?id=90433839012

[17] Okafor N, Okeke BC. Modern Industrial Microbiology and Biotechnology. 2da edición ed. London, England: CRC Press; 2018

[18] Boulton C, Quain D. Brewing Yeast and Fermentation. oxford: Blackwell Science Ltd; 2001.p. 424

[19] Hernández R, Fernández C, Baptista P. metodología de la investigación. 6ta edición ed. México: Mc Graw-Hill/Interamericana Editores, S. A; 2014

[20] Arellano-Plaza M, Herrera-López EJ, Díaz-Montaño DM, Moran A, Ramírez-Córdova JJ. Unstructured kinetic model for tequila batch fermentation. International Journal of Mathematics and Computers in Simulation. 2007;**1**:1-6 [Internet]. URL: https://www.researchgate.net/publication/233764715_Unstructured_Kinetic_Model_for_Tequila_Batch_Fermentation

[21] Ávila-Fernández Á, Galicia-Lagunas N, Rodríguez-Alegría ME, Olvera C, López-Munguía A. Production of functional oligosaccharides through limited acid hydrolysis of agave fructans.

Food Chemistry. 2011;**129**(2):380-385. DOI: 10.1016/j.foodchem.2011.04.088

[22] Fermentis by Lesaffre for beverages. Levadura activa seca para producción de tequila y mezcal [Internet]. 2017. URL: http://www.castlemalting.com/Publications/YeastProducts/Yeast-Fermentis-SafAle-BE-256_SP.pdf

[23] Liu X, Jia B, Sun X, Ai J, Wang L, Wang C, et al. Effect of initial PH on growth characteristics and fermentation properties of Saccharomyces cerevisiae. Journal of Food Science. 2015;**80**(4):M800-M808. [Internet]. DOI: 10.1111/1750-3841.12813

[24] García-Herrera EJ, de Jesús Méndez-Gallegos S, Talavera-Magaña D. El género agave spp. En México: principales usos de importancia socioeconómica y agroecológica. Revista Salud Pública y Nutrición. 2010;**5**(1): 109-129. URL: https://uiimseminario.files.wordpress.com/2014/05/09.pdf

[25] Rodríguez-GarayB, Lomelí-SenciónJA, Tapia-Campos E, Gutiérrez-Mora A, García-GalindoJ, Rodríguez-DomínguezJM, et al. Morphological and molecular diversity of Agave tequilana weber var. Azul and Agave angustifolia haw. Var. Lineno. Industrial Crops and Products. 2009;**29**:220-228. DOI: 10.1016/j.indcrop.2008.05.007

[26] Montañez-Soto J, Venegas-González J, Vivar-Vera M, Ramos-Ramírez E. Extracción, caracterización y cuantificación de los fructanos contenidos en la cabeza y en las hojas del Agave tequilana Weber AZUL. Bioagro. 2011;**23**(3):199-206. Available from: http://ve.scielo.org/scielo.php?script=sci_arttext&pid=S1316-33612011000300007&lng=es&nrm=iso

[27] Gil-Vega K, Díaz C, Nava-Cedillo A, Simpson J. AFLP analysis of Agave tequilana varieties. Katia Gil-Vega; Corina Díaz; Alejandro Nava-Cedillo. June Simpson. 2006;**170**(4):904-909. DOI: 10.1016/j.plantsci.2005.12.014

[28] Waleckx E, Mateos-Diaz JC, Gschaedler A, Colonna-Ceccaldi B, Brin N, García-Quezada G, et al. Use of inulinases to improve fermentable carbohydrate recovery during tequila production. Food Chemistry. 2011;**124**(4):1533-1542. DOI: 10.1016/j.foodchem.2010.08.007

[29] Hu W-S. Engineering Principles in Biotechnology [Internet]. London, England: ISTE Ltd and John Wiley & Sons; 2018 URL: https://catalog.loc.gov/vwebv/search?searchCode=LCCN&searchArg=2017016663&searchType=1&permalink=y

Chapter 6

Extrusion Processing of Biomass By-Products for Sustainable Food Production

Jordan Pennells, Ishka Bless, Pablo Juliano and Danyang Ying

Abstract

The sustainability of the food supply chain is gaining increasing attention in the quest to balance economic, environmental, and social dimensions. A key opportunity to enhance food system sustainability is by addressing food waste through upcycling strategies to generate higher value, functional foods. Extrusion is a food manufacturing technology that is emerging as a promising option for the incorporation of various types of biomass by-products, such as fruit and vegetable pomace, brewer's spent grain, bagasse, and oil press cake. In this chapter, we present an overview of the latest research conducted on incorporating biomass by-products into extruded food products, with an emphasis on the challenges and opportunities associated with this approach. A meta-analysis study was conducted regarding a key challenge for product quality when incorporating by-products, which is the reduction in radial expansion index of expanded snack and breakfast cereal products. To highlight future opportunities, two case studies illustrate successful examples of by-product incorporation for commercial extruded food products, while emerging protein sources from waste-consuming insects were also explored. Overcoming these challenges and leveraging opportunities can contribute to a more sustainable food system through the integration of by-products into value-added extruded foods.

Keywords: extrusion, by-products, food waste, upcycling, radial expansion index, optimisation, quality

1. Introduction

1.1 Chapter overview: from biomass to sustainable food products

There are a myriad of specialty biobased products that can be generated from biomass across the broad subject areas of biofuel, biochemical, biomaterial, pharmaceutical, food, and feed production. While biobased products have been around for centuries, recent advancements in technology have allowed for more efficient and cost-effective production, improving their functionality and economic viability. The focus of this chapter will be on the production of sustainable food products using biomass by-products, specifically manufactured through extrusion technology. It is clear from **Figure 1** that extrusion technology is linked to sustainability through the processing of food by-products into functional foods, in the effort to fight food waste. Biomass by-products that have been incorporated into extruded products include fruit and vegetable pomace, bagasse, oil press cake, brewer's spent grain (BSG), crop by-products such as bran or husks, and general food waste that forms the feed material for insect-based protein. The chapter outlines the latest research into by-products incorporated into extruded foods, detailing the major challenges facing this topic and sharing a couple of case studies of commercial ventures incorporating by-products into extruded food. This chapter forms a resource for researchers to study when developing a research questions and hypotheses within this field, and for industrials to consult when investigating opportunities to up-cycle biomass by-products into extruded foods across a range of applications including snack foods, breakfast cereals, and plant-based meats.

1.2 What is food system sustainability?

Food system sustainability is a complex topic that spans aspects all along the supply chain, from on-farm growth of crops by agricultural producers, ingredient preparation and product processing by food manufacturers, the selling of food products by retailers, and the consumption of food by consumers and the food service industry [1]. Sustainability can be addressed from multiple aspects, including sustainable agricultural practises to optimise crop yield with minimal inputs, the substitution of existing food ingredients with sustainable alternatives, optimised efficiency of food manufacturing processes, reducing food waste, and promoting social sustainability through food security and equity considerations.

A key opportunity to enhance food system sustainability is through addressing food loss and food waste. Food loss refers to a reduction in food production due to pre-consumer losses that occur all along this supply chain from harvest, post-production, storage, transportation, primary processing and wholesale. Recent estimates from the Food and Agriculture Organisation of the United Nations (FAO) stating that around 14% of global food production is wasted before reaching the retail market [2]. On the other hand, food waste refers to the disposal of food that is fit for human consumption. The United Nations Environment Programme (UNEP) estimates that an additional 17% of global food production is wasted in retail and consumer settings [3].

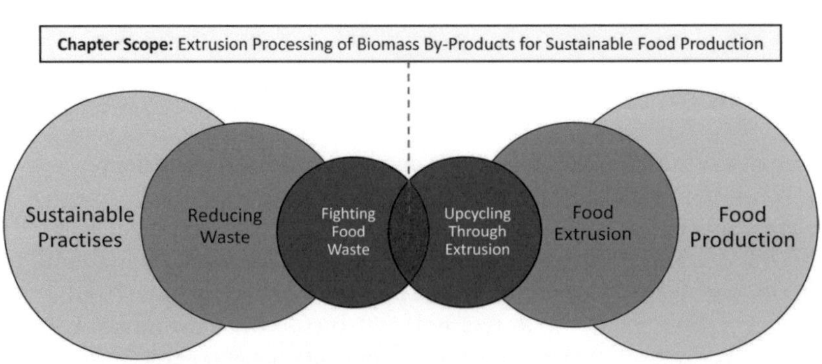

Figure 1.
Scope of the book chapter.

Although a high level of investment is dedicated toward agricultural research, greater than 90% of this investment is targeted at optimising crop productivity, while approximately 5% of investment is aimed toward research to reduce food losses [4]. This is a significant misjudgement, as greenhouse gas (GHG) emissions embedded in the wasted food product increase as the product moves along the supply chain [2]. Addressing food waste can have significant environmental, social, and economic benefits. It can help to reduce GHG emissions, conserve natural resources, and increase food security. Additionally, reducing food waste can also help to save money for both businesses and consumers.

1.3 Food waste hierarchy

Various options have been proposed to fight food waste, ranging from most to least preferable, including waste prevention, recycling, recovery, and disposal. Waste prevention involves strategies to avoid food waste from being generated in the first place, such as reducing the overproduction of perishable food products, optimising food manufacturing processes, and improving retail inventory management and storage practises. It also involves reduction strategies to limit the loss of food waste streams to worthless end point such as landfill, achieved through food recovery and redistribution approaches. Examples of this strategy include donating excess food to food bank organisations, marketing and selling imperfect-looking produce, or diverting food waste to animal feed applications. Waste recycling strategies involve the repurposing of food waste as an input into a subsequent process to create a product of higher value than the original waste stream. Examples of this strategy include the use of anaerobic digestion to produce biogas and biosolids, commercial composting to produce bio-fertiliser, or other bioconversion processes to generate valuable ingredients, chemicals, and materials [5]. Lastly, waste recovery aims to extract the residual energy stored within the biomass. **Figure 2** outlines the food waste hierarchy, which is presented to create an emphasis on moving food up the waste hierarchy.

Figure 2.
Food waste hierarchy developed by Stop Food Waste Australia.

1.4 Food by-product upcycling through extrusion

Based on the food waste hierarchy, by-product recycling into new food products is the second most preferable option behind waste prevention. A significant opportunity for waste upcycling is the incorporation of by-products into existing food manufacturing processes. There is a dual benefit for this approach, which includes: (1) Environmental sustainability benefits relating to reduced demand for existing food ingredients, which involves considerable resource use in terms of land, water, fertiliser, fuel, chemicals for pest and weed control, and fertile soil [6]; and (2) Health benefits by creating value-added food products that have a higher nutrient content, improved textural properties, and/or enhanced functionality.

Extrusion has emerged over the past decade as a processing option for food by-product valorisation, as demonstrated by a 2014 review article into current options for food waste valorisation only mentioning extrusion once [7]. Extrusion is a food manufacturing technology that is emerging as a promising option for the incorporation of various types of biomass by-products [8]. The extrusion process involves a rotating single or twin-screw conveyor enclosed within a heated barrel, which cooks the food material through the application of temperature, pressure, and mechanical shearing. The product is formed into the desired shape through the choice of the die size and shape. Extrusion is widely used to manufacture various food products, such as breakfast cereals, snacks, pasta, noodles, plant-based meat substitutes, and animal feeds. The key benefit of this technology is its functionality and versatility, which allows an extruder machine to perform multiple processing functions (i.e. mixing, heating, shearing, cooking, material expansion, and product shaping), all in a single, continuous process. The process can be fine-tuned by adjusting temperature, moisture content, feeding rate, screw configuration, and screw speed, which directly affect the functional, nutritional, textural and sensory attributes of the final product. The food ingredients involved in extruded food production can vary depending on the specific product type being manufactured. However, some common ingredients include starch-based flour (i.e. corn, rice, wheat) for expanded snacks, plant proteins (i.e. soy, pea, peanut) for plant-based meat analogues, and fibre-rich components (i.e. bran, hulls).

2. Biomass by-product sources incorporated in extrusion

In recent years, there has been growing interest in using alternative ingredients for extruded food production, such as food manufacturing by-products. These alternative ingredients are often rich in nutrients and have a lower starting cost compared to traditional ingredients. In fact, their utilisation may save costs in relation to waste management. Different types of biomass by-products have been incorporated into extruded foods, which include:

- Fruit and vegetable pomace [e.g., apple pomace (AP), carrot pomace, grape pomace]
- Industrial residues (e.g., BSG, sugarcane bagasse)
- Oil press cake (e.g., canola, rapeseed, hempseed)

- Emerging extrusion feedstocks (i.e. insects)

Each of these types of biomass by-products will be discussed in the following sections, with **Table 1** summarising relevant journal articles that study the incorporation of these ingredients in extruded food products.

By-product type	Citation	Research question	Key findings
Apple pomace	[9]	How does varying the content of AP (from 0% to 20%) within corn-based extruded snacks affect the content of dietary fibre and other health-promoting compounds, such as polyphenols, flavonoids, phenolic acids, flavonols, flavon-3-ols, and dihydrochalcones?	Up to 20% incorporation of AP led to significant increases in health-promoting compounds, included chlorogenic acid (up to 36 times), cryptochlorogenic acid (up to 4 times), catechin (up to 6 times), procyanidin (up to 3 times), phloridzin (up to 25 times) and epicatechin (up to 8 times).
	[10]	Can hydrochloric acid (HCl) pre-treatment of AP improve the interaction with starch during extrusion processing and increase the expansion ratio of direct-expanded products?	HCl reduced the cellulose, hemicellulose, and lignin contents, and increased the sugar content of modified apple pomace (MAP). In general, AP extrudates has a higher expansion ratio, although the MAP extrudate at 5 g/100 g had the highest expansion ratio of 3.87. Fourier Transform-Infrared (FTIR) Spectroscopy highlighted the 1035 cm^{-1} moiety, representing the C–O bonds between cellulose and lignin, as contributing to low expansion ratio of the extrudates.
	[11]	How does the addition of AP affect the total phenolic content, flavanols, total flavonoids, and antioxidant activity of yellow corn- based extruded ready-to-eat products?	The incorporation of AP increased the total phenolic content, flavonoids, and antioxidant activity of the extruded products. The maximum flavonoid content was generated at the die temperature of 160°C and feed moisture of 14 g/100 g.
	[12]	What is the effect of extrusion processing on the total phenolic content, antioxidant activity, textural, and functional properties of a blend of extruded products incorporating AP?	Extrusion processing at higher AP ratios and lower temperatures and screw speeds increased the total phenolic content and antioxidant activity. The extrudates had increased expansion ratio, brittleness, crispness, and water solubility index, but decreased hardness and water absorption index. The optimum conditions for the extrudates were 30% AP ratio, 25% moisture content, 132°C temperature, and 108 rpm screw speed.
	[13]	Can food industry co-products, such as fruit pomace and liquid whey, be converted into shelf-stable, nutrient-enriched extruded products using low-	Supercritical fluid extrusion generates extruded products that are rich in dietary fibre, phenolics, and antioxidants, while also exhibiting a low density of 0.21–0.35 g/cm^3. Replacing

By-product type	Citation	Research question	Key findings
		shear, low-temperature supercritical fluid extrusion?	water with liquid whey did not impact the textural properties of extrudate products.
	[14]	Can AP be incorporated into extruded snacks to improve their nutritional value and antioxidant content?	The incorporation of up to 20% of AP in extruded snacks increased the fibre content, phenolic content, and antioxidant capacity when compared to the products with no AP added. Although loss of phenolic compounds occurred during extrusion, the generation of thermally induced degradative products of phenolic origin compensated for that loss in terms of radical scavenging activity.
	[15]	How do extrusion parameters (die temperature and screw speed) and the addition of AP affect the quality of a corn flour-based extruded snack?	Screw speed had the greatest impact on the quality of extruded snacks, with an increase in screw speed leading to increased bulk density and decreased expansion ratio, while AP addition led to decreased product expansion. The optimal formulation contained 7.7% AP and was processed at a die temperature of 150°C and a screw speed of 69 rpm. Insoluble fibre in AP lowered expansion and adhesion to bubble structures punctured cells, reducing cell extensibility, in addition to damaging aerated bubbles do to its hydrophilic nature.
	[16]	What are the effects of different process conditions on the response variables (i.e. system parameters) of AP-soy flour-corn grits blends during extrusion?	Moisture content, temperature, screw speed, and apple pomace level all had significant effects on system parameters such as viscosity, specific mechanical energy, mass flow rate, torque, die pressure, and dough temperature during extrusion. Increasing the temperature resulted in decreased viscosity, die pressure, and specific mechanical energy. Increasing the level of apple pomace in the blend resulted in decreased viscosity, SME, and torque, and increased mass flow rate.
	[17]	Can extrusion modify the functionality of apple pomace?	Extrusion of apple pomace increased breakdown of cell wall structure and water solubility and increased the antioxidant activity during *in vitro* digestion.
	[18]	Can corn extrudates fortified with rosehip or AP improve the physico-chemical, sensory, and nutritional properties of extruded products?	The addition of fruit pomace to corn extrudates positively influenced the total polyphenol content and antioxidant activity of the product, particularly in samples enriched with rosehip pomace. While there was a

By-product type	Citation	Research question	Key findings
			slight decrease in some quality features, such as taste, flavour, structure, and colour, the extruded products obtained using fruit by-products were still well received by panellists and can be an excellent source of bioactive compounds in the human diet.
Carrot pomace	[19]	What is the impact of carrot pomace inclusion and moisture content on the quality of corn starch-based extrudates?	The study showed that the inclusion level of carrot pomace significantly impacted the extrudate quality, with the best expansion observed in extrudates with 5 g/100 g of carrot pomace inclusion and 15 g/100 g feed moisture.
	[20]	What are the effects of extrusion temperature and storage period on the microstructure, texture, and functional attributes of extruded snacks?	The increase in extrusion temperature resulted in an increase in *L*-value, *b*-value, and hardness of the extrudates, but a decrease in a-value, crispiness, β-carotene, and vitamin C. Storage for 6 months at 30°C and 50% relative humidity resulted in a decrease in L-value, b-value, crispiness, β-carotene, and vitamin C, and an increase in a-value and hardness of the extrudates.
	[21]	Can a low-cost, high-protein, and high-fibre snack be produced from composite flour made of rice, defatted soybean flour, carrot pomace powder, and cauliflower trimmings powder using extrusion processing?	The optimum extrusion conditions were 164°C die temperature, 313 rpm screw speed, and 85 g/100 g rice flour with a desirability value of 76.0%. The blend proportions under the optimum extrusion conditions had improved nutritional quality with 10.25 g/100 g protein and 0.84 g/100 g of fibre content.
	[22]	How much vegetable (carrot or broccoli) solids can be incorporated into extruded snacks with wet or dried vegetables and what is their effect on the extruded snack expansion?	Only ~3% (dry base) of wet vegetable (carrot or broccoli) can be incorporated into extrusion expanded snacks. Higher vegetable content (from 20% to 100%) in extruded snack is possible using dried vegetable powders. Increasing the amount of vegetable in the extruded snacks reduced the expansion.
	[23]	What is the effect of die temperature, feed rate, feed moisture, and carrot pomace incorporation on the quality of extruded products made from rice and gram flour?	Die temperature had the greatest effect on the quality of extruded products, followed by changes in feed moisture and carrot pomace level. The optimal incorporation level of carrot pomace powder was 5%, which resulted in extrudates with an overall acceptability score of 7.4. Feed rate was not significant for any of the response variables, indicating no impact on the quality of extruded products.

By-product type	Citation	Research question	Key findings
Berry pomace	[24]	Can hot air (HA) and microwave assisted hot air (MWHA) drying combined with extrusion be used to convert bilberry press cake into value-added extruded food products at different inclusion levels?	Different drying techniques to generate bilberry powders implied a difference in processing time (40% reduction with MWHA drying), but the retention of total phenolics and physical characteristics of extruded snacks were independent of drying techniques.
	[25]	Can nutritious products be generated by incorporating blackcurrant juice press residues into the extrusion process, while retaining advantageous textural and sensory properties?	Non-enzymatic press residue extruded with barley or oat flour exhibited superior properties in terms of extrudate expansion, lower hardness and density, and higher contents of fructose, glucose and fruit acids, which all contributed to positive sensory evaluation of texture, appearance, and flavour.
	[26]	How does the interplay between starch (waxy, regular) and pomace (blueberry, cranberry) impact the physicochemical characteristics of extruded products?	The addition of pomace to starch blends significantly decreased the expansion ratio (ER) of the extruded products as the level of pomace inclusion increased, with differences observed between different pomace types. Fourier-transform infrared (FTIR) and solid-state nuclear magnetic resonance (NMR) spectroscopy supported the conclusion that interactions occurred via hydrogen bonding, as opposed to covalent bonding.
	[27]	What is the effect of cherry pomace inclusion level and particle size on their expansion characteristics of corn starch extrudates?	The inclusion of 5% cherry pomace milled to the smallest particle size range (<125 μm) yielded extrudates with the highest expansion ratio, outperforming the control sample. The study found that particle size significantly affected the radial expansion ratio of extrudates, with the smaller particle size ranges yielding higher expansion ratio at all levels of pomace addition. Cherry pomace addition also significantly decreased the water absorption index and water solubility index of extrudates, while smaller particle ranges led to a higher water solubility index.
	[28]	How are the techno-functional and physical properties of ready-to-eat texturized cereal products affected by the incorporation of chokeberry pomace powder (CPP)?	Increasing the content of CPP decreased the cell pore size, expansion, and lightness of extrudates. The glucose release from extrudates during *in vitro* digestion decreased with higher levels of CPP inclusion. The dietary fibre content was unaffected by the addition of CPP. Anthocyanins were degraded by 70% in

By-product type	Citation	Research question	Key findings
			both CPP-starch extrudates, while phenolic acids and flavonols were fully retained.
	[29]	What is the effect of pomace chemical composition and physical properties, inclusion level, and extrusion processing parameters on the expansion of extrudates made from fruit (cranberry, blueberry, grape, and apple) pomace-corn starch blends?	Active interaction was exhibited between pomace and starch components through extrusion, such that the pomace type and inclusion level had a significant impact on the expansion of corn starch-based extrudates. The level of insoluble dietary fibre and material crystallinity correlates negatively with extrudate expansion, whereas soluble dietary fibre and sugar content presented a positive correlation with extrudate expansion at 30% pomace inclusion level.
	[30]	How does thermo-mechanical treatment change the chemical composition and physical properties of chokeberry pomace powder after extrusion processing at 90% loading?	Physicochemical transformation of the chokeberry pomace powder through extrusion led to a significant increase in water absorption index, anthocyanin and dietary fibre content, and glass transition temperature. However, non-significant change was observed in the water solubility index and sorption isotherms, with extrudates exhibiting a low expansion coefficient.
Tomato pomace	[31]	What is the effect of varied tomato pomace levels within extruded snack products on the total phenolic content and bioaccessibility, antioxidant capacity, and physicochemical properties?	Increasing levels of tomato pomace powder in extruded snacks resulted in increased content of individual phenolic compounds, as well as enhanced bioaccessibility of protocatechuic acid, chlorogenic acid, rutin, quercetin, and lycopene. However, snacks with higher levels of tomato pomace powder had lower expansion indices, water absorption index, and water solubility index, and higher hardness, redness, and yellowness values. Despite these outcomes, the extrudate with 10% tomato pomace inclusion had the highest overall acceptability.
	[32]	Can tomato pomace be incorporated into a corn and rice flour blend to create an acceptable and nutritious extruded snack food?	The fibre and protein content of tomato pomace incorporated extrudates were significantly higher, while sensory analysis showed the high desirability was the product with 25% tomato peel and 5% tomato seed, which was on par with the control sample.
	[33]	Can tomato pomace be incorporated into ready-to-eat extruded snacks to generate fibre and lycopene-rich products?	All four independent variables (pomace level, moisture content, screw speed and barrel temperature) had a significant impact on all the response

By-product type	Citation	Research question	Key findings
			variables investigated (specific mechanical energy, hardness, water solubility index, water absorption index, expansion index, bulk density, and colour). The optimum extrusion conditions were 10% tomato pomace addition, 14% moisture content, a screw speed of 300 rpm, and a barrel temperature of 170 C. The optimal lycopene content within the enriched extrudate was 2.73 ppm at the 10% tomato pomace addition level.
Other pomace types	[34]	Can grape pomace be effectively incorporated into extruded corn starch snack foods to improve their physicochemical quality and nutritional value?	Extrudates with the lowest feed moisture content (16%) had the highest expansion ratio (~3.8) and the lowest density (~0.11 g/cm^3). The extrudate samples with 5% grape pomace and 16% moisture content retained >74% of the total polyphenolic content when processed at all extrusion screw speeds.
	[35]	Can nutrient enriched extrudates be generated through the utilisation of cashew AP by-products?	Cashew AP can be used to biofortify, significantly increasing their protein and overall mineral content, in addition to improving their phenolic, flavonoid, and total antioxidant activities. However, incorporation of cashew AP had a negative impact on textural and physical properties of extrudates but had negligible impact on sensory evaluation. The extrudates made with 5% CAPP scored the highest for taste, crispness, odour, chewiness, colour, and surface character.
	[36]	How do different fractions of olive pomace, separated via centrifugation and freeze-dried into a powder, impact the physical properties of fibre and polyphenol enriched extruded products?	The addition of olive pomace fraction reduced the die pressure and specific mechanical energy during extrusion, resulting in lower radial expansion in the extruded product.
	[37]	How effectively can pineapple pomace be incorporated into extruded products as a source of dietary fibre?	Up to 10.5% of pineapple pomace can be added to extruded foods without affecting the hardness, yellowness, water absorption, and bulk density of the final product, but the addition of 21% of pomace negatively affected the characteristics of the extrudates.
Fruit / vege bagasse	[38]	What impact does processing temperature and moisture content on the colour and expansion ratio of extruded snacks formulated with cassava bagasse?	The moisture content and extrusion temperature significantly affected the physical and functional properties of snacks based on pre-gelatinized flours. Extruded snacks processed at 104°C and moisture of 16% exhibited a large

By-product type	Citation	Research question	Key findings
			expansion index, intermediate specific volume, and lighter yellow colour.
	[39]	What is the effect of extrusion temperature, moisture content, and naranjita bagasse content on the physical, sensory, and chemical properties of extruded snack foods?	The highest expansion index and the lowest penetration force were observed at high extrusion temperature and low moisture content, while the highest carotenoid content and colour difference (ΔE) were observed at high naranjita bagasse content. The optimal processing conditions were extrusion temperature of 125°C, moisture content of 23%, and naranjita bagasse content of 8.03%.
	[40]	How does orange bagasse, a low-cost, natural source of fibre, impact the physicochemical and structural properties of extruded snacks made with blue corn?	The incorporation of orange bagasse increased the crude fibre content and hardness of extruded snacks made from blue corn, in addition to decreasing the expansion index of the snacks, resulting in harder products with higher numbers of pores per area but of smaller sizes. The extruded starch lost its semicrystalline structure due to mechanical shearing and high temperature throughout the extrusion process.
Sugarcane bagasse	[41]	What impact does the addition of sugarcane bagasse fibre have on extrudate physicochemical characteristics?	The content of insoluble fibre decreased while soluble fibre increased during the extrusion process. High bagasse content and barrel temperature led to a decrease in expansion index values and an increase in water absorption and water solubility index values.
	[42]	How are the expansion characteristics of corn starch-based extruded products affected by the incorporation of sugarcane bagasse and AP by-products?	Inclusion of AP in corn starch-based extrudates produced higher initial and stable expansion indices, but also showed higher shrinkage than sugarcane bagasse extrudates, potentially due to the presence of sugars in the AP.
BSG	[43]	What is the effect of barrel temperature and BSG content on the physical properties of extruded snacks based on rice flour?	BSG reduced apparent density and volumetric expansion index of extruded snacks. SME input decreased significantly with an increase of BSG content, with the dilution of starch content leading to a reduction in melt viscosity. The frequency of structural ruptures was positively affected by barrel temperature, which is related to an increase in the number of internal pores of the extrudates and an increase in the crispness of extrudates.

By-product type	Citation	Research question	Key findings
	[44]	How does the incorporation of BSG impact the phenolic content, antioxidant capacity, arabinoxylan content and glycaemic index of extruded products?	The addition of BSG into extruded snacks increased the total phenolic content by 4 times, the DPPH radical scavenging activity by 19 times, and the antioxidant power by 5 times. The choice of base ingredients and extrusion conditions need to be optimised to produce low GI products more effectively with BSG incorporation.
	[45]	How do the extrusion processing parameters of moisture content, screw speed, and temperature profile impact the properties of barley based extrudates incorporating BSG?	The formulation water content and extrusion screw speed had the greatest impact on snack expansion. The addition of BSG led to an increase in the dietary fibre content of extruded snacks, while whey protein isolate addition led to extruded snacks with the highest protein content. Starch was required within the formulation to generate sufficient expansion and reduction in hardness for the extruded snacks.
	[46]	Can BSG be incorporated into extruded snack products to improve their nutritional and textural properties?	Increasing the level of BSG in the formulation led to a significant decrease in the sectional expansion index (SEI) of the extrudates, in addition to a decrease in the individual and total area of cells within the extrudate. BSG also created an increase in bulk density and phytic acid content.
	[47]	Can the addition of pectin resolve the expansion and textural problems of corn-based extruded snacks loaded with biomass by-products including BSG, sugar beet pulp, and AP?	While by-product addition lowered the expansion ratio and fracturability and increased the bulk density and hardness, the addition of pectin at 1% loading improved the expansion ratio of all extrudates incorporating by-products. However, no by-product incorporated extrudates reached the original expansion ratio of the corn-based control sample.
Oil press cake	[48]	What are the effects of defatted hemp cake addition on the physical properties of extruded snacks and process parameters optimise the quality of these products?	The level of defatted hemp cake addition and moisture content both decreased the expansion ratio and fracturability of the extruded snacks, while increasing their bulk density and hardness. In addition, the processing temperature significantly increased the hardness of extrudates.
	[49]	How does screw speed and semi-defatted sesame cake (SDSC) content affect the nutritional, textural, microstructural, and sensory properties of corn-based extrudates?	A 10% addition of SDSC resulted in a fourfold increase in protein content compared to the corn-based extrudates. Although, the addition of SDSC also reduced the sectional expansion and increased the puncture force of corn-based extruded snacks, which was

By-product type	Citation	Research question	Key findings
			hypothesised to be caused by the protein and fibre-rich sesame cake reinforcing the cell wall structure and increasing the resistance to deformation. It was also hypothesised that the introduction of fibre led to the premature rupture of gas cells during material expansion upon exiting the extrusion die Meanwhile, screw speed did not exhibit a significant impact on the puncture force and crispness work of extrudates. The product hardness was also increased due to the reduction porous cell size, which resulted in a less crispy extrudate.
	[50]	What are the optimal processing conditions for developing desirable extrudates in terms of physical, functional, textural, and sensory attributes by exploring the effect of virgin coconut oil (VCO) cake content, feed moisture and extrusion screw speed?	The optimal extrusion conditions were calculated to be 28.7% VCO cake addition, 14% feed moisture content, and a screw speed of 300 rpm, yielding a maximum desirability for extrudates of 0.88. A higher content of VCO cake generally led to a higher sensory acceptability and lower hardness, expansion ratio, peak viscosity, water absorption index, and water solubility index. A higher feed moisture content led to a lower sensory acceptability and water solubility index. A faster extrusion screw speed generally led to a higher expansion ratio, peak viscosity, and water absorption index.
	[51]	What is the impact of rapeseed press cake content, starch type, and extrusion temperature on the extruder response, rheological behaviour, expansion mechanisms, and technofunctional properties of extruded products?	The addition of rapeseed press cake from 0% to 70% increased the SEI of potato starch extrudates, although this relationship did not hold for waxy potato starch extrudates. It is hypothesised that the protein- and fibre-rich biopolymers within press cake by-products influences the rheological properties of the melted material, which impacts expansion behaviour upon exiting the extruder die through enhanced complex viscosity.
	[52]	What is the effect of varying concentrations of rapeseed press cake, moisture content and barrel temperature on the expansion behaviour of extruded starch blends and how do rheological properties influence this phenomenon?	For extrudates with 10% rapeseed press cake, the initial SEI of samples 10 seconds after exiting was in the very high range of 6–16, while the final SEI of extrudates dried at 40°C for 24 hours experienced severe shrinkage to reach ~3. For extrudates with 40% rapeseed press cake, the initial SEI was very low at ~1, but increased to ~3 for the final SEI due to due to time-dependent swelling. It was hypothesised that the high initial SEI was due to the presence

By-product type	Citation	Research question	Key findings
			of soluble fibres in the melted extrudate exiting the die, which entailed high water binding and holding capacity and enabled higher swelling due to more nucleation sites for water vaporisation and higher melt elasticity.
	[53]	How can extrusion be optimised to produce high-quality, protein-enriched, nutrient-dense ready-to-eat extrudates with defatted sesame flour (DSF)?	The experimental factors that had the greatest impact on the tested variables, in descending order, were the DSF content, moisture content, die temperature, and screw speed. The optimised extrudate had a protein content of 19.21 g/100 g. The optimal extrusion conditions were identified as 22% DSF content, 15.6% moisture content, a die temperature of 130°C, and screw speed of 250 rpm, with an overall sensory desirability of 73%.
	[54]	How does extrusion affect the canola meal?	Extrusion at high barrel moisture (36%) favoured protein aggregation resulting in lower extractable protein compared to extrusion at the lowest barrel moisture (24%). At lower barrel moisture contents (24% and 30%), a longer kneading block length increased extractable protein but this was not the case at 36% barrel moisture. Canola protein digestibility was improved upon extrusion at 30% barrel moisture but there was no significant change at lower (24%) or higher (36%) barrel moisture.
	[55]	How does extrusion affect the hempseed oil cake?	Extrusion enhanced the production of free polyphenols, flavonoids, and phenyl propionamide content, and α-glucosidase and acetylcholinesterase inhibition activities of hempseed oil cake.
	[56]	How does extrusion affect hempseed oil cake flour?	Extrusion at 40% moisture content and 400 rpm screw speed produced optimal overall characteristics and has potential to manufacture texturized protein meat analogue.

Table 1.
Summary of publications investigating the impact of incorporating fruit and vegetable pomace by-products into extruded food products.

2.1 Fruit and vegetable pomace

2.1.1 Apple pomace

AP refers to the fibrous solid residue remaining after pressing for juice or cider production, consisting of skin, core, and the flesh of the fruit. In the past, it has been considered a waste product with few uses outside of low-value applications such as pectin production, animal feed, soil composting [9]. The remainder of this by-product stream is typically sent to landfill. However, there has been a recent increase in interest to incorporate AP into extruded snack food products.

AP has been highlighted as a by-product resource of potential value, as it is rich in dietary fibre and phytochemicals such as phenolic acids and flavonoids [14]. A summarised analysis of the chemical components within AP showed that it consists of total dietary fibres (36.8%), fructose (16.0%), starch (14.0%), sucrose (8.4%), glucose (7.5%), water (7.3%), protein (3.7%), ash (1.9%), amino acids (1.8%), triterpenoids (1.6%), macro/microelements (0.6%), polyphenolic compounds (0.4%), and malic acid (0.03%) [57]. Isolated extracts and components from within AP have pointed to promising antioxidative, anti-inflammatory, antibacterial, and antiviral activities within medical studies, which may help address health issues such as diabetes, cardiovascular disease, and high cholesterol [57]. In addition to the pro-health properties of AP, its incorporation into extruded snacks at low levels (<10%) can improve the functionality of the extrudate products, such as enhanced expansion. However, above this critical level, the incorporation of AP can form highly dense and low expanded products [10]. Extrusion of AP increased breakdown of the cell wall structure, water solubility and the antioxidant activity during *in vitro* digestion [17].

2.1.2 Carrot pomace

Carrot pomace is a byproduct generated during the processing of carrot juice and is produced in significant quantities by the juice industry. Despite containing a high amount of beneficial nutrients, including bioactive compounds with antioxidant properties, carrot pomace is traditionally used as animal feed, which is a relatively low-value application for the material [19]. Meanwhile, the potential use of carrot pomace in food products has been investigated due to its rich content of dietary fibre, crude protein, iron, calcium, β-carotene, and vitamins such as thiamine, riboflavin, vitamin B-complex [58]. Studies have shown that extrusion can modify the functional properties of food by-products, including those of carrot pomace, by increasing water solubility, water holding capacity, and viscosity [59]. Shelf-stable powder was obtained from carrot pomace or cosmetically degraded carrots, with incorporation of carrot powder (3–100%) in extruded snacks enhancing their nutritional value [22].

2.1.3 Berry pomace

Various types of berries are utilised in industrial processing to extract juice, including bilberries, blueberries, raspberries, among others. In the juice industry, up to 30% of the original fresh berry weight may left over from the extraction process in the form of a press cake [60]. Currently, a large degree of press cake biomass is discarded as waste or used for low-value applications such as composting, despite containing valuable compounds such as polyphenolic antioxidants, vitamins, pectin, and lipids. The incorporation of dried berry press cake powders into extruded snack and breakfast cereal foods presents a significant opportunity to improve the flavour and nutritional value of novel products over existing products within the market [24]. However, challenges still need to be overcome relating to the loss of desirable physical properties of extruded products after the incorporation of berry pomaces, including reduced expansion index and increased density and hardness, in addition to maximising flavour and micronutrient retention [61].

2.1.4 Tomato pomace

Tomatoes are the second most consumed vegetable in the US behind the potato, including both raw and processed products such as a wholefood, paste, sauce, puree, juice, or soup [62]. Low value by-products of industrial processing include tomato seeds and peels, although they retain relatively high amounts of beneficial components such as lycopene, phenolic compounds, β-carotene, vitamin C, dietary fibre, and protein [63]. The utilisation of tomato processing waste is a crucial aspect of fighting food waste in the food manufacturing industry. Waste utilisation is an important activity to recover valuable bioactive components and prevent squandering of embedded resources [63]. In addition, spoilage of tomato waste creates an anaerobic environment that leads to the production of GHGs such as methane, creating a double negative to not utilising these waste resources.

2.2 Bagasse

Bagasse refers to the fibrous residue that is left over after pressing out the target compounds from crop products, which typically include sugars, starch, or other soluble bioactive compounds. Bagasse is typically discussed in the context of pressed sugarcane after sugar extraction but can also be related to starchy root vegetables such as cassava, and the fibre rich rind of some fruits. It is predominantly composed of cellulose, hemi-cellulose, and lignin, and has in the past been incorporated as a functional additive into food products such as bread, meat products, dehydrated gravies, puddings, and feed, and extruded snacks [41]. Cassava bagasse has been highlighted as difficult to further utilise due to its high moisture content, which promotes microbial contamination and requires energy intensive drying to prepare into a suitable feedstock condition. For the drying process to be economically viable, high value applications for the incorporation of this by-product resource must be demonstrated [38]. Bagasse is also discussed in the context of citrus fruit peel removed prior to juice extraction. The chemical composition varies between different species, but sits in the range of 50% cellulose, 10% pectin, 18% hemicellulose, and 18% lignin [64].

2.3 Brewer's spent grain

BSG is an abundant by-product generated during the wort filtration step of the brewing process. BSG has been shown to be a rich source of nutrients that remain after wort extraction, such as insoluble protein, cell wall fibre residue, and minerals [65]. The dietary fibre content present within the BSG may vary due to the type of malting process employed [45]. However, conventional applications have been limited to animal feed, with emerging applications in human nutrition products. The valorisation of BSG in extruded food not only reduces the environmental impact of the brewing industry but also offers a valuable source of dietary fibre and protein for human consumption.

DOI: http://dx.doi.org/10.5772/intechopen.111943

2.4 Oil press cakes

Oil press cake is a byproduct of oil extraction process from seeds or nuts. Oilseed species most commonly cultivated for oil extraction are rapeseed and sunflower, while cultivated for proteins is soybean, and cultivated for fibre is cotton [66]. The press cake has a high residual protein content and has been commonly used for animal feed applications. Further processing of the oil press cake can extract residual oil through solvent extraction or expeller pressing to form a defatted cake. Alternatively, press cake can either be used as a feed for food applications that require high protein contents. Oil press cakes are considered a cheap source of valuable components since modern protein extraction technologies enable recovery of target compounds and upcycling of this material within the food system as functional additives [48]. Extrusion has been found to enhance the protein digestibility of the canola seeds [54] and enhanced the production of free polyphenols, flavonoids, and phenyl propionamide content, and α-glucosidase and acetylcholinesterase inhibition activities of hempseed oil cake [55]. Extrusion of hempseed cake flour showed the potential to convert it into a meat-like textured product [56].

2.5 Research into extruded food incorporating by-products

Table 1 outlines the studies that investigate the incorporation of fruit and vegetable pomace into extruded food products, outlining the main research question investigated and the key findings presented within the study.

3. Challenges for by-product incorporation into extruded foods

There is a growing desire for organisations across the entire food supply chain to enhance the sustainability of their operations, alongside existing priorities for generating economic prosperity and social value. Depending on where each organisation sits along the food supply chain and their business operations, different opportunities and challenges will be present for by-product valorisation through extrusion processing.

3.1 Logistics and transport

For agricultural producers, opportunities exist to extract more value from an existing crop yield through collecting and distributing certain types of crop residues that still contain valuable components, such as stems, leaves, and products out of specification. In addition, opportunities exist for industrial food and biochemical manufacturers to turn existing 'waste' streams into additional value streams. However, challenges centre around logistical hurdles for sufficient by-product volume to satisfy downstream manufacturing requirements. Furthermore, large countries such as Australia suffer from the 'tyranny of distance' that significantly increases the transportation cost to aggregate these low value by-product resources [67]. These challenges must be addressed to ensure economic viability to justify a change in standard operating procedures for by-product producers. An idea proposed to address these challenges is investment into manufacturing hubs located at regional hotspots of waste generation, which presents the opportunity to develop new food manufacturing industry for a range of different product types and drive the transition to a circular economic model [68].

Additional processing steps (i.e. drying, grinding) are also required to remove water and minimise microbial activity, which can reduce subsequent transport costs, minimise food spoilage, and enhance the manufacturing viability of downstream products. Effective drying methods for different by-product types are required to be investigated to ensure to viability of energy and cost consumption [69]. Due to the distributed nature of by-product production, it is not feasible to perform these additional processing steps at the by-product generation site. Centralised collection and stabilisation points need to be investigated to address these challenges.

3.2 Feedstock variability

For food manufacturers looking to incorporate by-products into new products, opportunities are available to improve the nutritional profile, functionality, material costs, and consumer acceptance of their products. However, significant challenges still exist regarding the seasonal availability of by-product resources [70], the reliance on upstream by-product suppliers to provide relatively consistent material over time and into the future [71], and the impact of by-product incorporation on the functional properties of products. Further details of the challenges that are expected when approaching the opportunity of incorporating biomass by-products into extruded food products are discussed below.

3.3 Product safety

To enhance consumer acceptance of novel extruded food products that incorporate by-products, the presence of detrimental compounds in the by-product material must be recognised and tested for in the final product. A small fraction of compounds within food products have allergenic potential, such as glycinin and β-conglycinin, which are the main proteins that cause allergenicity in peanut and soybean-based products [72]. In addition, antinutritional compounds within the by-product sources (i.e. oil press cake) must be quantified both before and after processing. Suitable processing techniques are required to obtain novel products that are safe and allergen-free. Toxicity studies implemented within the development of novel food will be essential to ensure product safety and consumer acceptance [69].

Biological and enzymatic degradation is another important aspect that needs to be considered during transformation of biomass by-products intended for incorporation into food-grade products. Biological degradation refers to the breakdown of organic matter by microorganisms, with enzymatic activity enabling the breakdown of complex carbohydrates, proteins, and fats present in the biomass. By-products derived from agricultural crops or food processing often naturally contain enzymes and microorganisms that can lead to spoilage or undesirable changes in the final product, such as causing off-flavours, odours, or reducing the product shelf life. Managing biological and enzymatic degradation during by-product transportation and transformation is essential to ensure the production of safe and high-quality food products.

3.4 Product quality

When incorporating by-products into extruded food products, one of the challenges is ensuring consistent and desirable product quality. The extrudate porosity is an important property impacting the quality of products such as expanded snacks, breakfast cereals, and texturised vegetable protein. Porosity is a ratio of the volume occupied

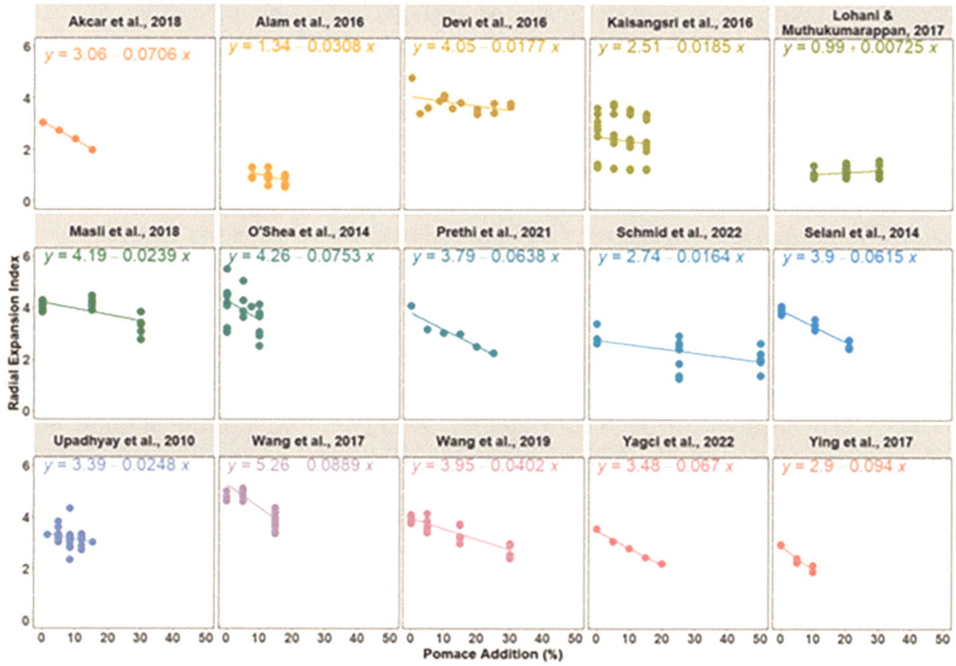

Figure 3.
The relationship between pomace inclusion and expansion ratio across various studies, with the citation for each study located in the title box above each panel.

by air within the external circumference of the extrudate piece. This is tightly linked to the extrudate expansion index, which refers to the increase in area (or puffing) of the extruded product as it exits the extruder die, relative to the area of the die orifice. Expansion ratio is an important parameter relating to product quality, as it impacts the texture (crispness), mouthfeel, and overall consumer acceptance of the product.

To quantify the relationship between by-product inclusion and expansion ratio in the literature, we performed a meta-analysis study to collect data on the impact of including pomace by-products on the radial expansion index (REI) of expanded snack products. A subset of the studies presented in **Table 1** were selected for quantitative analysis, based on their data availability and the quality of the experimental design. Multiple regression analysis was performed on the results of each study to visualise the relationship between pomace addition and REI. This meta-analysis across the published literature demonstrates that almost all the relevant studies present an inverse relationship between pomace level and REI. As such, pomace addition across different fruit and vegetable types decreases the expansion index of extruded products. **Figure 3** provides a summary of the relevant literature that studied the effect of AP incorporation into extruded snacks and the impact on extrudate physical properties.

4. Case studies for by-product incorporation into extruded food

4.1 Case study 1: Nutri V

Nutri V is a food manufacturing company that is aiming to promote healthy eating while also addressing on farm food waste during vegetable production. This start-up has partnered with Fresh Select, a major grower-packer supplying a range of vegetables to Coles supermarkets across Australia, to commercialise the production of vegetable powders and vegetable-based snacks (Nutri V 'goodies'). The vegetable powders can either be bought as is and consumed in a range of different product formats such as smoothies, bread, and pasta, or act as a feed ingredient for the extrusion process to generate the Nutri V 'goodies'. This technology was developed through collaboration with the Commonwealth Scientific and Industrial Research Organisation (CSIRO), Australia's national science agency and innovation catalyst, with the technology enabling retention of the natural colour and nutrients of the vegetable by-products. The vegetable by-products consist of out-of-specification products, which have been estimated at 20–30% of the overall crop yield for different vegetable types, in addition to vegetable stems and leaves that are not typically sold to consumers but are edible nonetheless. The vision of this venture is to provide greater benefit to agricultural producers (i.e. farmers) through shifting crop yield from the traditional range of 65–80%, to whole crop yield of 100% by collecting all the green biomass previously left on-farm. The complimentary value proposition of this venture is to encourage healthy eating for the Australian population by providing additional serves of vegetables in products that would not typically have nutritional value, such as traditional starch based ready-to-eat snacks.

The process to generate the vegetable fortified snacks begins with identify relevant crop by-products that do not meet aesthetic standard for wholesale, pre-processing of the fresh biomass through drying and grinding to yield a vegetable powder and stabilise the ingredients against microbial degradation, incorporation of vegetable powder into the extrusion process at approximately 10–15% loading alongside binding components (i.e. corn flour, potato starch), and finally flavour addition post-extrusion.

4.2 Case study 2: planetarians

Planetarians is a food technology company established in 2013 that aims to create food this is both healthy for people and good for the planet [73]. After initially launching a range of nutritionally complete drink products containing protein from oilseed cake by-products, which sold over one million servings, they pivoted into producing fibre and protein rich chip snacks from up-cycled sunflower oilcake in 2015. In 2017, they partnered with the pasta manufacturing company Barilla Group to produce high protein 'black pasta', once again incorporating oilseed cake by-products.

Finally, in 2021, they once again pivoted to enter the vegan meat space, utilising plant-based protein to produce 'whole cut' meat analogues through a high moisture extrusion process. A key point of difference within their production process is the incorporation of brewer's spent yeast, a by-product of the fermentation process to generate beer. Planetarians state that the incorporation of this cheap, readily available by-product resource provides a range of benefits including masking the beany after-taste of soy, adding umami flavour and meat-like colour, and increasing the protein content, quality, and functionality of the final product. In addition, yeast products have a Food and Drug Administration (FDA) status as Generally Recognised as Safe (GRAS), which facilitates convenient approval as a food-grade product. Recent capital raising has generated a $6.7 million seed II investment round to fund commercial production of its patented, plant-based protein product [74]. With verification of its technology at an industrial scale already complete, the company plans to use the capital to build a pilot facility and continue to increase production.

AB InBev, the largest beer manufacturer in the world, has committed to partnering with Planetarians for this venture, with the company stating that they are "excited to support new ways to bring circular solutions to the center of the plate though upcycled ingredients and alternative protein innovation" (Bernardo Novick, AB InBev). The vision for this partnership is to place its production line, which can be as compact as 3000 square feet, across AB InBev breweries all around the world.

5. Emerging opportunities in insect protein

Over the past decade, there has been growing interest in insect farming (e.g., raising and breeding insects as livestock) [75]. This attention is mostly attributed to their ability to efficiently convert low-value organic waste streams into high-quality products (e.g., protein, fat, chitin, and fertiliser) [76]. Compared with conventional livestock (e.g., pigs and cattle), research has also shown that insects produce lower GHG emissions and can be farmed with less land and water [77, 78]. Insect farming therefore provides a promising opportunity for both waste valorisation and sustainable food production.

5.1 Waste valorisation: insect-based bioconversion of agri-food waste and by-products into food

Insects can be farmed for food or feed. Currently, mealworms (*Tenebrio molitor*) and crickets (*Acheta domesticus*) are the main species farmed for human consumption. The nutritional composition of each insect varies due to factors such as rearing conditions, diet, and age. However, values typically sit between 15% and 25% protein, 4% and 13% fat and 2% and 4% fibre (made up mostly of chitin) (per 100 g edible portion) [79]. Both species also provide a source several vitamins and minerals, including iron, zinc, vitamin B_2 and vitamin B_{12} [80]. While commercial mealworm and cricket farming remains mostly reliant on mixed grain and soy-based feeds [81, 82], several agri-food waste and by-products have been explored as sole or supplementary feeding substrates substrates [76] (**Figure 4**). This includes:

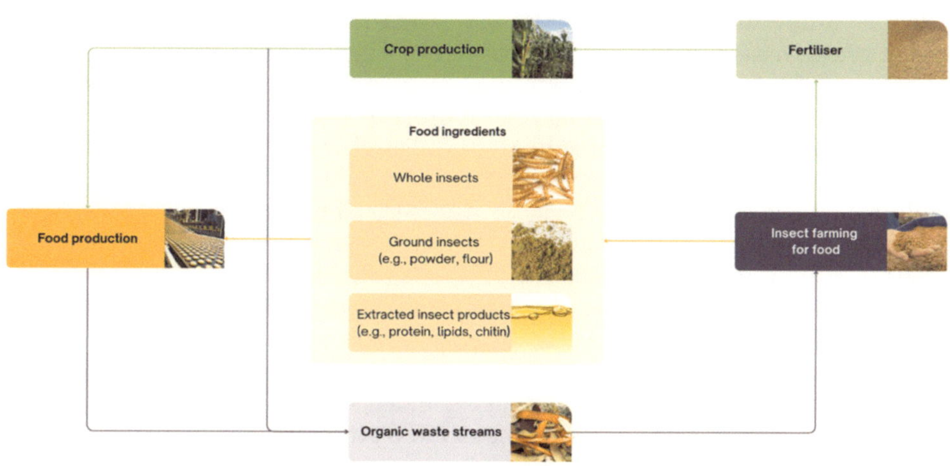

Figure 4.
Waste valorisation via insect-based conversion of agri-food waste and by-products into food.

- fruit and vegetable waste (e.g. peel, pomace, and leaves)
- BSG
- spirit distillers' grain
- beer yeast
- bakery remains (e.g. bread and cookies)
- pre-treated crop residues (e.g. wheat, rice and corn straw, rice bran and rice husk)

Despite clear potential for agri-food waste and by-products as feed for mealworms and crickets, several challenges remain. Firstly, while recent findings suggest that insects can be reared on organic waste streams, growth performance and survival rates vary. Mixed feeds (composed of more than one substrate) are therefore recommended to better meet physical and nutritional diet requirements [76, 83]. Local and year-round substrate availability should also be considered, alongside regulatory restrictions. For example, in regions such as Europe and Australia, insects are regarded as farmed animals and therefore subject to animal feed regulation. Potential feed sources such as catering waste are consequently prohibited from use [76, 84].

5.2 Incorporation of insects in extruded foods

Edible insects such as mealworms and crickets can be used in food to improve nutritional composition, substitute conventional protein sources, or contribute sensory attributes (e.g., flavour and texture). To prepare these species for human consumption, they typically undergo a series of processing steps including (1) inactivation/killing (generally by freezing, blanching, or steaming); (2) washing; (3) blanching; and (4) drying (generally by oven-drying or freeze-drying). Following drying, mealworms and crickets can be used whole, ground, or further processed to extract protein, lipids, and chitin.

Poor consumer acceptance remains a key barrier to the commercialisation of insect foods in Western food cultures, mainly driven by a lack of familiarity with insects in the context of food. However, research suggests that the use of non-recognisable insect ingredients (e.g. powder, flour, insect protein concentrate, texturized insect protein) could help to change perceptions [85]. Familiar foods such as bread, baked goods, pasta, and corn chips incorporating insect powder and flour are well-established among commercially available insect-based foods (e.g., Grillon Le Pain 'Crickbread', Chirps 'Chocolate Chirp Cookie mix', Sens 'Cricket Protein Pasta', Circle Harvest 'Cricket Corn Chips'). However, extrusion processing presents an emerging pathway for the expansion of non-recognisable insect ingredients in several product categories, including breakfast cereals, snack products and meat analogues. Several studies have already investigated applications for insect-ingredients in extruded food products, including protein enrichment in expanded snacks as well as soy substitution in meat analogues. Expanded snacks demonstrated acceptable sensory properties at 6–15% substitution [86–89] while 30–40% substitution was shown to provide meat-like properties and enhance flavour in meat analogues [90, 91]. Due to poor expansion properties, the high lipid and protein content of cricket and mealworm powders remains a key barrier to further enrichment of expanded snacks [86, 87]. However, some early products are already commercially available (e.g., Actually Foods 'Food puffs').

While the potential for insect-based extruded food products is promising, the nascent nature of the edible insect industry presents some challenges. Crucially, regulatory issues (related to approval of novel foods and a lack of clear industry standards) remain a key constraint to the farming and use of insects in food. Further research is also required to ensure safe farming, processing, storage and transport of insects and insect-based foods, as well as for management of allergenic risks (e.g., cross-reactive allergies and potential novel insect allergens) [92].

6. Summary and future perspectives

As awareness grows around the sustainability impacts of food waste, investigation into technologies that increase the value of by-product resources becomes more salient. Extrusion processing is one such technology that has been proposed to upcycle biomass by-products into higher value food across a range of different product types. This chapter explores the challenges and opportunities associated with the incorporation of by-products into extruded food products, in addition to case studies that share success stories of commercialised products that incorporate biomass by-products. Lastly, the emerging industry of insect protein is introduced as an opportunity for waste valorisation and sustainable food production.

Looking ahead, there are several important areas for future exploration in the field of extruded food development incorporating by-products. Firstly, addressing the logistical challenges related to the collection and transport of by-products is crucial. Investing in manufacturing hubs located strategically at regional waste generation hotspots could facilitate the transformation of low-value by-products into valuable resources, promoting a transition to a circular economic model. In addition, more focus needs to be place on strategies to overcome the issue of low radial expansion when incorporating by-products. Currently, the level of by-product addition is limited by this issue, which creates uncertainty around the value proposition of by-product incorporated snacks if their functional properties degrade above 5% addition.

Overall, addressing these challenges and taking advantage of emerging opportunities in this area will require enhanced collaboration and knowledge exchange between stakeholders across the food supply chain, including primary producers, post-production, primary food processing industry, waste management sector, consumer markets, and research institutions covering this topic. Extruded food manufacturers will have to integrate into existing activities in this space to overcome challenges include by-product transportation and logistics, food safety, and product acceptance.

Author details

Jordan Pennells[1*], Ishka Bless[2], Pablo Juliano[1] and Danyang Ying[1]

1 CSIRO Agriculture and Food, Werribee, Australia

2 School of Agriculture, Food and Wine, University of Adelaide, Adelaide, Australia

*Address all correspondence to: jordan_pennells@live.com.au

References

[1] Boxall N, King S, Kaksonen A, et al. Waste innovation for a Circular Economy: a summary report for the CSIRO Cutting Edge Science and Engineering Symposium. Clayton, Victoria. 2019

[2] FAO. The State of Food and Agriculture 2019. Moving forward on food loss and waste reduction. Rome. 2019

[3] United Nations Environment Programme. Food Waste Index Report 2021. Nairobi. 2021

[4] Stroosnijder S, Hetterscheid B, Castelein B. A Journey into the World's Food Systems in Search of Losses, Waste and Ways to Solve Them. The Netherlands: Wageningen University & Research; 2022

[5] Alexander S, Harris P, McCabe BK. Biogas in the suburbs: an untapped source of clean energy? Journal of Cleaner Production. 2019;**215**: 1025-1035. DOI: 10.1016/j.jclepro.2019.01.118

[6] Arsic M, O'Sullivan CA, Wasson AP, et al. Australia needs a national policy approach to successfully implement circular bioeconomy in agriculture and food systems. Australia: Australian Farm Institute; 2022

[7] Mirabella N, Castellani V, Sala S. Current options for the valorization of food manufacturing waste: a review. Journal of Cleaner Production. 2014;**65**: 28-41. DOI: 10.1016/j.jclepro.2013.10.051

[8] Leonard W, Zhang P, Ying D, Fang Z. Application of extrusion technology in plant food processing byproducts: an overview. Comprehensive Reviews in Food Science and Food Safety. 2020;**19**: 218-246. DOI: 10.1111/1541-4337.12514

[9] Gumul D, Ziobro R, Kruczek M, Rosicka-Kaczmarek J. Fruit waste as a matrix of health-promoting compounds in the production of corn snacks. International Journal of Food Science. 2022;**2022**:7341118. DOI: 10.1155/2022/7341118

[10] Dey D, Gu B-J, Ek P, et al. Apple pomace pretreated with hydrochloric acid exhibited better adherence with the corn starch during extrusion expansion. Carbohydrate Polymer Technologies and Applications. 2021;**2**:100089. DOI: 10.1016/j.carpta.2021.100089

[11] León EAV, Aldapa CAG, Rojas JA, et al. Phytochemical content and antioxidant activity of extruded products made from yellow corn supplemented with apple pomace powder. Food Science and Technology (Brazil). 2022;**42**:1-9. DOI: 10.1590/fst.91221

[12] Lohani UC, Muthukumarappan K. Effect of extrusion processing parameters on antioxidant, textural and functional properties of hydrodynamic cavitated corn flour, sorghum flour and apple pomace-based extrudates. Journal of Food Process Engineering. 2017;**40**: e12424. DOI: 10.1111/jfpe.12424

[13] Paraman I, Sharif MK, Supriyadi S, Rizvi SSH. Agro-food industry byproducts into value-added extruded foods. Food and Bioproducts Processing. 2015;**96**:78-85. DOI: 10.1016/j.fbp.2015.07.003

[14] Reis SF, Rai DK, Abu-Ghannam N. Apple pomace as a potential ingredient for the development of new functional

foods. International Journal of Food Science and Technology. 2014;**49**: 1743-1750. DOI: 10.1111/ijfs.12477

[15] O'Shea N, Arendt E, Gallagher E. Enhancing an extruded puffed snack by optimising die head temperature, screw speed and apple pomace inclusion. Food and Bioprocess Technology. 2014;**7**: 1767-1782. DOI: 10.1007/s11947-013-1181-x

[16] Singha P, Muthukumarappan K. Effects of processing conditions on the system parameters during single screw extrusion of blend containing apple pomace. Journal of Food Process Engineering. 2017;**40**:e12513. DOI: 10.1111/jfpe.12513

[17] Liu G, Ying D, Guo B, et al. Extrusion of apple pomace increases antioxidant activity upon in vitro digestion. Food & Function. 2019;**10**: 951-963

[18] Drozdz W, Tomaszewska-Ciosk E, Zdybel E, et al. Effect of apple and rosehip pomaces on colour, total phenolics and antioxidant activity of corn extruded snacks. Polish Journal of Chemical Technology. 2014;**16**:7-11. DOI: 10.2478/pjct-2014-0042

[19] Kaisangsri N, Kowalski RJ, Wijesekara I, et al. Carrot pomace enhances the expansion and nutritional quality of corn starch extrudates. LWT - Food Science and Technology. 2016;**68**:391-399. DOI: 10.1016/j.lwt.2015.12.016

[20] Dar AH, Sharma HK, Kumar N. Effect of extrusion temperature on the microstructure, textural and functional attributes of carrot pomace-based extrudates. Journal of Food Processing & Preservation. 2014;**38**:212-222. DOI: 10.1111/j.1745-4549.2012.00767.x

[21] Alam MS, Pathania S, Sharma A. Optimization of the extrusion process for development of high fibre soybean-rice ready-to-eat snacks using carrot pomace and cauliflower trimmings. LWT - Food Science and Technology. 2016;**74**:135-144. DOI: 10.1016/j.lwt.2016.07.031

[22] Ying D, Sanguansri L, Cheng L, Augustin MA. Nutrient-dense shelf-stable vegetable powders and extruded snacks made from carrots and broccoli. Food. 2021;**10**:2298

[23] Upadhyay A, Sharma HK, Sarkar BC. Optimization of carrot pomace powder incorporation on extruded product quality by response surface methodology. Journal of Food Quality. 2010;**33**:350-369. DOI: 10.1111/j.1745-4557.2010.00323.x

[24] Höglund E, Eliasson L, Oliveira G, et al. Effect of drying and extrusion processing on physical and nutritional characteristics of bilberry press cake extrudates. LWT. 2018;**92**:422-428. DOI: 10.1016/j.lwt.2018.02.042

[25] Mäkilä L, Laaksonen O, Ramos Diaz JM, et al. Exploiting blackcurrant juice press residue in extruded snacks. LWT - Food Science and Technology. 2014;**57**:618-627. DOI: 10.1016/j.lwt.2014.02.005

[26] Richter JK, Gu B-J, Ek P, et al. Potential interactions between starch and fruit pomace may impact the expansion ratio of direct expanded extrudates. Journal of Food Science. 2022;**87**:3513-3527. DOI: 10.1111/1750-3841.16240

[27] Wang S, Kowalski RJ, Kang Y, et al. Impacts of the particle sizes and levels of inclusions of cherry pomace on the physical and structural properties of

direct expanded corn starch. Food and Bioprocess Technology. 2017;**10**: 394-406. DOI: 10.1007/s11947-016-1824-9

[28] Schmid V, Mayer-Miebach E, Behsnilian D, et al. Enrichment of starch-based extruded cereals with chokeberry (Aronia melanocarpa) pomace: influence of processing conditions on techno-functional and sensory related properties, dietary fibre and polyphenol content as well as in vitro digestibility. LWT - Food Science and Technology. 2022;**154**: 112610. DOI: 10.1016/j.lwt.2021.112610

[29] Wang S, Gu B-J, Ganjyal GM. Impacts of the inclusion of various fruit pomace types on the expansion of corn starch extrudates. LWT - Food Science and Technology. 2019;**110**:223-230. DOI: 10.1016/j.lwt.2019.03.094

[30] Witczak T, Stępień A, Gumul D, et al. The influence of the extrusion process on the nutritional composition, physical properties and storage stability of black chokeberry pomaces. Food Chemistry. 2021;**334**:127548. DOI: 10.1016/j.foodchem.2020.127548

[31] Yagci S, Calıskan R, Gunes ZS, et al. Impact of tomato pomace powder added to extruded snacks on the in vitro gastrointestinal behaviour and stability of bioactive compounds. Food Chemistry. 2022;**368**:130847. DOI: 10.1016/j.foodchem.2021.130847

[32] Karthika DB, Kuriakose SP, Krishnan AVC, et al. Utilization of by-product from tomato processing industry for the development of new product. Journal of Food Processing & Technology. 2016;7:608

[33] Jabeen A, Naik H, Jan N, et al. Studying the effect of tomato pomace incorporation on physicochemical, nutritional and storage characteristics of corn-based extrudates using response surface approach. British Food Journal. 2022;**124**:3705-3723

[34] Bibi S, Kowalski RJ, Zhang S, et al. Stability and functionality of grape pomace used as a nutritive additive during extrusion process. Journal of Food Processing & Technology. 2017;**8**: 680-689

[35] Preethi P, Mangalassery S, Shradha K, et al. Cashew apple pomace powder enriched the proximate, mineral, functional and structural properties of cereal based extrudates. LWT - Food Science and Technology. 2021;**139**:110539. DOI: 10.1016/j.lwt.2020.110539

[36] Ying D, Hlaing MM, Lerisson J, et al. Physical properties and FTIR analysis of rice-oat flour and maize-oat flour based extruded food products containing olive pomace. Food Research International. 2017;**100**:665-673. DOI: 10.1016/j.foodres.2017.07.062

[37] Selani MM, Brazaca SGC, dos Santos Dias CT, et al. Characterisation and potential application of pineapple pomace in an extruded product for fibre enhancement. Food Chemistry. 2014; **163**:23-30

[38] Fiorda FA, Soares MS Jr, da Silva FA, et al. Physical quality of snacks and technological properties of pre-gelatinized flours formulated with cassava starch and dehydrated cassava bagasse as a function of extrusion variables. LWT-Food science and technology. 2015;**62**:1112-1119

[39] Ruiz-Armenta XA, de Jesus Zazueta-Morales J, Aguilar-Palazuelos E, et al. Effect of extrusion on the carotenoid content, physical and sensory properties

of snacks added with bagasse of naranjita fruit: optimization process. CyTA-Journal of Food. 2018;**16**:172-180

[40] Navarro Cortez RO, Gómez-Aldapa CA, Aguilar-Palazuelos E, et al. Blue corn (Zea mays L.) with added orange (Citrus sinensis) fruit bagasse: novel ingredients for extruded snacks. CyTA-Journal of Food. 2016;**14**:349-358

[41] Martínez-Bustos F, Viveros-Contreras R, Galicia-García T, et al. Some functional characteristics of extruded blends of fiber from sugarcane bagasse, whey protein concentrate, and corn starch. Food Science and Technology. 2011;**31**:870-878

[42] Masli MDP, Gu B-J, Rasco BA, Ganjyal GM. Fiber-rich food processing byproducts enhance the expansion of cornstarch extrudates. Journal of Food Science. 2018;**83**:2500-2510. DOI: 10.1111/1750-3841.14290

[43] Nascimento TA, Calado V, Carvalho CWP. Effect of Brewer's spent grain and temperature on physical properties of expanded extrudates from rice. LWT - Food Science and Technology. 2017;**79**:145-151. DOI: 10.1016/j.lwt.2017.01.035

[44] Reis SF, Abu-Ghannam N. Antioxidant capacity, arabinoxylans content and in vitro glycaemic index of cereal-based snacks incorporated with brewer's spent grain. LWT-Food Science and Technology. 2014;**55**:269-277

[45] Kirjoranta S, Tenkanen M, Jouppila K. Effects of process parameters on the properties of barley containing snacks enriched with brewer's spent grain. Journal of Food Science and Technology. 2016;**53**:775-783. DOI: 10.1007/s13197-015-2079-6

[46] Stojceska V, Ainsworth P, Plunkett A, İbanoğlu S. The recycling of brewer's processing by-product into ready-to-eat snacks using extrusion technology. Journal of Cereal Science. 2008;**47**:469-479. DOI: 10.1016/j.jcs.2007.05.016

[47] Ačkar Đ, Jozinović A, Babić J, et al. Resolving the problem of poor expansion in corn extrudates enriched with food industry by-products. Innovative Food Science & Emerging Technologies. 2018; **47**:517-524

[48] Jozinović A, AčkAr Đ, Jokić S, et al. Optimisation of extrusion variables for the production of corn snack products enriched with defatted hemp cake. Czech Journal of Food Sciences. 2017;**35**: 507-516

[49] Carvalho CWP, Takeiti CY, Freitas DDGC, Ascheri JLR. Use of sesame oil cake (Sesamum indicum L.) on corn expanded extrudates. Food Research International. 2012;**45**:434-443

[50] Shameena Beegum PP, Manikantan MR, Sharma M, et al. Optimization of processing variables for the development of virgin coconut oil cake based extruded snacks. Journal of Food Process Engineering. 2019;**42**: e13048

[51] Martin A, Osen R, Karbstein HP, Emin MA. Linking expansion behaviour of extruded potato starch/rapeseed press cake blends to rheological and technofunctional properties. Polymers (Basel). 2021;**13**:215

[52] Martin A, Osen R, Karbstein HP, Emin MA. Impact of rapeseed press cake on the rheological properties and expansion dynamics of extruded maize starch. Food. 2021;**10**:616

[53] Gojiya D, Davara P, Gohil V, Dabhi M. Process standardization for

formulating protein-augmented corn-based extrudates using defatted sesame flour (DSF): sesame oil industry waste valorization. Journal of Food Processing & Preservation. 2022;**e17203**

[54] Zhang B, Liu G, Ying D, et al. Effect of extrusion conditions on the physico-chemical properties and in vitro protein digestibility of canola meal. Food Research International. 2017;**100**: 658-664

[55] Leonard W, Zhang P, Ying D, et al. Effect of extrusion technology on hempseed (Cannabis sativa L.) oil cake: polyphenol profile and biological activities. Journal of Food Science. 2021; **86**:3159-3175

[56] Rajendra A, Ying D, Warner RD, et al. Effect of extrusion on the functional, textural and colour characteristics of texturized hempseed protein. Food and Bioprocess Technology. 2023;**16**:98-110

[57] Waldbauer K, McKinnon R, Kopp B. Apple pomace as potential source of natural active compounds. Planta Medica. 2017;**83**:994-1010. DOI: 10.1055/s-0043-111898

[58] Gull A, Prasad K, Kumar P. Effect of millet flours and carrot pomace on cooking qualities, color and texture of developed pasta. LWT - Food Science and Technology. 2015;**63**:470-474. DOI: 10.1016/j.lwt.2015.03.008

[59] Schmid V, Trabert A, Schäfer J, et al. Modification of apple pomace by extrusion processing: studies on the composition, polymer structures, and functional properties. Food. 2020;**9**:1-29. DOI: 10.3390/foods9101385

[60] Kryževičiūtė N, Kraujalis P, Venskutonis PR. Optimization of high pressure extraction processes for the separation of raspberry pomace into lipophilic and hydrophilic fractions. Journal of Supercritical Fluids. 2016;**108**: 61-68. DOI: 10.1016/j.supflu.2015.10.025

[61] Van Buggenhout S, Ahrné L, Alminger M, et al. Structural design of natural plant-based foods to promote nutritional quality. Trends in Food Science and Technology. 2012;**24**:47-59

[62] US Department of Agriculture. Potatoes and tomatoes are the most commonly consumed vegetables. 2020. https://www.ers.usda.gov/data-products/chart-gallery/gallery/chart-detail/?chartId=58340

[63] Lu Z, Wang J, Gao R, et al. Sustainable valorisation of tomato pomace: a comprehensive review. Trends in Food Science and Technology. 2019;**86**:172-187. DOI: 10.1016/j.tifs.2019.02.020

[64] Eaks IL, Sinclair WB. Cellulose-hemicellulose fractions in the alcohol-insoluble solids of Valencia orange peel. Journal of Food Science. 1980;**45**: 985-988

[65] Mussatto SI, Dragone G, Roberto IC. Brewers' spent grain: generation, characteristics and potential applications. Journal of Cereal Science. 2006;**43**:1-14. DOI: 10.1016/j.jcs.2005.06.001

[66] Ancuța P, Sonia A. Oil press-cakes and meals valorization through circular economy approaches: a review. Applied Sciences. 2020;**10**:7432

[67] Lyons B, De Daunton F. Recruitment and retention of a rural and regional workforce: a regional development perspective. Australia: Australian Farm Institute; 2022

[68] Hetherington J, Juliano P, MacMillan C, Loch A. Circular economy opportunities and implementation barriers for Australia's food, feed and fibre production. Farm Policy Journal. 2022. pp. 30-43

[69] Usman I, Saif H, Imran A, et al. Innovative applications and therapeutic potential of oilseeds and their by-products: an eco-friendly and sustainable approach. Food Science & Nutrition. 2023;**11**(6):2599-2609. DOI: 10.1002/fsn3.3322

[70] Araújo-Rodrigues H, Santos D, Campos DA, et al. Development of frozen pulps and powders from carrot and tomato by-products: impact of processing and storage time on bioactive and biological properties. Horticulturae. 2021;**7**:1-22. DOI: 10.3390/horticulturae7070185

[71] Lowe EA. Creating by-product resource exchanges: strategies for eco-industrial parks. Journal of Cleaner Production. 1997;**5**:57-65

[72] Nevara GA, Giwa Ibrahim S, Syed Muhammad SK, et al. Oilseed meals into foods: an approach for the valorization of oilseed by-products. Critical Reviews in Food Science and Nutrition. 2022:1-14. DOI: 10.1080/10408398.2022.2031092

[73] Planetarians. 2023. Planetarians. Available from: https://www.planetarians.com/ [Accessed: 14 Apr 2023]

[74] Prepared Foods. Planetarians Raises $6 million for commercial production of plant-based, climate friendly alternative protein. Prepared Foods. 2022. https://www.preparedfoods.com/articles/127951-planetarians-raises-6-million-for-commercial-production-plant-based-climate-friendly-alternative-protein

[75] Van Huis A. Prospects of insects as food and feed. Organic Agriculture. 2021;**11**:301-308

[76] Van Peer M, Frooninckx L, Coudron C, et al. Valorisation potential of using organic side streams as feed for Tenebrio molitor, Acheta domesticus and Locusta migratoria. Insects. 2021;**12**:796

[77] Oonincx DGAB, De Boer IJM. Environmental impact of the production of mealworms as a protein source for humans – a life cycle assessment. PLoS One. 2012;**7**:e51145

[78] Oonincx DGAB, Van Itterbeeck J, Heetkamp MJW, et al. An exploration on greenhouse gas and ammonia production by insect species suitable for animal or human consumption. PLoS One. 2010;**5**:e14445

[79] Orkusz A. Edible insects versus meat — nutritional comparison: knowledge of their composition is the key to good health. Nutrients. 2021;**13**:1207

[80] Payne CLR, Scarborough P, Rayner M, Nonaka K. A systematic review of nutrient composition data available for twelve commercially available edible insects, and comparison with reference values. Trends in Food Science and Technology. 2016;**47**:69-77

[81] Dobermann D, Swift JA, Field LM. Opportunities and hurdles of edible insects for food and feed. Nutrition Bulletin. 2017;**42**:293-308

[82] Sorjonen JM, Valtonen A, Hirvisalo E, et al. The plant-based by-product diets for the mass-rearing of Acheta domesticus and Gryllus bimaculatus. PLoS One. 2019;**14**:e0218830

[83] Lundy ME, Parrella MP. Crickets are not a free lunch: protein capture from scalable organic side-streams via high-density populations of Acheta domesticus. PLoS One. 2015;**10**:e0118785

[84] Lähteenmäki-Uutela A, Marimuthu SB, Meijer N. Regulations on insects as food and feed: a global comparison. Journal of Insects as Food and Feed. 2021;**7**:849-856

[85] Kröger T, Dupont J, Büsing L, Fiebelkorn F. Acceptance of insect-based food products in western societies: a systematic review. Frontiers in Nutrition. 2022;**8**:1186

[86] Azzollini D, Derossi A, Fogliano V, et al. Effects of formulation and process conditions on microstructure, texture and digestibility of extruded insect-riched snacks. Innovative Food Science & Emerging Technologies. 2018;**45**:344-353

[87] Igual M, García-Segovia P, Martínez-Monzó J. Effect of Acheta domesticus (house cricket) addition on protein content, colour, texture, and extrusion parameters of extruded products. Journal of Food Engineering. 2020;**282**:110032

[88] Ribeiro L, Cunha LM, García-Segovia P, et al. Effect of the house cricket (Acheta domesticus) inclusion and process temperature on extrudate snack properties. Journal of Insects as Food and Feed. 2021;**7**:1117-1129

[89] Ruszkowska M, Tańska M, Kowalczewski PŁ. Extruded corn snacks with cricket powder: impact on physical parameters and consumer acceptance. Sustainability. 2022;**14**:16578

[90] Cho SY, Ryu GH. Effects of mealworm larva composition and selected process parameters on the physicochemical properties of extruded meat analog. Food Science & Nutrition. 2021;**9**:4408-4419

[91] Smetana S, Larki NA, Pernutz C, et al. Structure design of insect-based meat analogs with high-moisture extrusion. Journal of Food Engineering. 2018;**229**:83-85

[92] FAO. Looking at Edible Insects from a Food Safety Perspective. Challenges and Opportunities for the Sector. Rome, Italy: FAO; 2021

Chapter 7

Hydrothermal Conversion of Lignocellulosic Biomass to Hydrochar: Production, Characterization, and Applications

Meegalla R. Chandraratne and Asfaw Gezae Daful

Abstract

The high moisture content poses a major technical barrier to using wet biomasses in thermochemical conversions. Hydrothermal conversions open efficient ways to convert wet biomass into carbonaceous products as an alternative to thermochemical methods such as pyrolysis, gasification, and combustion. Three types of hydrothermal conversions, hydrothermal carbonization (HTC), hydrothermal liquefaction (HTL), and hydrothermal gasification (HTG), use different operating conditions to convert wet biomass into distinct products: solid (hydrochar), liquid (aqueous soluble bio-oil), and gaseous fractions. Water plays a dominant role in hydrothermal conversions. HTC uses relatively mild conditions. HTL and HTG use subcritical and supercritical conditions, respectively. Conversion mechanisms and the effect of process parameters are also discussed in detail. The solid product hydrochar (HC) has properties comparable to biochar and activated carbon, hence a range of potential applications. Current and emerging applications of HC, including energy production and storage, soil amendment, wastewater treatment, carbon capture, adsorbent, and catalyst support, are discussed.

Keywords: bioenergy, hydrothermal carbonization, bio-oil, biofuel, thermochemical conversion, biomass composition, pyrolysis

1. Introduction

Over the past few decades, the world has seen a rapid increase in energy demand and consumption. Currently, fossil fuels (oil, natural gas, and coal) are the primary source of global energy supply. The combustion of fossil fuels releases greenhouse gases into the atmosphere, resulting in global warming, environmental pollution, and climate change. Due to higher prices and a finite supply of fossil fuels and to reduce greenhouse gas emissions and mitigate climate change, more attention has been given to finding a sustainable long-term alternative to fossil fuels. As a result, renewable energy resources (such as biomass, solar, wind, geothermal, and tidal) are gaining popularity. Biomass, available abundantly at a relatively low cost, is considered a promising alternative renewable energy source to substitute for fossil fuels. It is a clean, environmentally friendly, and carbon-neutral source of energy. Burning biomass or fossil fuel releases carbon dioxide (CO_2), a greenhouse gas. However, when biomass burns, it releases CO_2 back into the atmosphere that the plants have absorbed recently by photosynthesis during the growth process producing biomass. Thus, burning biomass does not add to the total CO_2 inventory of the earth and is considered carbon neutral. Nevertheless, the overall biomass chain, including planting, harvesting, processing, and transporting, needs to be considered for true carbon neutrality of biomass. In addition, burning biomass releases a variety of pollutants, including carbon monoxide (CO), nitrogen oxides (NOx), and particulate matter [1–4].

2. Lignocellulosic biomass

Biomass is defined as a plant or animal-based organic matter of recent origin. Biomass includes agricultural crop residues (residues of agricultural crops that are not harvested for commercial use, including stalks, leaves, husks, etc.), forestry and wood residues (residues associated with the production of timber in the forest, as well as the processing of timber into their final products), dedicated energy crops (herbaceous and short rotation woody energy crops), aquatic biomass (algae, plants, and microbes found in water), sewage sludge, digestate (remains of anaerobic digestion), industrial crops, animal, industrial, municipal and food waste [1, 5]. A detailed discussion on biomass sources, classification, composition, and analysis is available in [1, 5]. If improperly managed, these biomass residues and waste could lead to environmental issues, including water contamination, greenhouse gas emissions, pests and insects breeding, and foul odor. Converting biomass residues and wastes into bioenergy (a) reduces the burden on waste management, (b) energy generated reduces the dependence on fossil fuels, and (c) reduces the amount of decomposing waste and associated environmental issues [5–7].

Biomass is a complex heterogeneous mixture mainly of organic matter (and a small amount of inorganic matter). Biomass is also classified based on its chemical composition as carbohydrates, lignin, essential oils, vegetable oils, animal fats, natural resins (gums), etc. The most abundant biomass on earth is lignocellulosic biomass (LCB), which includes agricultural crop residues, forestry and wood residues, dedicated energy crops, industrial crops, and food waste. LCB is a complex mixture of biopolymers consisting of carbon (C), oxygen (O), and hydrogen (H), with total content reaches typically above 95%. The elemental composition of C, O, and H in plant biomass on a dry basis is 42–47%, 40–44%, and 6%, respectively. Plant biomass has a substantial amount of oxygen with a carbon-to-oxygen (C/O) ratio of almost one. Because of the higher level of oxygen content, the energy density of biomass is relatively low compared to fossil fuels. Plant biomass also contains macronutrients, micronutrients, trace elements, and other heavy metals, which are present in varying amounts, depending on the plant species and environment, summing all together up to about 4% [1, 4, 8–10].

The three major constituents of LCB are cellulose, hemicellulose, and lignin. These constitute the cell wall of plants. The most abundant polymer of the lignocellulosic plant cell wall is cellulose, which provides structural support that gives trees and wood their strength. It is a linear polymer chain composed of D-glucose (pyranose) units linked by β-1,4 glycosidic bonds. Cellulose has a high (300–15,000) degree of polymerization (DP) depending on the plant species. Cellobiose, glucose disaccharide, is the repeating unit of cellulose. Cellulose chains are grouped to form elementary fibrils, which are interlinked by hydrogen bonds and Van der Walls forces to form long microfibrils. Microfibrils represent the main component of the cell wall. Because of the high degree of hydrogen bonding, cellulose is highly stable and resistant to chemical attack. Interaction of hydroxyl (OH) groups present on the inner and outer surfaces of cellulose forms hydrogen bonds (intra- and intermolecular), which stiffens the chains and provides crystalline structure to cellulose. The interchain hydrogen bonds create crystalline (uniform and ordered) and amorphous (loose and disordered) regions of cellulose. The amorphous regions are more reactive than crystalline regions. The crystalline structure of cellulose leads to its mechanical strength and chemical stability and provides strength and toughness to leaves, roots, and stems. Cellulose is insoluble in almost any solvent [4, 5, 8, 11–13].

Hemicellulose is the second most abundant polymer in LCB. It is a branched heterogeneous structural polysaccharide composed of C5 and C6 sugars, including D-xylose, L-arabinose, D-mannose, D-galactose, and D-glucose units. Hemicellulose has a lower DP (80–200) than cellulose. Because of the branched structure, hemicellulose is amorphous and easy to hydrolyze by dilute acids, bases, and enzymes. Hemicellulose lateral chains form a tightly bound network with cellulose microfibrils through hydrogen bonds, which makes a highly rigid cellulose-hemicellulose-lignin matrix with the interaction of lignin *via* covalent bonds [4, 9, 11, 12].

Lignin is the third most abundant polymer in LCB. Lignin is an aromatic polymer composed of three phenylpropanoid monomers: guaiacyl, p-hydroxyphenyl, and syringyl. The phenylpropanoid monomers in lignin are linked in alkyl-aryl, alkyl-alkyl, and aryl-aryl ether bonds. Lignin, an amorphous and highly complex aromatic hydrophobic biopolymer, is the natural phenolic glue that tightly binds cellulose and hemicellulose. Lignin plays a cementing role for linkages between cellulose and hemicellulose to form a 3-D structure of lignin-polysaccharide complex in the cell wall, leading to a strong cell wall structure that makes it insoluble in water and provides mechanical strength to plants. Lignin provides sealing for a water-conducting system linking roots with leaves. The cross-linking between polysaccharides (hydrophilic and permeable) and lignin (hydrophobic and impermeable) creates vascular tissues for the efficient conduction of water in plants. Lignin forms a natural protective shield protecting cellulose and hemicellulose in plants and makes plants resistant to pathogens, oxidative stresses, and biodegradation by enzymes and microorganisms [5, 14, 15].

The key component in woody biomass is cellulose. Leaves and grasses are rich in hemicellulose, and lignin is the major component in shells. Hemicellulose is thermally least stable, and lignin is the most stable of all three. The cellulose, hemicellulose, and lignin content vary with the biomass falling in the range of 40–60%, 15–30%, and 10–25%, respectively. In addition to three major components, inorganic compounds and organic extractives, including fats, waxes, proteins, simple sugars, gums, resins, starches, phenolics, pectins, essential oils, and mineral compounds, are also present in biomass as nonstructural components, which are responsible for the smell, color, flavor, and natural resistance to decaying of some species. Inorganics present in LCB include nitrogen (N), phosphorus (P), potassium (K), magnesium (Mg), sulfur (S), calcium (Ca), chlorine (Cl), iron (Fe), boron (B), manganese (Mn), zinc (Zn), copper (Cu), molybdenum (Mo), nickel (Ni), selenium (Se), silicon (Si), sodium (Na), aluminum (Al), titanium (Ti), cadmium (Cd), and chromium (Cr). Woody biomass contains a high (~90%) amount of cellulose, hemicellulose, and lignin, whereas agricultural and herbaceous biomass contain more extractives and ash [5, 8, 11, 12].

Biomass analysis, including proximate and ultimate analysis, as well as higher heating value (HHV), is essential in understanding the behavior of biomass in energy applications. The proximate analysis determines the volatile matter (VM), fixed carbon (FC), ash (noncombustible solid residue), and moisture (M) content. VM is the vapors/gases released during heating. FC is the nonvolatile solid carbon that remains after devolatilization. Inorganic matter, including silica, calcium, iron, aluminum, potassium, sodium, magnesium, and titanium, are the main constituents of ash. High-ash content biomasses include agricultural residues, grasses, and straws. Contamination of biomass with dirt, soil, rock, and other impurities during collection and handling partly contributes to ash content. Moisture content has a significant impact, and high moisture content is a major concern in biomass conversion. The moisture content of some biomasses, such as water hyacinth, can be very high (> 90%). Moisture can affect the storage and handling of biomass. High moisture content significantly increases transportation and energy costs. It lowers the calorific value as the energy released is used in the evaporation of moisture, which is not recovered. Moisture drains much of the deliverable energy during conversion. The ultimate analysis expresses biomass composition in terms of major elements (C, H, O, S, and N) on a mass percent, dry, and ash-free basis. These are useful for performing mass balances on biomass conversion processes. Typically, the sulfur and nitrogen content of biomass is very low. The characteristics of raw biomass include high moisture content, low bulk density, low energy density (calorific value), poor grindability, compositional non-homogeneity, hygroscopic nature, higher biological degradation, and lower storability [9, 16–18].

3. Biomass conversion

The primary goal of biomass conversion is to break down the rigid and complex structure of biomass polymers into smaller and lower molecular weight (MW) compounds. Typically, the reaction time, temperature, and reaction medium influence the rate of destruction of the biomass polymeric structure. Chemical, biochemical, and thermochemical conversion processes convert biomass into end products such as chemicals, biofuels, or heat. Chemical conversion processes use chemicals to convert biomass into biofuels. Biochemical conversion processes utilize microorganisms or enzymes to convert biomass into products. Thermochemical conversion processes employ high temperatures in oxygen free, reduced, or normal oxygen environments. The thermochemical conversion approaches are applicable for a wide range of biomass types using different temperatures to break down the bonds of organic matter in a relatively short period. The advantages of thermochemical conversions over biochemical conversions include better conversion efficiency, handling a wide variety of feedstocks, shorter reaction times, and high energy efficiency. In most situations, the selection of the conversion process is based on the biomass feedstock available and the desired form of end products. Environmental standards, policy, economic conditions, and specific factors related to the project may influence the selection process. The moisture content of biomass is a primary aspect determining the conversion process. Thermochemical conversions, such as pyrolysis, gasification, or combustion, generally require relatively dry biomass with a moisture content of less than 30 wt.%, but moisture contents of 10 wt.% are preferred. The low energy density of wet biomass due to higher moisture content makes it unsuitable for these processes. The biochemical conversion process, a wet conversion process, is suitable for processing high moisture content biomass. Hydrothermal (HT) processing is a thermochemical conversion that involves applying heat (and pressure) in subcritical or supercritical water. It can process high moisture content (> 70 wt.%) biomass feedstock without dewatering or drying. Biochemical conversion consumes less energy compared to HT processing but requires more time. Subsequently, cost-effective HT processing has gained growing attention as a method suitable for processing wet biomass [1, 5–8, 19, 20].

4. Hydrothermal (HT) processing

Biomass with a moisture content of more than 30 wt.% needs to be dried before being suitable for some thermochemical processes such as pyrolysis, combustion, and gasification. Drying is a highly energy-intensive process and requires a large amount of energy. As a result, for biomass with high moisture content, the heat of moisture evaporation is higher than the heat available from biomass, becoming a net energy consumption. Wet biomass, typically with 70 wt.% or more water, can be converted using HT processing without energy-intensive drying [1, 5, 8].

HT processing applies heat and pressure in subcritical or supercritical water in a closed reactor. The biomass is surrounded by water during the reaction, and the presence of water speeds up biomass conversion. HT reactions are supported by water, which plays an active role as a solvent, reactant, and even a catalyst or catalyst precursor in the hydrolysis reactions. Due to the high ionic product in subcritical conditions, water shows both acidic and basic properties and behaves as a catalyst. As a result, the addition of acid or base catalyst can be avoided. Water behaves as a catalyst precursor due to relatively high concentrations of hydronium (H_3O^+) and hydroxide $(OH^-)^-$ ions resulting from the dissociation of water. In HT processing, water is used as a reacting medium. It is cheap, environmentally friendly, nontoxic, inherently present in wet biomass, and a better alternative to corrosive chemicals and toxic solvents. In HT processing, the reaction pressure is usually not controlled, but the temperature maintains the autogenic pressure corresponding to the saturation vapor pressure of water. It is an attractive process to convert wet biomass into three distinct product fractions, solid (HC), liquid (bio-oil/water), and a mixture of non-condensable gases. All three product fractions (solid, liquid, and gaseous) are formed at all temperatures in HT processing. However, reaction temperature (and pressure) determines the product distribution. The reaction shifts from solid products at low temperatures through liquid products at medium temperatures to gaseous products at high temperatures. The advantages of HT processing include the elimination of energy-intensive drying, high conversion efficiency, and relatively low operating temperature. Depending on the operating conditions, HT processes can be classified into three processes as hydrothermal carbonization (HTC), hydrothermal liquefaction (HTL), and hydrothermal gasification (HTG) [1, 5, 8, 21–23].

The physicochemical properties of water are strongly affected by pressure and temperature. Water exhibits gas-like behavior at or around the critical point with low density, low viscosity, high compressibility, and high diffusivity. Some physical and chemical properties of water at various conditions are shown in **Table 1**. Above the critical point, a single homogeneous fluid phase exists, and the density, viscosity, compressibility, and diffusivity of water are very sensitive to pressure and temperature changes. This unique property of water is a result of hydrogen bonding. When water is heated, hydrogen bonding in the water molecules becomes weak, allowing the dissociation of water into H_3O^+ and OH^- ions. The characteristics of water change

	Ambient	Subcritical water	Supercritical water
Temperature (°C)	0–100	100–374	> 374
Pressure (MPa)	0.003 (24°C)	0.1 (100°C)-22.1 (374°C)	> 22.1
Density (kg/m^3)	997	958 (101°C, 0.11 MPa) 692 (330°C, 30 MPa)	between gas-like and liquid-like densities 252 (410°C, 30 MPa)
Viscosity (mPa s)	0.89 (25°C, 0.1 MPa)	0.11 (250°C, 5 MPa) 0.064 (350°C, 25 MPa)	0.03 (400°C, 25 MPa) 0.07 (400°C, 50 MPa)
Dielectric constant	78.5 (25°C, 0.1 MPa)	27.1 (250°C, 5 MPa) 18.2 (330°C, 30 MPa) 14 (350°C, 20 MPa)	5.9 (400°C, 25 MPa) 10.5 (400°C, 50 MPa)
Ion product K_w (mol^2/L^2)	10^{-14} (25°C) 10^{-12} (100°C)	10^{-12} (100°C) 10^{-11} (300°C)	10^{-20} (400°C) 10^{-23} (550°C)
Heat capacity C_P (kJ/kg K)	4.22 (25°C, 0.1 MPa)	4.86 (250°C, 5 MPa) 10.1 (350°C, 25 MPa)	13 (400°C, 25 MPa) 6.8 (400°C, 50 MPa)
Thermal conductivity (W/m K)	0.608 (25°C, 0.1 MPa)	0.62 (250°C, 5 MPa)	0.16 (400°C, 25 MPa) 0.438 (400°C, 50 MPa)

Table 1.
Physical and chemical properties of water at various conditions [24–27].

from a polar solvent at ambient conditions to a nonpolar solvent at supercritical conditions. At ambient conditions, organic compounds, and gases are poorly miscible in water. But the high dielectric constant (about 80) makes water a good solvent for salts. The dielectric constant decreases rapidly with increasing temperature to about 27 (at 250°C and 5 MPa) and about 14 (at 350°C and 20 MPa). Under HT conditions, water displays less polar behavior due to decreasing dielectric constant and the miscibility of organic compounds is improved, which opens new reaction pathways. Temperature and pressure can be controlled to adjust reaction rates. Above the critical point, organic compounds are completely miscible. Below the critical point, organic compounds are miscible. When HT processing products are cooled down to ambient conditions, water, and organic compounds are separated again [23, 26, 28–30].

4.1 Hydrothermal carbonization (HTC)

HTC is carried out in compressed water at temperatures between 180 and 260°C under autogenous pressures (up to 4 MPa), and biomass feedstock in water is converted mostly to solids (HC). The residence time of HTC varies from minutes up to several hours. HTC aims to maximize the yield of HC. In addition, HTC produces bio-oil mixed with water in the liquid fraction (aqueous solution) and a small volume of gases (consisting of H_2, CH_4, and CO_2). The product distribution and characteristics primarily depend on the process conditions and the type of biomass feed. The reaction temperature remains the main process parameter even though both reaction time and temperature influence the physicochemical characteristics of products. Solids concentration has a relatively smaller influence. The characteristics of HC are significantly affected by the chemical composition of the feedstock. The liquid fraction contains valuable chemical compounds, including organic acids (acetic acid, formic acid, lactic acid, levulinic acid, and propionic acid), furan compounds (furfural, furfuryl alcohol, and 5-hydroxymethylfurfural (5-HMF)), esters, phenolic compounds, lignin fragments, amino acids, sugars, and some nutrients (nitrogen and phosphorus) that are formed *via* the biomass polymers degradation. Hemicellulose degradation temperature in subcritical water is usually around 180°C and cellulose decomposition in subcritical water usually starts above 200°C, while lignin decomposes above 220°C [5, 8, 21, 31–34].

During HT decomposition, the physical structure of biomass is altered through a series of reaction mechanisms such as hydrolysis, dehydration, decarboxylation, aromatization, polymerization, and condensation. These reactions do not run sequentially; instead, they show simultaneous and interconnected reaction paths and the mechanisms depend on the feedstock type. Hydrolysis has a lower activation energy than most decomposition reactions, which initiates biomass HT degradation [35–37]. Under HTC conditions, marked changes in water properties (such as dielectric constant, ionic product, and polarity) zcatalyze the hydrolysis of biomass. These property changes unlock different reaction pathways that are not possible in other conversion conditions such as pyrolysis, liquid water, or steam at atmospheric pressures. The dissociation of water into H_3O^+ ions facilitates hydrolysis reactions. As a result of property changes, under HTC conditions, water acts as a reagent for various reactions and as a solvent for a wide range of organic biomass compounds. In hydrolysis, biomass polymers (such as hemicellulose, cellulose, extractives, and lignin) react with water and break ester and ether (mainly β-1-4 glycosidic) bonds creating oligosaccharides and fragments of lignin that enter the liquid phase. Consequently, complete disintegration of the physical structure of biomass may occur. Rapid degradation and depolymerization of hemicellulose and cellulose chains following hydrolysis produces a wide range of fragments, including water-soluble oligomers (cellobiose, celotriose, cellotetraose, cellopentaose, cellohexaose, arabinose, etc.) and monomers (glucose and fructose). Hydrolysis of lignin fragments gives rise to phenolic compounds. Most extractables possess good water solubility and are eluted simultaneously [35–37].

Subsequently, two important reactions (dehydration and decarboxylation) occur simultaneously, reducing both the oxygen and hydrogen content of the biomass. The extractables increase dehydration and decarboxylation, leading to a condensation reaction. The reaction mechanism of the HTC process mainly involves dehydration, decarboxylation, and polymerization. Dehydration is favored at low temperatures (< 300°C) [38, 39]. The soluble hydrolysis products undergo further dehydration, which removes water from the biomass by eliminating hydroxyl (-OH) groups. It also releases more water into the reaction medium and carbonizes biomass by lowering the H/C and O/C ratios. Decarboxylation removes CO_2 from the biomass, eliminating carboxyl (-COOH) groups. Condensation and polymerization convert these compounds to larger molecules, which undergo further aromatization. The net result for LCB is highly aromatized fragments combining phenolic structures derived from the lignin dehydration with aromatization resulting from the carbonization of carbohydrates. The fragments undergo condensation and polymerization to form humic acid and bitumen-like material and partly precipitate to form HC [23, 40].

The main reason for a significant decrease in oxygen content could be the elimination of carboxyl groups, mainly from hemicellulose, cellulose, and extractives. Aromatization occurs because of dehydration and decarboxylation. In the process, hydroxyl and carboxyl groups in biomass are replaced by C=O and C=C bonds. Aromatic carbon structures formed are considered the building block of HC. The furfural compounds generated undergo hydrolysis, which further breaks them down into acids, aldehydes, and phenols. Due to the formation of organic acids such as acetic acid, lactic acid, formic acid, propionic acid, and levulinic acid pH of biomass and process water decrease as HTC proceeds, which further promotes hydrolysis and dehydration of small-chain polymers and monomers into much smaller fragments. The dehydration process of eliminating OH groups is sometimes called chemical dehydration. During the HTC of biomass, physical dehydration also occurs in addition to chemical dehydration. In physical dehydration, the reduction in the hydrophilic nature of biomass expels water out of the biomass matrix [35, 36].

The dehydration and decarboxylation of hydrolyzed products (fragments and extractable) lead to the formation of intermediate compounds. The 5-HMF generated from C6 sugars (D-fructose and D-glucose) and furfural generated from C5 sugars (D-xylose) are well-known dehydration intermediate compounds of sugars. The HTC process under acidic conditions allows the effective conversion of D-glucose to 5-HMF. D-glucose first isomerizes to D-fructose, and then undergoes dehydration to form 5-HMF. The 5-HMF ($C_6H_6O_3$) is C6 heterocyclic aldehyde containing aldehyde and alcohol functional groups. It has formyl (-CHO) and hydroxy methyl (-CH_2-OH) functional groups attached to 2 and 5 positions, respectively. The 5-HMF, one of the "top 10 biomass derived value-added chemicals" defined by the US Department of Energy, is a versatile platform chemical for a wide range of fuel and chemical products, including 2,5-furan dicarboxylic acid (FDCA), 2,5 dimethylfuran (DMF), 2 methyl furan, levulinic acid, formic acid, furfuryl alcohol, adipic acid, caprolactam, and maleic acid. The furfural ($C_5H_4O_2$) is C5 heterocyclic aldehyde containing aldehyde (-CHO) and two olefins (-CH=CH-) functional groups. Furfural can undergo various chemical reactions due to its aldehyde and olefin functional groups are considered a promising platform chemical that can be used to produce a range of chemical products, including DMF, tetrahydrofuran (THF), furan, furfuryl alcohol, 2-methyl furan, and levulinic acid [1, 35, 38, 39, 41, 42].

There are two pathways to HC formation: solid-to-solid conversion (primary char) and polymerization of aqueous phase intermediate compounds (secondary char). The carbon-rich intermediate compounds (such as furfural and 5- HMF) resulting from hydrolysis and dehydration undergo condensation, polymerization, and aromatization to produce bio-oil. After successive polymerization and aromatization, bio-oil converts into a solid product with or without auto nucleation to form secondary char [37]. A high concentration of HMF favors secondary char formation. The typical HTC temperatures are too low to convert lignin completely, and lignin forms primary char *via* solid-to-solid conversion. Sequential hydrolysis, dehydration, and isomerization producing furfurals and cleavage reactions yielding intermediate organic acids are thought to have resulted in secondary char. These dissolved intermediates can lead to the precipitation of the furfurals as a secondary organic phase, which polymerizes as microspheres. Both primary and secondary chars are called HC when they are formed by HTC, despite their different chemical structures [34, 43].

The solid fraction of the HTC is in the form of slurry mixed with an aqueous fraction, and it has to pass through a series of steps such as mechanical dewatering (compressing), filtering, and drying before it can be used as a fuel. HC has high hydrophobic and homogeneous properties and can be easily separated from the liquid fraction. HTC process removes part of oxygen from biomass during decarboxylation and dehydration. Consequently, by mechanical dewatering, the moisture content of solid fraction can be reduced to 50% (only 70–75% moisture content can be achieved in mechanical dewatering of wet biomass). This reduces the energy and time required for drying. Dried HC (moisture content <5%) pellets can be produced from the separated solid fraction, which can be used for energy production [21, 44].

Initially, the liquid fraction was considered a waste, hence higher costs for wastewater treatment and disposal. Recently, it has been considered a potential added resource of the HTC process as it contains various valuable organic and inorganic compounds, including organic acids, furan compounds, esters, phenolic compounds, lignin fragments, amino acids, sugars, and some nutrients. Mono sugars and chemicals can be recovered from liquid fraction. The gas fraction contains less CO and CO_2 and is less harmful [5, 8, 21, 31–34].

The inorganic elements (Ca, Mg, P, K, Na, S, and Fe) present in LCB remain in ash as oxide forms (CaO, MgO, P_2O_5, K_2O, Na_2O, SO_3, and Fe_2O_3) after combustion. Even though the ash content of LCB is usually minimal, these oxides can cause severe agglomeration, fouling, clinker formation, and corrosion during combustion, pyrolysis, and gasification. Inorganic compounds are very stable and may remain unchanged under HTC conditions, but the degradation of biomass polymers might release inorganics from the solid structure; hence some of the ash-forming elements can leach out from the biomass to the aqueous solution reducing the overall ash content of the HC. In addition, the HTC process can convert organic chlorine to inorganic chlorine reducing chlorine content in HC, which reduces the potential for corrosion and dioxin formation in combustion [21, 22].

Some of the advantages of the HTC process include (a) fewer emissions, low carbonization temperatures, and lower energy consumption than pyrolysis; (b) a greater variety of wet and dry feedstocks can be processed in an aqueous medium (no drying is required); (c) higher solid yields; (d) final product with lower ash content; and (e) inexpensive process. Feedstock material and processing conditions significantly influence the performance and properties of the HC. HC obtained from HTC has the following properties: (a) versatile properties make it suitable for a wide range of applications in the energy and environmental sectors, (b) oxygenated functional groups at the surface (OH, C=O, and COOH groups), (c) controlled porosity can be easily introduced using activation procedures, thermal treatments, etc., (d) uniform spherical micro-sized particles, (e) easily controlled surface chemistry and electronic properties *via* additional thermal treatment, and (f) specific physicochemical properties tailored for particular applications can be obtained by adding other components (such as inorganic nanoparticles) to biomass [1, 37–39].

HC is more energy dense, hydrophobic, and easily friable and has a lower H/C ratio and O/C ratio than the feedstock used. H/C and O/C ratios of selected biomasses and HC are given in **Table 2**. This is a result of reducing the oxygen and hydrogen content of the biomass through dehydration and decarboxylation, disrupting colloidal structures, and reducing the hydrophilic functional groups. A complex series of reactions, reducing hydrophilic functional groups, take place in HT medium, including the removal of hydroxyl groups through dehydration, removal of carboxyl and carbonyl groups through decarboxylation, and cleavage of many ester and ether bonds through hydrolysis [23, 40]. Specific characteristics, including H/C ratio, O/C ratio, elemental composition, porosity, conductivity, morphology, energy content, etc., are needed for applications in many fields such as solid fuel in soil amendment, adsorbent in water purification, power generation, and carbon capture. The characteristics of HC can be modified by combining HTC with other processes. The initial use for the HTC process was organic materials degradation, liquid and gaseous fuels,

Feedstock	Temperature, time, B/W	H/C**	O/C**	Fuel Ratio (FC/VM)	HHV (MJ/kg)	Ref.
Sewage sludge	220°C, 12 h, 1/7.5	2.07*, 1.77	0.52*, 0.36	0.12*, 0.22	9.26*, 5.97	[45]
Peanut shells	220°C, 12 h, 1/7.5	1.58*, 0.99	0.75*, 0.27	0.36*, 0.76	12.6*, 28.14	[45]
Orange peel	220°C, 12 h, 1/7.5	1.76*, 1.02	0.86*, 0.27	0.23*, 0.83	12.06*, 28.21	[45]
Fallen leaves	220°C, 12 h, 1/7.5	1.63*, 1.21	0.63*, 0.26	0.25*, 0.5	18.4*, 27.21	[45]
Rice straws	240°C, 1.5 h, 1/12	1.74*, 1.11	0.86*, 0.28	0.22*, 0.73	16.93*, 20.6	[46]
Waste textile	240°C, 1.5 h, 1/12	1.39*, 0.81	0.58*, 0.36	0.2*, 0.29	19.23*, 22.11	[46]
Waste wood	240°C, 1.5 h, 1/12	2.13*, 1.29	0.74*, 0.28	0.4*, 1.53	18.04*, 26.46	[46]
Wastepaper	240°C, 1.5 h, 1/12	2.07*, 1.2	0.9*, 0.29	0.16*, 0.53	14.28*, 24.05	[46]
Waste food	240°C, 1.5 h, 1/12	2.25*, 1.44	0.54*, 0.15	0.18*, 0.94	21.33*, 25.65	[46]
Hornbeam wood	225°C, 1 h, 1/20 250°C, 1 h, 1/20 275°C, 1 h, 1/20	1.51*, 1.17 1.08 1.02	0.69*, 0.57 0.49 0.47	0.17*, 0.32 0.95 1.36	17.29*, 18.45 19.91 20.46	[47]

*B/W — biomass to water ratio, * Values of biomass, ** atomic ratios.*

Table 2.
Properties of biomass and hydrochar.

and basic chemical production. In recent years, the technology gained research interest to produce solid HC and as a technique to synthesize nano- and micro-size carbon particles [1, 31, 32, 48].

The HC produced by HTC directly from LCB lacks porosity. Only a small porosity is developed after further carbonization at a higher temperature. This is due to HC being pre-carbonized material produced under autogenic pressures and temperatures between 160 and 200°C. The high surface area and porosity of HC are essential for most industrial applications, such as adsorption or catalysis. This would ensure efficient transport and diffusion throughout the material. Different techniques have been developed to improve porosity [1, 32].

4.2 Hydrothermal liquefaction (HTL)

HTL is carried out at temperatures between 260 and 350°C under autogenous pressures of 5–20 MPa (subcritical); biomass in compressed water is primarily converted to an aqueous fraction (bio-oil, bio-crude). Bio-oil is a complex mixture of organic compounds containing acids, alcohols, aldehydes, ketones, furans, sugars, amines, amides, esters, ethers, phenols, etc. Free radical and ionic reactions are considered to be leading reactions in HTL. The dissociation of water increases with an increase in temperature. The ionic product of water (K_w) 10^{-14} mol^2/L^2 at 25°C increases to 10^{-11} mol^2/L^2 at 300°C. At HTL conditions (with elevated temperature), mass transfer is improved because of accelerated mass-transfer-limited chemical reactions resulting from the lower viscosity of water. In HTL, water plays an important role as reaction media, helps in the dispersion of biomass, provides hydrogen, and stabilizes radicals. Water inhibits the polymerization of intermediate products, improving the quality and production of bio-oil but at the expense of HC yield [5, 26, 49–51].

The primary conversion of biomass during the HTL comprises three basic reaction mechanisms: depolymerization, decomposition, and recombination. Long-chain biomass polymers are depolymerized and decomposed into shorter-chain compounds. These short-chain compounds are highly reactive and may recombine (repolymerize) to form liquid, gaseous, and solid products. Some bio-oil components could originate from biomass feedstocks, while others could be converted through hydrolysis, depolymerization, decomposition, and recombination of reactive fragments. The important parameters for the depolymerization of long-chain polymer structures to shorter-chain hydrocarbons are temperature and pressure. Depolymerization liquefies macromolecules of biomass by breaking down their physical and chemical components and characteristics. Higher temperatures of the HTL process and desirable physicochemical properties of water help the depolymerization process to overcome the resistance of biomass. Water acts as a catalyst at HTL conditions. The three steps involved in decomposition are dehydration (loss of H_2O molecule), decarboxylation (loss of CO_2 molecule), and deamination (removal of amino acid content). The two mechanisms, dehydration and decarboxylation, facilitate the removal of oxygen from the biomass in the form of H_2O (eliminating the hydroxyl group) and CO_2 (eliminating the carboxyl group), respectively. Biomass macromolecules are hydrolyzed to form monomers and oligomers. When HTL occurs at lower temperatures, biomass undergoes hydrolysis to convert biopolymers into soluble intermediates such as 5-HMF, furfural, amino acids, and fatty acids. The hydrolysis of polysaccharides and proteins begins around 190°C. When temperature increases, subcritical water at HTL conditions breaks down the hydrogen bonds of the cellulose structure to form sugar monomers, which are rapidly degraded by different reactions, including isomerization, dehydration, hydrolysis, reverse aldol defragmentation, rearrangement, and recombination to reactive intermediates. The degradation products, such as furfurals, phenols, organic molecules, glycoaldehydes, and organic acids, are highly soluble in water. Recombination and repolymerization of light MW compounds to yield higher MW compounds occur due to excess oxygen or unavailability of hydrogen molecules [26, 52–55].

Higher temperatures of the HTL process provide the necessary activation energy to decompose lignin in the presence of water. In addition, desirable physicochemical properties, such as low dielectric constant, high ionic product, weak hydrogen bond, and high diffusivity at HTL conditions, catalyze the hydrolysis of biomass breaking down the hydrogen bonds of the cellulose structure to form sugar monomers *via* free radical reactions [56]. Hydrolysis and alkylation reactions rapidly convert lignin fragments into phenolic compounds, including catechols, phenols, guaiacols, etc. The phenolic compounds are then polymerized to form solid residues, which take place at a slower rate than hydrolysis reactions. Unconverted lignin forms char by solid-to-solid conversion [51]. Lignin fragmentation depends on the operating conditions and reactor design and/or is further converted into H_2, CO, CO_2, CH4, C_2H_6, and other low MW fragments. HTL process relies on the cleavage of ether or C–C bonds in lignin to form low MW fragments. Competition reactions hydrolysis and cleavage of the ether and C-C bond, demethoxylation, alkylation, and condensation occur during HTL of lignin for phenolic production [50].

The increase in temperature reduces the HC yield. At high temperatures, biomass polymers are degraded, and solid-to-solid conversion necessary for the formation of the HC does not occur or is reduced. The heating rate has been shown to affect the production of HC. Fast heating rates have resulted in higher bio-oil yield and lower HC yield; however, slow heating rates have resulted in a lower bio-oil and higher HC yield. The fast-heating rate significantly increased the reaction temperature, thereby inhibiting the formation of the char [51]. Bio-oil (liquid fraction) is the main product of HTL. The product distribution (gaseous, liquid, and solid) depends on the operating parameters of HTL, including biomass feedstock, temperature, heating rate, residence time, pressure, biomass-to-water ratio (B/W), particle size, and catalyst. The HTL process is endothermic at low temperatures but exothermic at higher temperatures. The advantages of HTL compared to pyrolysis are low operating temperature, high energy efficiency, low gases yield, and low tar yield [52].

The HTL temperature and pressure could impact the reaction directly (activation energy and reaction equilibria) and indirectly (solvent properties). During HTL, cellulose, hemicellulose, and lignin behave differently. Generally, biomass with high cellulose and hemicellulose tends to yield more bio-oil. But there are conflicting views on this. In addition to the three main components, extractives may also affect the yields of bio-oil and HC. Higher bio-oil yields have been obtained from hardwood samples than from softwood. Hardwood contains less lignin than softwood, hence, higher bio-oil yield [49, 50, 57]. Both temperature and lignin contents of wood were shown to have a marked effect on bio-oil yield. Maximum bio-oil production was obtained from wood with low lignin contents [58, 59].

Biomass particle size has negligible to minimal effects on the HTL process as sub/near supercritical water in HTL acts as a good heat transfer medium to overcome the heat transfer limitations, which makes the particle size of biomass a secondary parameter. Excessive biomass feedstock size reduction is not needed. Particle sizes between 4 to 10 mm have been reported as more suitable for the HTL process, which may vary with the operating conditions. No stipulated criteria for particle size selection are available to achieve maximum bio-oil yield during HTL [49, 55, 58].

Although many process parameters can influence the results of the HTL process, generally, it is accepted that temperature is a dominant factor that can alter the yield and properties of bio-oil. It is considered the key process parameter because it has the highest impact, as well as the parameter that can be directly controlled. A temperature range of 260–350°C would be viable for HTL of LCB [49, 52, 55, 58, 60]. Due to the increased fragmentation of biomass at higher temperatures, the effect of temperature on the bio-oil yield is synergetic. Biomass depolymerization occurs when the temperature is sufficiently large enough to overcome the activation energy of bond dissociation. The competing reactions hydrolysis, fragmentation, and repolymerization define the role of temperature during the HTL process. The dominant reaction during the initial stages of HTL is depolymerization. At later stages of HTL, repolymerization becomes active, leading to the formation of HC [49, 58]. Higher HTL temperatures enhance the reaction rates, as well as change the reaction mechanisms. Ionic reactions are favored at lower temperatures; the formation of radicals by homolytic bond breakage is promoted at higher temperatures. A highly diverse product spectrum usually results in free radical reactions, leading finally to gas formation [49, 61, 62]. Various authors have observed increased bio-oil yields with increasing HTL temperature. Due to reduced bio-oil yield at the maximum operating temperature, various optimum temperatures have been proposed by different authors for various biomasses. The optimum bio-oil yield is considered to occur in a specific temperature range which depends on the properties of biomass feedstock. The reduction of bio-oil yield at high temperatures is due to hydrolysis and repolymerization reactions. It can be assumed that the temperature range of 280–350°C would be suitable for the decomposition of biomass under HTL conditions. Intermediate temperatures usually produce higher bio-oil yields. The peak of bio-oil yield is usually around 300°C with an optimum range of 280–320°C have been reported. The final optimum HTL temperature varies with the biomass type. Temperature increase beyond optimum results in lower bio-oil yield and higher HC yield [49, 52, 55, 58, 60].

The heating rate is still a contentious parameter. Some researchers believe that the bio-oil yield and conversion rate mainly depend on the final temperature; the contribution of the heating rate is negligible. The perception of other researchers is that the heating rate is an important parameter in enhancing the bio-oil yield. The effect of heating rates on the product distributions in HTL is minimal compared to pyrolysis due to the better dissolution and stabilization of fragments in subcritical water. The undesirable breakdown of organic compounds and excessive polymerization of intermediates are reduced at faster heating rates. A positive correlation between heating rates and bio-oil yields has been obtained. Slow heating rates tend to yield more HC due to the increased secondary reactions. Secondary reactions are dominant at very high heating rates and yield more gases. In another study, it was found that the effect of the heating rate depends on the solvent. The heating rate is an important parameter in subcritical water but an optional parameter in supercritical alcohol. Moderate heating rates may be suitable to overcome heat transfer limitations, leading to extensive fragmentation and minimal secondary reactions [49, 52, 53, 58, 60].

The effect of residence time (reaction time) on product distribution during the HTL process has been examined by several researchers. Residence time has been observed to affect biomass conversion, bio-oil yield, and residue yield. Sufficient residence times are required to get high biomass conversions and bio-oil yield. The residence time may characterize the overall biomass conversion and the product compositions. For short residence time, bio-oil yield is lower due to incomplete biomass conversion. Alternatively, if the residence time is too long, the bio-oil may be decomposed and repolymerized, resulting in higher gaseous and solid fractions. A high repolymerization is not conducive to the yield and quality of bio-oil as degradation and repolymerization reactions convert some liquid organic fractions into other compounds. It is essential to inhibit the decomposition of intermediates into lighter products to obtain a higher bio-oil yield. Retention time can be regulated to control further cracking and repolymerization of intermediates. Improved yield and properties of bio-oil have been obtained with increasing residence times. However, a decrease in bio-oil yield has been observed at extended residence times (beyond the threshold limit). The type of biomass, its composition, type of catalysts, and operating conditions define the threshold limit of residence time. Two optimum residence times for maximum bio-oil yield have been reported depending on reaction temperature (120 min at 280°C and 5 min at 375°C), suggesting that shorter residence times are needed at higher reaction temperatures to reduce repolymerization of liquid fraction. In contrast, longer residence times can improve biomass conversion and bio-oil yield at lower reaction temperatures. The products obtained at different residence times of the same temperature can vary. Therefore, it is important to consider all factors and obtain the optimal residence time and temperature. Generally, bio-oil yield is maximum at optimum residence time, whereas gas yield and biomass conversion continue to increase. Residence time is a key parameter in HTL, which strongly affect the bio-oil yield [49, 50, 52–55, 58, 60].

In the HTL process, pressure is an important parameter for biomass decomposition. It alters the bio-oil yield and supports a single-phase system. HTL uses pressure to maintain a single-phase system and avoid large heat inputs required for phase change. A large heat supply is needed to maintain the temperature of two-phase systems. Pressure increases solvent density and penetrates effectively into larger molecules of biomass, resulting in improved disintegration into smaller fragments. Pressure, once reached supercritical condition, has no significant effect on bio-oil or gas yield. Higher pressures increase the density of the water; it may also cause the local solvent density to increase. This lead to a cage effect around the C–C bonds, which impedes the free radical reactions, inhibits the cleavage of C–C bonds, and ends up in low fragmentations, leading to a decrease in bio-oil yield. Within a specific range, pressure is positively correlated with bio-oil yield, but increasing the pressure above the upper limit shows no noticeable effect [24, 49, 52, 53, 55].

The influence of different solvents (such as subcritical and supercritical alcohols) on the bio-oil yield of LCB has been investigated. Alcohols have lower critical pressures and temperatures than water and significantly milder HTL reaction conditions could be used. Dielectric constants of alcohols are lower than water; hence, relatively high MW intermediates derived from cellulose, hemicelluloses, and lignin are expected to dissolve. Widely employed alcohols for biomass liquefaction have been ethanol and methanol. The biomass/water mass ratio is considered a vital parameter for the HTL process. Different authors have investigated the effect of water density on HTL bio-oil yield. The dilution of reactants, intermediates, and products during the reaction benefit all solvolytic conversions. Cross-reactions are minimized by dilutions and produce a more distinct product spectrum. Cross-reactions due to higher substrate concentrations inevitably lead to undesirable polymerization of the reaction products. Such processes have been observed for the HTL of biomasses. High biomass concentrations were shown to promote HC formation [24, 49, 52, 53, 55].

The lower oxygen and moisture content of HTL bio-oil promotes the higher calorific value and stability. The HTL bio-oil has lower H/C and O/C ratios, indicating a higher energy density than pyrolysis bio-oil. The decrease in the O/C ratio is higher than the H/C ratio indicating that more oxygen is removed as CO_2 or CO during HTL. As a result, HTL produces bio-oil with low oxygen content and improved quality [60].

4.3 Hydrothermal gasification (HTG)

At temperatures between 400 and 600°C and pressures of 23–45 MPa, biomass is primarily converted to a gas fraction (a mixture of non-condensable gases, including H_2, CO, CH_4, and CO_2) in a process known as HTG. It can produce syngas enriched with H_2. The biomass polymers decompose in supercritical water (SCW) above the critical point of water (374°C and 22.1 MPa) with enhanced conversion efficiency. The conversion rate of HTG is typically higher than 80%. The higher reaction temperatures of HTG cause reactions to progress faster, achieving complete decomposition of biomass, a distinctive feature of HTG compared to HTC and HTL. HTC and HTL often produce undesirable by-products that are occasionally dissolved in liquid fraction, and one of the problems of HTC and HTL is the difficulty in by-products posttreatment. Due to high conversion, only a small amount of organic compounds remain in the liquid fraction of HTG. Consequently, posttreatment of liquid fraction is not required or easily carried out [5, 8, 63].

When water changes from ambient conditions (25°C and 0.1 MPa) to supercritical conditions (400°C and 25 MPa), the properties of water change significantly and exhibit lower density, viscosity, ionic product, and dielectric constant. The dielectric constant at 400°C and 25 MPa is about six. Because of the decrease in the number of hydrogen bonds, water begins to display the behavior of a nonpolar solvent that can completely dissolve many organic compounds, hydrocarbons, and gases (such as CO_2, CH_4, H_2, and N_2). Due to the absence of phase boundaries, many rapid homogeneous reactions involving organic compounds occur at supercritical conditions. Inorganic polar compounds, usually soluble in subcritical water, are insoluble in SCW and easily separated from the reaction products. SCW has low density, low viscosity and exhibits gas-like properties. SCW combines both the dissolution of the liquid phase and the diffusion of the gaseous phase. The advantages of using SCW as the reaction medium in HTG include high diffusion coefficient, enhanced mass transfer, increased solvation properties, single-phase reaction environment in the reactor, and complete miscibility with different organics and gases, and product gas (syngas) does not have tar. It has a high heating value and prevents poisoning of catalyst (if used) and coke formation [6, 26, 49, 63, 64].

Conventional gasification is ineffective for wet biomasses (moisture content >80%), but HTG can be effectively employed when biomass is wet. In conventional gasification, syngas is partially diluted with nitrogen due to partial oxidation using air and contain tar. HTG converts biomass in the presence of water; hence, no dilution of syngas occurs with nitrogen. Syngas from HTG does not contain tars; if produced, they remain in the liquid fraction. HTG produces a high amount of H_2 and a very small amount of CO and char compared to conventional gasification. At low reaction temperatures or when the B/W is too high, HTG encounters a problem with the formation of tarry material. It can cause low gasification efficiency, reactor plugging, or the process water turns dark. At high heating rates, tarry material yield is decreased. This can be explained by assuming ionic reaction during tarry material production and free radical reaction during gasification [5, 8, 63].

During the HTG process, the biomass polymers cellulose and hemicellulose are converted into simple sugars (glucose, fructose, xylose, etc.). Some of the glucose is isomerized to fructose. Fructose produces 5-HMF and furan, which are then converted to alcohols, ketones, and organic acids. Hydrolysis products of small molecules in water are consequently gasified to produce H_2-rich syngas. Simultaneously, glucose and fructose are converted into glyceraldehyde, dihydroxyacetone, etc. Additionally, highly polymerized oligomers produced from the gasified small molecules and some intermediates eventually become liquid products. Phenols that are formed during the reaction are considered to be the final obstacle to complete biomass gasification. Inverse aldol condensation reaction converts xylose (a decomposition product of hemicellulose) to glyceraldehyde and methyl formate, producing propionic acid and acetic acid. Propionic acid and acetic acid are eventually gasified into small molecule gases (H_2 and CO). Propanoic acid may decompose in a second pathway and be gasified to ethane and CO_2. Simultaneously, dehydration of xylose produces furfural, which is decomposed in three proposed pathways. In the first pathway, furfural can be converted into water-soluble humic substances, which are gasified to CO and H_2, or in the second pathway, methyl cyclopentenolone, which is gasified to CO, CH_4, and H_2, or third pathway, gasified directly to CO, H_2, CH_4, and CO_2 [63, 65–67].

At low temperatures (< 500°C), biomass tends to produce low concentrations of H_2 in the gaseous fraction along with oil-based liquid faction. The liquid fraction contains a wide range of products, including acids, phenols, aldehydes, and furfurals. Ions (H^+ and OH^-) from SCW ionization support the cleavage of ring compounds to form simple molecules. At low-temperature HTG of biomass, ions are stable and ionic reactions dominate. In contrast, at high temperatures (> 500°C), free radicals are more stable than ions, and free radical mechanisms dominate in HTG of biomass. The gasification temperature and reactants influence free radical concentration. The most difficult biomass component to gasify is lignin. The efficient conversion of lignin in SCW is particularly important for the efficiency of the gasification process. In SCW, Lignin primarily dissociates into phenolics; phenolics decompose into gases [65, 66].

The operating conditions, including biomass characteristics, temperature, pressure, residence time, B/W, and catalyst, strongly influence the performance of HTG. The key parameters that affect the H_2 yield of the HTG process include temperature, pressure, B/W, and residence time. Short residence times are expected to degrade biomass successfully as the rate of hydrolysis and decomposition is relatively fast in the HTG process. For efficient destruction of biomass, residence time optimization is required. A large heat supply is needed to maintain the temperature of two-phase systems. HTG uses pressure to maintain a single-phase system and avoid large heat inputs required for phase change. By maintaining pressure higher than the supercritical pressure, the rate of hydrolysis and biomass dissolution can be controlled, which may enhance favorable reaction pathways for bio-oil or gas yield. In the supercritical region, pressure has a minimal effect on the properties of water. As a result, the effect of pressure on bio-oil or gas yield is minor or negligible [49, 58, 64, 68].

5. Applications of Hydrochar

HC is considered a valuable material for various agricultural, environmental, and industrial applications. The high versatility of HC allows for many applications, including soil amendments in agriculture, solid fuel in power generation, electrode materials in energy storage technologies, adsorbents in contaminant removal, and materials used as sensors and fuel cell catalysts [34]. HC could show different properties based on HT technology, process conditions, and feedstock used. Due to the formation of hydrocarbons on the surface, the surface area and porosity of HC are generally low, hindering its application. High surface area and porosity are vital for contaminant adsorbents and catalysts/catalyst support applications. The physicochemical properties of HC can be altered and improved through different activations. Chemical activation is achieved by impregnating HC with one or a mixture of chemical agents, followed by an activation process under a nitrogen flow. The most used chemical activating agents for the chemical activation process are potassium hydroxide (KOH), phosphoric acid (H_3PO_4), Zinc chloride ($ZnCl_2$), sodium hydroxide (NaOH), and potassium carbonate (K_2CO_3) [69]. Biochar has a high degree of carbonization, a highly aromatized carbon structure, lower H/C, and O/C ratios, with a strong anti-decomposition ability (environmentally more stable). But the cost of biochar production was higher. Alternatively, HC has higher H/C and O/C ratios than biochar and is mostly composed of aliphatic hydrocarbons; hence, its environmental stability is lower. The lower H/C and O/C ratios of biochar indicate a higher degree of aromaticity and maturation than HC [70].

Due to low bulk and energy densities, high moisture, and ash contents, untreated biomass is a poor-quality fuel. Biomass is usually pelletized to increase bulk density. But, long-term storage of biomass pellets causes moisture adsorption and biochemical and microbiological activities [71]. Granulated HC showed higher densities. HC densities in the range of 180 to 482% than raw biomass pellets made from four types of biomasses (pine sawdust, coconut fiber, coconut husk, and rice husk) have been obtained. HC pellets showed lower moisture uptake than raw biomass [72]. Consequently, HC has better fuel characteristics compared to biomass. HC needs to meet certain fuel characteristics, including energy density, combustion behavior, grindability, hydrophobicity, and thermal stability, to be effective as a replacement for coal. Its fuel-related characteristics are similar to lignite and can be used in power generation. But the chemical composition is different. HC has a significantly higher amount of volatiles than lignin. Also, the oxygen content of HC is higher than lignite. HC has a similar HHV as lignite. HC could be easily incorporated within existing coal-based processes. During the HTC process O/C ratio of the solid fraction is reduced, leading to increased HHV [28, 32, 37, 69, 73]. A high HHV (20.6–29.2 MJ/kg) of HC has been reported by many researchers as comparable to soft coal (20.93–33.5 MJ/kg) [51, 73, 74]. In addition to increased HHV, HC has lower volatile content (compared to biomass), which ensures better combustion. The low ash melting temperature of biomass makes burning more complicated. Due to the removal of many ash-forming minerals in liquid fraction, HC might have reduced mineral content. As a result, HC can achieve similar ash melting temperature as lignite. Only low-ash HC from certain biomasses (low ash) is suitable for power generation. High ash biomasses include sludge from wastewater treatment plants, agricultural residues, grasses, straws, etc. [28, 32, 37, 73].

The direct use of HC in agricultural and environmental applications might be complicated due to the presence of phenolic and organic acid compounds on the HC surface, which can cause negative plant and microbial responses. -Post-treatments, such as composting or anaerobic digestion, can reduce the toxicity of HC and make it suitable for soil application [70]. The factors, such as feedstock used, production process and process conditions, nature of the HC, morphological properties, and nature of the soil (loamy clay, fertile, sandy, or infertile), are important for HC application of crop improvement in the agricultural sector. The application of HC could result in either productive or counterproductive crop yield response. As a result of the low quantity of polar functional groups on the surface of the freshly produced HC, it shows hydrophobicity. However, due to oxidation by interacting with atmospheric oxygen, HC becomes more hydrophilic in nature by creating phenolic and carboxylic functional groups on the surface over time when mixed with soil. The water-holding capacity, cation exchange capacity, and nutrient retention capacity would increase significantly due to these functional groups on the surface [69]. The ability of HC to increase the nutrient supply for plants and decrease leaching losses makes it a good soil amendment to improve soil nutrient retention capacity. HC has a porous structure, low surface area, charged surface, and functional groups, including carbonyl, carboxyl, hydroxyl, and phenolic hydroxyl groups, to remedy the soils contaminated with heavy metals and organic compounds through adsorption [54]. Increased fertility has been achieved by the direct application of HC as a soil amendment, in addition to the formation of stable carbon sinks [75].

Reduced N_2O emission was observed in studying the effect of corn-based HC on the soil. HC was less effective as it is less stable than biochar. However, its production costs are less. Other studies have found that HC reduced N_2O emissions and contributed to NH_3, CO_2, and CH_4 emissions [69, 76]. The nutrient level of HC (especially plant biomass derived) is low; still, it can be used as fertilizer. HC can reduce loss from the surface runoff of fertilizers, enhancing fertilizer use. A decrease in total nitrogen, total phosphorus, nitrate, phosphate, and organic carbon has been reported using fertilizer containing an HC additive. HC addition improved fertilizer runoff and retention. Adding HC to soil could increase nutrient capacity and water retention in soil. Nutrients accumulated in the HC pores are released as needed. Adding HC to soil could change water aggregation, pH, cation, and anion exchange capacity. The addition of HC might improve the water-holding capacity of sandy soils. The colonization of the HC surface by fungi has been observed due to the hydrophobicity of the HC, which had a negative effect on water retention [76].

Increasing levels of HC in soil have been shown to deteriorate the growth of the Taraxacum plant. Some studies reported the negative effects of HC caused by nitrogen on crop yields. The absence of nitrogen migration in the first week of HC addition and the slow release of nitrogen with time has resulted in the nonavailability of nitrogen to the plants. Mixing HC into the soil for several weeks before planting has been suggested to overcome this. More research is needed to understand the ecotoxicological properties of HC and evaluate the impact of HC on soils. This will help lower the negative effects of HC in soil improvement for agricultural applications [69].

Energy storage becomes more critical with the increasing use of renewable energy. Carbon-based materials are widely utilized in capacitors. Thus, HC is an attractive material for energy applications. Its properties include surface area, electrical conductivity, tunable pore structure and size, easy accessibility, and strong mechanical properties. These are favorable for applications as supercapacitors and as anode and cathode material for batteries and fuel cells [69, 77, 78]. Due to surface area, polarity, porosity, aromaticity, and stability, HC has gained attention for electrochemical devices such as supercapacitors and batteries. Compared to the lower cycle stability, discharge/charge rate, and higher energy density of rechargeable batteries, HC-based supercapacitors usually exhibit higher cycle stability and power density [69]. Very high capacitance has been reported of HC derived from different sources and activated with KOH at different temperatures. Successful results have been reported of HTC nanospheres as anodes in Li^+ and Na^+ batteries. Their reversible capacity has reached up to 370 mA h/g at a 1C rate, which was better than that of traditional graphite electrodes. In another study, corn straw-based carbon spheres have been used as an anode in Li^+ batteries. The device showed excellent cycle stability with a specific capacity of 577 mA h/g after 100 cycles at 0.2C [77, 78]. Corncobs-based HC having a high surface area has been investigated as a carbon source. Sulfur loaded on HC achieved a discharge capacity of 1600 mA h/g and a reversible capacity of 554 mA h/g after 50 cycles [69]. Walnut shell-derived HC, activated HC/ZnO composites, and activated HC has been investigated for supercapacitors. The specific surface areas of 819 and 1073 m^2/g have been obtained for activated HC/ZnO composite and activated HC, respectively. The specific capacitance of activated HC/ZnO composite was 117.4 F/g at a current density of 0.5 A/g in KOH aqueous solution, which was found to be stable for 1000 cycles [79].

HC can also be used as an adsorbent to remove impurities in aqueous solutions. Depending on the raw material and manufacturing conditions, HC has a wide range of sorption properties [76]. The factors that determine the adsorption efficiency of HC include the specific surface area, pore structure, and surface functional groups. The functional groups on the surface give the HC a high chemical affinity and hydrophobicity, which has good potential adsorption applications [75, 78]. Relative to biochar, HC has a lower surface area. However, the adsorption capability of the HC is higher than biochar due to the abundance of oxygen-rich functionality and the presence of functional groups such as carbonyl, carboxyl, and hydroxyl groups on the surface [69]. Some of the weaknesses of HC include poor sorption properties compared to other adsorbers, high amounts of volatiles, low pore volume, low surface area, and negative surface charges, which repel negatively charged compounds such as phosphate. Despite the weaknesses, sorption properties for polar and nonpolar contaminants have been confirmed [76, 80]. HC subjected to chemical activation has better properties, which can be done during or after the HTC process. After chemical activation, enhanced sorption of heavy metals has been reported. Lanthanum activation has shown to be effective in phosphate removal as it can the negative surface charges of the HC. The results showed a maximum absorption of 61.5 mg/pg. [76, 81].

Heavy metal removal using HC has attracted many research interests. Different mechanisms, such as complexation, physical adsorption, precipitation, and electrostatic interactions, can be used to extract heavy metals. The research focused on heavy metal remediation using high-temperature HC has been reported. HC produced at high temperatures has improved surface area, less volatiles, increased ash content, and reduced functional groups. Higher adsorption of Cu by HC (compared to biochar) has been reported due to large quantities of functional groups found in HC. Removal of lead using HC derived from pine wood and rice husk has also been reported. The presence of oxygen functional groups on the HC surface was the primary factor influencing the strong removal capacity of heavy metals [51]. The comparison of the performance of switchgrass-based HC, KOH-activated HC, and activated carbon to remove Cu and Cd from the aqueous solution has shown close to 100% adsorption for Cu and Cd in 24 h by activated HC relative to the HC and activated carbon. The adsorption of lead from an aqueous solution using pinewood and rice husk-derived HC produced by HTL was favored at high temperatures as it is a physical endothermic process [69]. HC has effectively removed Cr (VI) from an aqueous solution. Low pH has been shown to give maximum adsorption efficiency. Pinewood sawdust-derived HC activated with H_2O_2 has been reported to have enhanced adsorption efficiency. The uptake of Pb^{2+} ions was 42 times higher [78, 82].

Adsorption of organic contaminants, including common inflammatory drugs (diclofenac sodium (DCF) and ketoprofen) and fungicides (diphenylamine), have been performed successfully using cellulose-derived carbon spheres. Successful removal of DCF from aqueous samples using KOH-activated HC produced from municipal woody and herbaceous pruning has been demonstrated [78, 82]. Absorption of pharmaceuticals (salicylic acid, flurbiprofen, and DCF) using orange peel-derived HC activated with H_3PO_4 has shown to be effective [83].

HC could serve as an adsorbent for capturing CO_2 to mitigate CO_2 quantities in the atmosphere. HC is economical, but for effective adsorption of CO_2, it needs post-treatments. The CO_2 adsorption potential of HC derived from woody and herbaceous pruning activated with KOH at higher temperatures was tested. The HC showed uptake of 84.5 mg CO_2/g. HC has excellent prospects as a cost-efficient and environmentally friendly material for CO_2 capturing, treating, and monitoring wastewater [78]. Sugarcane bagasse-derived HC and KOH-activated HC have been used to study CO_2 adsorption, which resulted in higher affinity for N_2 and CO_2 at 50°C by activated HC [69]. In another CO_2 adsorption study, silver fir sawdust-derived HC and KOH-activated HC were used in a pressure swing adsorption that resulted in 6.57 mmol/g of CO_2 adsorption at 5 bars with HC. The performance of HC was higher than activated HC and some traditional sorbents [84].

Dye removal by chemical and biological methods is effective, but they produce a lot of by-products. Natural physical adsorbents are environmentally safe, cheap, and abundant and can provide a better solution to this issue [48]. Activated carbon is the widely employed adsorbent for dye removal. The use of activated carbon is limited due to its high cost. HC has proven to be an efficient and economically viable adsorbent for the treatment of dye-contaminated water. Studies have shown that the adsorption performance of HC is most in favor of cationic dyes (methylene blue, rhodamine b, and methyl green) rather than anionic dyes (methyl orange and acid red 1). LCB-derived HC and sewage sludge have shown slightly lower adsorption capacity than commercial activated carbon. The same study reported a higher dye adsorption capacity of HC (128.6 to 160.5 mg/g) compared to biochar (12 to 130 mg/g). HC has been shown more effective in removing cationic dyes such as malachite green (10–40 mg/g) and methylene blue (15–45 mg/g) from the aqueous solution than anionic dye methylene orange (< 6.5 mg/g). HC could be used as an industrial adsorbent and a potential low-cost replacement for activated carbon after undergoing activation [51, 75]. Chemically activated HC could be used as a low-cost adsorbent to remove malachite green [85]. A maximum adsorption capacity of 34.9 mg/g of HC has been achieved in removing methylene blue dye from an aqueous solution using coffee husk-derived HC. In another study, the maximum adsorption capacity was 97 mg/g of HC in removing congo red dye using bamboo-derived HC [69]. A new chitosan-based adsorbent has been synthesized and used to remove methylene blue from wastewater. The adsorbent had outstanding reusability and a maximum adsorption capacity of 215.73 mg/g at 318.15 K based on Langmuir isotherm. Another novel chitosan-based adsorbent has been prepared and tested for methylene blue, methyl orange, and rhodamine B removal. The adsorbent had excellent reusability and ease of separating them from the solution using a magnetic field, and the maximum adsorption capacity for rhodamine B was 191.57 mg/g at 25°C. Carbon-coated polyacrylonitrile nanofibers have been used to adsorb methylene blue, and the adsorption capacity was 153.37 mg/g at room temperature. Even after 5 cycles, adsorption efficiency remained high [48].

Applications of HC in wastewater treatment have been extensively studied. HC has been shown to remove about 55% of gross pollutants from blended (50% of raw +50% of post HTL) wastewater. In the same study, ammonia, COD, nitrate, and phosphate removal efficiencies were 48, 53, 59, and 60%, respectively [21, 51, 86]. Sewage sludge-derived HC and KOH-activated HC has been used to remove orthophosphate (anions) and copper (cations) from wastewater, revealing that 97% of orthophosphates were removed through activated HC at 6 g/L, a higher adsorption capacity than HC [69]. HC is porous, has functional groups, and has a high content of micro- and macronutrients; therefore, it is a highly favorable and suitable environment for microorganisms to grow and can be used as an additive for anaerobic digestion. The addition of HC has been reported to increase methane yield in anaerobic digestion. HC has been shown to increase COD removal capacity in anaerobic digestion [51].

6. Conclusions

LCB, the most abundant biomass, represents a major carbon source. It is a sustainable and eco-friendly alternative renewable energy source. LCB is a complex heterogeneous mixture of three main biopolymers (cellulose, hemicellulose, and lignin) and a small amount of inorganic compounds and organic extractives. Biomass conversion breaks down the complex structure of biopolymers into smaller molecules. The selection of the biomass conversion process depends on several factors. Thermochemical conversion processes usually offer many advantages, but processes such as pyrolysis, gasification, or combustion, generally require dry biomasses. High moisture content is a major barrier and significantly impacts the biomass conversion process as it requires a large amount of energy to evaporate moisture.

In recent years, more attention has been given to HT processing as a method suitable for processing high moisture content biomass. It converts biomass with high moisture content in the presence of water, eliminating drying, and reducing energy costs. Water has multiple roles in HT conversions as a solvent, reactant, catalyst, and/or catalyst precursor. There are three HT processes, HTC, HTL, and HTG, based on the temperature, pressure, and residence time. The main products of HTC, HTL, and HTG are solid (HC), liquid (bio-oil/water), and gaseous (non-condensable) fractions, respectively. The conversion mechanisms of HTC differ from pyrolysis in that hydrolysis is the determining first step of HTC. Under HT conditions, the onset of carbonization takes place around 180°C.

Hydrolysis initiates the biomass degradation process in HTC as it has the lowest activation energy. Water catalyze and facilitates hydrolysis reactions. Degradation and depolymerization of hydrolyzed products of hemicellulose and cellulose produce a wide range of fragments, oligomers, and monomers; lignin fragments produce phenolic compounds. Subsequent dehydration and decarboxylation reduce oxygen and hydrogen content, leading to condensation reactions. Smaller molecules produced by dehydration and decarboxylation undergo condensation and polymerization, producing larger molecules; further, polymerization and aromatization produce highly aromatized fragments. Condensation and polymerization of these fragments form HC. The intermediate compounds, 5-HMF and furfural, are promising platform chemicals that can produce various chemical products. The process conditions and the biomass feedstock primarily determine the product distribution and characteristics.

The primary conversion mechanisms of HTL are depolymerization, decomposition, and recombination (repolymerization). Long-chain biopolymers undergo depolymerization and decomposition into short chains. Recombination of these highly reactive short-chain molecules produces liquid, gaseous, and solid products. High temperatures and physicochemical properties of water (acts as a catalyst) facilitate the depolymerization process. The three steps of decomposition are dehydration, decarboxylation, and deamination. Subcritical water at HTL conditions breaks down hydrogen bonds to form monomers. These are rapidly degraded to reactive intermediates by isomerization, dehydration, hydrolysis, reverse aldol defragmentation, rearrangement, and recombination. Recombination and repolymerization of short-chain compounds yield long-chain compounds. Under HTG conditions, biopolymers are converted to simple sugars. The simple sugars are gasified in different routes to produce a mixture of gases (CO, H_2, CH_4, and CO_2), the composition of which depends on the conversion route.

Further, research is required to investigate the reaction mechanisms, reaction kinetics and chemistry, integration of processes to make it more energy efficient, uses of process water, environmental implications of liquid and gas fractions, the effect of HC on soil applications, recovery of nutrients, heat transfer, energy, and heat recovery, etc.

Author details

Meegalla R. Chandraratne* and Asfaw Gezae Daful
Department of Chemical Engineering, Higher Colleges of Technology,
United Arab Emirates

*Address all correspondence to: mchandra@hct.ac.ae

References

[1] Chandraratne MR, Daful AG. Advances in bioenergy production using fast pyrolysis and hydrothermal processing. In: Samer M, editor. Biomass, Biorefineries and Bioeconomy. London: IntechOpen; 2022. pp. 269-289

[2] Güleç F, Riesco LMG, Williams O, et al. Hydrothermal conversion of different lignocellulosic biomass feedstocks – Effect of the process conditions on hydrochar structures. Fuel. 2021;**302**:121166

[3] Hodge BK. Alternative Energy Systems and Applications. Hoboken, NJ: John Wiley & Sons; 2017

[4] Carpenter NE. Chemistry of Sustainable Energy. Boca Raton, FL: Taylor & Francis; 2014

[5] Chandraratne MR, Daful AG. Recent advances in thermochemical conversion of biomass. In: Bartoli M, Giorcelli M, editors. Recent Perspectives in Pyrolysis Research. London: IntechOpen; 2022. pp. 145-176

[6] Okolie JA, Nanda S, Dalai AK, et al. A review on subcritical and supercritical water gasification of biogenic, polymeric and petroleum wastes to hydrogen-rich synthesis gas. Renewable and Sustainable Energy Reviews. 2020;**119**:109546

[7] Basu P. Biomass Gasification, Pyrolysis and Torrefaction: Practical Design and Theory. Second ed. Oxford: Elsevier; 2013

[8] Daful AG, Chandraratne MR. Biochar production from biomass waste-derived material. In: Hashmi S, Choudhury IA, editors. Encyclopedia of Renewable and Sustainable Materials. Oxford: Elsevier; 2020. pp. 370-378

[9] Daful AG, Chandraratne MR, Loridon M. Recent perspectives in biochar production, characterization and applications. In: Bartoli M, Giorcelli M, editors. Recent Perspectives in Pyrolysis Research. London: IntechOpen; 2022. pp. 327-344

[10] Sanchez J, Curt MD, Robert N, et al. Biomass resources. In: Lago C, Caldés N, Lechón Y, editors. The Role of Bioenergy in the Emerging Bioeconomy: Resources, Technologies, Sustainability and Policy. Oxford: Elsevier Inc.; 2019. pp. 25-111

[11] Dhyani V, Bhaskar T. Pyrolysis of biomass. In: Pandey A, Larroche C, Dussap C-G, et al., editors. Biofuels: Alternative Feedstocks and Conversion Processes for the Production of Liquid and Gaseous Biofuels. Oxford: Elsevier; 2019. pp. 217-244

[12] Wang S, Dai G, Yang H, et al. Lignocellulosic biomass pyrolysis mechanism: A state-of-the-art review. Progress in Energy and Combustion Science. 2017;**62**:33-86

[13] Sannigrahi P, Ragauskas AJ, Tuskan GA. Poplar as a feedstock for biofuels: A review of compositional characteristics. Biofuels, Bioproducts Biorefining. 2010;**4**:209-226

[14] Zheng Y, Shi J, Tu M, et al. Principles and development of Lignocellulosic biomass Pretreatment for biofuels. In: Li Y, Ge X, editors. Advances in Bioenergy. Vol. 2. Oxford: Elsevier Inc.; 2017. pp. 1-68

[15] Agbor VB, Cicek N, Sparling R, et al. Biomass pretreatment: Fundamentals toward application. Biotechnology Advances. 2011;**29**:675-685

[16] Wolf JP. Biomass combustion for power generation: An introduction. In: Rosendahl L, editor. Biomass Combustion Science, Technology and Engineering. Cambridge: Woodhead; 2013. pp. 3-8

[17] Jameel H, Keshwani DR. Thermochemical conversion of biomass to power and fuels. In: Cheng JJ, editor. Biomass to Renewable Energy Processes. Boca Raton, FL: Taylor & Francis; 2018. pp. 375-421

[18] Chen WH, Lin BJ, Lin YY, et al. Progress in biomass torrefaction: Principles, applications and challenges. Progress in Energy and Combustion Science. 2021;**82**:100887

[19] Williams CL, Dahiya A, Porter P. Introduction to bioenergy. In: Dahiya A, editor. Bioenergy - Biomass to Biofuels. London: Elsevier; 2015. pp. 5-36

[20] Adams P, Bridgwater T, Lea-Langton A, et al. Biomass conversion technologies. In: Thornley P, Adams P, editors. Greenhouse Gas Balance of Bioenergy Systems. London: Elsevier; 2018. pp. 107-139

[21] Kambo HS, Dutta A. A comparative review of biochar and hydrochar in terms of production, physicochemical properties and applications. Renewable and Sustainable Energy Reviews. 2015;**45**:359-378

[22] Zhao P, Shen Y, Ge S, et al. Clean solid biofuel production from high moisture content waste biomass employing hydrothermal treatment. Applied Energy. 2014;**131**:345-367

[23] Biller P, Ross AB. Production of biofuels via hydrothermal conversion. In: Luque R, CSK L, Wilson K, et al., editors. Handbook of Biofuels Production - Processes and Technologies. Duxford: Elsevier Ltd.; 2016. pp. 509-547

[24] Moller M, Nilges P, Harnisch F, et al. Subcritical water as reaction environment: Fundamentals of hydrothermal biomass transformation. ChemSusChem. 2011;**4**:566-579

[25] Uematsu M, Frank EU. Static dielectric constant of water and steam. Journal of Physical and Chemical Reference Data. 1980;**9**:1291-1306

[26] Toor SS, Rosendahl L, Rudolf A. Hydrothermal liquefaction of biomass: A review of subcritical water technologies. Energy. 2011;**36**:2328-2342

[27] Krammer P, Vogel H. Hydrolysis of esters in subcritical and supercritical water. Journal of Supercritical Fluids. 2000;**16**:189-206

[28] Kruse A, Dahmen N. Water - a magic solvent for biomass conversion. Journal of Supercritical Fluids. 2015;**96**:36-45

[29] Machmudah S, Diono W, Kanda H, et al. Hydrolysis of biopolymers in near-critical and subcritical water. In: Gonzalez HD, MJG M, editors. Water Extraction of Bioactive Compounds: From Plants to Drug Development. Oxford: Elsevier Inc.; 2017. pp. 69-107

[30] Lachos-Perez D, Torres-Mayanga PC, Abaide ER, et al. Hydrothermal carbonization and liquefaction: Differences, progress, challenges, and opportunities. Bioresource Technology. 2022;**343**:126084

[31] Heidari M, Dutta A, Acharya B, et al. A review of the current knowledge and challenges of hydrothermal carbonization for biomass conversion. Journal of the Energy Institute. 2019;**92**:1779-1799

[32] Titirici MM, Funke A, Kruse A. Hydrothermal carbonization of biomass. In: Pandey A, Bhaskar T, Stöcker M,

et al., editors. Recent Advances in Thermochemical Conversion of Biomass. Oxford: Elsevier; 2015. pp. 325-352

[33] Lam PS, Tooyserkani Z, Naimi LJ, et al. Pretreatment and Pelletization of Woody biomass. In: Fang Z, editor. Pretreatment Techniques for Biofuels and Biorefineries. Berlin: Springer-Verlag; 2013. pp. 93-116

[34] Paksung N, Pfersich J, Arauzo PJ, et al. Structural effects of cellulose on hydrolysis and carbonization behavior during hydrothermal treatment. ACS Omega. 2020;**5**:12210-12223

[35] Fang J, Zhan L, Ok YS, et al. Minireview of potential applications of hydrochar derived from hydrothermal carbonization of biomass. Journal of Industrial and Engineering Chemistry. 2018;**57**:15-21

[36] Sharma HB, Sarmah AK, Dubey B. Hydrothermal carbonization of renewable waste biomass for solid biofuel production: A discussion on process mechanism, the influence of process parameters, environmental performance and fuel properties of hydrochar. Renewable and Sustainable Energy Reviews. 2020;**123**:109761

[37] Fernández-Sanromán Á, Lama G, Pazos M, et al. Bridging the gap to hydrochar production and its application into frameworks of bioenergy, environmental and biocatalysis areas. Bioresource Technology. 2021;**320**:124399

[38] Titirici MM, White RJ, Falco C, et al. Black perspectives for a green future: Hydrothermal carbons for environment protection and energy storage. Energy & Environmental Science. 2012;**5**:6796-6822

[39] Yu KL, Lau BF, Show PL, et al. Recent developments on algal biochar production and characterization. Bioresource Technology. 2017;**246**:2-11

[40] Funke A, Ziegler F. Hydrothermal carbonization of biomass: A summary and discussion of chemical mechanisms for process engineering. Biofuels, Bioproducts Biorefining. 2010;**4**:160-177

[41] Shao Y, Ding Y, Dai J, et al. Synthesis of 5-hydroxymethylfurfural from dehydration of biomass-derived glucose and fructose using supported metal catalysts. Green Synthesis and Catalysis. 2021;**2**:187-197

[42] Mathew AK, Abraham A, Mallapureddy KK, et al. Lignocellulosic biorefinery wastes, or resources? In: Bhaskar T, Pandey A, Mohan SV, et al., editors. Waste Biorefinery: Potential and Perspectives. Amsterdam: Elsevier B.V.; 2018. pp. 267-297

[43] Lucian M, Volpe M, Gao L, et al. Impact of hydrothermal carbonization conditions on the formation of hydrochars and secondary chars from the organic fraction of municipal solid waste. Fuel. 2018;**233**:257-268

[44] Hitzl M, Corma A, Pomares F, et al. The hydrothermal carbonization (HTC) plant as a decentral biorefinery for wet biomass. Catalysis Today. 2015;**257**:154-159

[45] He C, Zhang Z, Ge C, et al. Synergistic effect of hydrothermal co-carbonization of sewage sludge with fruit and agricultural wastes on hydrochar fuel quality and combustion behavior. Waste Management. 2019;**100**:171-181

[46] Lin Y, Ge Y, Xiao H, et al. Investigation of hydrothermal co-carbonization of waste textile with

waste wood, waste paper and waste food from typical municipal solid wastes. Energy. 2020;**210**:118606

[47] Ercan B, Alper K, Ucar S, et al. Comparative studies of hydrochars and biochars produced from lignocellulosic biomass via hydrothermal carbonization, torrefaction and pyrolysis. Journal of the Energy Institute. 2023;**109**:101298

[48] Ababneh H, Hameed BH. Chitosan-derived hydrothermally carbonized materials and its applications: A review of recent literature. International Journal of Biological Macromolecules. 2021;**186**:314-327

[49] Singh R, Prakash A, Balagurumurthy B, et al. Hydrothermal liquefaction of biomass. In: Pandey A, Bhaskar T, Stöcker M, et al., editors. Recent Advances in Thermochemical Conversion of Biomass. Oxford: Elsevier; 2015. pp. 269-291

[50] Cao L, Zhang C, Chen H, et al. Hydrothermal liquefaction of agricultural and forestry wastes: State-of-the-art review and future prospects. Bioresource Technology. 2017;**245**:1184-1193

[51] Ponnusamy VK, Nagappan S, Bhosale RR, et al. Review on sustainable production of biochar through hydrothermal liquefaction: Physico-chemical properties and applications. Bioresource Technology. 2020;**310**:123414

[52] Biswas B, Kumar J, Bhaskar T. Advanced hydrothermal liquefaction of biomass for bio-oil production. In: Pandey A, Larroche C, Dussap C-G, et al., editors. Biofuels: Alternative Feedstocks and Conversion Processes for the Production of Liquid and Gaseous Biofuels. Oxford: Elsevier Inc.; 2019. pp. 245-266

[53] Gollakota ARK, Kishore N, Gu S. A review on hydrothermal liquefaction of biomass. Renewable and Sustainable Energy Reviews. 2018;**81**:1378-1392

[54] Fan Y, Hornung U, Dahmen N. Hydrothermal liquefaction of sewage sludge for biofuel application: A review on fundamentals, current challenges and strategies. Biomass and Bioenergy. 2022;**165**:106570

[55] Mishra RK, Kumar V, Kumar P, et al. Hydrothermal liquefaction of biomass for bio-crude production: A review on feedstocks, chemical compositions, operating parameters, reaction kinetics, techno-economic study, and life cycle assessment. Fuel. 2022;**316**:123377

[56] Yiin CL, Bin OE, SSM L, et al. A review on potential of green solvents in hydrothermal liquefaction (HTL) of lignin. Bioresource Technology. 2022;**364**:128075

[57] Bhaskar T, Sera A, Muto A, et al. Hydrothermal upgrading of wood biomass: Influence of the addition of K2CO3 and cellulose/lignin ratio. Fuel. 2008;**87**:2236-2242

[58] Akhtar J, Amin NAS. A review on process conditions for optimum bio-oil yield in hydrothermal liquefaction of biomass. Renewable and Sustainable Energy Reviews. 2011;**15**:1615-1624

[59] Zhong C, Wei X. A comparative experimental study on the liquefaction of wood. Energy. 2004;**29**:1731-1741

[60] Fan Q, Fu P, Song C, et al. Valorization of waste biomass through hydrothermal liquefaction: A review with focus on linking hydrothermal factors to products characteristics. Industrial Crops and Products. 2023;**191**:116017

[61] Williams PT, Onwudili J. Composition of products from the supercritical water gasification of glucose: A model biomass compound. Industrial and Engineering Chemistry Research. 2005;**44**:8739-8749

[62] Yoshida T, Oshima Y, Matsumura Y. Gasification of biomass model compounds and real biomass in supercritical water. Biomass and Bioenergy. 2004;**26**:71-78

[63] Matsumura Y. Hydrothermal gasification of biomass. In: Pandey A, Bhaskar T, Stöcker M, et al., editors. Recent Advances in Thermochemical Conversion of Biomass. Oxford: Elsevier; 2015. pp. 251-267

[64] Gemechu ED, Kumar A. The environmental performance of hydrogen production pathways based on renewable sources. In: Ren J, editor. Renewable Energy Driven Future-Technologies, Modelling, Applications, Sustainability and Policies. London: Elsevier Inc.; 2021. pp. 375-406

[65] Wang Q, Zhang X, Cui D, et al. Advances in supercritical water gasification of lignocellulosic biomass for hydrogen production. Journal of Analytical and Applied Pyrolysis. 2023;**170**:105934

[66] Nanda S, Mohammad J, Reddy SN, et al. Pathways of lignocellulosic biomass conversion to renewable fuels. Biomass Conversion and Biorefinery. 2014;**4**:157-191

[67] Goodwin AK, Rorrer GL. Reaction rates for supercritical water gasification of xylose in a micro-tubular reactor. Chemical Engineering Journal. 2010;**163**:10-21

[68] Kersten SRA, Potic B, Prins W, et al. Gasification of model compounds and wood in hot compressed water. Industrial and Engineering Chemistry Research. 2006;**45**:4169-4177

[69] Masoumi S, Borugadda VB, Nanda S, et al. Hydrochar a review on its production technologies. Catalysts. 2021;**11**:939

[70] Masek O. Biochar in thermal and thermochemical biorefineries - production of biochar as a coproduct. In: Luque R, Lin CSKI, Wilson K, et al., editors. Handbook of Biofuels Production - Processes and Technologies. Duxford: Elsevier Ltd.; 2016. pp. 655-671

[71] Lehtikangas P. Storage effects on pelletised sawdust, logging residues and bark. Biomass and Bioenergy. 2000;**19**:287-293

[72] Liu Z, Quek A, Balasubramanian R. Preparation and characterization of fuel pellets from woody biomass, agro-residues and their corresponding hydrochars. Applied Energy. 2014;**113**:1315-1322

[73] Maniscalco MP, Volpe M, Messineo A. Hydrothermal carbonization as a valuable tool for energy and environmental applications: A review. Energies. 2020;**13**:4098

[74] Hoekman SK, Broch A, Robbins C. Hydrothermal carbonization (HTC) of lignocellulosic biomass. Energy and Fuels. 2011;**25**:1802-1810

[75] Nachenius RW, Ronsse F, Venderbosch RH, et al. Biomass Pyrolysis. In: Marin GB, West DH, Li J, et al., editors. Advances in Chemical Engineering, Chemical Engineering for Renewables Conversion. Vol. 42. Oxford: Elsevier Inc.; 2013. pp. 75-139

[76] Czerwińska K, Śliz M, Wilk M. Hydrothermal carbonization process: Fundamentals, main parameter

characteristics and possible applications including an effective method of SARS-CoV-2 mitigation in sewage sludge. A review. Renewable and Sustainable Energy Reviews. 2022;**154**:111873

[77] Kruse A, Funke A, Titirici MM. Hydrothermal conversion of biomass to fuels and energetic materials. Current Opinion in Chemical Biology. 2013;**17**:515-521

[78] Adolfsson KH, Yadav N, Hakkarainen M. Cellulose-derived hydrothermally carbonized materials and their emerging applications. Current Opinion in Green and Sustainable Chemistry. 2020;**23**:18-24

[79] Li Y, Liu X. Activated carbon/ZnO composites prepared using hydrochars as intermediate and their electrochemical performance in supercapacitor. Materials Chemistry and Physics. 2014;**148**:380-386

[80] Fang J, Gao B, Chen J, et al. Hydrochars derived from plant biomass under various conditions: Characterization and potential applications and impacts. Chemical Engineering Journal. 2015;**267**:253-259

[81] Dai L, Wu B, Tan F, et al. Engineered hydrochar composites for phosphorus removal/recovery: Lanthanum doped hydrochar prepared by hydrothermal carbonization of lanthanum pretreated rice straw. Bioresource Technology. 2014;**161**:327-332

[82] Xia Y, Yang T, Zhu N, et al. Enhanced adsorption of Pb(II) onto modified hydrochar: Modeling and mechanism analysis. Bioresource Technology. 2019;**288**:121593

[83] Fernandez ME, Ledesma B, Román S, et al. Development and characterization of activated hydrochars from orange peels as potential adsorbents for emerging organic contaminants. Bioresource Technology. 2015;**183**:221-228

[84] Gallucci K, Taglieri L, Papa AA, et al. Non-energy valorization of residual biomasses via HTC : CO2 capture onto activated Hydrochars. Applied Sciences. 2020;**10**:1879

[85] Hammud HH, Shmait A, Hourani N. Removal of malachite green from water using hydrothermally carbonized pine needles. RSC Advances. 2015;**5**:7909-7920

[86] Arun J, Varshini P, Prithvinath PK, et al. Enrichment of bio-oil after hydrothermal liquefaction (HTL) of microalgae C. vulgaris grown in wastewater: Bio-char and post HTL wastewater utilization studies. Bioresource Technology. 2018;**261**:182-187

Chapter 8

Biogas Production and Process Control Improvements

Moses Jeremiah Barasa Kabeyi, Oludolapo Akanni Olanrewaju and Joseph Akpan

Abstract

Biogas is a sustainable energy produced from biodegradable organic matter through anaerobic digestion. Biogas mainly contains methane (CH_4) and carbon dioxide (CO_2) and various contaminants, such as water vapor (H_2O), ammonia (NH_3), hydrogen sulfide (H_2S), methyl siloxanes (MSs), nitrogen (N_2), oxygen (O_2), halogenated volatile organic compounds (VOCs), carbon monoxide (CO) and hydrocarbons whose presence largely depends on the source of biogas. The value of biogas is enhanced further by the production of organic manure as substitutes for chemical fertilizers in crop production. The vast range of applications makes biogas an attractive sustainable energy choice. Bioenergy generation capacity, mainly from biogas globally, has shown promising worth over non-renewable energy options; hence, its production has expanded rapidly during the previous two decades, from 42.38 Exajoules/year in 2000 to about 57.74 Exajoules/year in 2023, an increase of about 36.24%. This study presents the biogas production and potential improvement measures that include factors influencing the process and quality of biogas produces such as C/N (carbon nitrogen ratio, the pH, dilution of feedstock, composition and nutritive value of feedstock, residence period, mixing and stirring, temperature, presence of toxicants, loading time and redox conditions). The main barriers to biogas production and use include lack of technical know-how, lack of subsidies and incentives for biogas production and use. Access to appropriate and affordable biogas technologies, the development of a comprehensive policy on the construction and operation of bio-digesters as well as utilization of biogas and sales of biogas energy products and services are important for an increasing adoption of biogas.

Keywords: bioenergy, biogas production, feedstock for biogas production, biogas process improvement, low carbon technologies, sustainable energy development

1. Introduction

Anaerobic digestion is an attractive biomass-to-energy and waste management technology due to the significant presence of feedstock in form of the organic wastes, food wastes and energy crops, and technology maturity [1]. Biogas is an important renewable energy resource whose production and use can aid in the mitigation of greenhouse gas (GHG) emissions by substituting fossil fuels and promoting the utilization of biodegradable plant and animal waste feedstocks [2, 3]. With the expected depletion of coal, gas, and oil within the next decade and a half, renewable energy sources like biogas are becoming increasingly important [4]. Another motivation for the use of biogas as an energy resource is the international treaties or agreements that commit countries to reduce their carbon footprint like agenda 21 and Kyoto Protocol [5, 6]. These mechanisms advocated for the transition to renewable and low-carbon sources of energy to reduce global greenhouse gas emissions, particularly from the energy sector which is dominated by fossil fuel sources of energy [7, 8].

Biogas is a viable option to the worldwide energy dilemma and has enormous promise as a sustainable and renewable energy source for commercial, industrial, and residential use [9, 10]. Because of rising environmental concerns and measures in response to concerns over greenhouse gas emissions and climate change, there has been growth in interest in using biomass resources as a renewable feedstock for electricity generation, fuel production, chemical processing, and hydrogen production [11, 12], resulting in increasing energy generation from organic waste production as it reduces the threat to climate change [13]. Electricity generation, thermal applications like cooking, heating, and lighting, and biofuel production are the primary uses for biogas. Biogas is used to produce over 7000 megawatts (MW) of electricity annually [14, 15].

Many developing countries still face the challenge of providing modern energy services to their populations e.g., India alone has about 450 million people with no access to modern energy for electricity [16, 17], yet they have huge biomass resources [18]. There is substantial potential for biomass to play a leading role in global energy generation is evidenced by the 9.7% annual growth rate of biofuels since 1990 [19, 20]. In this context, the role of biogas energy within this huge biomass supply chain cannot be overemphasized. Beyond providing food for all of humanity, agriculture is the primary source of income for more than two-thirds of the world's population. Additionally, bout 82% of the global population is directly or indirectly involved in smallholder agriculture and other industries, making this economic activity the mainstay of many countries [16, 21]. This shows that biogas is important as a renewable and sustainable energy resource for the energy transition and that agriculture and other carbon transformative processes should boost its production and consumption [22, 23].

Biogas has a significant role to play in the ongoing energy transition as a renewable and sustainable energy resource because of its high energy conversion potential and the widespread availability for power generation, industrial applications as a renewable process feedstock and for thermal energy applications [24–28]. Besides biogas production, the anaerobic digestion (AD) process can act as a waste treatment and disposal method while the digestate produced is a valuable organic fertilizer that can replace chemical fertilizers in sustainable agriculture [29, 30]. Anaerobic digestion is an important process in global carbon cycle which produces biogas through biodegradation of organic matter that releases 590–800 million tons of methane into the atmosphere annually which can be mitigated through controlled anaerobic digestion that produces useful biogas [25, 31, 32]. Numerous different organic feedstocks are anaerobically digested to produce biogas, which can then be upgraded to produce biomethane which is an important substitute of natural gas. The sustainable technique of turning biomass waste into biogas can lead to advantages including less carbon emissions, improved organic waste management, and increased resource use efficiency [25, 32].

Biogas has the potential to contribute to sustainable fuel for multiple applications as a fuel characterized by a high methane content, which is generated through the process of anaerobic digestion of organic substrate under carefully regulated biochemical process conditions [33, 34]. Indigestible carbohydrates, proteins, and lipids are not present in the biomass substrate utilized for biogas production, which can complicate or slow down the digestion [35, 36]. Biogas can also be used as a feedstock for production of biomethane, carbon dioxide, hydrogen, and various biofuels for a wide range of applications including generation of heat and power and feedstock for various industrial processes [20, 37].

In the period preceding the 1800s, man predominantly relied on biomass as its primary source of energy [38, 39]. With the beginning of the industrial revolution, the energy landscape underwent a significant change that was typified by a predominate reliance on fossil fuels, mainly coal, and later petroleum derivatives like natural gas and diesel. A major increase in oil prices was present between the early 1970s to the late 1980s. This situation made it necessary to investigate and promote ecologically friendly, sustainable, and renewable energy sources. During this time, biogas became one of the most popular choices [40–42]. According to [43], in order to mitigate environmental degradation and reduce greenhouse gas emissions, it is imperative to engage in environmental conservation practices and adopt proper waste management strategies. Rural communities, comprising a significant proportion of the populace in numerous developing countries and primarily involved in subsistence agriculture, possess the capacity to generate biogas like in India, where approximately 70% of the populace resides in rural regions [17, 44], whereas on a global average basis, the enormous energy potential of biogas is largely unrealized. A sizeable section of the population works in small-scale agriculture in rural areas of many developing countries. These people have the opportunity to use biogas technology to meet a significant amount of their energy needs [45, 46].

There is considerable potential of biogas as a sustainable energy source, both currently and in the foreseeable future, is significant. Despite the significant potential of biogas to make a substantial contribution to the achievement of sustainable development goals and the global transition toward sustainable energy due to its minimal greenhouse gas emissions, widespread availability, and access to raw materials, its incorporation into the global energy and electricity mix has been constrained [20, 44]. Therefore, it is crucial to initiate a change in the current circumstances [7, 47–49]. The overall objective of this study is to investigate the technological progress and possible uses of biogas in promoting a sustainable worldwide transition and attaining the targets specified in the sustainable development goals. The primary objective of this study is to evaluate the feasibility of incorporating biogas into the existing power grid and other energy applications as a means of reducing emissions. Furthermore, this study presents a comprehensive strategy for the transformation of biogas into electrical energy within the context of a large-scale energy transition. This measure will contribute to mitigating the dual challenge presented by the increasing levels of greenhouse gas emissions and the consequent climate change, in accordance with the emissions and climate goals outlined in the Paris Agreement. This research investigates and evaluates different alternatives, approaches, and feasible technologies to produce electricity using biogas, with a specific emphasis on its implementation in agricultural and industrial contexts. This chapter provides discussion on the potential, prerequisites, and obstacles linked to the production of biogas and the processes to improvement. Furthermore, this study offers an examination of the biogas potentials that can be employed for the purpose of generating heat and electricity [31, 41, 42, 46, 50].

1.1 Problem statement

The heightening need for food and energy resulting from the growth of the global population, particularly in developing countries, has intensified the strain on energy production and consumption. This strain is currently associated with the release of greenhouse gases and the subsequent impact on climate change [16]. Although biogas has a great deal of promise to supply clean energy to both rural and urban populations, the prevalence of inoperable and abandoned biogas facilities has raised concerns about the sustainability of the technology. This poses questions concerning the dependable creation, distribution, and use of biogas fuel [17]. It is technically possible to convert this waste into valuable energy sources, thereby generating additional revenue [34, 51, 52].

Sustainable electricity generation is crucial. Hence, sustainable farm-level biogas electricity generation requires appropriate infrastructure design and selection due to susceptibility to biogas system failures [53, 54]. Another thing to consider in such system design is the natural decomposition of agricultural waste that results in the emission of substantial amounts of methane, a potent greenhouse gas, into the atmosphere. Take for instance, in the year 2015, the exclusive contribution of livestock manure to methane emissions in the United States of America amounted to approximately 10%. However, a mere 3% of the total livestock waste underwent recycling via anaerobic digestion [34]. This presents the main issue in utilizing as it lies in its unsteady production value and variations in quality. These factors can potentially disrupt the generation process or hinder the effectiveness of biogas applications, resulting in reduced reliability [55].

This study presents the biogas production, process improvements parameters, and its contribution toward promoting sustainable energy generation for local applications, such as in agriculture, as well as its potential for facilitating the transition of grid electricity toward renewable energy sources. This study focuses on the examination of optimal and sustainable methods for producing biogas. It also proposes sustainable approaches for generating electricity from biogas and producing biofuels such as biohydrogen, methanol, and syngas [42, 56, 57].

1.2 Rationale of biogas production

The main challenges facing biogas production and use as an energy resource include high equipment cost and a lack of governmental incentive programs in many countries, low reliability or lack of guarantee of long term performance of biogas plants due to technological challenges, unpredictable investment environment, limited biogas distribution and storage capacity for some countries like Sweden and low cost of commercial fertilizers as well as low cost of fossil fuels like gas and oil in some countries [58]. A study by Huang et al. [2] on household biogas digester use in rural China showed that there was rapid growth in the early part of the century in the use of household biogas digesters but the operations were never smooth because out of 1743 households interviewed, 42% adopted household biogas digesters, but on average they worked for just 6.66 months each year which is an indicator of underutilization of the digestors.

Both developed and developing countries face challenges with the utilization of significant organic waste and pressure to replace fossil energy resources like natural gas, oil, and coal with renewable energy sources. Anaerobic digestion of organic biomass is now a mature technology while biogas in its raw form or cleaned and upgraded forms have multiple applications as a bioenergy substitute of fossil fuel sources. The production and utilization of biogas have also been growing with capacity growing by more than double between 2009 and 2022 [59]. Methane fermentation occurs naturally in the process of organic matter decay in oxygen-deficient environments like swamps and in landfills leading to the natural emission of methane to the atmosphere as a greenhouse gas. Controlled biogas production can be used to reduce these natural emissions [21, 60].

The human population is steadily growing leading to increased demand for food and energy which simultaneously aggravates the environmental challenges. Substituting fossil fuels with renewable energy alternatives is currently a major global issue of the 21st century and is a key sustainable development objective. Implementation of biogas technologies will significantly transform costly, socially sensitive issues and environmentally damaging fossil fuel dependence, environmental pollution, GHG emissions, and waste conversion into profitable options like electricity generation, heat production and production of biofertilizers. As well as the provision of green vehicular fuel substitute of fossil gas. A national biogas production development program can significantly lead to development of new enterprises, boost income for rural communities and create new jobs to improve people's socio-economic wellbeing [59].

The implementation of biogas technology enables the harnessing of renewable energy through the utilization of crop and animal waste. The energy source exhibits significant potential for utilization in electricity generation and as a viable thermal energy solution for residential structures. The ultimate outcome of this process can be efficiently harnessed as a valuable fertilizer in rural agricultural environments and households, leading to a reduction in costs related to waste management and disposal [43]. Biogas energy can improve rural people's lives and economies. Many small-holder farmers in developing nations burn biomass for disposal, even though it can be used as a fertilizer, and with biogas technology, an impoverished organic fertilizer that are more affordable can be made by smallholder farmers [61]. Designing biogas systems so that they can reliably supply most or 100% of their energy demands requires defining a number of factors [45]. The challenges limiting full use of biogas that should be overcome, include lack of national generating technologies, need to clean biogas before use, the challenge economic feasibility that requires incentives, the lack penalties for possible environmental damages from biogas schemes [58, 62].

2. Biogas production

The capacity of biogas plants globally has been on a growth trajectory in many countries because of factors like rising cost of fossil fuels, the easy and affordable supply of biomass feedstock, and growing concerns about emissions and the threat of a global climate emergency. The primary waste sources used in the production of biogas include, but are not limited to, home wastes such food, produce, and fruit as well as animal wastes like dung and bird droppings. Public moist wastes from markets, cafes, restaurants, and food outlets are also frequently used in this procedure, along with biodegradable trash from businesses with high moisture content and degradability. The use of anaerobic digestion to produce biogas acts as a catalyst for improving a country's energy mix while also significantly advancing the preservation of natural resources and defending the environment [30, 32, 48, 49, 63–65].

Under optimal conditions, a specific group of bacteria can generate biogas through anaerobic processes having a calorific value ranging from 21 to 24 MJ/m^3, thereby establishing it as an environmentally sustainable energy resource [22]. Biodegradation of organic matter generates between 590 and 800 million tons of methane per year, which is then released into the atmosphere without human involvement. Biogas, a byproduct of biomass decomposition, can be recovered and converted into usable energy by use of such devices [6, 22]. The composition of biogas can range from 50 to 70% methane and 30 to 50% carbon dioxide, depending on the substrate employed and the management and regulation of the production process. Nitrogen and hydrogen sulfide are among the extra components that are present. Gas distribution systems can be fitted with enriched biogas from larger plants. The metabolic process that ultimately produces methane and other gases from biomass relies on the action of a suite of bacteria that exists as at least three bacterial communities employed in biogas production systems [22, 66]. The utilization of anaerobic digesters usually occurs in the mesophilic (20–40°C) and thermophilic (above 40°C) temperature ranges [22, 66], depending on the substrate fed into the digester, biogas typically consists of 30–60% methane and 30–50% CO_2. Hydrogen gas is also present in biogas although in much smaller amounts.

Large biogas plants can generate biogas for supply into gas supply networks or mains, with a typical heating value of 21–24 MJ/m^3 or 6 kWh/m^3. This makes it useful for a wide range of applications, including cooking, heating, lighting, and electricity generating [22, 67, 68]. Over a long period of time, biogas has been used as fuel product anaerobic digestion of waste for sanitation. It is also typical to use biogas plant digested waste as a fertilizer [29, 69].

2.1 Global development of biogas energy resources

About 0.25% of the world's energy came from biogas in 2011, while biogas accounted for 27% the global biofuel market [35, 70]. Income and standard of living in rural areas can be improved through biogas production and use in two ways by selling electricity derived from to the grid and increasing agricultural productivity by application of organic manure/digestate. Not only may biogas be used for the aforementioned purposes, but it can also be converted into diesel fuel to power appliances in places like houses and classrooms [70, 71]. Biogas can be upgraded to biomethane, and utilized more efficiently as a transportation fuel, a natural gas replacement in commercial and residential settings, and as a feedstock for natural gas distribution networks. Feedstock for greenhouses and a raw ingredient for chemical fuel production, carbon dioxide can be derived from biogas [72]. In 2005, there were around 16 million small household biogas digesters in use worldwide, with China and India accounting for the great majority. A total of 16 million tons of firewood were replaced by biogas in India in 1996, and it met 4% of China's energy requirements at the time. Around 7 million biogas digesters did the same in the USA [70]. The efficiency of using biogas can be greatly increased through cogeneration, which simultaneously produces heat and power. Excess electricity can be sold to the grid to help stabilize the grid and lessen the consequences of global warming by displacing the generation of energy from fossil fuels [47].

Europe has been a leader in the development of biogas technologies from urban trash. More than 14% of Europe's municipal solid waste is converted into biogas by the region's more than 70 active anaerobic digesters [73, 74].

2.2 Challenges of biogas development

The technology for producing biogas from biomass is currently fully developed, but there are several obstacles preventing its adoption. The obstacles include general ignorance about biogas and its application among the very people who can benefit most and lack of a regulatory framework to support the installation of biodigesters and lack of necessary incentives [75, 76]. Statistics show that many biogas digesters are underutilized, with some even laying unused soon after being built [74]. Lack of technical management skills, together with economic and cultural factors, contributes to this. In 2002, during the World Summit on Sustainable Development, it was acknowledged that many countries may benefit from using renewable energy to alleviate poverty [2, 77].

The global biogas production capacity by volume is over 59 billion m^3 biogas or about 35 billion m^3 methane equivalent, but the European Union alone accounts for about half of the total capacity. This shows that many developing and some developed countries have unsatisfactory contribution and have a long way to go in biogas production and use compared with developed counterparts [76]. The biggest barrier to the widespread use of biogas technology is the high expenses incurred in digester operation and maintenance. Complex biogas digester designs and a lack of technical knowledge are further obstacles [67, 78]. The key issues with the adoption of technology of biogas are discussed below:

i. Poor maintenance and management

The efficient operation of a biogas system requires the adoption of suitable management practices for the digester and zero grazing units for animals to achieve maximum efficiency. However, there are a few competing zero grazing management standards that make it difficult to manage the digester effectively. Farmers typically place more importance on the health of their farms than on keeping the digester well maintained. Due to this situation, there is a restricted amount of biogas produced, which may make it more difficult for the farmer to return his costs [67]. Studies show that lack of knowledge on digester maintenance is a leading challenge for biogas production in rural China and other countries, and technical training and availability of technicians would help alleviate the challenge [2].

ii. Lack or low technological awareness

There is a significant knowledge gap among biogas users in terms of technology and feasible applications. It is challenging to run, maintain, and service the biogas digesters and biogas energy systems because of the widespread lack of knowledge and skills. When the needs for biogas are not met to their full potential, production suffers [2, 78].

iii. Lack of feedstock

Studies also show that there is pervasive perception that biogas technology is only ideal for people having sufficient animal and agricultural waste for them to venture into biogas energy production. Therefore, lack of existing supply chain for biogas feedstock limits some small-scale farmers from investing in biogas production. The provision of reliable information on potential feedstock may assist potential investors in decision making [75].

iv. High cost of Installation

The availability of trained workers hinders the widespread implementation of biogas technology. The lack of readily available specific biogas installation supplies necessitates the development of finding alternative materials in the market [79]. A study on challenges biogas adoption in south Africa showed that several respondents are limited by cost of necessary building materials and lack of initial capital to purchase and install biodigesters [75].

v. Systems failure

Biogas digesters frequently malfunction, and even the ones that do work do not always deliver optimal results. This misunderstanding has given biogas technology a bad name among eco-activists as not being sustainable [80].

vi. Lack or poor post-installation support.

Post-installation support for biogas systems has lagged demand because of a dearth of qualified experts and competent artisans. Service typically stops at the end of the "12-month guarantee period" after installation that is sometimes advertized. The farmers and operators are technologically illiterate, so they have no idea how to keep the systems working efficiently [2].

vii. Lack of standards

Biogas technology operates freely without any laid-out standards to regulate it. This makes it hard to ensure quality control measures are undertaken.

viii. Availability of cheaper and convenient alternatives

Many rural communities are accustomed to using traditional sources of energy like firewood, charcoal kerosene, which is readily available freely in most cases and is considered convenient to use compared to biogas preparation and related investment and operational costs [75].

ix. Limited awareness and capacity building

Awareness among the public is a major challenge in many countries leading to limited adoption of biogas technology in many owing to limited institutional and technical capacity in these countries that would have actively promoted the adoption resulting in its lack of popularity as an energy source [67, 81, 82]. For this technology to gain massive adoption, especially at local or community levels, proper awareness and capability programs are necessary.

x. Lack of economies of scale and sufficient demand

Households in rural areas having significant feedstock for biogas production are often small with limited energy demand making investment in biogas production unattractive as most of the biogas produced may not be used. Biodigesters running with no biogas consumption leads to emission of highly potent methane which is a greenhouse gas. Connection to grids may create extra demand to sustain higher levels of biogas production [83].

There is need for governments and policy developers to provide a comprehensive policy and legislative environment that facilitate biogas digester production, installation and maintenance as well as other financial and non-financial incentives to encourage biogas production and uses [2]. Biodigester acquisition and use is primarily influenced by socio-economic factors like availability of household labor resources, family, or household financial income, and availability and type of feedstock for biogas production. Digestors require labor input of labor for operation and maintenance and provide the energy demand for biogas, hence lack of it has a negative impact. Families with higher income may prefer alternative sources of energy which also negatively affects biogas production and use [2, 84, 85].

2.3 Benefits of biogas production and use

There are several socioeconomic and environmental advantages with biogas production and use [83], these benefits include.

i. Boost for decentralized generation.

The biogas plants deliver biogas for heating homes and farms, digested manure to boost food production rate; delivery and export excess power to the grid via decentralized generation which improves the economic position farmers [83].

ii. Job creation

Biogas production creates jobs in the supply chain right from equipment manufacturers, dealers and farm works which helps improve the social and economic conditions of local people [83].

iii. Better crop yields and sustainable agriculture

Digestate from biodigesters is proven biofertilizer whose application to fields reduces water needs and improves crop yields while saving cost incurred in buying expensive and polluting chemical fertilizers [83].

iv. Biogas forms the basis of energy independence.

By substituting the fossil fuels with biogas, the oil importing countries and regions acquire some energy self-sufficiency which saves their foreign exchange reserves and a cleaner environment. Biogas production and use is more reassuring for many countries since biodegradable biomass resources are abundant in all countries and regions of the world. By increasing decentralized generation as opposed to centralized energy systems, supply interruptions are easier to manage locally while enhancing the stability of the main grid hence higher reliability of energy supply [13].

v. Reduction in greenhouse gas emissions

Disposal of agricultural wastes by means of open fires and natural decomposition releases huge quantities GHG and other pollutants to the atmosphere which contributes to global warming and environmental pollution. This is avoided or minimized by controlled biodegradation in digester for biogas production, while producing a clean energy resource to substitute polluting fossil fuels. Since the global warming potential of methane is about 25 times higher than CO_2 it will be scandalous to freely release tons of methane to the atmosphere, when this can be avoided by controlled biodegradation [13, 54]. Controlled biogas production and its use encourages waste accumulation and digestion which avoids methane emissions to the atmosphere from uncontrolled bio digestion of biomass in fields [83].

vi. Carbon dioxide sequestration potential by the anaerobic fermentation process

Biogas use has the possibility to uptake and sequester carbon dioxide to meet the climate target of below 1.5°C temperature rise. Methane from biodigesters has about 30% carbon dioxide. By upgrading biogas to biomethane, this CO_2 is removed and can be used in multiple applications like feedstock for food and biomaterials production [13].

vii. Water purification through anaerobic digestion implementation

Anaerobic digestion can be used to treat effluents with a high biological oxygen demand (BOD) and reduce BOD levels. This is the oxygen demand for microbial metabolism of organic materials (including trash). The BOD levels in wastewater from the dairy industry are 25–40 times greater than those seen in wastewater from the average household. Through bio digestion, between 70 and 90% of the BOD can be removed cost-effectively compared to aerobic systems [13].

viii. Health benefits

Domestic scale biogas production contributes to reduce respiratory and eye diseases that are linked smoke and fumes from firewood and charcoal, in addition to a better living environment of households. Wastewater from animal waste like dung and domestic wastewater sources, can be connected to toilets then to biodigesters for better disposal [13].

i. Reduced cost of waste disposal

The production of biogas offers solutions, for the treatment and management of digestible animal and plant waste leading to sterilization and production of useful manure and sewage with significant volume and mass reduction which effectively reduces the cost of waste treatment and disposal [30, 64, 86]. According to the French Agency for Environment and Energy Management (ADEME), using manure as raw for anaerobic digestion reduces methane emissions from the storage and use of manure, and biogas reduces fossil greenhouse gas emissions (with an effective reduction in greenhouse gases of more than 751,000 tons of CO_2 equivalent). The global warming potential of carbon dioxide emissions from biogas is smaller than that of methane emissions from fossil fuels. Therefore, biogas production and use reduce greenhouse gas emissions in accordance with the Kyoto protocol and Paris agreements. The negative effect is that reduced use of manure increases use of chemical fertilizers for fertility or nutrient replenishment to the soil [13, 80].

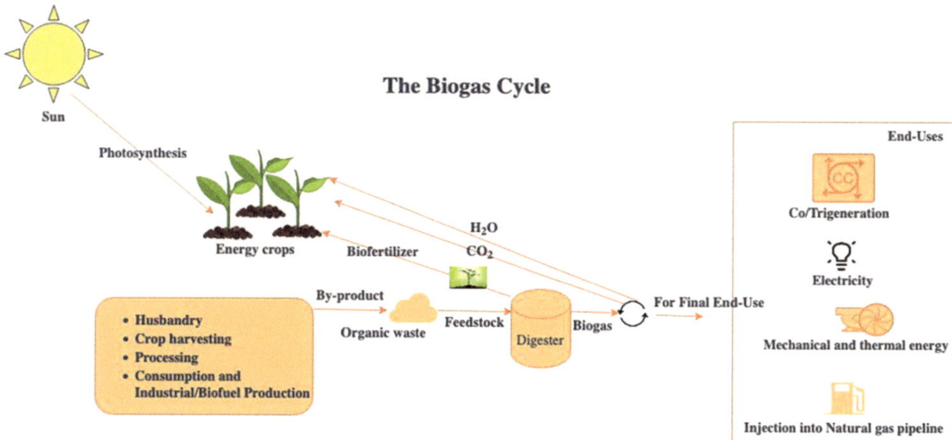

Figure 1.
The biogas cycle.

2.4 The biogas cycle

Biogas is generated by anaerobic digestion of biomass with the most exploited feedstock being agricultural wastes like manure, poultry litter, hay, and straw, stillage from ethanol production, activated sludge from wastewater treatment plants, etc. [87]. The biogas cycle is demonstrated in **Figure 1**.

Figure 1 shows the biogas cycle which consists of biodegradation of biomass mainly from energy crops and wastes from animal husbandry, biofuel process products, crop harvesting waste, industrial waste and food consumption wastes which are fed to anaerobic digester whose main products are biogas and biofertilizer which is taken back to the soils to support biomass regeneration [88]. The CO_2 released in combustion is absorbed by the vegetation through photosynthesis to form biomass. The organic nitrogen rich residual sludge from the digester is used as fertilizer for the soil [87]. Biogas can be treated and enriched and used as natural gas, but can also be combusted for production of thermal energy or electricity through different generation prime movers [89, 90].

The anaerobic digesters have been used widely to supply biogas by many households [87, 91]. The new trend in biogas production and utilization is the biorefinery concept which mainly uses renewable biomass as energy source and combines with production of chemicals, like plastics, solvents, and synthetic fuels [79, 89, 91]. A typical example is the Danish Bioethanol Concept which comprises the ethanol production from lignocellulosic biomass while biogas is produced from stillage and cellulose waste. Then residual cellulose waste is also recycled after wet oxidation for additional conversion into biogas [39, 87, 92]

2.5 Biogas production process

Different types of organic matter can be digested anaerobically to produce biogas. The final step in the chain of chemical and biological processes used to break down organic matter for the purposes of producing biogas and managing garbage is called anaerobic digestion [73]. The production of biogas involves the controlled breakdown of organic material into smaller molecules by several anaerobic bacteria [73, 77].

Chemical reactions such as hydrolysis, acidogenesis, acetogenesis, and methanogenesis occur during anaerobic digestion, which produces biogas from biomass [35, 73].

Biogas is produced by the microbial action in the digester soon after biomass is prepared and fed after undergoing preparation and a gradual fermentation process; the biomass is introduced into the reactor, where microbial activity within the digester rapidly generates biogas. Consequently, the process is a consequence of bacterial consumption of organic matter, specifically proteins, carbohydrates, and lipids/fats. This microbial digestion gives rise to the production of gases, predominantly methane and carbon dioxide. The process of biogas production encompasses several distinct stages, namely pretreatment, hydrolysis, acidogenesis, acetogenesis, and methanogenesis. Prior to introducing feedstock into the digester for the purpose of anaerobic degradation, the initial stage of biogas production involves feedstock processing or pretreatment. The pretreatment of feedstock is an essential and pivotal stage to mitigate instances of failure, enhance the production and enhance the quality of digestate, among other advantageous outcomes.

The bio digestion process for *biogas* production involves several key stages, namely the *pretreatment* stage of raw materials, anerobic digestion, purification of raw biogas, final biogas utilization, and *post-treatment* of *digestate*. Impurities like sand are, stones, plastics, etc. removed from the feedstock at the pretreatment stage, and the total solids (TS) concentration and temperature of the *feedstock* moderated for optimum bio-digestion [93]. The digester constitutes the core or the main element of the biogas production plant whose size and performance characteristics should be selected based on specific process and output conditions. At the terminal stage of the biogas production, the residue and slurry are further treated for intended application e.g. biofertilizer [21, 23, 94]. The process flowchart of a biogas plant process is shown in **Figure 2**.

From **Figure 2**, it is noted that the process of biogas production starts with feedstock collection and pretreatment before it is fed into the anerobic digester. The main products of the anaerobic digestion process are raw biogas and digestate. Biogas can be purified and stored ready for distribution and consumption. The digestate in slurry form is prepared or conditioned and applied in irrigation or controlled discharge as an organic fertilizer. Digested can also be used as a solid fertilizer in addition to slurry form. The various stages of biogas production are discussed as follows.

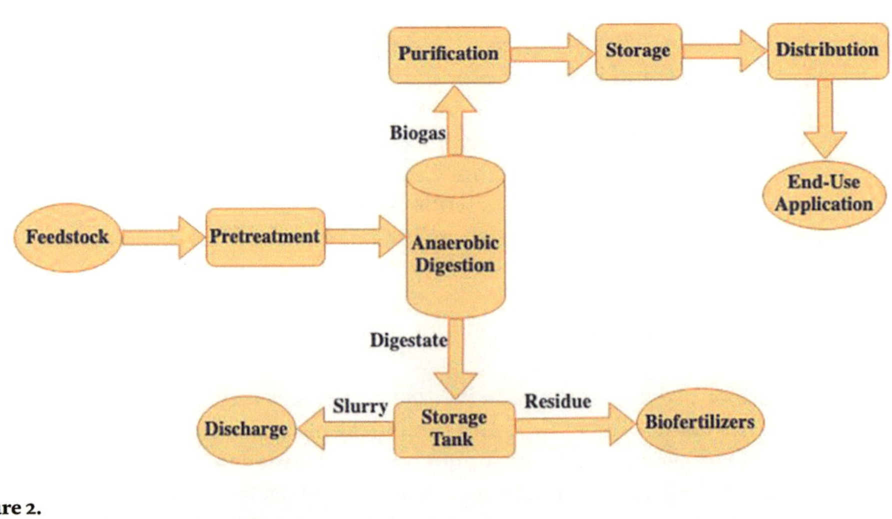

Figure 2.
Biogas plant operation process and products.

2.5.1 Pretreatment stage

Pretreatment is implemented as a strategy to enhance the breakdown of substrates, thereby enhancing the overall efficacy of the process. The employed pretreatment methods can be classified into four primary categories: chemical, mechanical, thermal, and enzymatic processes. The aforementioned techniques are designed to expedite the process of decomposition, although it is crucial to acknowledge that they do not necessarily lead to augmented biogas generation [10]. First, the feedstock is washed, then it is macerated, screened, and finally pressed (however, these stages can change depending on the feedstock). Mechanical parts can survive longer if they are free of contaminants like plastic, and magnetic traps are used to effectively eliminate magnetic pollutants. Glass, eggshells, pottery, bones, and sand are among nonmetallic impurities that must be removed during the preparation step. These pollutants do not decompose during digestion and hence accumulate as deposits at the digestor's base [72].

Enzymes are required for hydrolyzing lignocellulose, which is necessary for digestion. An economic and technical barrier to producing biogas from lignocellulosic waste is its complicated structure. Strong and compact molecular bonds are formed in lignocellulose due to the presence of cellulose, hemicellulose, and lignin. The effectiveness of pretreatment is critical for maximizing lignocellulose's potential as a biogas source. Pretreatment, in general, hastens reactions, boosts biogas yields, and creates a plethora of novel substrates [10, 95].

Technological advancements in pretreatment have enhanced biogas yields from lignocellulosic feedstocks, resulting in lower methane emissions into the atmosphere. Pretreatment is crucial because it increases biomass digestion and provides more biogas than without it by facilitating microbial breakdown of lignocellulose and its polymers, especially cellulose and hemicellulos. It has been shown that autoclaving and microwave heating can cause the hydrolysis of several non-biodegradable chemicals found in municipal garbage. Biomass pretreatment achieves this by partially or completely breaking down the feedstock to yield fermentable sugars, reducing the lignin barrier, and smoothing out the crystalline structure of cellulose [10, 30, 88, 89].

Table 1 demonstrates the wide range of pretreatment options available, each with its own set of benefits and drawbacks, and cost benefit implications. While pretreatment enhances the efficacy of the process and the biogas yield, it also introduces difficulties such as higher energy input, increased operating and maintenance expenses, and the introduction of inhibitory chemicals.

AD have become increasingly popular in producing renewable and sustainable energy from waste. Although synergies do exist between various technologies, the configurations to couple need further development [96].

2.5.2 Processes in anaerobic digestion

i. Hydrolysis

Hydrolysis is a chemical reaction in which water is broken down into OH^- anions and H^+ cations. In the presence of an acidic catalyst, hydrolysis breaks down the massive biomass polymers present in the substrate [33]. The proteins, carbohydrates, and lipids, all of which are massive organic polymers, are broken down into their component simple sugars, fatty acids, and amino acids to make up biomass [33, 73, 77]. During the process of hydrolysis, fermenting bacteria, such as Bactericides, Clostridia, and Bifidobacterial, efficiently decompose biopolymers, including carbohydrates,

	Technology	Merits	Demerits
1	Feedstock milling	• Does not produce any process inhibitors. • Increased methane (5–25%)	• High energy requirements • High maintenance cost
2	Feedstock Extrusion	• Extrusion increases the surface area to volume ratio	• Leeds to increase energy consumption. • Increases equipment maintenance cost
3	Steam pretreatment/ steam explosion	• Increased cellulose fiber reactivity. • Can make use of process energy/ biogas	• Risk of producing inhibitors (e.g., furfural and HMF) • Reduces digestible biomass due to lignin condensation. • More process energy input and reduced net energy output. • Precipitation phenomena
4	Hot water treatment	• Solubilized hemicellulose and lignin products are present in lower concentrations. • Reduced risk of producing inhibitors like furfural and HMF) • Increased enzyme accessibility	• High heat demand and hence energy consumption • Not effective at all temperature ranges
5	Microwave	• Leads to more biogas production by 4–7%	• Increased process energy input • Increases labor and maintenance cost
6	Diluted or strong acid pretreatment	• Used to solubilize hemicellulose component. • Methanogens can adapt to inhibiting compounds.	• Acids are expensive. • Can form inhibiting compounds. • Causes corrosion problems.
7	Alkaline pretreatment	• Used to solubilize hemicellulose and some lignin. • Increased methane production	• Can produce inhibitors. • Can cause high alkali concentration in reactor

Table 1.
Pretreatment methods and applications [10].

proteins, and lipids, into soluble forms such as sugars, fatty acids, and amino acids [97]. Hydrolysis produces acetate and hydrogen, which are put to good use by methanogens in the final stages of anaerobic digestion. Methane formation by acidogenesis necessitates further hydrolysis product breakdown, as these are still quite large molecules [45, 98].

ii. Acidogenesis

Anaerobic activity of acetogens produces mostly organic acids, alcohols, hydrogen gas, and hydrogen sulfide throughout this process [33, 97]. In an acidic environment generated by the fermentative bacteria, hydrolysis products are degraded by acidogenic microorganisms to produce ammonia, carbon dioxide, hydrogen sulfide, carbonic acid, shorter-volatility fatty acids, alcohol, and other trace products based on the substrate composition and products of hydrolysis [97]. Acidogenesis byproducts are too large to be ideal for methane production. Thus, they engage in acetogenesis [67].

iii. Acetogenesis

The process of acetogenesis produces acetate from acetic acid [99]. Through anaerobic digestion, acetic acid, hydrogen, and carbon dioxide are generated during acidogenesis. The results of acetogenesis and other processes are digested to the point where methanogens can produce methane [67].

iv. Methanogenesis

Methane and carbon dioxide are created in the final stage, called methanogenesis, thanks to the activity of acetoclastic methanogens (AM) and carbon dioxide (CO_2) reducing methanogens (CM), respectively. Finally, anaerobic bacteria called methanogens like *Methanosarcina barkeri*, *Methanosaeta concilii*, and *Methanococcus mazei* complete the methanogenesis process by producing methane from the byproducts of acetogenesis and other intermediate products of acidogenesis and hydrolysis [33, 97, 100]. The first three stages of methanogenesis (as shown in Eqs. (1) and (2)) involve the utilization of CO_2 and acetic acid (CH_3 COOH):

$$CO_2 + 4\,\mathrm{H}_2 \rightarrow CH_4 + 2\mathrm{H}_2\mathrm{O} \tag{1}$$

$$CH_3\mathrm{COOH} \rightarrow CH_4 + CO_2 \tag{2}$$

While (CO_2) might theoretically be turned into water and methane, at this time the acetic acid pathway is the primary mechanism for methane generation. Therefore, (CO_2) and (CH_4) are produced as the primary products of anaerobic digestion via the acetic acid pathway [67]. H_2 and CO_2 gas combinations are transformed into biogas, which is composed of 60–70% CH_4 and 40–60% CO_2 [16]. The four steps of biogas production, together with associated activities and microorganisms, are depicted in **Figure 3**.

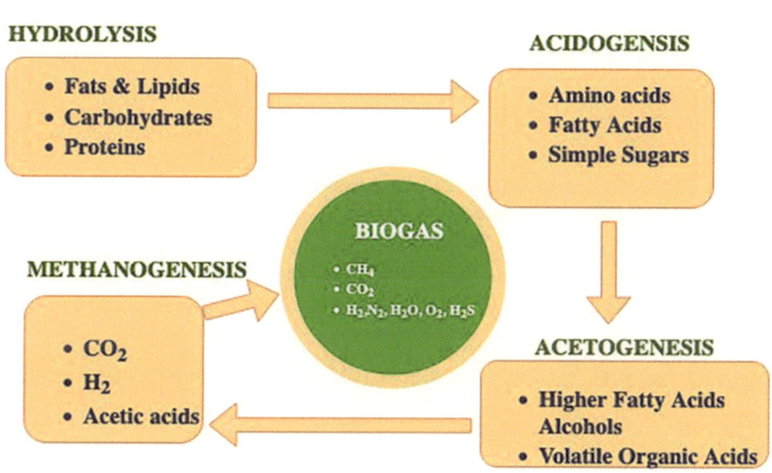

Figure 3.
Summary of the biogas formation process.

Figure 3 shows the main stages in biogas production being hydrolysis, acidogenesis, acetogenesis, and methanogenesis as the four primary stages in anaerobic digestion. Maximal biogas production requires careful management of the four main stages.

2.5.3 Factors influencing digester's efficiency and performance

There are many factors that influence anerobic digestion and biogas production, e.g., biogas potential of feedstock, nature of the substrates, inoculums, digestion pH, temperature, digester loading rate, hydraulic retention time (HRT), carbon/nitrogen (C:N) ratio, presence of volatile fatty acids (VFA), presence inhibitory substances, etc. [101].

i. Feedstock

Any biodegradable waste that has fats, proteins, carbohydrates, cellulose, and hemicelluloses is suitable as feedstock in anaerobic digestion. However, the various feedstocks have different theoretical chemical oxygen demand (COD) and the potential biogas. They have different or varied biochemical structure and rate of its biodegradability leading to differences in productivity. The factors in feedstock that influence biogas productivity are BOD, volatile solids, carbon/nitrogen (C:N) ratio, COD, volatile solids (VS) content, and presence of inhibitory substances [78, 101].

ii. Temperature

The bio-digestive process can be accelerated within tolerable limits by heating process. The same holds true for the processes involved in making biogas. The microorganisms engaged in anaerobic digestion perform both thermophilic and mesophilic digestion. Temperatures between 45 and 80°C are ideal for thermophiles, while those between 25 and 40°C are ideal for mesophiles, yet the Psychrophilic bacteria operate below 25°C [67]. A decrease in methane concentration and a rise in CO_2 might impair the heating value of the gas produced if global warming is not controlled. The best number to use for reliable and productive biogas production depends on the organism being used [94, 101].

iii. Digester instrumentation

Anaerobic digesters require continual monitoring of production levels and parameters to ensure optimal process management and, consequently, output is achieved by detecting and responding to changes in biomass quantity, anomalies, and the health of microorganisms [100]. The amount of biomass still to be decomposed and the amount of time spent doing so should be indicated by the gases produced. Knowing when to add new feedstock and the digester's efficiency can be gauged with this information. Maintaining optimal microbial digestion requires monitoring gas generation in a system where biomass is added regularly [67]. By monitoring the gas produced, issues in the digester, like a shift in pH or temperature, can be identified and remedied [4].

iv. The pH

The pH is an important parameter in the operation of the digester, and the changes in pH value is is different at different states of the digestion. Biogas yield is optimized at pH range of between 6.5 and 7.2 [101, 102]. At a pH of about 5, acid-forming bacteria are at their most productive, while methane-forming bacteria thrive at a pH value of 6.2 or higher [33]. The bacteria population used in the bio methane process does best between 6.8 and 7.2, but may survive in a higher pH range [103]. When the pH of the digester falls outside of that range, the methane-producing bacteria deteriorates, providing an unideal environment for the microorganisms responsible for methane synthesis to survive. With a continuous decrease in pH below 6.2, the biogas productivity is reduced due to the accumulation of VFA, as they are toxic to the methane-producing bacteria. For the digester to generate the required biogas, the microorganism activeness is required as well as regular attention to the pH regulation [99, 104].

v. C/N ratio

The C/N ratio determines the suitability of biomass for anaerobic digestion. Where the C/N ratio is high indicates low nitrogen content needed for microbial growth which leads to low uptake of nitrogen by methanogens needed for protein production. The result is wastage of carbon and low biogas production [101, 105]. The overall carbon to nitrogen ratio of the substrate. Although very high C/N ratios foster population expansion of methanogens, which has little to no influence on the carbon residue in the substrate and results in poor methane output, the optimal ratio for anaerobic digestion is between 20 and 30. Overabundant ammonia reduces methane generation and poses a threat to methanogenic bacteria when the carbon to nitrogen ratio is too low [99].

vi. Toxicants/Inhibitors

Inhibitors (i.e., which are toxic to the anaerobic digestive process) of methanogenesis, such as antibiotics and other residues, result in less methane synthesis and a rise in the concentration of volatile acids. Avoiding a high nitrogen to carbon ratio is important since it increases the likelihood of creating hazardous circumstances for the bacteria [33].

The most common inhibitory substances in substrates for the anaerobic digestion process are (CO_2), (NH_3), long chain fatty acids (LCFA), H_2S. Methanogens are more sensitive to toxic material than other groups of bacteria. Short-chain fatty acids being the main inhibitors for methanogens. The concentration of ammonia is important but needs control with less than 200 mg/l being suitable for biogas generation. Nitrogen is a nutrient, hence C:N ratio of 30:1 improves biogas productivity by supplying nutrients to micro-organisms for the microbial activity. The methanogenic phase is also inhibited by the effect of ammonia which is rich in swine waste, poultry waste, swine waste, and high proteinaceous sludge. It is the free NH_3 and not ammonium ion (NH_4^+) that is responsible for inhibitory effect. Studies show that ammonia with a concentration of 1.77–14 g/l of total ammonia nitrogen can reduce the biogas yield by about 50%. The inhibition by ammonia is influenced by the temperature and the pH. Changes in the temperature and pH can lead to process dynamics disturbance leading to as much as 30% loss in biogas yield. Thermophiles are more sensitive to LCFA inhabitation compared to mesophiles. LCFA are found slaughterhouse waste, agricultural residues, olive oil, food waste, wastewater, etc. which act as inhibitory substances in the process of anaerobic digestion [101].

vii. Hydraulic retention time (HRT)

The hydraulic retention time (HRT) refers to the time taken by the biodegradable material in the bioreactor. Factors influencing the *HRT include the* temperature inside the digester, technologies applied, and the type of feedstock used. The recommended HRT for mesophilic digester is 10–40 days while the thermophilic digester is about 14 days. A very short retention time may have the bacteria washed out of the digester before getting multiplied which leaves the digester in a state of standstill, while longer retention time increases the volume of the reactor to handle the same feedstock. To reduce the retention time and reactor volume, an optimum loading rate should be maintained for optimizing methane production, with 2–3 weeks being regarded as generally optimum for lignocellulosic material to degrade and produce biogas [101].

viii. Co-digestion

The performance of a digester can significantly be enhanced by co-digestion which helps supply the necessary missing nutrients to micro-organisms for higher efficiency. Common materials for co-digestion includes leftover foods, cow manure, vegetable wastes, and fruit. Co-digestion with properly selected feedstock materials can significantly increase biogas yield [101].

ix. Redox conditions

The digestor's Redox conditions dictate the degree and velocity of contaminant biodegradation. Aerobic biodegradation rates are typically substantially higher than anaerobic biodegradation rates for most pollutants. While oil that has settled into anaerobic sediments may be quite tenacious, petroleum-based hydrocarbons that penetrate the aerobic zone of waters in lakes and rivers are susceptible to microbial destruction. Highly reduced hydrocarbons, like the alkanes, have low molecular weights and require oxygen to degrade. Methane and other alkanes cannot undergo anaerobic degradation. Hexadecane ($C_{16}H_{34}$), a high-molecular-weight alkane, may be present in some conditions. These chemicals rarely degrade and are often found in petroleum-contaminated areas. However, heavily chlorinated molecules like perchloroethene (PCE) resist oxygen decomposition. Anaerobic environments can bio transform these substances [106].

For a biodigester, the methanogenic bacteria require redox condition between −300 and − 330 mV for optimal performance [107]. Redox condition is important for fermentative H_2 production process in which anaerobic bacteria are crucial and rate limiting. The efficiency with which substrates are metabolized, proteins and other storage resources are synthesized, and metabolic waste products are excreted, are all impacted by the Redox state. Pyruvate undergoes an oxidation-reduction reaction (AB) that produces VFAs and H_2 under acidic circumstances. Methane and CO_2 are produced by mesophilic bacteria during neutral processes, while solventogenesis occurs during basic operations. The repression of mesophilic bacteria indirectly promotes H_2 producers within the system. The anaerobic bacteria operate below pH 6, but MB optimum range is between 6.0 and 7.5. The pH range of 5.5–6.0 is ideal to avoid both methanogenesis and solventogenesis. Good H_2 is realized at a pH of about 6 compared to near-neutral. However, highly acidic pH (<4.5) inhib-its H_2 production as it inactivates AB. Hydrogenase enzyme activity gets inhibited by maintaining low or high pH beyond optimum range [108].

x. Mixing rate

The biogas production plants need an appropriate mixing system to provide favorable conditions for microbial community, for process stability and efficiency in anaerobic digestion [96, 109]. The mixing intensity or shear rate and the length of time that shear rate is applied to a mixer determines the degree of mixing realized. Application of uniform shear rate is an important strategy to achieve stability of bio digestion process. In some studies, higher mixing intensity of 67 rpm for 5 min h^{-1} led to 15–18% higher production output compared 10 rpm and 30 rpm without generating instability in terms of VFA accumulation and dead zones. A higher mixing rate reduces dead zones to less than 2%. Therefore, mixing significantly affects the biogas production rates but the impeller design is the principal factor. The best results were achieved by using a large diameter impeller running at medium mixing rate in anaerobic reactor [109, 110].

xi. Inoculum

The inoculum is used to reduce the start-up time of biodigestion by seeding of the which accelerates biogas formation. By use of inoculum, about 80% biogas can be obtained within the first 10 days of digestion period [101].

xii. Moisture

The substrate dilution or moisture content affects the process of anaerobic digestion with best results in terms of methane yield achieved under 60–80% humidity [89, 101].

xiii. Organic loading rate (OLR)

The OLR is the rate at which organic matter is added to the digester volume in relation to either time or the substrate's biological conversion capacity. The organic loading rate is directly promotional to the quantity of volatile solids loaded in the digester and hence it influences the biogas yield, with more methane production for lesser the OLR. If the reactor has higher concentration VFA's, then it implies that it is overloaded [101]. The loading rate of a digester is the daily volume of volatile solids given to the digester, and methane production are favorable with high loading rates [33].

xiv. Biological oxygen demand (BOD) and chemical oxygen demand (COD):

The BOD and the COD affect biogas production too. The biochemical oxygen demand determines the amount of oxygen needed by the microbes to decompose the substrate. If the BOD is high, then a more rapid organic degradation is achieved which leads to more biogas production. The COD measures all organic and inorganic biodegradable matter in feedstock or substrate. The value of COD of an organic waste helps predict the theoretical methane yield from the substrate. From experiments, if 0.5 l/gm of COD is removed, the approximate methane production is about 0.35 l/g [101, 111].

2.6 Energy requirements for biogas process

The production of biogas necessitates the addition of heat to keep the substrate at an appropriate temperature and the constant stirring of the substrate, either by hand or with an electric motor-driven stirrer [99].

Therefore, process heat energy requirement is determined by the relation in Eq. (3)

$$Q_t = M * C * (T_2 - T_1) \tag{3}$$

where
Q_t = Overall heat required to heat the slurry in Kilojoule (KJ);
M = Mass of the slurry in (Kg);
C = The specific heat capacity of slurry and is expressed in KJ/Kg°C.;
T_2 = Desired digester slurry temperature in °C;
T_1 = Temperature of the digester charge/slurry in °C,
M = Digester size (V) in m^3× density of slurry (ρ) Kg/m^3; Density of slurry (ρ) = (Density of water + substrate density dung)/2; Specific heat of slurry = (specific heat of water (4.2KJ/Kg°C) + specific heat of the substrate e.g., cow dung (2.8KJ/Kg°C)/2 = 3.5 Kg/Kg°C.

The digestor effluent contains waste heat that can be recovered and used to produce the necessary heat energy. The slurry's input temperature may rise by a few degrees Celsius as a result. This results in a potential 50% reduction in the total heat required by thermophilic digesters [31, 112]. External heating is most effective for digesters of a large or medium size and requires approximately twice as energy intensive as central heating. A total of 850–1000 W/m^2 K^{-1} is the typical range for space heating, and 300–400 W/m^2 K^{-1} is the typical range for space cooling [112].

Parameters such as microbe species, feedstock pretreatment, and biogas purification processes, substrate composition, substrate properties, and ideal reactor conditions must be managed to produce biogas profitably. Further investigation is required and have been actively a high interest area toward the enhancement of biogas production and biogas utilization through the application of engineering, biology/biotechnology, and technical innovations in order to achieve cost-effective biogas production [10, 113].

3. Biogas feed stocks

A wide range of biowaste can be used as substrates for biogas production through anaerobic digestion. Significant volumes of lignocellulosic waste are generated from agricultural, municipal, and other activities. The most commonly used materials for biogas production are animal manure and slurry, waste water and sewage sludge, municipal solid waste, and food processing and consumption generated waste [10]. The most used substrates for biogas production are pig and cattle manure, and poultry litter which generally compete with traditional manure [10, 87, 114]. Sewage sludge generated from wastewater treatment plants is the main feedstock for biogas production in places like Gävleborg region and Sweden while the use manure, food waste and industrial waste have been on the rise, although limited. Horse manure was also widely used as a substrate during co-digestion. Policy measures like ban on the landfilling of organic waste and competing applications like manure use can shift the demand for biogas feedstock selection [114, 115].

The quantity and composition of biogas produced by biomass digestion is a function of depending on the substrate used and process parameters control. Different types of feedstocks can be applied for biogas production ranging from animal wastes, agricultural residues, and energy crops [59]. The various feedstock used have different characteristics and hence biogas potential and hence may require different type and level of pretreatment and process conditions during anaerobic digestion [2, 78]. Feedstock biomass can take the form of a solid, a slurry, or a liquid (either concentrated or diluted). Many different materials can be used as feedstocks, including animal manure, straw, biowaste from homes and businesses, waste from bioethanol and biodiesel production, waste from markets and restaurants, sewage sludge from wastewater treatment plants, and energy crops like maize, silage, grass, sorghum, cereals, and sugar beets. The high lignin concentration of wood makes it unsuitable for anaerobic digestion and biogas production [78, 79, 88, 116].

Several different kinds of biomass garbage can be converted into biogas. Litter from farms, city dumps, and other sources can all contribute to the production of lignocellulosic garbage. Common forms of waste that are utilized as feed stock include animal manure, slurry, sewage sludge, municipal solid waste, food waste, and agricultural waste [10]. The biomass feedstock used to create biogas consists of a wide variety of components, including lipids, proteins, carbohydrates, cellulose, and hemicellulose. Co-substrates are added to alter the organic composition and boost biogas generation. Most co-substrates consist of organic industrial waste, food scraps, and municipal biowaste. Although lipids generate more biogas overall, carbohydrates and proteins convert more quickly [10]. Both biogas potential and yield are affected by the feedstock and co-substrates used in production, in addition to the process parameters [60, 84, 86].

There are many substrates that are suitable for anaerobic digestion, which can be characterized and classified into four main groups, namely energy crops, agricultural residues, food production wastes, and organic wastes [30, 117]. In the European countries, about 70% bio digestion feedstock comes from the agricultural sector like manure and agricultural residues. According to the European Biogas Association, agricultural wastes account for 40–60% in biogas production in Germany, Cyprus, Denmark, France, Italy and Poland [59]. The residual digestate is a feasible material for use as organic fertilizers, with huge optimization potential leading to increased farm production and profitability [117]. The various feedstocks for biogas production are discussed below.

3.1 Crop residues for biogas

Crop residues, a byproduct of crop husbandry, can be put to several different applications. Combustion, composite manufacture, bio-digestion, and animal feed are just some of the ways that waste materials like these can be repurposed. In terms of surplus agricultural products, India yearly produces around 500 Mt. (million tons) [16]. Nearly 70 million metric tons of organic waste are generated annually in the United States of America. Examples of agricultural waste include spoiled food and water [34]. Agricultural crop residues are a serious concern of farmers and cause significant contributors to GHG generation. Primary agricultural residues remain in fields as by-products after harvesting cereal grain straws, wheat (Triticum), barley, rice (Oryza), corn stovers, stalks, leaves, etc. On the other hand, secondary agricultural residues are products of agricultural resource processing like bagasse, sunflower husks, nutshells etc. Two sustainability issues associated with the use of agricultural wastes for biogas production are the potential competition with feeding in animal husbandry and the probable depletion of organic matter in the soil and nutrients in farmlands [59].

GHG, including CH_4 and N_2O are produced when crop waste is burned on-site. In terms of GHG from crop residues, maize crop residues produce the most (in gigagrams of CO2 equivalent). GHG emissions can be lowered and items like fertilizer can be more profitable if we can find better applications for these wastes [16]. **Figure 4** shows that most carbon dioxide emissions from crop residue combustion are caused by the burning of maize residues as found in [16].

Figure 4 shows the main crop residues used for biogas production led by maize, rice, wheat, and sugarcane wastes. The potential for making biofuels from cellulosic waste is enormous. This includes materials like bagasse, energy crops, agricultural wastes, and even sewage. Cellulose, hemicellulose, and lignin are the three primary organic components of lignocellulose [10]. Cellulose is the primary structural component that gives plant cell walls their mechanical stability. Macroscopic hemicellulose is composed of repeating polymers of pentoses and hexoses. Three aromatic alcohols (coniferyl alcohol, sinapyl alcohol, and p-coumaryl alcohol) are synthesized within lignin. Composition of lignocellulose differs greatly depending on both the source and the time of year. Cellulose is a linearly connected polymer that contains many β-1,4-glycosidic linkages. Crystalline and amorphous elements coexist throughout the structure [10].

By applying temperatures in the range of 320°C and pressures of 25 MPa, crystalline cellulose can be converted into amorphous cellulose. Cellulose is the most common organic component on Earth, making up more than 25% of plant material. Hemicellulose is an important part of biomass because of the dynamic nature of its structure and the vast variety of polymers it contains, such as pentoses like xylose and arabinose, hexoses like mannose, glucose, and galactose, and sugar/uronic acids like glucuronic, galacturonic, and methylgalacturonic acid. Xylan makes up roughly 90% of the hemicellulose structure, though this value varies with the kind and origin of the feedstock being processed. Hemicellulose requires a wide variety of enzymes in order to be thoroughly degraded into free monomers [10]. Hemicellulose is comprised of multiple sugar units arranged in polymer structures that can be readily hydrolyzed. It possesses shorter side chains and a comparatively lower molecular weight when compared to cellulose. Hemicellulose enhances the overall compactness of the

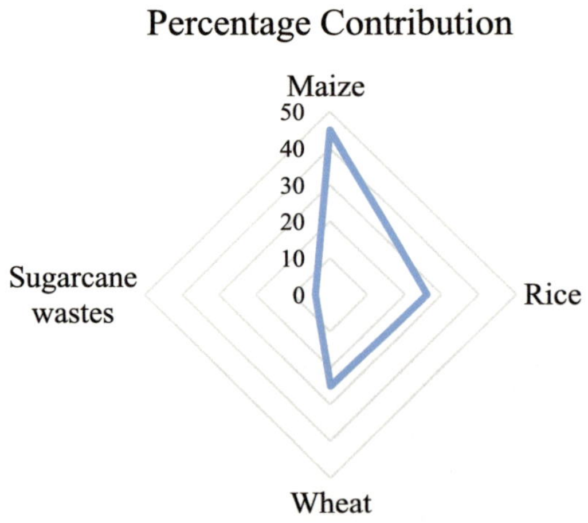

Figure 4.
Leading crop residues in terms of CO_2 gigagram equivalent GHG emissions.

cellulose-hemicellulose-lignin network through its interconnection of lignin and cellulose molecules. The solubility range of hemicellulose molecules is typically observed to be between 150 and 180°C, with variations influenced by environmental conditions such as acidity, neutrality, and alkalinity [10, 118–120].

Lignin is a naturally occurring heteropolymer found in the cell wall, composed of three phenylpropane-based units, namely p-coumaryl alcohol, coniferyl alcohol, and sinapyl alcohol, which are interconnected through linkages. The structure of lignin exhibits resistance to microbial degradation and oxidative stress. Lignin exhibits limited solubility in water, resulting in reduced degradability and hindered biogas production. At approximately 180°C in a neutral setting, the dissolution of lignin and hemicellulose in water occurs. The solubility of lignin in acidic, neutral, or alkaline environments is determined by the presence of the phenylpropane-based unit within its molecular structure. Lignin constitutes approximately 30–60% of the composition of wood, whereas agricultural residues and grasses typically contain lignin in the range of 5–30%. The agricultural produce primarily consists of hemicellulose [120].

The composition and structure of lignin have been found to have a positive impact on the hydrolysis process, resulting in an increase in the generation of biogas. Nevertheless, the presence of higher lignin content in the feedstock has been observed to decrease the efficiency of degradation [10, 120].

3.2 Energy crops

Energy crops encompass a wide range of plant species that can be classified into two main categories: herbaceous and woody. Herbaceous energy crops include grass, *maize* (*Zea mays* L.), and *raps* (*Brássica nápus* L.), while woody energy crops consist of species such as *willow* (*Salix*), *poplar* (*Populus*), and *oak* (*Quercus*). However, it is important to note that woody crops require a specific delignification pretreatment digestion process to be utilized effectively for energy purposes. Lignin, due to its resistance to degradation in anaerobic digestion, presents a favorable characteristic for gasification, incineration, or composting processes when considering woody substrates [86]. Herbaceous energy crops have many important characteristics making them suitable for anaerobic digestion e.g., high solar energy conversion efficiency leading to high yields, need low agrochemical inputs, and have got low nutrient and water requirement due to their extensive rooting system, to hold onto fertilizers and water, and they also have low moisture content at harvest time. Biogas can be made cost effective and high yielding by using crops with perennial growth habits with low establishment costs and fewer field operations and crops providing high biogas output as a feedstock source [59, 121].

3.3 Animal residues

Animal waste is a valuable energy resource containing biodegradable nutrients and renewable energy. However, by practice, most animal waste is deposed off, collected in lagoons or left to decompose in the open land causing environmental hazard [37]. Generally, the term "animal waste" refers to feces, urine, dung, or other excrement, urea, digestive emission, urea, or similar substances emitted by farm animals like livestock, poultry, or fish [88, 89, 122]. Animal waste like manure emit methane, ammonia nitrous oxide, volatile organic compounds, hydrogen sulfide, and particulate matter, which can cause serious environmental concerns and health problem [4, 61].

The primary sources of animal waste commonly encountered include dairy shed effluent, mainly comprising of wash water, urine, dung, residual milk, and waste feed. Additionally, poultry litter, which encompasses water, manure, spilled feed, feathers, and bedding material, is another significant source [21, 60]. Moreover, renderings from slaughterhouses and other byproducts originating from livestock finishing operations contribute to the overall volume of animal waste. Dairy manure and poultry litter are widely recognized as prominent examples of animal wastes that are extensively be used in large-scale biogas energy production [60]. In the past, significant volumes of bovine excrement were subjected to treatment processes and subsequently utilized as a form of agricultural fertilizer, or alternatively, were disposed of through the practice of spreading onto farmland. Nevertheless, numerous countries have implemented rigorous environmental regulations and legislations to regulate the emission of odors, as well as the pollution of surface and groundwater, soil contamination, and nutrient management. These regulations have prompted the adoption of alternative approaches that discourage the disposal of animal manure. Consequently, there is an increasing emphasis on utilizing animal manure in biomass-based production chains, driven by the incentives provided by these regulations [79, 121].

The challenge with poultry waste is high phosphorus (P) run-off to surface water which creates issues like odor and taste problems in drinking water, because of excess algae growth arising from excess phosphorus. For this reason, local land applications of poultry litter are restricted to protect water quality. Poultry waste on the other hand has high methane potential making it an efficient feedstock for biogas production. Slaughterhouses and fish processing factories also generate significant digestible waste for biogas production [78, 121].

3.4 Industrial and process wastes

Increased demand and consumption of food and industrial produce leads to large scale generation of waste. There is need for proper management of organic waste to minims harm to the environment like as climate change, ecosystem damage, and resource depletion [59]. About one-third of food produced for human consumption globally is discarded as waste, which is approximately 1.3 billion tons of waste per year. Anaerobic digestion of waste prevents the spread of pathogens and environmental degradation arising when waste is carried by to runoff water to basins and drain into the oceans. Anaerobic digestion surpasses thermal treatments technologies for bakery wastes such bread, biscuits, donuts, and pizza dough, as wastes like bakery garbage promotes methane production because it includes easily degradable carbs and vitamins [20, 59].

Implementation of anaerobic technologies started with wastewater treatment hence Industrial wastewater is a major source of aquatic pollution that endangers surrounding environments and ecosystems. Anaerobic digestion of industrial waste is more advantageous than aerobic wastewater treatment since it produces useful biogas, and substantially lowers energy requirement, and generates less sludge that requires disposal Brewery wastewater, cassava (Mánihot esculénta Crantz) starch wastewater, palm oil mill effluent, biodiesel wastewater, and bioethanol wastewater (vinasse) have demonstrated significant biogas potential [59].

3.5 Biogas potential of various feedstocks

Different biomass feedstock has different biogas potential in terms of both quality and quantity. **Table 2** illustrates gas yields and methane contents for various substrates after 10–20 days of retention at 30°C [37, 61].

Substrate	Gas Yield (L/kg Vs*)	Methane Content (%)
Pig manure	340–550	65–70
Cow manure	90–310	65
Poultry droppings	310–620	60
Wheat straw	200–300	50–60
Rye straw	200–300	59
Barley straw	250–300	59
Oats straw	290–310	59
Corn straw	380–460	59
Flax	360	59
Hemp	360	59
Grass	280–550	70
Elephant grass	430–560	60
Sunflower leaves	300	59
Agricultural waste	310–430	60–70
Fallen leaves	210–290	58
Algae	420–500	63

Table 2.
Methane content and gas yield for different biomass [37, 101].

As can be seen in **Table 2**, there is a wide range in both the quantity and quality of biogas produced from various forms of biodegradable biomass feedstocks. According to **Table 2**, biogas can be produced from a variety of sources, including algae, cow manure, pig manure, agricultural solid wastes, and grass. Among the many potential sources of biogas, algae have the highest output, followed by sugar beets, elephant grass, sewage sludge, clover, elephant grass, and maize straw. There is not much potential for biogas production from materials like reed, sugarcane bagasse, rice seed coat, or rice seed straw. Therefore, a farmer must carefully select his biomass feedstock, or the plants and animals he raises, to achieve maximum biogas output [100].

3.6 Estimation of biogas potential

According to [100, 123], biogas can be used to generate heat, the cost of which can be roughly calculated using Eq. (4) as shown:

$$E_{pi} = E_m^{cd} * \eta_i \tag{4}$$

where:

E_{pi} – Production of the energy (heat only, electricity only or heat and electricity in co-generation), [GWh/year].

η_i –efficiency of energy conversion process.

E_m^{cd} – Possible production of total biogas energy.

Total biogas energy potential from livestock dung can be determined using the following relation in Eq. (5):

$$E_m = \Sigma(N_{i.n} * M_{i.n} * oDM_{i.n} * CH_{4i.n}) * NCV * 365 * 10^{-6} \quad (5)$$

where:

E_m - Expected production from livestock manure available in the region, [GWh/year].

Ni.n--Animals of the same category (cattle, pigs, poultry etc.) in the region, [number of animals].

Mi.n - Manure outcome of each category of animals, [kg of manure/animal/day].

$oDM_{i.n}$ -Organic dry matter content in the manure for each category of animals, [%].

$CH_{4i.n}$ -Methane outcome, [Nm^3 CH4/kg organic dry weight of the animal manure].

NCV – Net Calorific Value [kWh/Nm^3].

3.7 Biogas production and fertilizer application

The composition of the feedstock and the stability of the biogas plant's operations are two factors that influence the biogas yield, biogas quality, status of digestate, and stability of plant operations [124], and the quality of the sludge from anaerobic digestion, that can be used as an organic fertilizer [22], to improve soil quality, boost crop yields, and aid in the advancement of sustainable agriculture [61]. Biodigester slurry can boost agricultural output by 10–20% by enhancing soil fertility and productivity. It is advised that 5 tons of the digestate be applied per hectare on farms that do not use irrigation, and that 10 tons be applied per hectare on farms that do use irrigation [125].

The chemical and nutrient composition of the biodigester substrate and digestate is determined by the efficacy of the process and the feedstock composition. Very little preparation is needed before spraying the farm with the substrate or digestate or utilizing it as bedding for the animals [34]. Digester slurry is a 25% more effective fertilizer than just putting manure on the farm. This suggests that increasing agricultural yield by employing digestate in this way. In comparison to applying manure directly, biogas slurry makes it easier for plants to absorb the nutrients they need [61].

Slurry from a biodigester can be used as a fertilizer because of its high nutritional content and low environmental impact compared to synthetic alternatives. The digestate can be used as a foliar fertilizer on the farm if it is mixed with water and sprayed on the right leaves. Slurry is collected from the digester via a ditch as gas pressure increases after the biogas plant is refueled with fresh feedstock. Overflow and back pressure can be avoided in a biodigester system with regular ditch emptying by the operator [126].

Digestate consists of water and any unreacted feedstock or co-substrates. Digestate management is necessary for monitoring and controlling biogas production. A farm can either use the digestate as fertilizer or store it in a lagoon or storage tank. The wastewater from a biogas plant that has not been filtered for coarse fiber reduces bio digestion rates and gas production. Bedding and compost can be made from the screened-out material. After fibers have been extracted, the liquid organic residue is known as filtrate. Dry matter nitrogen, phosphorous, and potassium levels in manure filtration concentrate range from 3 to 4.5%. The crops can be sprayed with nutrient-rich sewage water.

The filtrate can be refined into a liquid concentrate and a solid product known as filter cake. This is especially important when bringing garbage from off-site locations into the digester's sphere of impact. Plants' basic nutrients, nitrogen, phosphorous, and potassium, may be found in abundance in these wastes. Thus, it is important to think about how this would affect the final disposal site's overall nutrient management strategy. Technology for recovering nutrients to control digestate nutrient levels [127].

As a result, biogas generation can be a more economical alternative to costly chemical fertilizers by boosting agricultural output and yields while saving money for farmers and contributing to long-term energy security. Farmers can earn revenue by selling the digestate or filtrate [30, 60, 65, 66].

4. Conclusion

Anaerobic digestion is a biological process where biodegradable matter is degraded to by the action of microorganisms in the absence of oxygen to produce biogas, which is a clean fuel from renewable feedstocks with potential use in heat and electricity generation. Factors affecting bio digestion include type of raw materials, temperature, pH, mixing, level preparation and pretreatment, residence time, redox effect, hydraulic pressure, etc. The efficiency of anaerobic digestion can be evaluated in terms of bio-methane potential, gross energy yields, process energy demand, net energy yield, and energy efficiency. Energy crops are the most used substrate for biogas production.

Concerns about greenhouse gas emissions and depletion of fossil fuels have increased demand for sustainable renewable sources. Biogas production is one of the innovative and viable for production of energy from biowastes and biomaterials, with an extra benefit of producing biofertilizer to support sustainable agriculture. Huge quantities of biodegradable waste are generated by humanity globally, which can pose a serious threat to human, animal health, and their environment. There are many wastes disposal and treatment methods employed to avoid environmental pollution. Among the many options, anaerobic digestion is a popular non-conventional energy source used to produce renewable energy from biodegradable biomass.

Biogas is a product of the anaerobic digestion process with many applications as in generation of renewable energy. The main component of biogas with energy value is methane, but has impurities like moisture, carbon dioxide, siloxanes, hydrogen sulfide, siloxanes, hydrocarbons, oxygen, ammonia, oxygen, carbon monoxide and nitrogen whose presence is undesirable as they reduce the calorific value of biogas and create operational problems in the energy systems. This necessitates biogas cleaning and application of multi-stage technologies to produce upgraded biogas called biomethane. Biomethane gas is a flexible and easy to store fuel with similar properties and applications to natural gas with no need to modify any equipment settings for natural gas devices and equipment.

The factors influencing the process and quality of biogas produces include the C/N ratio, the pH, dilution of feedstock, composition and nutritive value of feedstock, residence period, mixing and stirring, temperature, presence of toxicants, loading time and redox conditions. The main challenges facing biogas production and use include lack of technical training on biogas production and use, lack of subsidies and incentives. Therefore, besides easy access to appropriate and affordable biogas technologies, there is a need for developing a comprehensive policy on construction and operation of biodigesters as well utilization of biogas and sale of biogas energy products and services. Biogas education and exposure, availability of simple and affordable biodigesters and facilitating legal and policy environment are key tools needed to increase the adoption of biogas technology and its sustainable use.

Raw or cleaned biogas which is free from harmful contaminants has significant energy value and applications. The multifunctional uses of biogas produce include the production of biofuels and renewable energies, reduction in greenhouse gas emissions mainly methane and carbon dioxide, reduction in nuisances resulting from odors and flies associated with uncontrolled biodegrading of biomass, economic or financial savings for farmers, sustainable waste treatment and recycling of organic waste, minimization of water and air pollution, rural poverty alleviation, better of social conditions of rural or remote settlements like gender, balance as women who bear the responsibility of looking for firewood are relived of the struggle and conservation of forests and vegetation cover due reduced use of firewood and charcoal.

Acknowledgment

The authors are grateful to the Durban University of Technology for providing the resources to undertake this study.

Funding

The article publication charges were paid by the Durban University of Technology.

Conflict interest

The authors declare that they have no conflict of interest.

Availability of data

All the data used for the study have been declared.

Author details

Moses Jeremiah Barasa Kabeyi*, Oludolapo Akanni Olanrewaju and Joseph Akpan
Industrial Engineering Department, Durban University of Technology, Durban, South Africa

*Address all correspondence to: moseskabeyi@yahoo.com

References

[1] Ugwu SN, Enweremadu CC. Enhancement of biogas production process from biomass wastes using iron-based additives: Types, impacts, and implications. Energy Sources, Part A: Recovery, Utilization, and Environmental Effects. 2022;**44**(2):4458-4480. DOI: 10.1080/15567036.2020.1788675

[2] Huang X, Wang S, Shi Z, Fang L, Yin C. Challenges and strategies for biogas production in the circular agricultural waste utilization model: A case study in rural China. Energy. 2022;**241**:122889. DOI: 10.1016/j.energy.2021.122889

[3] Kabeyi MJB, Olanrewaju AO. Environmental impact of energy resources. In: Presented at the International Conference on Industrial Engineering and Operations Management. Manila, Phillipines: IEOM Society; 2023. p. 612 [Online]. Available from: https://ieomsociety.org/proceedings/2023manila/612.pdf

[4] Kabeyi M, Olanrewaju O. Slaughterhouse waste to energy in the energy transition with performance analysis and design of slaughterhouse biodigestor. Journal of Energy Management and Technology. 2022;**6**(3):188-208. DOI: 10.22109/jemt.2021.292954.1309

[5] Kabeyi MJB, Olanrewaju OA. Energy and environment: Oportunities and challenges for the energy transition. In: Presented at the Fifth European Conference on Industrial Engineering and Operations Management. Rome, Italy: IEOM Society; July 26-28, 2022. 2022. p. 427 [Online]. Available from: https://ieomsociety.org/proceedings/2022rome/427.pdf

[6] Kabeyi MJB, Olanrewaju OA. The role of electrification of transport in the energy transition. In: Presented at the Fifth European Conference on Industrial Engineering and Operations Management. Rome, Italy: IEOM Society; July 26-28, 2022. 2022. p. 426 [Online]. Available from: https://ieomsociety.org/proceedings/2022rome/426.pdf

[7] Sahota S et al. Review of trends in biogas upgradation technologies and future perspectives. Bioresource Technology Reports. 2018;**1**:79-88. DOI: 10.1016/j.biteb.2018.01.002

[8] Abanades S et al. A critical review of biogas production and usage with legislations framework across the globe. International Journal of Environmental Science and Technology. 2021;**19**:3377-3400. DOI: 10.1007/s13762-021-03301-6

[9] Kumar R, Jilte R, Ahmadi MH. Electricity alternative for e-rickshaws: An approach towards green city. International Journal of Intelligent Enterprise (IJIE). 2018;**5**(4):333-344. DOI: 10.1504/IJIE.2018.10016762

[10] Achinas S, Achinas V, Euverink GJW. A technological overview of biogas production from biowaste. Engineering. 2017;**3**(3):299-307. DOI: 10.1016/J.ENG.2017.03.002

[11] Pasternak G. Chapter 9 - Electrochemical approach for biogas upgrading. In: Aryal N, Ottosen LDM, Kofoed MVW, Pant D, editors. Emerging Technologies and Biological Systems for Biogas Upgrading. Massachusetts, United States: Academic Press; 2021. pp. 223-254

[12] Stevens DJ. Hot gas conditioning: Recent progress with larger-scale biomass gasification systems.

In: National Renewable Energy lab. (NREL), Colorado, USA; August 2001. [Online]. Available from: https://www.osti.gov/biblio/786288-hot-gas-conditioning-recent-progress-larger-scale-biomass-gasification-systems-update-summary-recent-progress

[13] Tagne RFT, Dong X, Anagho SG, Kaiser S, Ulgiati S. Technologies, challenges and perspectives of biogas production within an agricultural context. The case of China and Africa. Environment, Development and Sustainability. 2021;**23**(10):14799-14826. DOI: 10.1007/s10668-021-01272-9

[14] Herz G, Reichelt E, Jahn M. Design and evaluation of a Fischer-Tropsch process for the production of waxes from biogas. Energy. 2017;**132**:370-381. DOI: 10.1016/j.energy.2017.05.102

[15] Kabeyi MJB. Project and program evaluation consultancy with terms of reference, challenges, opportunities, and recommendations. International Journal of Project Management and Productivity Assessment (IJPMPA). 2020;**8**(2):47-68. DOI: 10.4018/IJPMPA.2020070103

[16] Sarkar S et al. Management of crop residues for improving input use efficiency and agricultural sustainability. Sustainability, Review. 2020;**12**(9808):1-24. DOI: 10.3390/su12239808

[17] Palit D et al. The trials and tribulations of the village energy security programme (VESP) in India. Energy Policy. 2013;**57**(2013):407-417. DOI: 10.1016/j.enpol.2013.02.006

[18] IEA. Renewables Information: Overview. International Energy Agency. Paris, France; 2021. Available from: https://www.iea.org/reports/renewables-information-overview. [Accessed: March 19, 2023]

[19] IEA. Global Energy Review 2020. International Energy Agency. Paris, France; 2020. Available from: https://www.iea.org/reports/global-energy-review-2020/renewables; [Accessed: Decemeber 16, 2020]

[20] Kabeyi MJB. Challenges of implementing thermal powerplant projects in Kenya, the case of Kipevu III 120MW power station, Mombasa Kenya. [Masters], Department of Education Management, University of Nairobi, Nairobi, 5866. 2012 [Online]. Available from: http://erepository.uonbi.ac.ke:8080/xmlui/handle/123456789/11023

[21] Kabeyi MJB, Olanrewaju OA. Technologies for biogas to electricity conversion. Energy Reports. 2022;**8**(Supplement 16):774-786. DOI: 10.1016/j.egyr.2022.11.007

[22] Bond T, Templeton MR. History and future of domestic biogas plants in the developing world. Energy for Sustainable Development. 2011;**15**(4):347-354. DOI: 10.1016/j.esd.2011.09.003

[23] Kabeyi MJB, Olanrewaju OA. Bagasse electricity potential of conventional sugarcane factories. Journal of Energy. 2023;**2023**:1-25. DOI: 10.1155/2023/5749122

[24] Machado PG, Teixeira ACR, Collaço FMA, Mouette D. Review of life cycle greenhouse gases, air pollutant emissions and costs of road medium and heavy-duty trucks. WIREs Energy and Environment. 2021;**10**(4):e395. DOI: 10.1002/wene.395

[25] Kabeyi MJB, Oludolapo AO. Preliminary Design of a Bagasse Based Firm Power Plant for a sugar factory. In: Presented at the South African Universities Power Engineering Conference (SAUPEC), North West

University, South Africa, 27-28 January 2021. 2021. p. 104 [Online]. Available from: https://ieeexplore.ieee.org/abstract/document/9377242

[26] Kabeyi MJB, Oludolapo AO. Central versus wellhead power plants in geothermal grid electricity generation. Energy, Sustainability and Society. 2021. Art no. ESSO-D-20-00011R4;**11**(7):1-23. DOI: 10.1186/s13705-021-00283-8

[27] Kabeyi MJB, Olanrewaju OA. Geothermal wellhead technology power plants in grid electricity generation: A review. Energy Strategy Reviews. 2022;**39**:100735. DOI: 10.1016/j.esr.2021.100735

[28] Kabeyi MJB. Geothermal electricity generation, challenges, opportunities and recommendations. International Journal of Advances in Scientific Research and Engineering (IJASRE). 2019;**5**(8):53-95. DOI: 10.31695/IJASRE.2019.33408

[29] Lebuhn M, Munk B, Effenberger M. Agricultural biogas production in Germany - from practice to microbiology basics. Energy, Sustainability and Society. 2014;**4**(1):10. DOI: 10.1186/2192-0567-4-10

[30] Kabeyi MJB, Oludolapo AO. Optimization of biogas production for optimal abattoir waste treatment with bio-Methanation as solution to Nairobi slaughterhouses waste disposal. In: Presented at the Second African International Conference on Industrial Engineering and Operations Management. Harare, Zimbabwe: (IEOM) Society; 5-7 December 2020. 2020. p. 083 [Online]. Available from: http://ieomsociety.org/harare2020/papers/83.pdf

[31] Kabeyi MJB, Oludolapo AO. Development of a cereal grain drying system using internal combustion engine waste heat. In: Presented at the 11th Annual International Conference on Industrial Engineering and Operations Management. Singapore: IEOM Society; March 7-11, 2021. 2021 [Online]. Available from: http://www.ieomsociety.org/singapore2021/papers/188.pdf

[32] Kabeyi MJB, Oludolapo AO. The potential of power generation from municipal solid waste. In: Presented at the Second African International Conference on Industrial Engineering and Operations Management. Harare, Zimbabwe: IEOM Society; 5-7 December 2020. 2020. p. 081 [Online]. Available from: http://ieomsociety.org/harare2020/papers/81.pdf

[33] Arogo J, Ignosh J, Bendfeldt E. Biomethane production technology. [Online]. Available from: https://www.pubs.ext.vt.edu/content/dam/pubs_ext_vt_edu/442/442-881/BSE-240.pdf

[34] Tanigawa S. Biogas: Converting waste to energy. 2017 [Online]. Available from: https://www.eesi.org/files/FactSheet_Biogas_2017.09.pdf

[35] Das S, Sherpa MT, Najar IN, Thakur N. Biomethane. Environmental Sustainability. 2020;**3**(4):453. DOI: 10.1007/978-981-13-8307-6_5

[36] Soeder DJ, Borglum SJ. 7 - energy economics. In: Soeder DJ, Borglum SJ, editors. The Fossil Fuel Revolution: Shale Gas and Tight Oil. Amsterdam, Netherlands: Elsevier; 2019. pp. 213-247

[37] Energypedia. Biogas plant used for power generation.png. Energypedia. Available from: https://energypedia.info/wiki/Electricity_Generation_from_Biogas [Accessed: December 16, 2020]

[38] Moriarty P, Honnery D. Energy accounting for a renewable energy

future. Energies. 2019;**12**:1-16. DOI: 10.3390/en12224280

[39] Kabeyi MJB, Olanrewaju OA. Performance analysis and evaluation of ethanol potential of Nzoia sugar company ltd. Energy Reports. 2022;**8**(16):787-799. DOI: 10.1016/j.egyr.2022.11.006

[40] Biogas for better life: An African initiative, a cost-benefit analysis of national and regional integrated biogas and sanitation programs in Sub-saharan Africa. 2007

[41] Kabeyi MJB, Oludolapo AO. Dual cycle cogeneration plant for an operating diesel powerplant. In: Presented at the 11th Annual International Conference on Industrial Engineering and Operations Management. Singapore: IEOM Society; March 7-11, 2021. 2021 [Online]. Available from: http://www.ieomsociety.org/singapore2021/papers/200.pdf

[42] Kabeyi MJB, Oludolapo AO. Fuel from plastic wastes for sustainable energy transition. In: Presented at the 11th Annual International Conference on Industrial Engineering and Operations Management. Singapore: (IEOM) Society; March 7-11, 2021. 2021 [Online]. Available from: http://www.ieomsociety.org/singapore2021/papers/199.pdf

[43] Chen B, Hayat T, Alsaedi A. Biogas Systems in China. Berlin: Springer; 2017. pp. 151. DOI: 10.1007/978-3-662-55498-2

[44] Kabeyi MJB, Olanrewaju OA. A techno-economic assessment of diesel to gas power plant conversion. In: Presented at the 12th Annual Istanbul International Conference on Industrial Engineering and Operations Management. Istanbul, Turkey: IEOM Society; March 7-10, 2022. 2022. p. 406 [Online]. Available from: https://ieomsociety.org/proceedings/2022istanbul/406.pdf

[45] Kariuki LN. Indoor carbon monoxide emission and particulates produced from combustion of carbon based fuels and their health implications on rural households of Manyatta division, Embu Districts, Kenya. [Msc thesis] Department of environmental Sciences, Kenyatta University. 2009

[46] Kabeyi MJB, Oludolapo AO. Performance analysis and modification of a slaughterhouse waste biogas plant for biogas and electricity generation. In: Presented at the 11th Annual International Conference on Industrial Engineering and Operations Management. Singapore: IEOM Society; March 7-11, 2021. 2021 [Online]. Available from: http://www.ieomsociety.org/singapore2021/papers/203.pdf

[47] Kabeyi MJB, Olenwaraju AO. Managing sustainability in electricity generation. In: Presented at the 2020 IEEE International Conference on Industrial Engineering and Engineering Management (IEEM), Singapore, 14-17 December 2020. 2020. IEEM20-P-0406

[48] Kabeyi MJB, Olanrewaju OA. Relationship between electricity consumption and economic development. In: Presented at the International Conference on Electrical, Computer and Energy Technologies (ICECET), Cape Town-South Africa, 9-10 December 2021. 2022 [Online]. DOI:10.1109/ICECET52533.2021.9698413

[49] Kabeyi MJB. Sustainability in the energy transition to renewable and low carbon grid electricity generation and supply. Frontiers in Energy Research. 2022;**9**:1-45. DOI: 10.3389/fenrg.2021.743114

[50] Kabeyi MJB, Oludolapo AO. Performance analysis of diesel engine Powerplant and development cogeneration plant for KIPEVU III

power plant. In: Presented at the Seventh International Conference on Advances on Clean Energy Research. Barcelona, Spain; 20-22 April 2022. pp. 800-814. DOI: 10.1016/j.egyr.2022.11.064

[51] Kabeyi MJB, Olanrewaju OA. Fuel from plastic wastes for sustainable energy transition. In: Presented at the 11th Annual International Conference on Industrial Engineering and Operations Management. Singapore: IEOM Society; March 7-11, 2021. 2021 [Online]. Available from: http://www.ieomsociety.org/singapore2021/papers/199.pdf

[52] Kabeyi MJB, Olanrewaju OA. Review and design overview of plastic waste-to-pyrolysis oil conversion with implications on the energy transition. Journal of Energy. 2023;**2023**:1821129. DOI: 10.1155/2023/1821129

[53] Kabeyi M, Olanrewaju O. Performance analysis of a sugarcane bagasse cogeneration power plant in grid electricity generation. In: Presented at the 11th Annual International Conference on Industrial Engineering and Operations Management. Singapore: IEOM Society; March 7-11, 2021. 2021 [Online]. Available from: http://www.ieomsociety.org/singapore2021/papers/201.pdf

[54] Kabeyi MJBK, Olanrewaju OA. The use of smart grids in the energy transition. In: 2022 30th Southern African Universities Power Engineering Conference (SAUPEC), 25-27 Jan. 2022. 2022. pp. 1-8. DOI: 10.1109/SAUPEC55179.2022.9730635 [Online]. Available from: https://ieeexplore.ieee.org/abstract/document/9730635

[55] Farooque M, Leo A, Rauseo A, Wang J-Y. Efficient and ultra-clean use of biogas in the fuel cell - the DFC experience. Energy, Sustainability and Society. 2015;**5**(1):11. DOI: 10.1186/s13705-015-0041-0

[56] Kabeyi MJB, Oludolapo AO. Performance analysis of an open cycle gas turbine power Plant in Grid Electricity Generation. In: Presented at the 2020 IEEE International Conference on Industrial Engineering and Engineering Management (IEEM), Singapore, Singapore, 14-17 December 2020. DOI: 10.1109/IEEM45057.2020.9309840

[57] Kabeyi MJB, Oludolapo AO. Performance analysis of diesel engine power plants for grid electricity supply. In: Presented at the 31ST Annual Southern African Institution for Industrial Engineering Conference, South Africa, 5th – 7th October 2020. 2020. p. 4423 [Online]. Available from: https://www.saiie.co.za/system/files/2021-11/SAIIE31%20Conference%20Proceedings.pdf

[58] Salomon KR, Silva Lora EE. Estimate of the electric energy generating potential for different sources of biogas in Brazil. Biomass and Bioenergy. 2009;**33**(9):1101-1107. DOI: 10.1016/j.biombioe.2009.03.001

[59] Kulichkova GI et al. Plant Feedstocks and their biogas production potentials. The Open Agriculture Journal. 2020;**2020**(14):219-234. DOI: 10.2174/1874331502014010219

[60] Kabeyi MJB, Olanrewaju OA. The potential of grid power generation from municipal solid waste for Nairobi city. In: Presented at the Second African International Conference on Industrial Engineering and Operations Management. Harare, Zimbabwe: IEOM Society; 5-7 December 2020. 2020. p. 081 [Online]. Available from: http://ieomsociety.org/harare2020/papers/81.pdf

[61] Energypedia. Biogas basics. Energypedia. Available from: https://energypedia.info/wiki/Biogas_Basics [Accessed: December 23, 2020]

[62] Ogunjuyigbe ASO, Ayodele TR, Alao MA. Electricity generation from municipal solid waste in some selected cities of Nigeria: An assessment of feasibility, potential and technologies. Renewable and Sustainable Energy Reviews. 2017;**80**:149-162. DOI: 10.1016/j.rser.2017.05.177

[63] Kabeyi MJB, Olenwaraju AO. Development of a biogas plant with electricity generation, heating and fertilizer recovery systems. In: Presented at the Second African International Conference on Industrial Engineering and Operations Management. Harare, Zimbabwe: IEOM Society; December 7-10, 2020. 2020 [Online]. Available from: http://www.ieomsociety.org/harare2020/papers/82.pdf

[64] Kabeyi MJB. Investigating the challenges of bagasse cogeneration in the kenyan sugar industry. International Journal of Engineering Sciences & Research Technology. 2020;**9**(5):7-64. DOI: 10.5281/zenodo.3828855

[65] Kabeyi M, Olanrewaju O. Slaughterhouse waste to energy in the energy transition with performance analysis and design of slaughterhouse biodigestor. Journal of Energy Management and Technology. 2022;**6**(3):188-208. DOI: 10.22109/jemt.2021.292954.1309

[66] Kabeyi MJB, Oludolapo AO. Development of a biogas plant with electricity generation, heating and fertilizer recovery systems. In: Presented at the Second African International Conference on Industrial Engineering and Operations Management, IEOM Society International. IEOM Society; 5-7 December 2020. 2020. p. 082 [Online]. Available from: http://ieomsociety.org/harare2020/papers/82.pdf

[67] Bond T, Templeton M. History and future of domestic biogas plants in the developing world. Energy for Sustainable Development. 2011;**15**(4):347-354

[68] Kabasci S. Optimizing utilization of biogas: Combined heat and power delivers greatest benefits. Power Engineering International; 2009;**17**(7). [Online]. Available from: https://www.powerengineeringint.com/decentralized-energy/on-site-renewables/optimizing-utilization-of-biogas-combined-heat-and-power-delivers-greatest-benefits/

[69] Barasa KMJ, Oludolapo AO. Biomethane production and applications. In: Sevcan A, editor. Anaerobic Digestion - Biotechnology for Reactor Performance and Environmental Sustainability. Rijeka: IntechOpen; 2023. p. 45. Ch. 9

[70] Hahn H. Economic and ecological evaluation of biogas plant configurations for a demand oriented biogas supply for flexible power generation. In: Nelles M, editor. 10 Rostock bioenergy forum Proceedings. 2016. p. 490

[71] Kabeyi M, Olanrewaju O. Diesel powerplants: Design and operation and performance enhancements. In: Presented at the Fifth European Conference on Industrial Engineering and Operations Management. Rome, Italy: IEOM Society; July 26-28, 2022. 2022. p. 425 [Online]. Available from: https://ieomsociety.org/proceedings/2022rome/425.pdf

[72] Jain S et al. Global Potential of Biogas. London, United Kingdom: World Biogas Association; 2019. p. 56 [Online]. Available from: https://www.worldbiogasassociation.org/wp-content/uploads/2019/09/

WBA-globalreport-56ppa4_digital-Sept-2019.pdf [Accessed: November 23, 2021]

[73] Amigun B, Blottnitz HV. Investigation of scale economies for African biogas installations (in 0196-8904). Energy Conversion and Managment. 2007;**48**(12):3090-3094. DOI: 10.1016/j.enconman.2007.05.009

[74] Scriba C. Biogas Plant Project at Folkecenter Controlled by C-Control/plus: Project: Control System for a Test and Exhibition Biogas Plant at Folkecenter. Sweden: Nordvestjysk Folkecenter for Vedvarende Energi; 1999. p. 1999

[75] Muvhiiwa R, Hildebrandt D, Chimwani N, Ngubevana L, Matambo T. The impact and challenges of sustainable biogas implementation: Moving towards a bio-based economy. Energy, Sustainability and Society. 2017;7(1):20. DOI: 10.1186/s13705-017-0122-3

[76] Patinvoh RJ, Taherzadeh MJ. Challenges of biogas implementation in developing countries. Current Opinion in Environmental Science & Health. 2019;**12**:30-37. DOI: 10.1016/j.coesh.2019.09.006

[77] Amigun B, Sigamoney R, Blottnitz HV. Commercialisation of biofuel industry in Africa: A review. Renewable and Sustainable Energy Reviews. 2006;**12**(2008):690-711. DOI: 10.1016/j.rser.2006.10.019

[78] Kabeyi MJB, Olanrewaju OA. Electricity and gas potential of abattoir waste. In: Presented at the 12th Annual Istanbul International Conference on Industrial Engineering and Operations Management. Istanbul, Turkey: IEOM Society; March 7-10, 2022. 2022. p. 403 [Online]. Available from: https://ieomsociety.org/proceedings/2022istanbul/403.pdf

[79] Kabeyi MJB, Olanrewaju OA. Cogeneration potential of an operating diesel engine power plant. Energy Reports. 2022;**8**(16):744-754. DOI: 10.1016/j.egyr.2022.10.447

[80] Kabeyi MJB, Olanrewaju OA. Biogas production and applications in the sustainable energy transition. Journal of Energy. 2022;**2022**(8750221):43. DOI: 10.1155/2022/8750221

[81] Cucchiell F, D'Adamo I, Gastaldi M. Biomethane: A renewable resource as vehicle fuel. Resources. 2017;**6**(58):1-13

[82] United Nation Economic and Social Commission for the Asia and the Pacific, Recent Developments in Biogas Technology for Poverty Reduction and Sustainable Development, United Nation Economic and Social Commission for the Asia and the Pacific, Beijing. 2007

[83] Saidmamatov O, Rudenko I, Baier U, Khodjaniyazov E. Challenges and solutions for biogas production from agriculture waste in the Aral Sea basin. 2021;**9**(2):1-15. DOI: 10.3390/pr9020199

[84] Khan MU et al. Current status of biogas upgrading for direct biomethane use: A review. Renewable and Sustainable Energy Reviews. 2021;**149**:111343. DOI: 10.1016/j.rser.2021.111343

[85] Kabeyi MJB, Olanweraju OA. Sustainability assessment for non-combustible renewable power generation. In: Presented at the 12th Annual Istanbul International Conference on Industrial Engineering and Operations Management. Istanbul, Turkey: IEOM Society; March 7-10, 2022. 2022. Paper 429 [Online]. Available

from: https://ieomsociety.org/proceedings/2022istanbul/429.pdf

[86] Kabeyi MJB. Potential and challenges of bagasse cogeneration in the Kenyan sugar industry. International Journal of Creative Research Thoughts. 2022. Art no. IJCRT_218740;**10**(4):379-526. DOI: 10.1729/Journal.30042

[87] Sonia D, Venko B. Biogas as a source of energy and chemicals. In: Venko B, editor. Biorefinery Concepts, Energy and Products. Rijeka: IntechOpen; 2020. Ch. 2

[88] Kabeyi M, Olanrewaju O. Development of a cereal grain drying system using internal combustion engine waste heat. In: Presented at the 11th Annual International Conference on Industrial Engineering and Operations Management. Singapore: IEOM Society; March 7-11, 2021. 2021 [Online]. Available from: http://www.ieomsociety.org/singapore2021/papers/188.pdf

[89] Kabeyi MJB, Oludolapo AO. Development of a biogas plant with electricity generation, heating and fertilizer recovery systems. In: Presented at the Second African International Conference on Industrial Engineering and Operations Management. Harare, Zimbabwe: IEOM Society; 5-7 December 2020. 2020. p. 082 [Online]. Available from: http://ieomsociety.org/harare2020/papers/82.pdf

[90] Kabeyi MJB, Olanrewaju O. Diesel to gas engine power plant conversion: A review and preliminary design for an operting power plant. Journal of Energy Management and Technology. 2023;7(2):103-115. DOI: 10.22109/jemt.2022.292982.1312

[91] Kabeyi MJB, Olanrewaju O. Optimum biodigestor design and operations. In: Presented at the Fifth European Conference on Industrial Engineering and Operations Management. Rome, Italy: IEOM Society; July 26-28, 2022. 2022. p. 424 [Online]. Available from: https://ieomsociety.org/proceedings/2022rome/424.pdf

[92] Kabeyi MJB, Olanrewaju OA. Performance analysis and electricity potential for Nzoia sugar factory. Energy Reports. 2022;**8**(16):755-764. DOI: 10.1016/j.egyr.2022.10.432

[93] Liu X, Yan Z, Yue ZB. 3.10 - Biogas. In: Moo-Young M, editor. Comprehensive Biotechnology. 2nd ed. Burlington: Academic Press; 2011. pp. 99-114

[94] Adnan AI, Ong MY, Nomanbhay S, Chew KW, Show PL. Technologies for biogas upgrading to biomethane: A review. Bioengineering. 2019;**6**(4):1-23. DOI: 10.3390/bioengineering6040092

[95] Kabeyi MJB, Oludolapo AO. Preliminary design of a bagasse based firm power plant for a sugar factory. In: Presented at the South African Universities Power Engineering Conference. SAIEE, North West University, South Africa, 27-28 January 2021. 2021. p. 104 [Online]. Available from: https://ieeexplore.ieee.org/abstract/document/9377242

[96] Beegle JR, Borole AP. Energy production from waste: Evaluation of anaerobic digestion and bioelectrochemical systems based on energy efficiency and economic factors. Renewable and Sustainable Energy Reviews. 2018;**96**:343-351. DOI: 10.1016/j.rser.2018.07.057

[97] Kour D et al. Technologies for Biofuel Production: Current Development, Challenges, and Future Prospects. Cham: Springer; 2019

[98] Kabeyi MJB, Olanrewaju O. Slaughterhouse waste to energy in the

energy transition with performance analysis and design of slaughterhouse biodigestor, (in en). Journal of Energy Management and Technology. 2022;**6**(3):176-196. DOI: 10.22109/jemt.2021.292954.1309

[99] Karlsson A, Björn A, Yekta SS, Svensson B. Improvement of the Biogas Production Process. Linköping, Sweden: Linköping University; 2014 [Online]. Available from: http://www.diva-portal.org/smash/get/diva2:776575/FULLTEXT01.pdf

[100] Karekezi SW, Kithyoma S. The Potential for Small and Medium Scale Renewable in Poverty Reduction in Africa: The Role of Renewable in Poverty Reduction Nairobi. 2007

[101] Bhajani SS, Pal SL. Review: Factors affecting biogas production. Ijraset Journal For Research in Applied Science and Engineering Technology. 2022. Art no. IJRASET40192;**10**(II):79-89. DOI: 10.22214/ijraset.2022.40192

[102] Kabeyi MJB, Olanrewaju OA. Electricity planning models. In: Characteristics and Applications Presented at the Seventh North American International Conference on Industrial Engineering and Operations Management. Orlando, Florida, USA: IEOM Society; June 12-14, 2022; 2022 [Online]. Available from: https://ieomsociety.org/proceedings/2022orlando/396.pdf

[103] Ghimire PC. Missions report on selection of biogas plant design and formulation of quality control framework framework and certification procedures for biogas constructors. SNV Netherlands Development Organisation 2009. p. 1-30. Available from: https://bibalex.org/baifa/en/resources/document/338060

[104] Kemausuor F, Adaramola MS, Morken J. A review of commercial biogas systems and lessons for Africa. Energies. 2018;**11**:1-21. DOI: 10.3390/en11112984

[105] Kabeyi MJB, Olanrewaju OA. Conversion of diesel and petrol engines to biogas engines as an energy transition strategy. In: Presented at the Fourth African International Conference on Industrial Engineering and Operations Management, Nsukka, Nigeria, April 5-7, 2022. 2022. p. 448 [Online]. Available from: https://ieomsociety.org/proceedings/2022nigeria/448.pdf

[106] Maier RM, Gentry TJ. Chapter 17 - microorganisms and organic pollutants. In: Pepper IL, Gerba CP, Gentry TJ, editors. Environmental Microbiology. 3rd ed. San Diego: Academic Press; 2015. pp. 377-413

[107] Bahaa S, Gerald K, Martin W, Johann F. Working fluids for low-temperature organic Rankine cycles. Energy. 2007;**32**(2007):1210-1221. DOI: 10.1016/j.energy.2006.07.001

[108] Mohan SV, Mohanakrishna G, Srikanth S. Chapter 22 - biohydrogen production from industrial effluents. In: Pandey A, Larroche C, Ricke SC, Dussap C-G, Gnansounou E, editors. Biofuels. Amsterdam: Academic Press; 2011. pp. 499-524

[109] Singh B et al. Enhancing efficiency of anaerobic digestion by optimization of mixing regimes using helical ribbon impeller. Fermentation. 2021;**7**(4):251 [Online]. Available from: https://www.mdpi.com/2311-5637/7/4/251

[110] Mei R, Narihiro T, Nobu MK, Kuroda K, Liu W-T. Evaluating digestion efficiency in full-scale anaerobic digesters by identifying active microbial populations through the lens of microbial activity. Scientific Reports. 2016;**6**(1):34090. DOI: 10.1038/srep34090

[111] Kabeyi MJB, Olanrewaju OA. Sugarcane molasses to energy conversion for sustainable production and energy transition. In: Presented at the 12th Annual Istanbul International Conference on Industrial Engineering and Operations Management. Istanbul, Turkey: IEOM Society; March 7-10, 2022. 2022. p. 405 [Online]. Available from: https://ieomsociety.org/proceedings/2022istanbul/405.pdf

[112] Han R et al. Review on heat-utilization processes and heat-exchange equipment in biogas engineering. Journal of Renewable and Sustainable Energy. 2016;**8**(032701). DOI: 10.1063/1.4949497

[113] Stucki M, Jungbluth N, Leuenberger M. Life Cycle Assessment of Biogas Production from Different Substrates, Final Report. Federal Department of Environment, Transport, Energy and Communications, Federal Office of Energy: Bern; 2011

[114] Arfan M, Wang Z, Soam S, Eriksson O. Biogas as a transport fuel—A system analysis of value chain development in a Swedish context. Sustainability. 2021;**13**(8):1-21. DOI: 10.3390/su13084560

[115] Calbry-Muzyka A, Madi H, Rüsch-Pfund F, Gandiglio M, Biollaz S. Biogas composition from agricultural sources and organic fraction of municipal solid waste. Renewable Energy. 2022;**181**:1000-1007. DOI: 10.1016/j.renene.2021.09.100

[116] IRENA. Biogas for road vehicles: Technology brief. Abu Dhabi. 2018 [Online]. Available from: https://www.irena.org/publications/2017/Mar/Biogas-for-road-vehicles-Technology-brief

[117] Lindorfer H, Frauz B. Chapter 6 - biogas biorefineries. In: Pandey A, Höfer R, Taherzadeh M, Nampoothiri KM, Larroche C, editors. Industrial Biorefineries & White Biotechnology. Amsterdam: Elsevier; 2015. pp. 271-294

[118] Sweet MS, Winandy JE. Influence of degree of polymerization of cellulose and hemicellulose on strength loss in fire-retardant-treated southern pine. Holzforschung. 1999;**53**(3):311-317. DOI: 10.1515/HF.1999.051

[119] Garrote G, Dominguez H, Parajo J. Hydrothermal processing of lignocellulosic materials. Holz als roh- und werkstoff. 1999;**57**(3):191-202

[120] Grabber JH. How do lignin composition, structure, and cross-linking affect degradability? A review of cell wall model studies. Crop Science, Conference Paper. 2005;**45**(3):820-831. DOI: 10.2135/cropsci2004.0191

[121] Balaman ŞY. Chapter 1 - introduction to biomass—Resources, production, harvesting, collection, and storage. In: Balaman ŞY, editor. Decision-Making for Biomass-Based Production Chains. Massachusetts, United States: Academic Press; 2019. pp. 1-23

[122] Kabeyi MJB, Olanrewaju OA. Performance analysis and modification of a slaughterhouse waste biogas plant for biogas and electricity generation. In: Presented at the 11 th Annual International Conference on Industrial Engineering and Operations Management. Singapore: IEOM Society; March 7-11, 2021. 2021 [Online]. Available from: http://www.ieomsociety.org/singapore2021/papers/203.pdf

[123] Karanja GM, Kiruiro EM. Biogas production, KARI Technical Note No. 10, Embu, Kenya. 2003

[124] Saravanan K, Suganthi R, Chinnaiyan VK. Review on biogas from

organic waste. In: 2014 International Conference on Green Computing Communication and Electrical Engineering (ICGCCEE). IEEE Xplore; 6-8 March 2014. 2014. pp. 1-4. DOI: 10.1109/ICGCCEE.2014.6922445

[125] Macharia D, Musungu WN. Institutional Biogas Installer Manual. Nairobi, Kenya: IT Power Eastern Africa; 2008. p. 29 [Online]. Available from: https://www.wisions.net/files/uploads/IT%20Power_Installer%20Biogas%20Manual_SEPS_SC048.pdf

[126] Energy Management Assistance Program (ESMAP). The Power of Dung: Lessons Learned from On-farm Biodigester Programs in Africa. Washington DC: World Bank; 2019 [Online]. Available from: http://documents1.worldbank.org/curated/en/468451557843529960/pdf/The-Power-of-Dung-Lessons-Learned-from-On-Farm-Biodigester-Programs-in-Africa.pdf

[127] AgSTAR. Anaerobic digester/Biogas system operator guidebook. in A guidebook for operating anaerobic digestion/biogas systems on farms in the United States. United States Environmental Agency (EPA), EPA 430-B-20-003. 2020 [Online]. Available from: https://www.epa.gov/sites/production/files/2020-11/documents/agstar-operator-guidebook.pdf

Chapter 9

Depolymerization of Native Lignin into Vanillin, Vanillic Acid, and Other Related Compounds via Alkaline Aerobic Oxidation: Reaction Mechanisms and Process Control Using Organic Cations

Yuki Hirano, Takashi Hosoya and Hisashi Miyafuji

Abstract

Vanillin is one of the platform chemicals in industry, which is used not only as a traditional fragrance but also as a raw material for medicines and polymer materials. Industrial vanillin production by alkaline aerobic oxidation of lignin is carried out with degradation of lignosulfonate at temperatures around 170°C under pressurized air in the presence of Cu^{2+} as a catalyst. However, this method has problems such as low vanillin yields and by production of sulfur-containing wastewater, and various studies have been conducted to solve them. Research on the mechanism of the vanillin formation reaction by aerobic oxidation of lignin and the search for a method to control this reaction has also been conducted. In this chapter, we review relevant studies from the above perspective, mainly those conducted by the authors' research group.

Keywords: tetrabutylammonium ion, cyclic polyether, bio-based aromatics, model compound, organic cation

1. Introduction

Currently, environmental issues, such as the depletion of fossil resources and global warming, are of worldwide concern, which has raised growing interest in renewable and environmentally friendly biomass resources as an alternative to fossil resources. Among the various types of biomasses, wood is a promising resource because of its huge stocks and inedibility. Wood is composed mainly of three components: cellulose, hemicellulose, and lignin. Among these, lignin constitutes 20–35% of the wood [1]. Because of its aromatic nature, lignin is expected as bio-based source of low molecular weight (MW) aromatics that are indispensable in the chemical industry [2–8].

The production of low MW aromatics from lignin requires conversion methods that facilitate effective cleavage of interunit linkages in lignin polymer. Among many methods currently suggested, alkaline aerobic oxidation is a promising way from the viewpoint that lignin is effectively degraded by nontoxic molecular oxygen [9–21]. Aerobic oxidation of lignosulfonate in waste liquor from sulfite pulping of softwood has been put into practical use as industrial vanillin (4-hydroxy-3-methoxybenzaldehyde) production [22–27].

Vanillin is a versatile compound that is useful as a source of medicines and polymer materials, as well as a traditional fragrance [24–30]. The vanillin production from lignosulfonate is usually carried out in aqueous alkali, that is, aqueous NaOH solution, under compressed air in the presence of a Cu^{2+}-based catalyst. This method was a major industrial vanillin production method until 1990s. However, the yield of vanillin is usually less than 10% based on the original sodium lignosulfonate [25, 31]. In addition to this relatively low yield of the target compound, high temperature and air pressure, typically 160–200°C/0.5–2.0 MPa, required for the process and management of wastewater containing sulfur compounds are drawbacks of this biomass-based vanillin production [25]. Recent vanillin production is, therefore, substituted almost with petrol-based ones, that is, formylation of guaiacol [27].

In light of recent environmental concerns, however, the production method of vanillin derived from biomass should be continuously studied and improved as it could be a breakthrough in establishing a method to produce low MW aromatic compounds from wood. To promote the adoption of the lignin-based vanillin production method in the chemical industry, it is necessary to address some of the shortcomings of this method described above. One solution to the sulfur-containing waste problem is to use native lignin in wood, as well as lignosulfonate as raw material for vanillin. Developing methods that provide the target product in good yield without the use of transition metals or high-temperature and high-pressure conditions would be another issue to tackle. Furthermore, mechanistic knowledge of alkaline aerobic oxidation as the foundation of the process development is essential for realizing these goals.

This chapter presents the results of fundamental and applied studies on vanillin production from lignin, conducted mainly in the authors' research group, from the above perspective. In addition to vanillin, other related oxidation products, for example, vanillic acid, are also produced in the aerobic oxidation of lignin. The formation of these compounds will also be briefly summarized.

2. Alkaline aerobic oxidation of lignin in the presence of organic cations

In vanillin production via alkaline aerobic oxidation of lignin, simple alkaline solution, such as aqueous NaOH solution, is generally employed as a reaction medium. Alternatively, our research group developed a new alkaline medium, that is, tetrabutylammonium hydroxide (TBAH) for lignin degradation. The chemical structure of tetrabutylammonium ion, Bu_4N^+, is presented in **Figure 1**. TBAH is a salt of Bu_4N^+ and OH^- and its aqueous solution exhibits strongly alkaline nature, as aqueous NaOH solution does. In this respect, TBAH is similar to NaOH, but a distinguishing characteristic of TBAH is that it possesses the bulky organic cation.

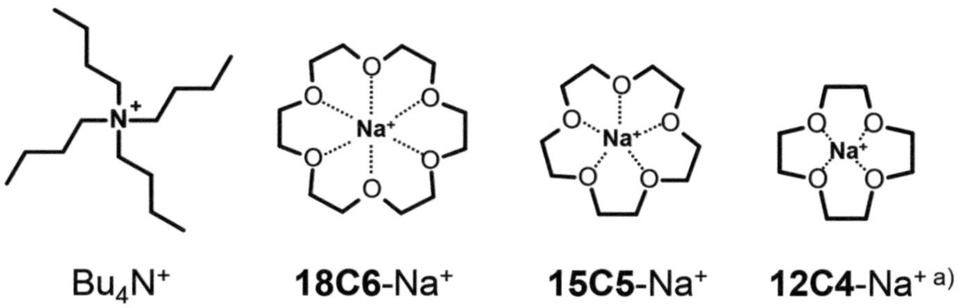

Figure 1.
Organic cations employed in our studies [32, 33]. [a] *In the case of* **12C4**-*Na^+, a complex cation in which Na^+ is bound by two molecules of* **12C4** *is also reported [34].*

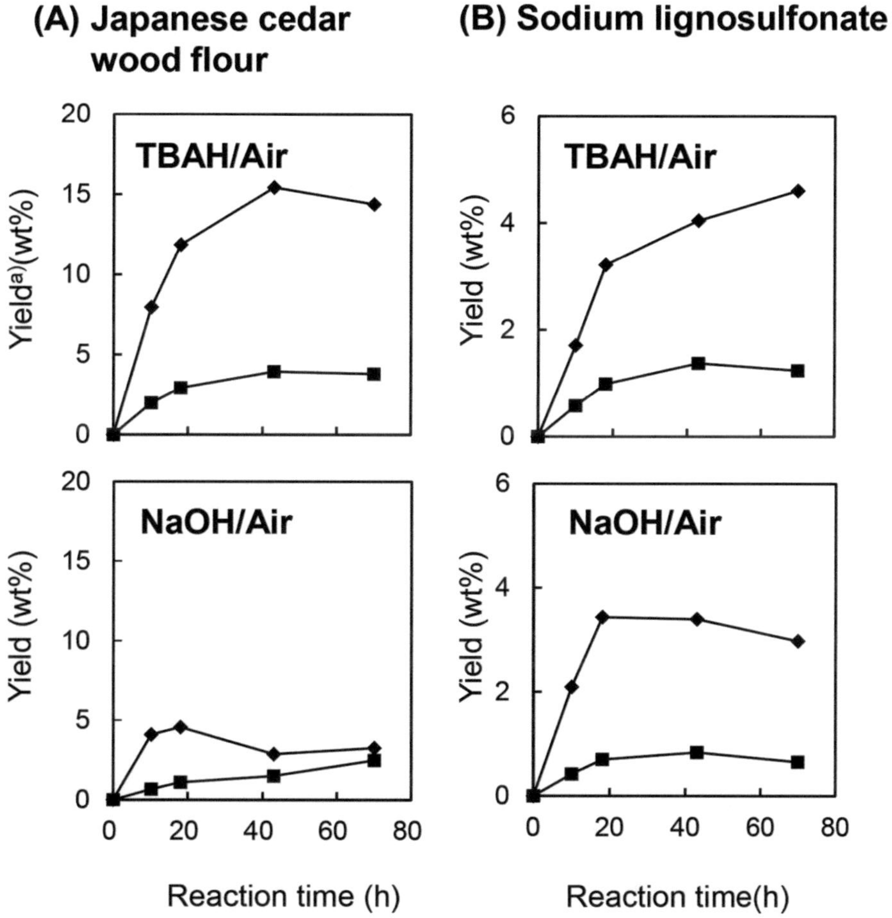

Figure 2.
Formation behavior of vanillin (♦) and vanillic acid (■) from Japanese cedar wood flour (A) and sodium lignosulfonate (B), during their alkaline aerobic oxidation under air in a sealed system (volume of the gas phase: 8 mL). Reaction conditions: sample 14 mg/ reaction solution (1.25 mol/L aqueous TBAH solution or aqueous NaOH solution 2.0 mL)/ 120°C. [a] *Yield from the wood flour is based on its Klason lignin amount [33].*

Figure 2 shows the changes with time in yields of vanillin and its related substance, vanillic acid (4-hydroxy-3-methoxybenzoic acid); when sodium lignosulfonate or Japanese cedar (*Cryptomeria japonica*) wood flour is oxidatively degraded at 120°C under air in the aqueous NaOH solution or aqueous TBAH solution with their concentration being 1.25 mol/L. In the vanillin production from the sodium lignosulfonate, the raw material for current lignin-based vanillin production [33], it has been indicated that the -SO_3^- moieties in the raw material play important roles in the vanillin formation [35, 36]. This point can be confirmed by comparing the yield of vanillin in the NaOH medium in **Figure 2** between wood flour and sodium lignosulfonate. In other words, in the case of the NaOH medium, the yields of vanillin and vanillic acid are comparable between sodium lignosulfonate and the wood flour (vanillin yield: 3–5 wt%; vanillic acid yield: 1–3 wt%). Considering significant degradation of the original lignin structures during its isolation process, it can be concluded that sodium lignosulfonate gives the target compounds in relatively high yields. This result would be attributed to the presence and the positive effect of the -SO_3^- moiety.

On the other hand, higher yields of vanillin and vanillic acid are obtained in the TBAH medium, regardless of the presence of the -SO_3^- moiety, as shown in **Figure 2**. Also, the wood flour exhibited much higher vanillin yield in TBAH than sodium lignosulfonate did in the same medium. This result is interesting from the perspective of reducing sulfur-containing waste during the vanillin production; efficient vanillin production from the wood flour, an originally sulfur-free material, was achieved in TBAH, which leads to the establishment of processes that are not bothered by the sulfur-containing waste. It should be also emphasized that the vanillin production was carried out without any transition metal additives. The employed reaction temperature, 120°C, was much lower than those generally adapted in current vanillin production processes.

The vanillin production shown in **Figure 2** takes a long reaction time, over 43 h, to reach the maximum product yield. The current process solves this problem by using pressurized air and high temperatures to promote the oxidation. **Table 1** summarizes the maximum yields of vanillin and vanillic acid and the time to reach them for the oxidation of Japanese cedar wood flour in various reaction media. It becomes clear that the addition of NaOH(s) to the reaction system and replacing the air in the reaction system with pure O_2 significantly shortens the reaction time and increases the yields of the products [16, 20]. Finally, the yields of vanillin and vanillic acid under optimum conditions are 23.2, 1.2 wt%, respectively, and the time to reach of this yield is 4.0 h. Thus, in the TBAH media, it is possible to improve the reaction efficiency at low temperature of 120°C without pressurizing the reaction system, which is advantageous compared to the current process with harsh reaction conditions. For vanillic acid, on the other hand, the compound tends to form in relatively low yields under the conditions in which vanillin production is facilitated. This may suggest that the production pathways of vanillin and vanillic acid are in trade-off relationship.

The above yields of vanillin and vanillic acid are remarkable in the history of lignin chemistry. One of the most important reactions in lignin chemistry is alkaline nitrobenzene oxidation (AN oxidation). This is an oxidative degradation of a lignin-containing sample such as wood and isolated lignin by nitrobenzene to give *p*-hydroxybenzaldehydes reflecting the substitution pattern of the lignin sample, mostly vanillin in the case of softwood lignin. Despite its more than 60-year history, AN oxidation has been one of the most efficient lignin degradation methods in use today and the product yield in the AN oxidation has been employed as a benchmark to measure the efficiency of other oxidation methods. **Table 1** shows that the yields of vanillin and vanillic acid in the TBAH+NaOH/O_2 system are comparable to those of

Medium/oxidant	$[OH^-]$ (mol/L)	Time (h)[b]	Temperature (°C)	Yield (wt %)[a]		
				Vanillin	Vanillic acid	Total
NaOH/air	4.0	96	120	6.2	2.4	8.6
NaOH/nitrobenzene[c]	2.0	2.5	170	27.2	1.2	28.4
TBAH/air	1.25	43	120	15.4	3.9	19.3
TBAH+NaOH[d]/air	3.75	20	120	18.9	2.2	21.1
TBAH+NaOH/oxygen[e]	3.75	4.0	120	23.2	1.2	24.4
NaOH +**18C6**[f]/air	4.0	96	120	16.1	2.3	18.4
NaOH +**15C5**[f]/air	4.0	96	120	15.2	2.3	17.5
NaOH +**12C4**[f]/air	4.0	72	120	16.1	2.3	18.4
NaOH +1,4-dioxane[f]/air	4.0	96	120	5.5	3.4	8.9
NaOH + **TEG**[f]/air	4.0	96	120	8.9	5.0	13.9
NaOH + **TRG**[f]/air	4.0	96	120	5.4	9.2	14.6

[a]*The yield is based on the Klason lignin content of the wood flour.*
[b]*The reaction time is the one in which the yield of vanillin becomes maximum.*
[c]*The alkaline nitrobenzene oxidation (AN oxidation) was carried out under the established conditions.*
[d]*In the TBAH+NaOH medium, $[Bu_4N^+]$ is set to 1.25 mol/L with $[OH^-]$ being adjusted to the presented value by adding NaOH(s) to the TBAH solution.*
[e]*"Oxygen" means the experiment is carried out under pure O_2.*
[f]*The reaction medium is prepared by adding the ether compounds to the NaOH solution, where* **18C6, 15C5,** *and* **12C4** *are the polycyclic ethers with different cavity sizes (see* **Figure 1***), and* **TEG** *and* **TRG** *are tetraglyme and triglyme (non-cyclic polyethers with different chain lengths), respectively.*

Table 1.
The yield of vanillin and vanillic acid from Japanese cedar wood flour after its oxidation in several oxidation systems [16, 32, 33].

AN oxidation. These results indicate that the TBAH-based alkaline aerobic oxidation is a highly efficient oxidation method and exploit the potential of the raw material to produce the target compounds.

The high yield of vanillin achieved in the aforementioned TBAH is clearly due to the presence of the bulky tetrabutylammonium ion, compared to Na^+. This led us to the idea of producing vanillin under the presence of various bulky organic cations. Namely, we added polycyclic ethers, that is, 18-Crown-6 (1,4,7,10,13,16-hexaoxacyclooctadecane, **18C6**), 15-Crown-5 (1,4,7,10,13-pentaoxa-cyclopentadecane, **15C5**), and 12-Crown-4 (1,4,7,10-tetraoxacyclododecane, **12C4**) to an aqueous NaOH solution and studied the alkaline aerobic oxidation behaviors of Japanese cedar wood flour in these reaction media. These crown ethers capture Na^+ in the aqueous NaOH solution, forming three types of complex cations **18C6**-Na^+, **15C5**-Na^+, and **12C4**-Na^+ as shown in **Figure 1** [37]. It is expected that these organic cations would exhibit enhancing effects on the vanillin production, which is similar to those observed for Bu_4N^+.

The aerobic oxidation of Japanese cedar wood flour was conducted in a 4.0 mol/L aqueous NaOH solution in the presence of 2.0 mol/L of the three crown ethers. As shown in **Table 1**, the vanillin yield under the three crown ethers (**18C6, 15C5, 12C4**) is 15.2–16.1 wt%, which is clearly higher than that in the pure NaOH system. In contrast, in the aerobic oxidation using equimolar amounts of tetraglyme (**TEG**) and triglyme (**TRG**) and non-cyclic analogs of **15C5** and **12C4** respectively, the vanillin yields are 8.9 and 5.4 wt%, nearly the same as that obtained in the pure NaOH system. This can be explained by the fact that cyclic polyethers generally exhibit much stronger complexing ability toward metal cations than their corresponding non-cyclic polyethers [38], suggesting that the increased vanillin yield due to the addition of the crown ethers results from the formation of the complex cations in **Figure 1**. Moreover, the increase in vanillin yield by cyclic polyethers seems to depend highly on their Na^+ accommodation ability. For instance, the addition of 1,4-dioxane, which has a considerably smaller ring size than the three crown ethers, results in a vanillin yield of 5.5 wt% (**Table 1**). This is likely because the cavity size of 1,4-dioxane is too small to accommodate Na^+.

For vanillic acid, another major product, the yield from Japanese cedar wood flour in the presence of complex cations is 2.3 wt%, regardless of the types of the crown ethers. This value is nearly identical to that obtained in the absence of the complex cations (2.4 wt%). This phenomenon of minimal influence of the complex cations on the production of vanillic acid is also observed in the aforementioned aerobic oxidation of Japanese cedar wood flour in the presence of Bu_4N^+. Thus, it can be stated that the complex cations formed from Na^+ and the crown ethers and Bu_4N^+ exert quite similar effects, significantly increasing the yield of vanillin while having little impact on the formation of vanillic acid. In the following section, we will elaborate on this issue from the perspective of reaction mechanisms. Moreover, the addition of **TEG** and **TRG**, which are non-cyclic polyethers, does not substantially increase the vanillin yield, but it increases the yield of vanillic acid. For instance, the addition of **TRG** significantly increases the yield of vanillic acid from 2.4 to 9.2 wt% (**Table 1**). Although there are few clues to explain these results rationally at this point, they represent intriguing phenomena when considering the potential for further reaction control by complex cations.

3. Mechanism of vanillin formation from lignin and the roles of organic cations

Bu_4N^+ and the complex cations between Na^+ and the polycyclic ethers are found to facilitate the vanillin production by alkaline aerobic oxidation of lignin. However, the mechanisms underlying the improved vanillin yield caused by the presence of the organic cations are not clear. We, thus, investigated the role of organic cations (Bu_4N^+ and the crown ether-based complex cations) in the vanillin formation mechanisms, by analyzing the behaviors of lignin model compounds **LM1** and **LM2** shown in **Figure 3** under the alkaline aerobic oxidation conditions. These models carry a β-O-4 linkage, which is the most abundant linkage in lignin, but are different in that **LM1** has the phenolic OH group, while **LM2** does not.

LM1 with the phenolic OH group was examined as a model for the phenolic end group of lignin [16]. As shown in **Figure 3**, an enol ether intermediate **EE** is formed in the initial stage of the oxidation via the formation of a quinone methide **QM** structure from **LM1** and the following elimination of C_γ position as HCHO (Pathway A1 and A2). The second step involves the oxidation of the $C_\alpha = C_\beta$ moiety of **EE** finally leading to vanillin (Pathway A3). The mechanisms in the oxidation of **EE** are not clear at the moment, but vanillin formation from **EE** via a dioxetane intermediate was suggested in several studies on O_2 pulp bleaching conditions [40, 41]. For the effect of Bu_4N^+ in the TBAH medium, Bu_4N^+ stabilizes **EE** probably by coordinating with the side chain of **EE**, which results in slower vanillin formation from **LM1** in TBAH

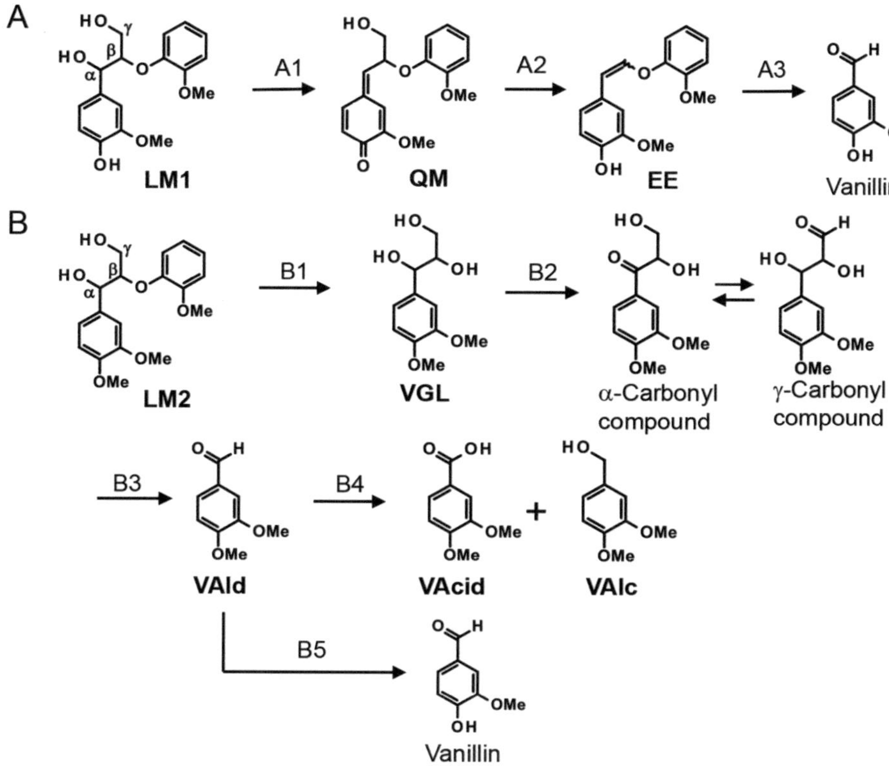

Figure 3.
Reaction pathways involved in alkaline aerobic oxidation of **LM1** *(A) and* **LM2** *(B) [16, 33, 39].*

compared to that in NaOH. It is, therefore, likely that Bu_4N^+ does not have positive effects on the vanillin production from the phenolic end of lignin.

Experiments using **LM2**, a model for the middle β-O-4 phenylpropane unit of lignin, have suggested the following reaction mechanisms [16, 39]. First, the cleavage of the β-ether of **LM2** via the neighboring group participation of C_α-OH occurs to give a veratryl glycerol intermediate **VGL** (Pathway B1). This intermediate is then oxidized to an α-carbonyl compound, which is in equilibrium with the corresponding γ-carbonyl compound through the keto-enol tautomerization (Pathway B2). A retro-aldol reaction proceeds from the γ-carbonyl compound, leading to the formation of veratraldehyde (**VAld**) via Pathway B3. On the other hand, another retro-aldol pathway starting from the α-carbonyl is less significant [39]. There are two possible fates of **VAld** with one being disproportionation to produce veratric acid (**VAcid**) and veratryl alcohol (**VAlc**) (Pathway B4) and the other being conversion of the 4-OMe group to OH to produce vanillin (Pathway B5).

Comparison of the degradation behaviors of **LM2** in the TBAH and NaOH system revealed that TBAH promoted the cleavage of the β-ether in the initial stage of the reaction (Pathway B1) and increased the selectivity of the vanillin production from **VAld** in the final stage (Pathway B5). The former promotion of the β-ether cleavage can be explained by the fact that the OH^- activity in the aqueous TBAH solution is higher than that of the aqueous NaOH solution, even when their OH^- concentration is the same. On the other hand, the latter promotion of vanillin formation from **VAld** is due to the interaction of Bu_4N^+ with **VAld** in the aqueous TBAH solution, as indicated by the ^1H NMR analysis of **VAld** in the presence of Bu_4N^+ salt [16]. In other words, **VAld** is in the cage made of Bu_4N^+ in the aqueous TBAH solution, which inhibits the progress of the disproportionation (Pathway B4). This results in relative increase in the selectivity of the vanillin forming pathway B5.

The influence of the complex cations on the reactivity of **LM2** is confirmed by evaluating the additive effect of crown ethers (**18C6**, **15C5**, and **12C4**) on the reactivity of **LM2** in the aqueous NaOH solution. When the complex cations are present in the reaction system, a decrease in the yields of **VAcid** and **VAlc** is observed, along with an increase in vanillin yield, similar to the previously confirmed Bu_4N^+ system. Thus, it is inferred that both complex cations and Bu_4N^+ essentially increase the vanillin yield through the same mechanism, namely, the suppression of the disproportionation pathway of **VAld** due to the cage effect. Interestingly, in the case of the non-cyclic ethers such as **TEG** and **TRG**, and 1,4-dioxane, a cyclic ether with a much smaller ring size, no significant increase in vanillin yield is observed in the model experiments using **LM2**. This suggests that the formation of complex cations shown in **Figure 1**, and the suppression of the disproportionation pathway (Pathway B4) are closely related. In addition to the similar effects of Bu_4N^+ and the complex cations as mentioned above, the complex cations exhibit several specific effects that are not observed with Bu_4N^+ [32]. For instance, in the case of the complex cation systems, the inhibition of the oxidative degradation of the **VGL** side chain to the aldehyde group is observed probably by the stabilization of **VGL** by its interaction with the complex cation.

The above phenomena observed for **LM1** and **LM2** suggest that vanillin formation from middle phenylpropane units in lignin is promoted in the presence of the organic cations, with the reaction at the phenolic end being rather suppressed. In other words, the formation of vanillin from the middle unit plays an important role in the formation of vanillin under the alkaline aerobic oxidation conditions. **Figure 4** shows a possible pathway for vanillin formation from the middle unit of lignin based on the study of the reaction mechanisms of **LM2**. First, the β-ether of the middle β-O-4 unit undergoes alkaline hydrolysis to form the glycerol end (pathway I). The glycerol end is oxidized to a C_α-aldehyde end (pathway II), and finally, vanillin is released from this end (pathway III). In an actual lignin molecule, the middle units are considered to be more abundant than the phenolic ends. It is, thus, likely that the suppressing effect of Bu_4N^+ on the vanillin production from the phenolic end is canceled out, and the overall vanillin yield was increased in the organic cation-containing media.

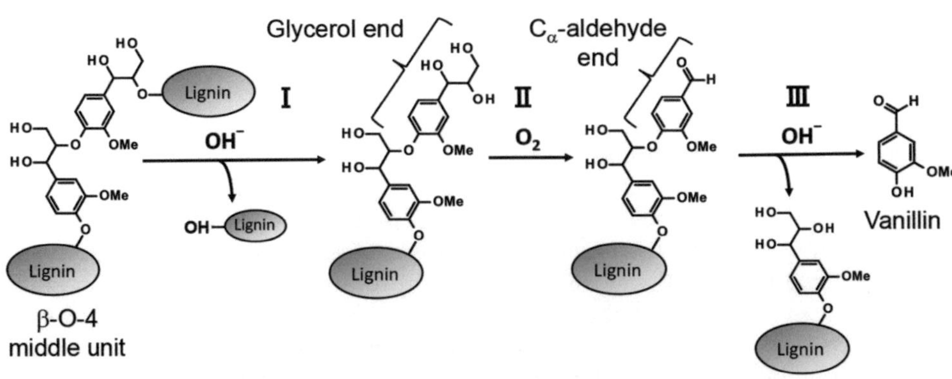

Figure 4.
Vanillin formation from a β-O-4 middle unit of lignin polymer [16].

The model compounds **LM1** and **LM2** discussed so far are designed to mimic the β-O-4 structure in lignin. However, the compound has a much more simplified structure than the actual lignin. Therefore, when applying the reaction pathway based on the degradation behavior of **LM2** (**Figure 3**) to actual lignin molecules, careful consideration is required. When comparing the vanillin production behavior from **LM2** and wood flour, a certain correspondence can be recognized between the oxidation behaviors of actual lignin and the model compounds in that the increase in vanillin yield from the wood flour observed in the presence of organic cations is reproduced even in the case of the model compound. Thus, it is likely that the suppression of the disproportionation by the cage effect of the organic cations plays a certain role also in the vanillin production from actual lignin. In the case of actual lignin, however, **VAld**, which was produced from **LM2**, never forms, and instead the C_α-aldehyde end group shown in **Figure 4** is formed as a vanillin precursor. The rates of vanillin elimination from this C_α-aldehyde end and the conversion reaction of **VAld** to vanillin (conversion of the 4-position methoxy group to a hydroxy group) are expected to differ significantly, which may affect the degree of suppression of disproportionation by the cage effect. Also, in the C_α-aldehyde end linked to the actual lignin side chain, more complex side reactions not observed with **VAld** might proceed. According to these ideas, future studies using model compounds that more accurately mimic the actual C_α-aldehyde end are needed to more precisely elucidate the mechanism of increased vanillin yield by the cage effect.

4. Conclusions

This chapter presented high-yield vanillin production from lignin (mainly native softwood lignin) by the alkaline aerobic oxidation in the presence of organic cations. When conducting alkaline aerobic oxidation of lignin in the presence of bulky organic cations such as Bu_4N^+ and the complex cations formed between the crown ethers and Na^+, a significant increase in the yield of vanillin from the lignin samples is observed. In this organic cation-containing system, the alkaline aerobic oxidation allows for the efficient production of vanillin at a temperature of 120°C, employing native lignin in the wood flour as the starting material. This vanillin production process is more advantageous than the current process using lignosulfonate in that it can achieve a high vanillin yield comparable to that of AN oxidation and does not produce sulfur-containing effluent.

One of the vanillin production pathways in the alkaline aerobic oxidation of lignin involves the elimination of vanillin from the C_α-aldehyde structure, which is produced by the oxidative degradation of non-phenolic intermediate units in lignin. It has been found that the aforementioned organic cations improve the selectivity of this vanillin elimination reaction. Research on such reaction mechanisms and further mechanistic insights into the function of the organic cations are expected to lead to the development of more rational reaction control methods to improve the efficiency of the vanillin production processes.

Acknowledgements

This work was supported by a Grant-in-Aid for Scientific Research (C) (No. 21K05715 and 23K05343) from the Japan Society for the Promotion of Science.

Author details

Yuki Hirano, Takashi Hosoya and Hisashi Miyafuji*
Graduate School of Life and Environmental Sciences, Kyoto Prefecture University, Kyoto, Japan

*Address all correspondence to: miyafuji@kpu.ac.jp

References

[1] Ralph J, Brunow G, Boerjan W. Lignins. Encyclopedia of Life Science. 2007:1-10

[2] Liu X, Bouxin FP, Fan J, Budarin VL, Hu C, Clark JH. Recent advances in the catalytic depolymerization of lignin towards phenolic chemicals: A review. ChemSusChem. 2020;**13**:4296-4317

[3] Jing Y, Dong L, Guo Y, Liu X, Wang Y. Chemicals from lignin: A review of catalytic conversion involving hydrogen. ChemSusChem. 2020;**13**:4181-4198

[4] Sun Z, Fridrich B, de Santi A, Elangovan S, Barta K. Bright side of lignin depolymerization: Toward new platform chemicals. Chemical Reviews. 2018;**118**:614-678

[5] Schutyser W, Renders T, Van den Bosch S, Koelewijn SF, Beckham GT, Sels BF. Chemicals from lignin: An interplay of lignocellulose fractionation, depolymerisation, and upgrading. Chemical Society Reviews. 2018;**47**:852-908

[6] Tarabanko VE, Tarabanko N. Catalytic oxidation of lignins into the aromatic aldehydes: General process trends and development prospects. IJMS. 2017;**18**:2421

[7] Gillet S, Aguedo M, Petitjean L, Morais ARC, da Costa Lopes AM, Łukasik RM, et al. Lignin transformations for high value applications: Towards targeted modifications using green chemistry. Green Chemistry. 2017;**19**:4200-4233

[8] Rinaldi R, Jastrzebski R, Clough MT, Ralph J, Kennema M, Bruijnincx PCA, et al. Paving the way for lignin valorisation: Recent advances in bioengineering, biorefining and catalysis. Angewandte Chemie (International Ed. in English). 2016;**55**:8164-8215

[9] More A, Elder T, Jiang Z. Towards a new understanding of the retro-aldol reaction for oxidative conversion of lignin to aromatic aldehydes and acids. International Journal of Biological Macromolecules. 2021;**183**:1505-1513

[10] Zhu Y, Liao Y, Lv W, Liu J, Song X, Chen L, et al. Complementing vanillin and cellulose production by oxidation of lignocellulose with stirring control. ACS Sustainable Chemistry & Engineering. 2020;**8**:2361-2374

[11] Wang J-X, Asano S, Kudo S, Hayashi J. Deep delignification of Woody biomass by repeated mild alkaline treatments with pressurized O2. ACS Omega. 2020;**5**:29168-29176

[12] Tarabanko VE, Kaygorodov KL, Vigul DO, Tarabanko N, Chelbina YV, Smirnova MA. Influence of acid prehydrolysis on the process of wood oxidation into vanillin and pulp. Journal of Wood Chemistry and Technology. 2020;**40**:421-433

[13] Rawat S, Gupta P, Singh B, Bhaskar T, Natte K, Narani A. Molybdenum-catalyzed oxidative depolymerization of alkali lignin: Selective production of vanillin. Applied Catalysis A: General. 2020;**598**:117567

[14] Paananen H, Eronen E, Mäkinen M, Jänis J, Suvanto M, Pakkanen TT. Base-catalyzed oxidative depolymerization of softwood Kraft lignin. Industrial Crops and Products. 2020;**152**:112473

[15] Jeon W, Choi I-H, Park J-Y, Lee J-S, Hwang K-R. Alkaline wet oxidation of

lignin over Cu-Mn mixed oxide catalysts for production of vanillin. Catalysis Today. 2020;**352**:95-103

[16] Hosoya T, Yamamoto K, Miyafuji H, Yamada T. Selective production of bio-based aromatics by aerobic oxidation of native softwood lignin in tetrabutylammonium hydroxide. RSC Advances. 2020;**10**:19199-19210

[17] Cabral Almada C, Kazachenko A, Fongarland P, Da Silva Perez D, Kuznetsov BN, Djakovitch L. Oxidative depolymerization of lignins for producing aromatics: Variation of botanical origin and extraction methods. Biomass Conversion and Biorefinery. 2020;**12**:3795-3808

[18] Abdelaziz OY, Ravi K, Mittermeier F, Meier S, Riisager A, Lidén G, et al. Oxidative depolymerization of Kraft lignin for microbial conversion. ACS Sustainable Chemistry & Engineering. 2019;7:11640-11652

[19] Schutyser W, Kruger JS, Robinson AM, Katahira R, Brandner DG, Cleveland NS, et al. Revisiting alkaline aerobic lignin oxidation. Green Chemistry. 2018;**20**:3828-3844

[20] Maeda M, Hosoya T, Yoshioka K, Miyafuji H, Ohno H, Yamada T. Vanillin production from native softwood lignin in the presence of tetrabutylammonium ion. Journal of Wood Science. 2018;**64**:810-815

[21] Hosoya T, Okamoto D, Miyafuji H, Yamada T. Production of vanillin and vanillic acid by aerobic oxidation of polyethylene glycol (PEG)-modified glycol lignin in tetrabutylammonium hydroxide. Lignin. 2021;**2**:9-18

[22] Pacek AW, Ding P, Garrett M, Sheldrake G, Nienow AW. Catalytic conversion of sodium lignosulfonate to vanillin: Engineering aspects. Part 1. Effects of processing conditions on vanillin yield and selectivity. Industrial and Engineering Chemistry Research. 2013;**52**:8361-8372

[23] Ding P, Garrett M, Loe Ø, Nienow AW, Pacek AW. Generation of hydrogen gas during the catalytic oxidation of sodium lignosulfonate to vanillin: Initial results. Industrial and Engineering Chemistry Research. 2012;**51**:184-188

[24] Fache M, Boutevin B, Caillol S. Vanillin production from lignin and its use as a renewable chemical. ACS Sustainable Chemistry & Engineering. 2016;**4**:35-46

[25] Hocking MB. Vanillin: Synthetic flavoring from spent sulfite liquor. Journal of Chemical Education. 1997;**74**:1055

[26] Vu TT, Lim Y-I, Song D, Hwang K-R, Kim D-K. Economic analysis of vanillin production from kraft lignin using alkaline oxidation and regeneration. Biomass Conversion and Biorefinery. 2023;**13**:1819-1829

[27] Havkin-Frenkel DV. In Kirk-Othmer Encyclopedia of Chemical Technology. New York: Wiley; 2018. pp. 1-12

[28] Zhang H, Yong X, Zhou J, Deng J, Wu Y. Biomass vanillin-derived polymeric microspheres containing functional aldehyde groups: Preparation, characterization, and application as adsorbent. ACS Applied Materials & Interfaces. 2016;**8**:2753-2763

[29] Llevot A, Grau E, Carlotti S, Grelier S, Cramail H. From lignin-derived aromatic compounds to novel biobased polymers. Macromolecular Rapid Communications. 2016;**37**:9-28

[30] Bjørsvik H-R, Liguori L. Organic processes to pharmaceutical chemicals based on fine chemicals from Lignosulfonates. Organic Process Research and Development. 2002;**6**:279-290

[31] Vidal JP. Vanillin. In: Kirk-Othmer Encyclopedia of Chemical Technology. Hoboken: Wiley; 2006. DOI: doi.org/10.1002/0471238961

[32] Hosoya T, Kawase K, Hirano Y, Ikeuchi M, Miyafuji H. Alkaline aerobic oxidation of native softwood lignin in the presence of Na^+-cyclic polyether complexes. Journal of Wood Chemistry and Technology. 2022;**42**(1):1-14

[33] Yamamoto K, Hosoya T, Yoshioka K, Miyafuji H, Ohno H, Yamada T. Tetrabutylammonium hydroxide 30-hydrate as novel reaction medium for lignin conversion. ACS Sustainable Chemistry & Engineering. 2017;**5**:10111-10115

[34] Buschmann H-J. Stability constants and thermodynamic data for complexes of 12-Crown-4 with alkali metal and alkaline-earth cations in methanol solutions. Journal of Solution Chemistry. 1987;**16**:181-190

[35] Tomlinson GH, Hibbert H. Studies on lignin and related compounds. XXIV. The formation of vanillin from waste sulfite liquor. Journal of the American Chemical Society. 1936;**58**:345-348

[36] Tomlinson GH, Hibbert H. Studies on lignin and related compounds. XXV. Mechanism of vanillin formation from spruce lignin sulfonic acids in relation to lignin structure. Journal of the American Chemical Society. 1936;**58**:348-353

[37] Izatt RM, Pawlak K, Bradshaw JS, Bruening RL. Thermodynamic and kinetic data for macrocycle interactions with cations and anions. Chemical Reviews. 1991;**91**:1721-2085

[38] Tuemmler B, Maass G, Voegtle F, Sieger H, Heimann U, Weber E. Open-Chain Polyethers. Influence of aromatic donor end groups on thermodynamics and kinetics of alkali metal ion complex formation. Journal of the American Chemical Society. 1979;**101**:2588-2598

[39] Hirano Y, Izawa A, Hosoya T, Miyafuji H. Degradation mechanism of lignin model compound during alkaline aerobic oxidation: Formation of vanillin precursor from β-O-4 middle unit of softwood lignin. Reaction Chemistry & Engineering. 2022;**7**:1603-1616

[40] Imai A, Yokoyama T, Matsumoto Y, Meshitsuka G. Significant lability of Guaiacylglycerol β-Phenacyl ether under alkaline conditions. Journal of Agricultural and Food Chemistry. 2007;**55**:9043-9046

[41] Gierer J. Chemistry of delignification part 2: Reactions of lignins during bleaching. Wood Science and Technology. 1986;**20**:1-33

Chapter 10

Lactic Acid Production from Lignocellulosic Biomass

Asfaw Gezae Daful, Marie Loridon and Meegalla R. Chandraratne

Abstract

This chapter presents bio-based lactic acid production process from lignocellulosic biomass. Bio-based chemicals can replace the chemicals that we usually get from petroleum-based resources, and they are used to produce cleaners, solvents, adhesives, paints, plastics, textiles, and many other products. Lactic acid is one of such candidates of bio-based chemicals with important applications in various industrial sectors such as the chemical, pharmaceutical, food, and cosmetics industries, where its demand is steadily increasing. It is also an essential building block for numerous commodity and intermediate-biobased chemicals making it as a suitable alternative to their fossil-derived counterparts. The bioconversion process of transforming lignocellulosic biomass into lactic acid consists of four primary stages. Initially, pretreatment is performed to enable the utilization of all C_5 and C_6 sugars by the selected microorganism. These sugars are then hydrolyzed and fermented by a suitable microorganism to produce either L- or D-lactic acid, depending on the desired stereochemistry. Finally, the lactic acid is separated and purified from the fermentation broth to obtain a purified product. The promising method for the industrial production of bio-based lactic acid will be of continuous simultaneous saccharification and fermentation in a gypsum-free process using $Mg(OH)_2$ as neutralizer, followed by reactive distillation for purified lactic acid production. The *cradle-to-gate* life cycle assessment model for the biobased lactic acid production process indicated that the about 80–99% of the environmental burdens of most of the environmental impact categories can be reduced compared with its equivalent fossil-based lactic acid, making biobased lactic acid environmentally superior to the fossil-based lactic acid.

Keywords: lactic acid, biomass, bioconversion, life cycle assessment (LCA), environmental impacts

1. Introduction

Lactic acid (LA), the simplest hydroxy carboxylic acid with IUPAC name of 2-hydroxypropionic acid and a molecular formula $CH_3CH(OH)COOH$, is a bulk chemical with two optically active *enantiomers* (D(−) and L(+)), depicted in **Figure 1**, which has wide industrial applications in the food, chemical, pharmaceutical and health care industries [1–4]. The L(+) isomer is preferentially used for food and pharmaceutical

 IntechOpen

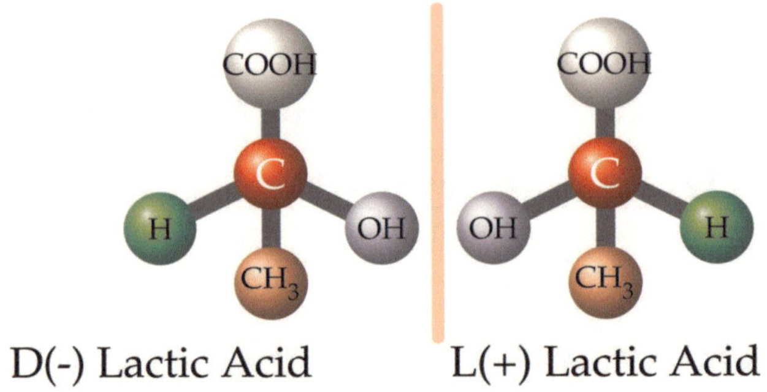

Figure 1.
Two stereoisomers (D(−) and L(+)) of lactic acid.

applications, being the only lactic acid isomer produced in the human body [5]. Lactic acid can be used as building block for a multiple of commodity and intermediate chemicals [3, 6, 7], such as acrylic acid, lactate esters, 1,2-propanediol, pyruvic acid, acetaldehyde, 2,3-pentanedione, ethanedioic acid, and polylactic acid (PLA), which is a biodegradable plastic suitable for packagings. Different applications of lactic acid can be either as final product or as feedstock for other processes like in food and cosmetics industries. PLA may be a suitable alternative to oil-derived plastics and has already a global market demand of approximately 10^5 metric tonnes per annum and is anticipated to grow by at least 28% per annum until 2025 [1, 8]. Low-molecular-weight PLA bioplastics can replace polymers such as polyethylene and polyethylene terephthalate, while high-molecular-weight PLA can replace polystyrene.

Lactic acid can be produced either by anaerobic fermentation [3, 9–11] from various sources, including whey, starch crops, sugar crops, and lignocellulosic materials, or by chemical synthesis [5, 10, 12–15] from fossil resources. The chemical synthesis pathway [5] produces an optically inactive *racemic* mixture of the L and D isomers, while the anaerobic fermentation pathway generally yields optically pure L or D isomer, depending on the microorganism used [1].

2. Production of lactic acid

Microbial fermentation and chemical synthesis are the two industrially important methods for lactic acid production as depicted in **Figure 2**. A *racemic* mixture of DL-lactic acid is produced by a chemical synthesis from fossil sources, while an optically pure D(−) or L(+) lactic acid is produced using microbial fermentation. Furthermore, microbial lactic acid fermentation offers several key advantages over chemical synthesis. First, the utilization of renewable feedstocks reduces dependence on non-renewable petrochemical resources and contributes to a more sustainable and environmentally friendly production process. Second, microbial fermentation enables the production of optically pure lactic acid, as either D(−) or L(+) enantiomers. Third, fermentation processes generally operate under milder conditions compared to chemical synthesis. They require lower energy input in terms of temperature and pressure, resulting in reduced energy consumption and associated costs. Finally, it can utilize a wide range of feedstocks, including sugars from various biomass sources, agricultural by-products, and even waste streams. This flexibility allows for the utilization of

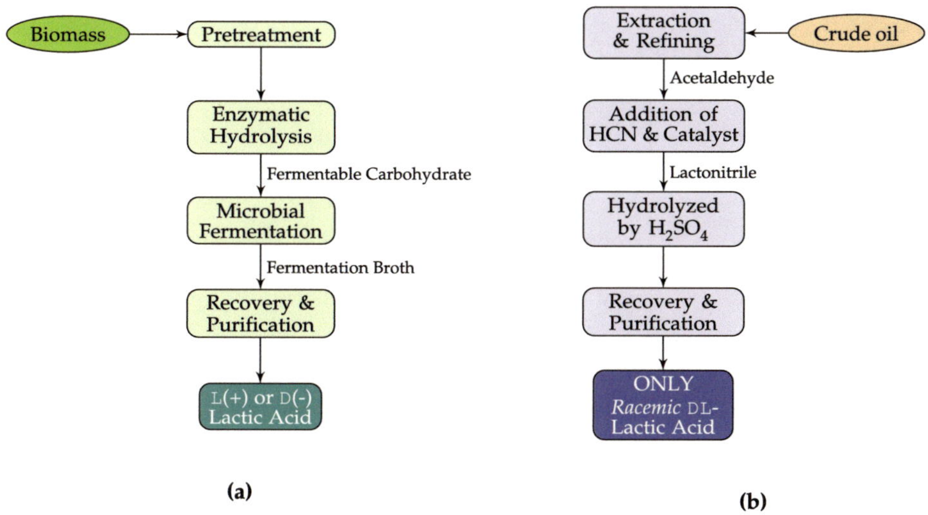

Figure 2.
An overview of LA production processes: (a) microbial fermentation from renewable recourses and (b) chemical synthesis from non-renewable recourses.

diverse and readily available substrates, contributing to a more sustainable and cost-effective production process.

2.1 Renewable resources for lactic acid

Lactic acid's demand has been rising due to its increment in a wide range of applications, aiming on sustainability the utilization of low-cost non-food biomass materials for its production. Given its availability, and low cost compared to refined sugars, lignocellulosic biomass is a promising source for sustainable lactic acid production. Nevertheless, due to its complexity, it is still challenging to commercially use lignocellulose for lactic acid synthesis due to the recalcitrance of the major constituents of lignocellulose for bioconversions [10, 16, 17]. The bioconversion processes for producing lactic acid from lignocellulosic biomass, depicted in **Figure 2a**, includes four main steps [18]: (a) pretreatment [19]—breaking down the structure of the lignocellulosic matrix, (b) enzymatic hydrolysis—depolymerizing lignocellulose to fermentable sugars, such as glucose and xylose, by means of hydrolytic enzymes, (c) fermentation—metabolizing the sugars to lactic acid using homolactic acid strains [20–22] and (d) separation and purification of lactic acid—purification of lactic acid to meet the standards of commercial applications.

2.2 Pretreatment

Lignocellulosic biomass are composed of cellulose, $(C_6H_{10}O_5)_n$, (40–50%), hemicelluloses, $(C_5H_8O_4)_m$, (25–35%) and lignin $[C_9H_{10}O_3(OCH_3)_{0.9-1.7}]_x$, (15–20%) in an intricate structure where the components are rigidly associated through non-covalent bonds and covalent cross-linkages along with minor amounts of other compounds such as proteins, ash, and pectin. The highly polymerized phenolic lignin encrusted on the crystallized cellulose and hemicellulose polymer matrix causes challenges in the conversion processes as shown in **Figure 3**. The main bottleneck in utilizing such

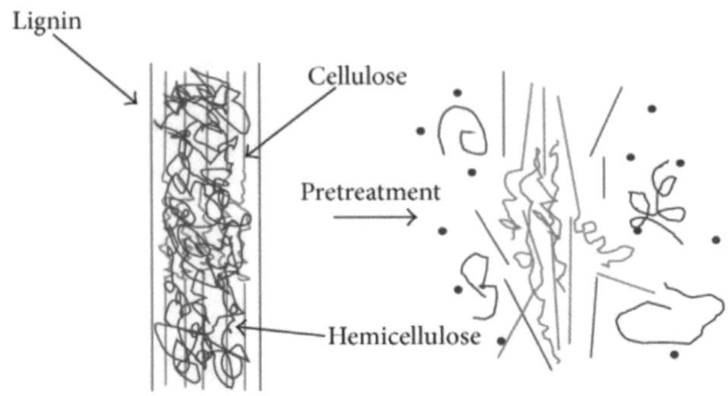

Figure 3.
Schematic of the role of pretreatment in the conversion of biomass into its constitutes, cellulose, hemicellulose, and lignin.

lignocellulosic biomass for bioproduct production is its complex, hierarchical, and recalcitrant nature. There are various pretreatment techniques that are classified as physical, chemical, physicochemical, and biological processes. The pretreatment step is required to increase the accessibility of the carbohydrate polymers by enzymes or bacteria/organism, where the enzymes depolymerize the polymer present in the lignocellulose, making them available for conversion of biomass to biofuels and value-added chemicals *via* fermentation. The size of the biomass feedstock is reduced first by milling to increase its surface area so as to increase hydrolysis efficiency. The main purposes of pretreatment are to condition the cellulose to increase its susceptibility to cellulase enzymatic digestion and hydrolyze the hemicellulose structural polysaccharides to oligomeric and monomeric sugars. Given the highly crystalline structure of cellulose that is embedded in a matrix of polymers-lignin and hemicellulose, pretreatment [18, 19, 23] is a very decisive step to overcome the recalcitrance of the lignocellulosic biomass to separate the cellulose from the matrix polymers and to make it more accessible for enzymatic hydrolysis.. Various types of pretreatment have been developed, including physical (mechanical, milling, extrusion), physicochemical (steam explosion pretreatment, hydrothermolysis, wet oxidation, ammonia fiber expansion), chemical (dilute acid, alkaline, ionic liquids), and biological methods (microorganisms, enzymes). In this chapter, steam explosion [24–27] pretreatment will be discussed, which is one of the viable technologies and has been explored extensively in the recent years.

2.2.1 Steam explosion

Steam explosion is one of the most commonly employed and effective pretreatment methods, which is typically a combination of both mechanical forces and chemical effects applied to lignocellulosic biomass. Steam explosion is a process in which lignocellulosic biomass is treated with steam at a temperature of 160–250°C under high pressure (0.69–1.5 MPa), for a very short period of time, and the lignocellulosic biomass then undergoes an explosive decompression, which causes a estrangement in its rigid structure [25, 28]. Steam is driven out of the reactor through a nozzle by inductive force after condensing at high pressure and "*wetting*" the lignocellulosic material. At this point, various processes take place such as the condensed moisture evaporates as a result of pressure reduction, lignocellulosic matrix desegregation

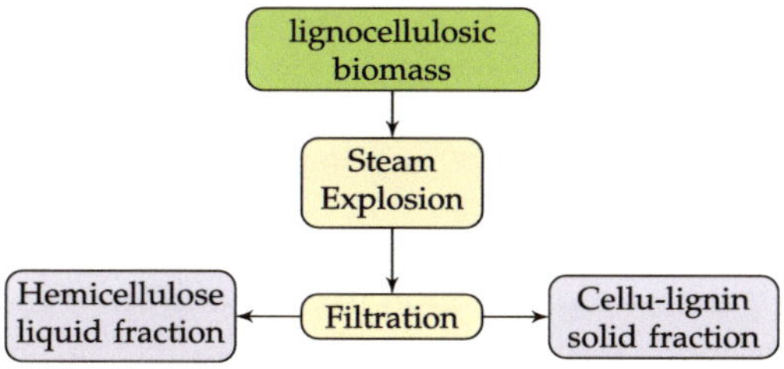

Figure 4.
Schematics of steam explosion pretreatment of lignocellulose biomass and separation of the hemicellulose liquid and cellu-lignin solid fractions by filtration.

occurs, and breaking down *inter-* and *intra*-molecular linkages [25–27, 29]. The cellulose bundles are defibrillated by the sudden release of pressure, which makes it easier for enzyme hydrolysis and fermentation to access the cellulose. In addtion, the high-temperature treatment causes the acetyl groups to hydrolyze, and the acetic acid that is then released catalyzes the hydrolysis of polysaccharides, primarily hemicelluloses, in a process known as *auto-hydrolysis* [25]. The solid cellulose and lignin network, cellu-lignin fraction, will be separated from the dissolved hemicellulose solution by filtration, as shown schematically in **Figure 4**.

2.3 Fermentation and purification processes

The hemicellulose liquid fraction and the cellu-lignin solid fraction are separated by filtration following the steam explosion treatments. For the purpose of removing sugars from the solid residues, the cellu-lignin solid fraction is washed. The hemicellulose liquid fraction (mainly xylose and other C_5 sugars) and the cellu-lignin solid fraction (mainly C_6 sugars) are used to produce lactic acid using various process methods. Lactic acid is produced from the liquid fraction of hemicellulose by simple fermentation, while cellulose and any remaining hemicellulose are hydrolyzed to produce fermentable monosaccharides from the cellu-lignin solid fraction to produce lactic acid using selected strains, which is known as a simultaneous saccharification and fermentation (SSF) [1, 9]. Acid-tolerant lactic acid bacteria were recently reported that can grow and ferment at pH 5.0 and at higher temperatures up to 60°C that are optimal conditions for cellulose hydrolysis by fungal enzymes. Such new isolates produce L(+)-lactic acid as the main fermentation product during the SSF processes with very low levels of cellulase [1, 30].

Lactic acid bacteria, particularly *Lactobacillus* spp., can ferment glucose (C_6 sugars) derived from the cellu-lignin solid fraction, which contains mainly C_6 sugars with some amounts of xylose (C_5 sugar) that is unfermentable by most lactic acid bacteria.

Lactococcus lactis IO-1 [31] and *Enterococcus mundtii* [32, 33] are the two common strains fermenting xylose to L(+) lactic acid, while *Lactobacillus pentosus* strains are for fermenting C_6 sugars [5, 10]. *Escherichia coli* is a candidate able to ferment all biomass-derived C_5 and C_6 sugars into L(+) lactic acid with high yields and optical purity, with minimal nutrient requirements, see Eq. (1) *Escherichia coli B*, WL204 [22], unlike the heterofermentative pathway which produces both acetic and lactic acids, is homofermentative genetically engineered strain to produce acetic acid free lactic acid with a yield of 90–97%, and an optical purity of 99.5% from xylose fermentation [1].

$$\underset{\text{Xylose}}{1.2\ C_5H_{10}O_5} \rightarrow \underset{\text{Lactic Acid}}{2\ C_3H_6O_3}$$
$$\underset{\text{Glucose}}{C_6H_{12}O_6} \rightarrow \underset{\text{Lactic Acid}}{2\ C_3H_6O_3} \quad (1)$$

The fermentation requires a stringent control of temperature at 35–45°C and pH of 5–6.5 [34], where the lactic acid production itself can drop the pH, which will have an inhibitory effect on the metabolic activities of the strains [35]. Appropriate bases and salts are added to neutralize the lactic acid to reduce the undesirable consequences of lactic acid accumulation in the industrial operations [3, 35]. The neutralization of lactic acid during fermentation has significant drawbacks, because it necessitates additional steps to recover the undissociated lactic acid from its respective salt so as to use the neutralizing agent again. Calcium hydroxide, $Ca(OH)_2$, and sulfuric acid, H_2SO_4, are commonly used in the fermentation process as a conventional neutralizing agent and in liberating lactic acid from calcium lactate, as shown in Eqs. (2) and (3), respectively [1].

$$2CH_3CHOHCOOH + Ca(OH)_2 \rightarrow (CH_3CHOHCOO^-)_2Ca^{2+} + 2H_2O \quad (2)$$

$$(CH_3CHOHCOO^-)_2Ca^{2+} + H_2SO_4 \rightarrow 2CH_3CHOHCOOH + CaSO_4 \quad (3)$$

Another lactic acid production processes without the formation of gypsum are either using acid-tolerant strains able to ferment both C_5 and C_6 sugars like *Bacillus coagulans* isolates [30], or using $Mg(OH)_2$ as neutralizing agent that forms Mg-*Lactate*.

The formed Mg-*Lactate* will react with organic amine that is miscible in water, triethylamine (R_3N), Eq. (4), forming both a $Mg(OH)_2$ crystal and a *triethylamine-LA complex*, (R_3N-LA) in a SWAP reactor [36], as depicted in **Figure 5**. The $Mg(OH)_2$ crystal will be filtered and recycled back to fermentation and the *triethylamine-LA complex* will be thermally decomposed to release LA where triethylamine will be recycled back to the SWAP reactor, Eq. (5).

$$Mg - Lactate + R_3N \rightarrow Mg(OH)_2 \downarrow_{crystal} + R3N - LA - Complex \quad (4)$$

$$R_3N - LA - Complex \rightarrow R_3N + Lactic\ Acid \quad (5)$$

The low volatility and the presence of nonvolatile impurities and other organic acids in the fermentation broth make the recovery of lactic acid challenging.

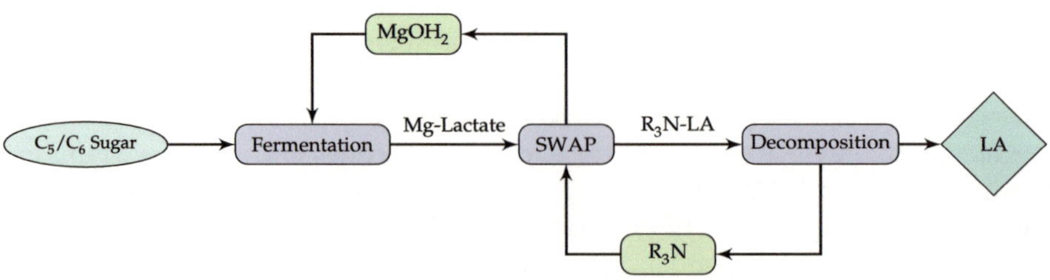

Figure 5.
Schematics of the neutralization and recycle of $Mg(OH)_2$ from the SWAP reactor and R_3N from decomposition of the triethylamine-LA complex. Adopted from a patent, EP1991517B1 [36].

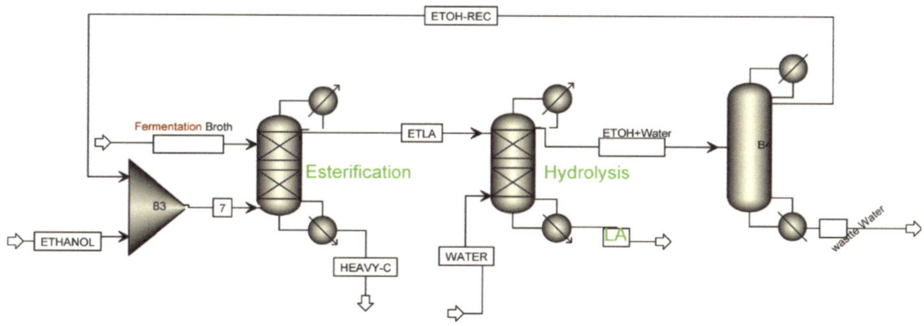

Figure 6.
Process flow sheet for the purification of crude LA from the fermentation broth via reactive distillation columns [1].

There are various lactic acid recovery processes from fermentation broth, such as nanofiltration [12, 34], adsorption [37], solvent extraction [38], electrodialysis [39, 40], membrane separation [41], ion-exchange resins [42], and reactive distillation [43, 44]. Reactive distillation [45, 46] was found to be promising at the industrial scale, with some cost advantages [43], and thus, two-step reactive distillation processes are modeled in the present work. Esterification and hydrolysis are the two simultaneous catalyzed reversible reactions that are required for lactic acid recovery by reactive distillation as given in Eq. (6). As shown in **Figure 6**, lactic acid from the fermentation broth is first converted to ethyl lactate (ETLA) in the esterification column using ethanol and this ethyl lactate is then removed from the column by distillation. Pure lactic acid, (99 wt.% L(+) lactic acid) is then recovered by hydrolysis of the ester (ethyl lactate) with water, in a hydrolysis reactive distillation column. The ethanol that is the top product from the hydrolysis reactive distillation will be purified by normal distillation and recycled (ETOH-REC) to the esterification column [1, 46].

$$\underset{\text{Lactic acid}}{C_3H_6O_3} + \underset{\text{Ethanol}}{C_2H_5OH} \underset{\text{Hydrolysis}}{\overset{\text{Esterification}}{\rightleftharpoons}} \underset{\text{Ethyl Lactate}}{C_3H_5O_3C_2H_5} + \underset{\text{Water}}{H_2O} \quad (6)$$

2.4 Lactic acid from non-renewable resources

The commercial process for chemical synthesis is based on lactonitrile [24, 47], $CH_3CH(OH)CN$ as shown in **Figure 2b**. Lactic acid can be chemically produced through various routes, including the conversion of renewable resources and petrochemical-derived materials. One of the commonly used chemical synthesis methods involves the production of lactic acid from acetaldehyde *via* lactonitrile intermediates.

$$\text{Crude oil} \rightarrow \text{Ethane} \rightarrow \text{Ethylene} \rightarrow \text{acetaldehyde} \quad (7)$$

The process begins with the extraction of crude oil, followed by refining to obtain ethane. Ethane is then converted into ethylene through processes such as steam cracking or catalytic cracking. Ethylene is further processed to produce acetaldehyde (CH_3CHO). This can be achieved through the oxidation of ethylene or by direct hydration of acetylene. Acetaldehyde is a key raw material for the subsequent steps in lactic acid synthesis. Acetaldehyde is reacted with hydrogen cyanide (HCN) in the presence of a base catalyst under high-pressure conditions. This catalytic reaction results in the formation of lactonitrile (acetaldehyde cyanohydrin), Eq. (8) [12]. Lactonitrile is then recovered, purified through distillation, and hydrolyzed using sulfuric acid (H_2SO_4). This hydrolysis reaction converts lactonitrile into lactic acid and ammonium sulfate byproduct as shown in Eq. (9). To further process the lactic acid, Eq. (10) involves esterification with methanol (CH_3OH), leading to the formation of methyl lactate ($CH_3CHOHCOOCH_3$). Methyl lactate is subsequently recovered, purified *via* distillation, and hydrolyzed with acidified water in Eq. (11). This hydrolysis step results in the production of lactic acid and the regeneration of methanol. Methanol is separated through distillation and can be recycled for further use. It is important to note that the chemical synthesis route described above typically yields a *racemic* mixture of lactic acid, which consists of equal amounts of D(−)-lactic acid and L(+)-lactic acid. However, in certain applications, such as the production of polylactide, optically pure L(+)-lactic acid is often preferred.

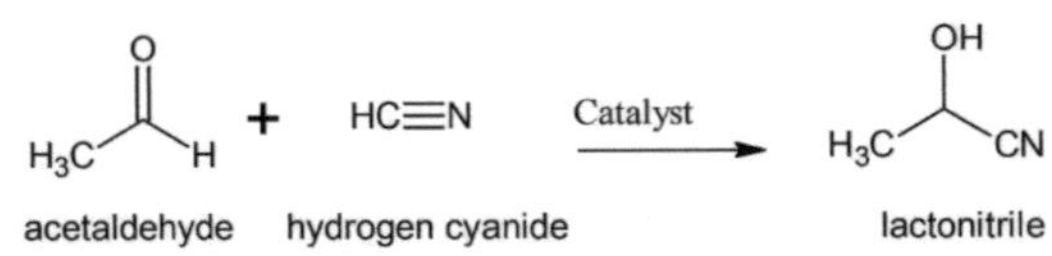

(8)

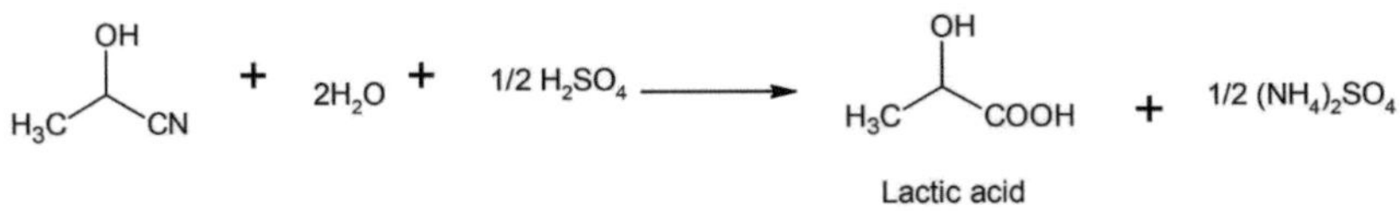

(9)

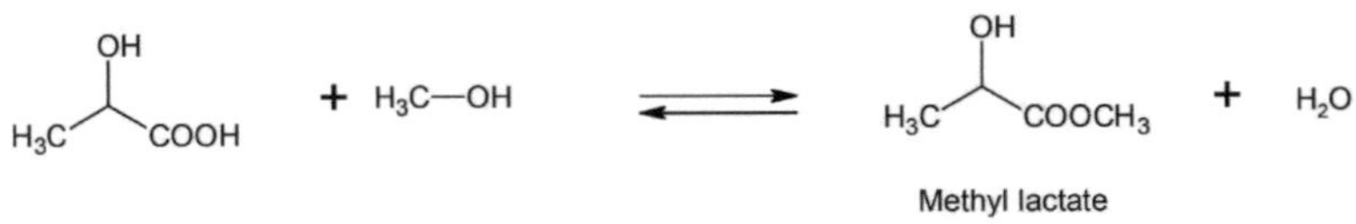

(10)

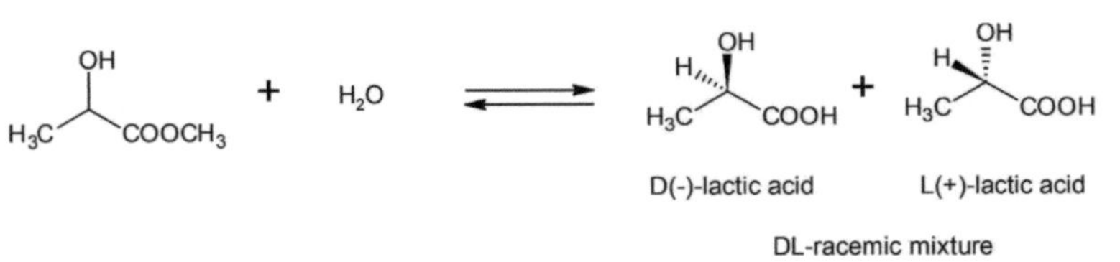

(11)

3. Environmental impact assessment

The environmental impact assessment of lactic acid production from various biobased feedstocks have been studied and the environmental benefits compared with fossil-based lactic acid production [1, 48, 49]. Environmental assessment was made using the standard Life Cycle Assessment for determining the potential environmental impacts of biobased lactic acid production processes. Environmental life cycle assessment, a standardized methodology [50], considers the full life cycle of the analyzed biochemical production systems and a broad range of environmental impact indicators, as shown in **Table 1**. A *cradle-to-gate* life cycle assessment is applied considering all steps across production systems, feedstock production, and lactic acid production

Description	Symbol
Abiotic depletion potential measured in kg of *substance* eq.	ADP
Acidification potential measured in kg of SO_2 eq.	AP
Freshwater eutrophication potential measured in kg of PO_4^{-3} eq.	FWEP
Marine eutrophication potential measured in kg NO_3^- eq.	MEP
Global warming potential measured in kg of CO_2 eq.	GWP
Ozone layer depletion potential measured in kg of CFC-11 eq. (chlorofluorocarbon)	ODP
Human toxicity potential measured in kg of 1,4-DB eq. (DB:dichlorobenzene)	HTP
Fresh water aquatic ecotoxicity potential measured in kg of 1,4-DB eq	FWAETP
Marine aquatic ecotoxicity potential measured in kg of 1,4-DB eq.	MAETP
Terrestrial ecotoxicity potential measured in kg of 1,4-DB eq.	TETP
Photochemical oxidation potential measured in kg of C_2H_4 eq.	POCP
Metal depletion potential measured in kg Fe eq.	MDP
Ionizing radiation potential measured in kBq U_{235} eq.	IRP
Particulate matter formation potential measured in kg PM_{10} eq.	PM
Water depletion potential measured in m^3 water used	WDP
Agricultural land occupation measured in m^2a	ALO
Urban land occupation measured in m^2a	ULO
Natural land transformation measured in m^2	NLT

Table 1.
Environmental impact categories considered for comparison of biobased lactic acid and fossil based lactic acid production for the LCA study.

as a stand-alone plant or biorefinery processes integrated with existing plants. The potential environmental impacts of a lignocellulosic biomass-based lactic acid process have been investigated by means of a life cycle assessment [50]. A life cycle assessment is carried out for biobased lactic acid from lignocellulosic biomass and compared with fossil-based lactic acid, as described in Sections 2.1 and 2.4, respectively. Similar studies were conducted on comparing environmental impacts of a biodegradabale plastic polylactic acid (PLA) compared with fossil based PLA [51–53]. The life cycle assessment complies with ISO [54, 55] standards, as shown in **Figure** 7, consisting of four stages: goal and scope definition, inventory analysis, impact assessment, and interpretation.

Ögmundarson et al. [48] quantified the environmental impacts of biochemical production of lactic acid using different biobased feedstocks like corn, corn stover, and macroalgae (Laminaria sp.) and discussed the potential for reducing impact. Despite differences in assumptions and data availability across the *cradle-to-gate* life cycle assessment model for the biobased lactic acid production process, several research works show there is an environmental savings that leads to the global sustainable developments [1, 48, 56]. Replacing fossil-based lactic acid with its biobased lactic acid reduces about 80–99% of the environmental burdens of most of the environmental impact categories [1], making biobased lactic acid is environmentally friendlier than its fossil-based equivalent in most of the impact categories.

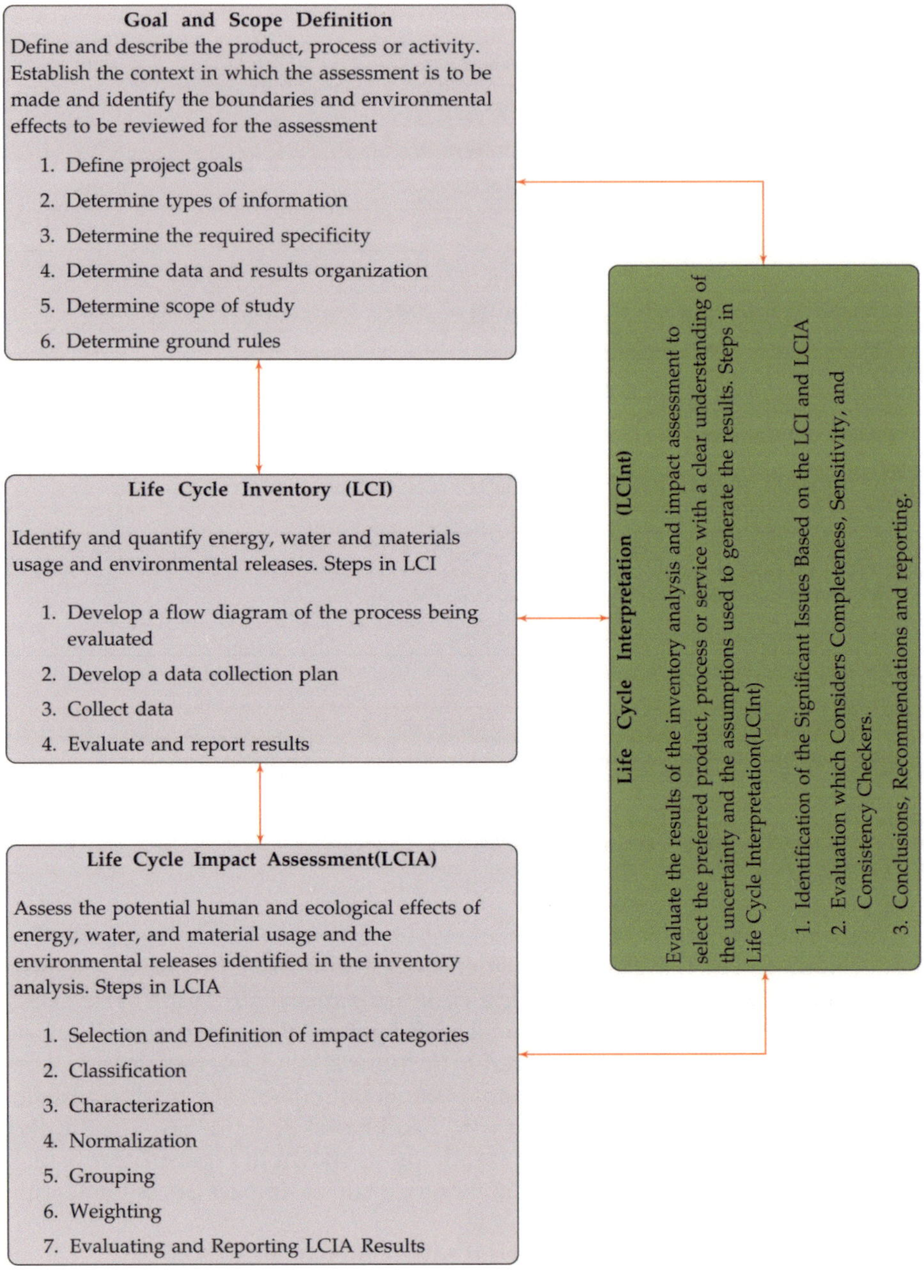

Figure 7.
Phases and guidelines of LCA, according to ISO 14040 series.

4. Conclusion

Lactic acid has important applications in various industrial sectors such as the chemical, pharmaceutical, food, and cosmetics industries, where its demand is steadily increasing. Furthermore, lactic acid is an essential building block for numerous commodity and intermediate biobased chemicals like acrylic acid, lactate esters,

1,2-propanediol, pyruvic acid, acetaldehyde, 2,3-pentanedione, ethanedioic acid, and polylactic acid, making it as a suitable alternative to their oil-derived counterparts. The bioconversion processes of converting lignocellulosic biomass into lactic acid need to overcome the following points or challenges: (1) effective pretreatment process to make all C_5 and C_6 sugars available for the selected microorganism that ferment them into stereospecific (L- or D-) lactic acid, (2) selection of suitable microorganism along with their required substrates and fermentation conditions, and (3) employing the separation and purification of lactic acid from the fermentation broth. The promising method for the industrial production of lactic acid will be of continuous simultaneous saccharification and fermentation in a gypsum-free process using $Mg(OH)_2$ as neutralizer, followed by reactive distillation for purified lactic acid production. The *cradle-to-gate* life cycle assessment model for the biobased lactic acid production process indicated that the about 80–99% of the environmental burdens of most of the environmental impact categories can be reduced compared with its equivalent fossil-based lactic acid, making biobased lactic acid environmentally superior to the fossil-based lactic acid.

Author details

Asfaw Gezae Daful[1]*, Marie Loridon[1] and Meegalla R. Chandraratne[2]

1 Department of Chemical Engineering, Higher Colleges of Technology, Abu Dhabi, United Arab Emirates

2 Department of Chemical Engineering, Higher Colleges of Technology, Ruwais, United Arab Emirates

*Address all correspondence to: agezae@hct.ac.ae; agezae@gmail.com

References

[1] Gezae Daful A, Görgens JF. Techno-economic analysis and environmental impact assessment of lignocellulosic lactic acid production. Chemical Engineering Science. 2017;**162**:53-65

[2] Gezae Daful A, Haigh K, Vaskan P, Görgens JF. Environmental impact assessment of lignocellulosic lactic acid production: Integrated with existing sugar mills. Food and Bioproducts Processing. 2016;**99**:58-70

[3] Abdel-Rahman MA, Tashiro Y, Sonomoto K. Recent advances in lactic acid production by microbial fermentation processes. Biotechnology Advances. 2013;**31**(6):877-902

[4] Chao G, Cuiqing M, Ping X. Biotechnological routes based on lactic acid production from biomass. Biotechnology Advances. 2011;**29**(6): 930-939

[5] Castillo Martinez FA, Balciunas EM, Salgado JM, Domínguez González JM, Converti A, Oliveira RPDS. Lactic acid properties, applications and production: A review. Trends in Food Science and Technology. 2013;**30**(1):70-83

[6] Päivi M-A, Simakova Irina L, Tapio S, Dmitry YM. Production of lactic acid/ lactates from biomass and their catalytic transformations to commodities. Chemical Reviews. 2014;**114**(3):1909-1971

[7] Dusselier M, Van Wouwe P, De Smet S, De Clercq R, Verbelen L, Van Puyvelde P, et al. Toward functional polyester building blocks from renewable glycolaldehyde with Sn cascade catalysis. ACS Catalysis. 2013;**3**(8):1786-1800

[8] Cellulac. Green Cleaner Bio-Chemicals. Available from: http://cellulac.co.uk/en/ market/ [Accessed: May 20, 2023]

[9] Wang Y, Tashiro Y, Sonomoto K. Fermentative production of lactic acid from renewable materials: Recent achievements, prospects, and limits. Journal of Bioscience and Bioengineering. 2015;**119**:10-18

[10] Abdel-Rahman MA, Tashiro Y, Sonomoto K. Lactic acid production from lignocellulose-derived sugars using lactic acid bacteria: Overview and limits. Journal of Biotechnology. 2011;**156**(4): 286-301

[11] Rathin D, Michael H. Lactic acid: Recent advances in products, processes and technologies- A review. Journal of Chemical Technology & Biotechnology. 2006;**81**(7):1119-1129

[12] Pal P, Sikder J, Roy S, Giorno L. Process intensification in lactic acid production: A review of membrane based processes. Chemical Engineering and Processing: Process Intensification. 2009;**48**:1549-1559

[13] Narayanan N, Roychoudhury PK, Srivastava A. L (+) lactic acid fermentation and its product polymerization. Electronic Journal of Biotechnology. 2004;**7**:167-179

[14] John RP, Madhavan NK, Pandey A. Fermentative production of lactic acid from biomass: An overview on process developments and future perspectives. Applied Microbiology and Biotechnology. 2007;**74**:524-534

[15] Environmental factsheet PLA. Environmental factsheet: Polylactic acid. 2015

[16] Mazzoli R, Bosco F, Mizrahi I, Bayer EA, Pessione E. Towards lactic acid bacteria-based biorefineries.

Biotechnology Advances. 2014;**32**:1216-1236

[17] Benjamin Y, García-Aparicio MP, Görgens JF. Impact of cultivar selection and process optimization on ethanol yield from different varieties of sugarcane. Biotechnology for Biofuels. 2014;7:60

[18] Taherzadeh MJ, Karimi K. Pretreatment of lignocellulosic wastes to improve ethanol and biogas production: A review. International Journal of Molecular Sciences. 2008;**9**(9):1621-1651

[19] Sreenath HK, Moldes AB, Koegel RG, Straub RJ. Lactic acid production by simultaneous saccharification and fermentation of alfalfa fiber. Journal of Bioscience and Bioengineering. 2001; **92**(6):518-523

[20] Tamakawa H, Ikushima S, Yoshida S. Efficient production of l-lactic acid from xylose by a recombinant candida utilis strain. Journal of Bioscience and Bioengineering. 2012;**113**(1):73-75

[21] Ying W, Ali A-RM, Yukihiro T, Yaotian X, Takeshi Z, Kenji S, et al. l-(+)-lactic acid production by co-fermentation of cellobiose and xylose without carbon catabolite repression using enterococcus mundtii QU 25. RSC Advances. 2014;**4**:22013-22021

[22] Zhao J, Xu L, Wang Y, Zhao X, Wang J, Garza E, et al. Homofermentative production of optically pure l-lactic acid from xylose by genetically engineered escherichia coli b. Microbial Cell Factories. 2013; **12**:57

[23] Mosier N, Wyman C, Dale B, Elander R, Lee YY, Holtzapple M, et al. Features of promising technologies for pretreatment of lignocellulosic biomass. Bioresource Technology. 2005;**96**(6): 673-686

[24] Wasewar KL. Separation of lactic acid: Recent advances. Chemical and Biochemical Engineering Quarterly. 2005;**19**:159-172

[25] Rocha GJM, Gonçalves AR, Oliveira BR, Olivares EG, Rossell CEV. Steam explosion pretreatment reproduction and alkaline delignification reactions performed on a pilot scale with sugarcane bagasse for bioethanol production. Industrial Crops and Products. 2012;**35**(1):274-279

[26] Oliveira FMV, Pinheiro IO, Souto-Maior AM, Martin C, Gonçalves AR, Rocha GJM. Industrial-scale steam explosion pretreatment of sugarcane straw for enzymatic hydrolysis of cellulose for production of second generation ethanol and value-added products. Bioresource Technology. 2013; **130**:168-173

[27] Agbor VB, Cicek N, Sparling R, Berlin A, Levin DB. Biomass pretreatment: Fundamentals toward application. Biotechnology Advances. 2011;**29**:675-685

[28] Kumar P, Barrett DM, Delwiche MJ, Stroeve P. Methods for pretreatment of lignocellulosic biomass for efficient hydrolysis and biofuel production. Industrial and Engineering Chemistry Research. 2009;**48**(8):3713-3729

[29] Kaar WE, Gutierrez CV, Kinoshita CM. Steam explosion of sugarcane bagasse as a pretreatment for conversion to ethanol. Biomass and Bioenergy. 1998;**14**:277-287

[30] Patel MA, Ou MS, Harbrucker R, Aldrich HC, Buszko ML, Ingram LO, et al. Isolation and characterization of acid-tolerant, thermophilic bacteria for

effective fermentation of biomass-derived sugars to lactic acid. Applied and Environmental Microbiology. 2006;**72**: 3228-3235

[31] Joshi DS, Singhvi MS, Khire JM, Gokhale DV. Strain improvement of lactobacillus lactis for d-lactic acid production. Biotechnology Letters. 2010; **32**(4):517-520

[32] Ali A-RM, Yukihiro T, Takeshi Z, Kenji S. Improved lactic acid productivity by an open repeated batch fermentation system using *Enterococcus mundtii* QU 25. RSC Advances. 2013;**3**: 8437-8445

[33] Abdel-Rahman MA, Tashiro Y, Zendo T, Hanada K, Shibata K, Sonomoto K. Efficient homofermentative l-(+)-lactic acid production from xylose by a novel lactic acid bacterium, enterococcus mundtii {QU} 25. Applied and Environmental Microbiology. 2011;**77**:1892-1895

[34] Isabel González M, Álvarez S, Riera F, Álvarez R. Lactic acid recovery from whey ultrafiltrate fermentation broths and artificial solutions by nanofiltration. Desalination. 2008;**228**: 84-96

[35] Hongo M, Nomura Y, Iwahara M. Novel method of lactic acid production by electrodialysis fermentation. Applied and Environmental Microbiology. 1986; **52**:314-319

[36] Krieken JV. Method for preparing an organic amine-lactic acid complex. November 14: EP Patent 1,991,517. 2012

[37] Hábová V, Melzoch K, Rychtera M, Sekavová B. Electrodialysis as a useful technique for lactic acid separation from a model solution and a fermentation broth. Desalination. 2004;**162**: 361-372

[38] Wee Y-J, Yun J-S, Lee YY, Zeng A-P, Ryu H-W. Recovery of lactic acid by repeated batch electrodialysis and lactic acid production using electrodialysis wastewater. Journal of Bioscience and Bioengineering. 2005;**99**(2):104-108

[39] Sena B, İnci İ, Hasan U. Adsorption of lactic acid from model fermentation broth onto activated carbon and amberlite ira-67. Journal of Chemical & Engineering Data. 2011;**56**(5):1751-1754

[40] Udachan IS, Sahoo AK. A study of parameters affecting the solvent extraction of lactic acid from fermentation broth. Brazilian Journal of Chemical Engineering. 2014;**31**: 821-827

[41] Gudena K, Rangaiah GP, Lakshminarayanan S. Modeling and analysis of hybrid reactive stripper-membrane process for lactic acid recovery. Industrial & Engineering Chemistry Research. 2013;**52**(8): 2907-2916

[42] González MI, Álvarez S, Riera F, Álvarez R. Purification of lactic acid from fermentation broths by ion-exchange resins. Industrial & Engineering Chemistry Research. 2006; **45**(9):3243-3247

[43] Chien-Yuan S, Cheng-Ching Y, I-Lung C, Jeffrey DW. Plant-wide economic comparison of lactic acid recovery processes by reactive distillation with different alcohols. Industrial & Engineering Chemistry Research. 2013;**52**(32):11070-11083

[44] Mo L, Shao-Tong J, Li-Jun P, Zhi Z, Shui-Zhong L. Design and control of reactive distillation for hydrolysis of methyl lactate. Chemical Engineering Research and Design. 2011;**89**(11): 2199-2206

[45] Taylor R, Krishna R. Modeling reactive distillation. Chemical Engineering Science. 2000;**55**(22): 5183-5229

[46] Gezae Daful A. Simulation of reactive distillation: Comparison of equilibrium and nonequilibrium stage models. International Journal of Chemical Engineering. 2012;**6**(10): 1231-1239

[47] Wee YJ, Kim JN, Ryu HW. Biotechnological production of lactic acid and its recent applications. Food Technology and Biotechnology. 2006; **44**(2):163-172

[48] Ogmundarson O, Sukumara S, Laurent A, Fantke P. Environmental hotspots of lactic acid production systems. GCB Bioenergy. 2020;**12**(1):19-38

[49] Helmes RJK, Lopez-Contreras AM, Benoit M, Abreu H, Maguire J, Moejes F, et al. Environmental impacts of experimental production of lactic acid for bioplastics from ulva spp. Sustainability. 2018;**10**(7):1-15

[50] Guinée JB, Heijungs R, Huppes G. Economic allocation: Examples and derived decision tree. LCA Methodology. 2004;**9**:23-33

[51] Riofrio A, Cornejo M, Baykara H. Life cycle and environmental impact evaluation of polylactic acid (pla) production in ecuador. The International Journal of Life Cycle Assessment. 2022; **27**:834-848

[52] Ghomi ER, Khosravi F, Ardahaei AS, Dai Y, Neisiany RE, Foroughi F, et al. The life cycle assessment for polylactic acid (pla) to make it a low-carbon material. Polymers (Basel). 2021;**13**:11

[53] Tecchio P, Foschia M, Talon O, Dejonghe S. Lactic acid production by pla-waste chemical recycling: Environmental impact evaluation. Semantic Scholar. 2013

[54] ISO. ISO 14040 – environmental management life cycle assessment - principles and framework. 2006

[55] ISO. ISO 14044 – environmental management life cycle assessment - requirements and guidelines. 2006

[56] Li Y, Bhagwat SS, Cortés-Peña YR, Ki D, Rao CV, Jin Y-S, et al. Sustainable lactic acid production from lignocellulosic biomass. ACS Sustainable Chemistry & Engineering. 2021;**9**(3): 1341-1351

Chapter 11

Opportunities and Challenges of Harnessing Biomass Wastes for Decentralized Heat and Energy Generation and Climate Mitigation via Fluidized-bed Gasification Pathway

Henok Mekonnen Mossissa, Million M. Afessa, Derese T. Nega, Muhammad Ahsan Amjed and A. Venkata Ramayya

Abstract

Biomass wastes offer immense potential as a renewable energy source, holding the promise to replace fossil fuels for heat and energy generation, in particular for decentralized power production. Furthermore, the utilization of biomass promotes circular economy by enabling the conversion of local resources into useful products and energy. However, the conversion of biomass into end-use products and heat/ energy is a complex process with multiple pathways, such as fluidized bed gasification, a well-established and efficient method for converting coal and biomass into heat. Despite its merits, this process is currently limited to industrial applications and encounters certain limitations and obstacles. Notably, the low energy density of biomass wastes and downstream pipe contamination from tar and polycyclic aromatic hydrocarbon (PAH) growth poses significant technological challenges. Nonetheless, a roadmap has been developed to guide the widespread adoption of fluidized bed gasification of biomass for decentralized power generation and climate mitigation. This book chapter delves into the opportunities and challenges of fluidized bed gasification as a viable option for decentralized power generation and climate mitigation through biomass waste conversion. The significance of well-crafted policies supporting renewable energy sources and optimizing fluidized bed gasifiers to achieve desirable end products are also emphasized.

Keywords: fluidized bed gasification, biomass wastes, GHG emission reduction, decentralized heat and power, climate change mitigation, circular economy

1. Introduction

Currently, a substantial portion of the world's energy supply is derived from fossil fuels, whose reserves are unsustainable. High carbon fuel consumption and concomitant greenhouse gas emissions are currently the most pressing and well-considered challenges [1]. To address these encounters, renewable energy seems quite promising. Among all, biomass wastes are one of the renewable energy sources which can potentially substitute fossil fuels for heat and energy generation [2]. As biomass waste is virtually everywhere, it is an excellent resource for distributed heat and power generation, which reduces dependence on fossil fuels and central power generation, which is hugely challenged by global energy politics. It also promotes a circular economy as local resources can be converted to useful products and energy [3]. It can also be coupled with carbon capture, and bio-remediation through special biomass sources to have a greater impact on the environment [4]. Also, the utilization of biomass waste such as municipal and agro-industrial waste as feedstocks in large quantities resolves the concerns associated with waste management, aiding to curb environmental pollution and severe health effects [5].

There are various biomass-to-end-use (bio-based products and heat/power) conversion pathways. Thermo-chemical and biochemical conversion of biomass is the major conversion route yielding a wide range of products and subsequent applications. Thermochemical conversion is a widely employed means of biomass conversion through combustion, pyrolysis, gasification, and liquefaction [6]. Among those pathways, gasification is gaining increased attention for its better conversion efficiency, accommodation of a variety of feedstocks, and yield of a variety of products for versatile applications [7]. Fluidized bed gasification is a very matured and efficient coal and biomass conversion pathway mostly limited to industrial applications to date. Due to high solid-to-gas contact and excellent heat transfer in the fluidized bed, it is considered very suitable for controlling operating conditions (such as reaction temperature, residence time, and solid gas heat transfer) and the end product distribution compared to other gasification and thermo-chemical conversion pathways [8]. In addition, it is also suitable for a wide range of feedstocks. **Figure 1** shows a typical gasification process with respect to the operating temperature. Even though fluidized bed gasification is one of the most promising technologies used for the thermo-chemical conversion of coal and biomass materials, it comes up with challenges and limitations [10]. Particularly, the low energy density of biomass waste and challenges like syngas quality. In terms of heating value and the amount of contaminants like tar, particulates, and heavy metals present in the syngas, the downstream processes like upgrading and cleaning have a direct impact on investment and operational costs. In addition, transporting raw biomass residue over a long distance is not feasible technically or economically because of the very low volumetric energy density of biomass, which may require more fuel energy than it can produce [3].

In a frontier to design optimized fluidized bed gasifiers for typical end-products, experimental and computer models have been extensively employed. Experimental works are highly limited to lab-scale reactors and fail to represent actual-scale gasification processes due to cost and complex processes which makes the experiment difficult. Although computer models are widely employed to simulate the fluidized bed phenomena, it comes with severe limitations as fluidized bed gasification involved complex multi-phase, multi-step, multi-scale processes, which are very expensive to be dealt with in great detail at the same time.

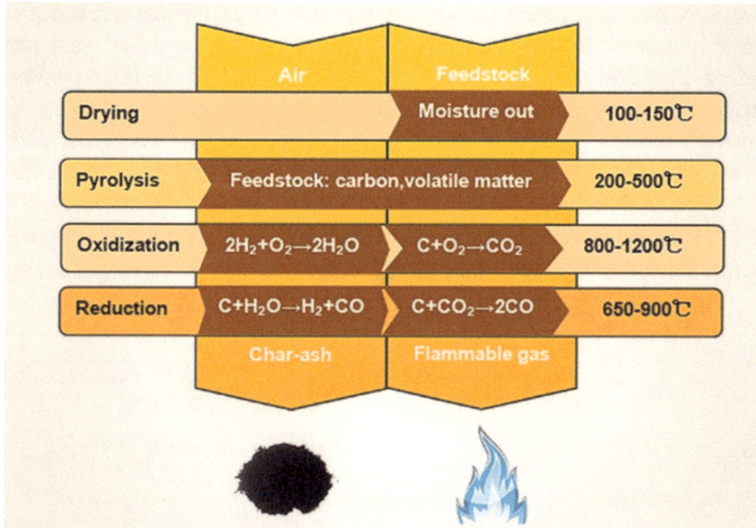

Figure 1.
Typical steps in gasification [9].

In addition, due to a lack of well-formulated strategies and policies towards renewable energy, biomass waste is highly under-exploited, and often times it is dumped and burned in the open air, which has been a common trend in many developing countries. A comprehensive approach combining the technical, economic, strategic, and policy frameworks should be followed to address the existing challenges.

In this book chapter, the opportunities and challenges of the existing fluidized bed gasification technology as a potential candidate for biomass to heat and energy conversion pathways will be discussed.

2. Fluidized bed gasification

2.1 Biomass wastes conversion via fluidized bed gasification

There are several Biomass waste conversion routes to valuable products or heat/power production as summarized in **Figure 2**. The choice of technology depends on the desired end-use, the nature of biomass, resource availability, and other considerations like the techno-economic and environmental aspects [11]. The main technologies employed so far are categorized under thermochemical conversion, biochemical conversion, and extraction leading to a wide range of applications such as heat and electricity generation, bio-oil, hydrogen, and various synthetic chemicals production [8]. Thermochemical conversion is among the widely exploited routes given its versatility in accommodating a wide range of feedstocks, design options, and a wide range of final products and application domains.

A prevailing thermochemical conversion method is biomass gasification as portrayed in red arrows in **Figure 2**, which typically uses a fixed or fluidized bed reactor to convert biomass into gaseous fuels at low to moderate temperatures [12]. Due to its several advantages over fixed bed gasifiers, fluidized bed gasification has garnered increasing attention over the others, in the biomass to heat/power conversion. High mixing and reaction rates, accommodation of various biomass feedstock, and its potential for scaling are the main winning points [13]. The gasification technologies are summarized in the **Table 1** with their typical features, advantages, and limitations.

Gasification usually produces a gas mainly composed of CO and H_2 with energy values between 5 and $20MJ/Nm^3$ depending on the biomass characteristics, operating

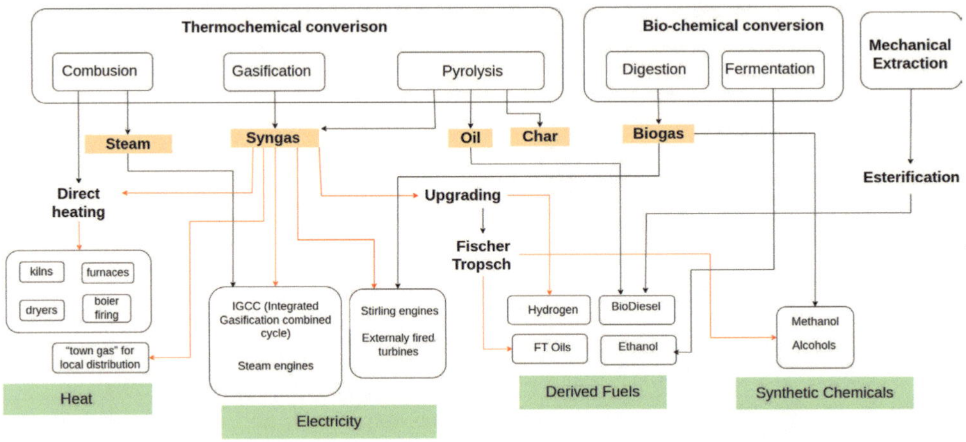

Figure 2.
Biomass conversion routes emphasizing gasification pathways with red arrows (adapted from [6–8]).

Interms of gasifier design	
Bubbling bed	***Features***
	The bed is a two-phase region of bubbles and the emulsion. The fluidizing gas is generally kept in continuous mode, entering through the bottom and exiting through the top [10].
	Pros → moderate investment cost—intermediate control [8]—can handle a wide range of feedstock types and gasifying agents (Air, Steam, and O_2) [8]—a wide range of scale, $10 - 100MW$ [8]—carbon conversion efficiency $> 90\%$ [8]—Intermediate tar yield [8]
	Cons → low thermal throughput ($1.2 - 1.6MW/m^2$) [8]—feedstock size affects bed hydrodynamics and fine particles are subject to elutriation [8]
	Application → direct heating—combined heat and power (CHP) using internal combustion (IC) engine or gas turbine and heating boilers—biomass-derived fuels via Fischer-Tropsch process [14]
Circulating fluid bed	***Features***
	the solids move in a cycle characterized by thorough mixing and high residence times within the solid circulation loop [10].
	Pros → can handle a wide range of feedstock types, sizes, and gasifying agents (air, steam, and O_2) [8]—highly scalable with $> 20MW$ [8]—carbon conversion efficiency $> 90\%$ (better than BFB) [8]—intermediate tar yield [8]—high thermal throughput ($5 - 7MW/m^2$) [8]

Interms of gasifier design	
	Cons → high investment cost and intermediate control [8] ***Application*** → IGCC/IGFC [15]—methanol synthesis [16]
Dual/Twin Fluidized bed	***Features***
	Circulating hot bed material is heated in a separate fluidized bed reactor by combustion of residual biomass char [17]
	Pros → better carbon conversion efficiency [18]
	Cons → complex construction [10]
	Application domain → some commercial combined heat and power (CHP) [18]—gas supply—hydrogen production—synthesis of liquid fuels—other industries [19]
Entrained bed	***Features***
	Powdered fuel ($0.75mm$) is injected into the reactor chamber along with the gasifying agent [10]. It operates at high gasification temperatures (1300–1500°C) [8]
	Pros → carbon conversion up to 100% [8]—very low tar yield [8]
	Cons → suitable for very fine feedstocks and air gasification [8] [20]—suitable for high plant capacity only ($>100MW$) [8]—high investment cost and complex operation control [8, 10]
	Application → suitable for IGCC (integrated gasification combined cycle) plants [10]
Updraft	***Features***
	Fixed bed type. The fuel is fed in from the top, and the gas generated also emerges from the reactor via the top [10]
	Pros → low investment cost [8]—very easy control
	Cons → sensitive to feedstock type and size unlike fluidized beds but better than downdraft [8]—very high tar yield [8]—suitable for plants upto $20MW$ only [21]—low power throughput ($1-2MW/m^2$) [8]
Downdraft	***Features***
	Fixed bed type. Biomass is fed in from the top and drops downwards, while air is injected from one side [10]
	Pros → Low investment cost [8]—very easy control—produces low tar gas [22]—has a carbon conversion efficiency of 93–96% [8]
	Cons → very sensitive to feedstock size and type [8]—suitable for low moisture content [22]—suitable for small-scale plants up to $5MW$ [21]—low power throughput ($1-2MW/m^2$) [8]
Interms of heating	
Direct heating (auto-thermal)	***Features***
	Gasifying agent is supplied along with the fuel creating an oxygen-deprived environment.
	Pros → reduced operation cost—reduced capital cost
	Cons → NA
Indirect heating (allothermal)	***Features***
	Heat is supplied using heat exchanger

Interms of gasifier design	
	Pros → better cold gas efficiency than direct heating [23]—higher gas quality (heating value of $12 - 20MJ/Nm^3$)—lower tar and char ***Cons*** → requires external heating usually using an external steam source [24]
Plasma	***Features***
	Inert gas is passed through the high-energy electric arc and heated to 1500–5000° C. Organic components are converted to gas while inorganic components are converted to virtuous slag when the biomass comes in contact with the plasma arc [25]
	Pros → can be used with organic MSW (municipal solid waste), and other wastes such as paper, plastics, glass, metals, textiles, wood, rubber, etc. [10]
	Cons → still under study not yet in commercial stage
Interms of operating pressure	
Atmospheric	***Features***
	Operates at operating pressure
	Pros → a more economical design
	Cons → NA
Pressurized	***Features***
	Gasification pressure between 5 and $40bar$
	Pros → higher energy conversion efficiency—higher exergy efficiency [24]
	Cons → higher operating cost
Interms of gasifying agent used	
Air gasification	
	Pros → low cost [21]—better plant safety [26]
	Cons → produces a low heating value fuel gas, generally from 4 to $7MJ/Nm^3$ [27] [21]
Steam gasification	***Pros*** → produces a gaseous secondary energy carrier with a calorific value typically in the range of $12 - 14MJ/Nm^3$ [28]
	Cons → NA
Oxygen gasification	***Pros*** → produces a higher heating value fuel gas, $12 - 28MJ/Nm^3$ [29]—direct oxygen-blown gasification suitable for large-scale Bio-SNG production [24]
	Cons → high cost of using O_2 [30] [21]
Combined gasification	***Features***
	steam–air, steam-oxygen, steam–air-oxygen and oxygenated air
	Pros → can make a tradeoff between cost and performance
	Cons → NA

Table 1.
Summary of features of gasification systems.

conditions, and gasifying agent used [2]. Integrated with a gas turbine, boiler, or steam turbine it can be a viable biomass-to-energy conversion component in heat, power, and combined heat and power systems. Syngas can also be an alternative fuel in modified internal combustion engines for electricity generation. Small and medium-scale Combined Heat and Power (CHP) systems also known as 'co-generation' are used to convert biomass into electricity while extracting waste heat, a promising application for commercial buildings such as hospitals, schools, or office building blocks but can also be used for decentralized power generation in remote and rural areas [31]. For existing coal plants, burning solid biomass in traditional power plants alongside coal, in a process known as 'co-firing', is a cost-effective, more efficient, and clean option with only minor technical adjustments. Caputo et al. [11] calculated and compared the overall system efficiencies for ï¬,uidized bed combustor with steam turbine cycle and fluidized bed gasifier with combined gas-steam cycle in a black-box model using literature data. The system with the fluidized bed gasifier has an efficiency between 36 and 45% while the former has an efficiency between 25 and 28% for a power scale between 5 and 50MW.

2.2 Fluidized bed performance indexes and important process parameters

The knowledge of performance indexes of any energy conversion system is crucial to determine plant size, investment cost, techno-economic feasibility, and environmental impact. The overall plant performance is the sum of the upstream, downstream, and gasification in-bed performances considering the gasification reactor as the powerhouse of the plant. The upstream process may require biomass feedstock treatment and in the downstream, there are auxiliary components such as gas cleaning units, boilers, engines, and turbines, depending on the plant's application. The commonly used performance indexes in fluidized bed gasification are gas yield, the high heating value of gas, cold gas efficiency, carbon conversion efficiency, and thermal efficiency [31, 32]. All performance indexes are attributed to the in-bed gasification process, except the thermal efficiency is calculated for the whole plant.

Gas yield

$$\gamma = \left(\frac{Q_{syngas}}{Wb(1 - X_a)}\right) \tag{1}$$

where Q_{syngas} is the flow rate of syngas (producer gas) (Nm^3/h), Wb is the mass flow rate of biomass (kg/h), X_a is the ash content in the feed (on a dry basis).

High heating value

$$HHV\left(MJ/Nm^3\right) = \sum HHV_i x_i \tag{2}$$

where x_i (vol%) and HHV_i (MJ/Nm^3) represent the volumetric percentage and the higher heating value of each component in the dry product gas (mainly CO, H_2, and CH_4 in gasification).

Cold gas efficiency

$$\eta_{CG} = \frac{M_{syn} \times LHV_{syn}}{M_{biomass} \times LHV_{biomass}} \tag{3}$$

where M denotes mass.

$$LHV\left(MJ/Nm^3\right) = \sum LHV_i x_i \tag{4}$$

where x_i (vol%) and LHV_i (MJ/Nm^3) represent the volumetric percentage and the higher heating value of each component in the dry product gas (mainly CO, H_2, and CH_4 in gasification).

Carbon conversion efficiency:

$$\eta_C = \frac{C_{syngas}}{C_{bio}} \times 100\% \tag{5}$$

where C_{syngas} is the total carbon amount in the syngas, (kg) and C_{bio} is the total carbon amount in solid fuel, (kg).

Thermal efficiency

$$\eta_{th} = \frac{P_{NET}}{M_{biomass} \times LHV_{biomass}} \tag{6}$$

where P_{NET} is the net thermal power output of the plant, $M_{biomass}$ is the biomass feed rate and $LHV_{biomass}$ is the lower heating value of the biomass.

The performance indexes of a gasification system are affected by a number of factors [33] such as biomass characteristics, temperature, pressure, residence time, catalytic effects (of catalytic bed material and ash), and type and ratios of gasifying agents used.

2.2.1 Biomass characteristics

Biomass residues are products of forestry, agricultural, municipal, and industrial waste and have significant variations in their physical, chemical, and morphological characteristics [34]. Biomass is characterized by its elemental composition (ultimate analysis), moisture content, fixed carbon content, ash content, heating value, density, porosity, and thermal conductivity. The heating value and composition of the product gas depend on the biomass type along with other process parameters. Also, the amount of tar, particulates (ash and elutriated char), and other impurities like heavy metals, are concerns of careful choice of gasification technology and in-bed process optimization and/or cost of downstream cleaning of gas. In the case of lignocellulosic biomass, the main representative building blocks are cellulose, hemicellulose, and lignin. Since these constituent species follow distinct kinetic pathways in the devolatilization step, their proportion in the biomass affects the product distribution. Therefore, the performance of a biomass-based conversion system is hugely dependent on biomass characteristics. Gonzalez et al. [32] studied the effect of biomass characteristics on different performance indexes using 10 different biomass residues. The results suggest a positive correlation between VM, C content, and HHV_{bio} to the combustible gas concentration, calorific value, gas yield, and energy yield of the product gas, while the H_2 concentration is more favored with the H/O ratio of biomass. The reactivity of biomass is affected by the inorganic content present, which in turn affects the carbon conversion efficiency [34]. Also, alkaline is found in some lignocellulosic biomass, which reacts with bed material causing agglomeration [35]. In addition, the amount of ash in the biomass can impact the plant's operating cost due to stringent gas cleaning requirements in some applications [34].

2.2.2 Gasifying agent

The choice of gasifying agent affects the performance and economic aspects of fluidized bed gasification. Syngas heating values typically range between 4 and 7, 10–18, and $12 - 28MJ/Nm^3$ when air, steam, and oxygen are used as gasifying agents respectively [29]. In another study, a simple directly heated fluidized bed air gasification delivers syngas having low heating value ($4 - 6MJ/Nm^3$) and high tar content ($10 - 40g/Nm^3$) [13]. Alternatively, syngas with higher heating values ($10 - 40g/Nm^3$) can be obtained when oxygen and steam are mixed as gasifying agents for a similar gasifier design [36]. It is also demonstrated that a medium heating value ($10 - 15MJ/Nm^3$) gas can be produced in a dual bed gasifier using steam and air [37].

The high amount of nitrogen dilutes the gas resulting in a considerably lower heating value of yield gas when air is used as a gasifying agent. Steam gasification has the advantage of maximizing hydrogen production via water gas shift reaction ($H_2O + CO = > CO_2 + H_2$). But the resulting yield gas has lower quality in other performance indexes. As per a report on steam gasification in a circulating fluidized bed there was a decrease in heating value, gas yield, carbon conversion and an increase in tar yield [29]. Oxygen can be an excellent gasification agent as it results in high gas yield, a high heating value of gas, and less tar yield due to high reaction rates. The limitation is the high cost of oxygen production and operation costs [12]. Oxygen is a by-product of green hydrogen production via electrolysis. As a result, the availability of such plants within a reasonable distance from gasification plants can create an affordable supply of oxygen given that there are no other competing interests such as medical use of oxygen.

2.2.3 Equivalence ratio

The equivalence ratio[1] is one of the most important operating parameters. Higher ER favors gas yield but decreases CO, CH_4, and H_2 production and increases CO_2 production [29]. The higher the ER the higher the gas yield as it promotes oxidation and carbon conversion. The heating value of the gas is related to the amount of CO, H_2, and CH_4 [31]. High ER results in lower heating value product gas because more CO_2 and N_2 dilute the combustible gas species. Previous studies demonstrate that lower ER is desirable but too low ER means reduced temperature and the optimum value of ER, usually between 0.2 and 0.4 [38] needs to be maintained. If tar is a concern in the application of the producer gas, a higher ER is desirable (0.3–0.4 [38]) so that the reaction operates at a higher temperature, which favors tar cracking [33]. The response surface method which is a statistical technique that employs regression analysis based on mathematical relations is used to optimize values of input factors like temperature and equivalence ratio for optimal system performance [39].

2.2.4 Steam to biomass ratio

Steam is used as a gasifying agent for improved gas yield, LHV, and carbon conversion efficiency [33]. Water gas shift ($CO + H_2O \rightarrow H_2 + CO_2$) reaction is

---

[1] Equivalence ratio is defined as the ratio of actual air to fuel ratio to stoichiometric air to fuel ratio in this chapter's context

favored for the SBR range of 1.35–4.04, which increases the H_2 and CO_2 fraction in the producer gas [33]. Due to the increased yield of H_2, steam gasification is very suitable for H_2 production. When the SBR increases more than the optimum, the reaction temperature reduces due to too much low-temperature steam [40].

2.2.5 Temperature

Bed temperature is one of the predominant parameters affecting the reactions in the gasification process [35]. The chemical kinetics of multi-phase and multi-step reactions in the gasification process is governed by the Arrhenius law of rate constant, which defines the temperature dependence of reactions. As a result, temperature plays the main role in deciding the output gas distribution and the gasifier performance. In a fluidized bed, the temperature remains almost constant due to the high thermal inertia of the bed material. The wide range of operating temperature in fluidized bed gasifiers is between 700 and 1000°C [41]. Pooya et al. [35], studied the effect of operating temperature between 650 to 1050°C, on the gasification of two different biomass in a BFB and the increasing temperature is in favor of gas yield, HHV_{gas}, carbon conversion efficiency, and cold gas efficiency. A higher reaction temperature is in favor of high carbon conversion and tar cracking, which means lower tar and char in the producer gas [33]. However, the gasification temperature is limited by agglomeration and sintering of bed material and ash [13]. Therefore, applications, such as gas engines, turbines, fuel cells, and conversion of gas for the synthesis of fuels or chemicals, need extensive and costly gas cleaning [42] as the temperature is usually kept below 850°C in typical fluidized bed gasifiers.

2.2.6 Bed material

In fluidized bed gasification, the bed material is used as a mixing and heat transfer medium and desirably has high thermal inertia to maintain a fairly uniform temperature in the bed [43]. It has high thermal inertia enough to maintain a nearly constant temperature throughout the bed. The secondary effect of bed material could be acting as a catalyst in the case of catalytic beds, which influences the reaction rates of typical reactions involved in turn affecting the species distribution of yield gas. Some literature reported insignificant tar produced when the catalyst Rh/CeO2/SiO_2 is used in low-temperature gasifications [29, 44]. Gallucci et al. [4] conducted a lab-scale assessment of four different catalytic bed materials (olivine, k-feldspar, kaolinite, and calcite) on their potential for emission reduction of heavy metals and pollutants from PABR (plant-assisted bio-remediation) plant biomass. Several papers reported that Dolomite is an effective catalyst for tar-cracking and enhances gas yield [45, 46].

2.3 Biomass potential via gasification

Biomass resource potentials are large enough to deliver about a quarter (i.e. $200 - 300EJ$) of the world's future energy supply [3] during the century. The share of biomass in the global energy mix has grown from 4% in 2004 to 7.7% in 2013 (which is about 65% share among the renewables) [2], and 10% in 2018 (two-thirds in developing countries) [47]. The sum effect of factors like biomass availability, logistics, technologically feasible options, and policies is what decides the extent of effectiveness of the use of biomass as a competitive energy resource [48]. The primary step for the efficient development of a biomass conversion system is the quantification and energy-potential characterization of the available biomass with respect to the desired application [49]. Jaswinder et al. [48], studied the biomass potential, challenges, technological options, and government policies towards promoting biomass use in decentralized power generation in the Indian context.

2.3.1 Technical potential of biomass

Estimation of the biomass potential that can be utilized for heat power and biomass-derived fuels is a very important step. The potential of biomass is estimated from the annual main production (P_i) and the product-to-residue ratio (R_j) when agro-industrial wastes are considered for utilization [49]. The subscript i denotes the main biomass while the subscript j denotes the residue from its main biomass source more than a single residue can be produced from the main biomass source. The moisture content (MC_j) and low heating values (LHV_j) of the biomass define the theoretical energy potential. Availability factor A should be taken into account in order to estimate the technical energy potential of a biomass (QT_{ij}). In a black-box approach, the gasification system potential can be calculated considering the yearly operational hours H, gasification reactor efficiency ($\eta_{gasific}$), and the generator efficiency (η_{gen}). The heating values and other important characteristics of biomass residues used in thermochemical conversion are summarized in **Table 2**.

Biomass potential from annual main biomass production: $B_{ij} = P_i.R_j$.
Theoretical energy potential of each biomass: $Q_{ij} = B_{ij}.(1 - MC_j).LHV_j$.
The technical energy potential of each biomass: $QT_{ij} = Q_{ij}.A$.
Gasification potential of each biomass: $QG_{ij} = QT_{ij}.H.\eta_{gasific}.\eta_{gen}$.
Regional energy potential: $\overline{Q} = \sum Q_{ij}$, $\overline{QT} = \sum QT_{ij}$, $\overline{QG} = \sum QG_{ij}$.

Hiloidhari et al. [63] used this methodology to estimate India's biomass potential but not limited to gasification only. 686 MT [64] of biomass production with 34% surplus, and can roughly produce 23 GW [29] of power equivalent to the 17% of India's total primary energy demand. The case of India's biomass potential is reported in various papers.

Following the same methodology a 2017 paper [48] estimated India's biomass potential to be $30GW$ of electricity from all the surplus crops considering availability between 28% and 48% from the surplus residue of $223MT$, rice husk, bagasse, and sawdust being the most abundant ones. The annual operation hours are considered as $6570h/year$ whereas the average lower calorific values, lowest thermal efficiency, and average energy requirements for selected feedstocks (Rice, Wheat, Coarse cereals, Total Cereal, Cotton, Sugarcane) are considered. The installed capacity of bio-energy in India as of 2017 was only $5GW$, which is about 17% of its calculated potential.

3. Opportunities

3.1 GHG emission reduction and climate mitigation

According to the International Energy Agency, by 2019 fossil fuels were responsible for 32.1 gigatonnes of CO_2 emissions [65]. Additionally, it predicted that by 2030,

Biomass	Proximate analysis (wt, %)				Ultimate analysis (wt, %, db)					HHV (MJ/kg)	Refs.
	VM	FC	MC	Ash	C	H	O	N	S		
Distilled grain	—	—	13.84^{db}	5.84^{db}	49.93	7.26	36.45	5.31	1.04	27.2	[50]
Olive kernel			4.59	3.46	48.59	5.73	44.06	1.57	0.57	20	[51]
Corn cobs			7.1	5.34	46.3	5.3	42.19	0.57	0	17.9	[52]
	71.21	16.11	9.71	2.97	40.22	4.11	42.56	0.39	0.04	16.65^{a}	[43]
Sunflower stalks			40	3	52.9	6.58	35.9	1.38	0.15	20.8	[52]
Rapeseed stalks			5.86	3.95	45.52	5.53	48.37	0.58	0	16.8	[53]
Corn stalks			0	45.53	6.4	6.15	41.11	0.78	0.13	17.8	[52]
Pine				0.37	52.1	6.36	41	0.07	0.05		[8]
Pine saw dust	82.29	17.16	^{db}	0.55	50.54	7.08	41.11	0.15	0.57	20.54^{a}	[54]
Oak				1.29	49.9	5.98	42.6	0.21	0.05		[8]
barley straw				4.95	42.9	5.53	45.5	0.56	0.25		[8]
MSW				12	47.6	6	32.9	1.2	0.3		[8]
Sewage sludge			85	37.5	32.6	4.5	18.9	4.38	1.69		[8]
Cotton gin trash	71.20^{db}	15.78^{db}		13.02^{db}	39.3	5.43	40.49	1.44	0.34		[55]
Saw dust	76.1	8.9	14.6	0.4	44.96	5.83	45.5	3.1	0.61	17.12^{db}	[35]
	70.4	17.9	10.4	1.3	46.2	5.1	35.4	1.5	0.06	18.81^{a}	[35]
Empty fruit bunch	79.34	8.36	7.8	4.5	43.52	5.72	48.9	1.2	0.66	15.22^{db}	[35]
Cedar wood	80–82	18–20	^{db}	0.3	51.1	5.9	42.5	0.12	0.02	19.26^{a}	[56]
Olive oil residue	76	19.4	9.5	4.6	50.7	5.89	36.97	1.36	0.3	21.2^{a}	[57]
Rice husk	73.8	13.1	12.3	0.8	45.8	6	47.9	0.3		13.36^{a}	[58]
Rice straw	65.23	16.55	5.58	12.64	38.61	4.28	37.16	1.08	0.65	14.4^{a}	[59]
Spruce wood pellet	74.2	17.1	8.4	0.3	49.3	5.9	44.4	0.1		18.5^{a}	[60]
Coffee husk	74.3	14.3	10.4	1	46.8	4.9	47.1	0.6	0.6	16.54^{a}	[58]
Coffee ground	71.8	16.7	10.5	1	52.97	6.51	36.62	2.8	0.05	22^{a}	[61]
Sugarcane bagasse	67–70	28.7–30.7	^{db}	1.26	48.58	5.97	38.94	0.2	0.05	19.05^{a}	[44]
Wheat straw	75.8	18.1	^{db}	6.1	46.1	5.6	41.7	0.5	0.08	17.2^{a}	[62]

^{a}*given in lower heating value (LHV),* ^{db}*given in dry basis.*

Table 2.
Some of the Biomass residues used in thermochemical conversion pathways.

emissions would be around 42 gigatonnes with the current global trend unless climate agreements are implemented with stringent measures [65]. It is critical to abide by the 2015 Paris Agreement that states global temperature rise should not exceed above 2°C from the pre-industrial levels in order to combat the potentially catastrophic global climate change threat [66]. In accordance with this aim, the EU under the 2030 climate and energy framework plans GHG emissions reduction by 40% through increasing the share of renewable energy by 27% [67]. Resource and environmental sustainability measures primarily involve renewable energy production [6].

Burning of biomass is regarded as carbon neutral (also written in EU legislation [2]) because carbon is re-absorbed during the growth of biomass through photosynthesis where CO_2 and H_2O are converted to glucose and O_2 as depicted in **Figure 3**. A briefing on EPRS(European Parliamentary Research Service) discussed the opportunities and challenges of biomass use for electricity and heat with regard to GHG emission, resource availability, environment, and human health [2]. It outlined that biomass thermochemical conversion for heat and power generation has as high as up to 70% savings in GHG emission in some cases, but is influenced by factors like feedstock type, transportation, and conversion efficiency. In order to compute CO_2 savings and effective emission, the IPPC European Directive put CO_2 emission factors of $0.43kg/KWhCO_2$ for European energy mix for electricity generation (40% coal, 30% gas, 30% non-fossil) and $0.23kg/KWhCO_2$ for European fuel mix for thermal applications (50% gas, 40% oil, 10% coal) [68]. **Table 3** shows the commitment of some exemplary countries, EU, and worldwide in terms of the total installed bio-fuel capacity.

Biomass has been utilized for millennia and simply utilizing it does not guarantee sound benefits over other alternatives. In fact, the traditional use of biomass is very

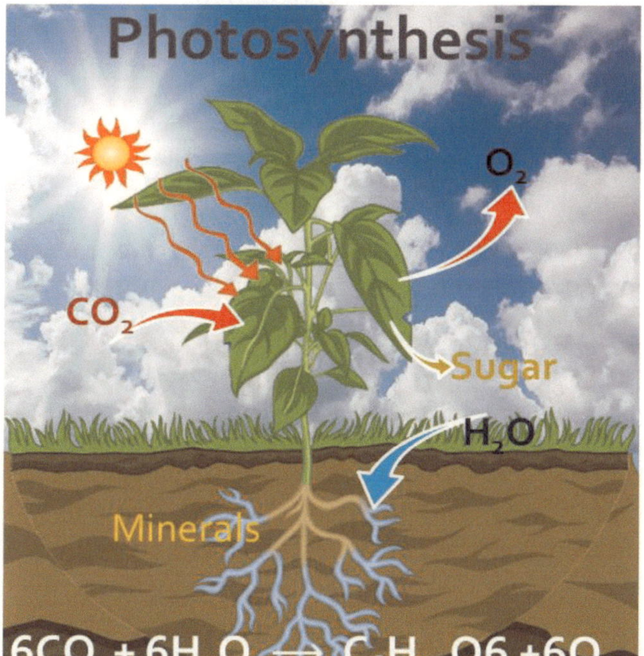

Figure 3.
Carbon cycle via photosynthesis.

Country	Total bio-fuel capacity	Year	Refs
UK	8.8 TWh (25% of total renewables energy	2019	[69]
Denmark & Finland	15% of total electricity production		[70]
Sweden, Austria, Estonia, Belgium, Italy, and Brazil	6 to 8% of total electricity production		[70]
Germany	49.1 GW	2014	[48]
China	17.8 GW	2018	[71]
US	16.2 GW	2018	[71]
India	10.2 GW	2018	[71]
Japan	4.0 GW	2018	[71]
EU	42 GW	2018	[71]
Global	93 GW (433 TWh production)	2014	

Table 3.
Bio-fuel installed capacities of countries, EU and global.

inefficient and can cause adverse effects on air quality and human health [2]. According to a 2017 World Energy Outlook special report [72], a third of the world's population (2.5 billion people) still traditionally serves with solid biomass, which is estimated to cause 2.8 million premature deaths annually due to indoor air pollution. Clean and efficient utilization of biomass with proven technologies like fluidized bed gasification is a promising technological solution for achieving GHG emission reduction and climate mitigation.

Biomass residues constitute municipal waste significantly, especially in rural and developing countries. The global municipal waste production in 2016 is 2.01 billion tonnes which is projected to grow to 2.2 billion in 2025 and 3.4 billion in 2050, according to World Bank statistics [73]. Out of which the trash-derived biomass accounts for 44% [26].

Direct and inefficient burning of biomass for heat and the production of charcoal is a common bio-energy utilization trend in developing countries. Wood, straws, cow dung, and other biomass residues are used for cooking, space heating, and lighting which accounts for about $30.7EJ$ and another 20 to 40% for informal sectors including charcoal production [74].

Potential deployment levels of bioenergy by 2050 could be in the range of 100 to $300EJ$ [3]. In sub-Saharan Africa alone, biomass has an outstanding potential of about $15{,}000MW$ from just 30% of residues from agricultural crops and forest logging residues [75]. However, there are large uncertainties in this potential such as market and policy conditions, and strong dependence on the rate of improvement in the agricultural sector too [3].

3.2 Distributed heat and power generation from local resource

Biomass conversion via fluidized bed gasification is a very good candidate for distributed power and heat generation. Biomass conversion is way more cost-effective when it is done locally than transported a long distance. In 2015, 72% of energy consumption for space heating and 64% for hot water comes from fossil fuels, and biomass share is about 12% [76]. Waste biomass for heating can fit here well to reduce dependency on fossil fuels. Once commercial, biomass gasification can be a cutting-edge technology for distributed energy generation. This in turn contributes to local income opportunities and green and sustainable development. Developing the local economy by adding value to local resources and reducing dependency on imported sources of energy access in many sustainable development schemes [3].

Biofuels are garnering attention as a promising renewable energy source and a business opportunity for rural communities by cutting fossil fuel demand and reducing greenhouse gas emissions [77]. Affordability, reliability, and sustainability of energy production in rural areas can be guaranteed by decentralized, small-scale systems owned by the local community [1].

As an interesting approach for biomass energy conversion, fluidized bed gasification technology has attractive features for distributed heat and power applications such as [3, 78]:

- Clean and sustainable conversion of local resources resulting in emission reduction and related incentives and government subsidies.
- Ability to handle a wide range of variety of biomass residues
- Techno-economically competitive small and medium-scale applications by avoiding hurdles related to biomass residue processing, storage, and transportation
- Fossil fuel substitute and independent energy access aside central grid
- Deployment of small and medium enterprises in the supply chain in agriculture and forestry

3.3 Circular economy integrating waste management and bio-products supply chain

A circular economy principle is based on closing the life cycle of a renewable resource [79] by extending its value through exploiting its waste. Local farmers can generate additional income by providing biomass fuels for small local power plants [3]. The circular and bio-economy structures are complementary in terms of maintainability and resource efficiency objectives [26]. The global bio-economy plans place a premium on the sustainable management of organic resources to ensure asset viability and biomass sustainability [80]. Such a circular bio-economy will demand the development of biorefineries in addition to enhanced sustainability measures [81]. PABR (Plant assisted bioremediation) plants are being researched for thermo-chemical conversion via catalytic fluidized bed gasification, which has the advantage of biomass being recycled in a sustainable way that complies with the circular economy principles [4] [2]. Several studies have demonstrated that comparable syngas quality can be produced from PABR plants in an FBG and the heavy metals are left in

---

[2] PABR (Plant assisted bioremediation): is a technique to recover soil contaminated by heavy metals and pollutants using plants' potential to extract them

the ash with the aid of catalytic beds [4], which maximizes the positive impact of the plant during its lifecycle.

3.4 Benefits attributed to carbon taxes and incentives

A carbon tax would give stakeholders in the biomass conversion projects, an incentive in return to efficiently reduce emissions of carbon dioxide and other greenhouse gases. Economically, biomass, as a renewable source of energy, attracts various tax benefits from many governments [82]. The steps taken by the EU such as decarbonization (20% GHG emission reduction) and significant renewable energy usage (20% increase) are among the most exemplary with a commitment of 20% upsurge in energy effectiveness by 2020 from 1990 levels [83]. Article 4 of the Directive (EU) 2018/2001 of the European Parliament and of the Council of 11, lists the support schemes for energy from renewable sources [84].

4. Existing challenges

4.1 Technical/technological bottlenecks

One of the key challenges in developing commercial advanced waste gasification technologies is to improve the quality of the syngas produced, making it suitable for a range of applications, including energy generation in gas engines or turbines, hydrogen production, or chemical feedstock [10]. Compared to traditional fuels, syngas from biomass gasification has a low heating value. During gasification, various compounds are released, such as tars, heavy metals, halogens, and alkaline compounds, which can cause both environmental and operational issues. Generally, the acceptable limit for tar is around $50 - 100\ mg/Nm^3$. While the amount of tar is important for engine applications, the dew point temperature of the tar is more crucial [13]. The heavy polyaromatic hydrocarbons primarily determine the dew point temperature of the total tar, even in small quantities.

As per recent research, both primary and secondary tar removal/reduction techniques are being widely used for advanced waste gasification technologies. Secondary methods like thermal or catalytic tar cracking and mechanical methods such as the use of cyclones, ceramic, fabric, or electrostatic filters, and wet scrubbers have been found to be highly effective in most cases, although they may not be economically feasible and can be particularly complex when very low tar content is required [85]. In-bed methods, which include the adequate selection of main operating parameters, use of proper bed additives or catalysts, and gasifier design and process optimization, are gaining more attention for waste gasification since they can significantly reduce the need for downstream cleanup [86].

Some in-bed tar treatment techniques are [13]:

- **Secondary air**: Secondary air in the freeboard contributes to more oxidation and reforming of water-soluble heterocyclic tars.

- **High-temperature gasification**: A recent experimental study on rice husk gasification [87] showed that non-catalytic high-temperature fluidized bed gasification yields higher heating values ($HHV = 3.6MJ/Nm^3$ and

$LHV = 3.2MJ/Nm^3$), higher carbon conversion efficiency (91.6%), higher thermal efficiency (75%), higher gas yield ($2.7m^3/kg$) and reduced tar + SPM (solid particulate matter) ($0.33g/Nm^3$). But the upper ceiling of the working temperature (900^0C in this case) is limited by the sintering of ash problem as rice husk has high ash content (18–20% in this case).

- **Narrowly optimized process**: 2–3 ring Light PAH and Heavy PAH require a narrow range of operating conditions of ER, gasification agents, catalysts, and biomass throughput. In a pilot-scale two-stage circulated fluidized bed process using an oxygen-steam gasifying agent, the tar yield was minimized to as low as $1.83g/Nm^3(db)$, in the optimum operating condition ($SBR = 0.15$ & $ER = 0.3$) [12].

- **In-bed catalysts:** In-bed catalysts such as calcined limestone, dolomite, and olivine sand and, less frequently, Ni-based or other metallic catalysts influence the tar cracking/reforming reactions.

In addition to tars, unwanted products in the produced gas (dust, fry ash, tars, ammonia, sulfur compounds, and others) pose a serious problem in the heating or power generation components of the plant and may need to undergo intensive gas cleaning which results in higher plant investment and operation costs [10]. Maximizing gasifier performance and obtaining high-quality fuel is also a very important consideration as it affects the overall technical and economic feasibility of the biomass conversion system. Still, further development and optimization are required to maximize CCE, thermal efficiency, and CGE and minimize tar, char, and ash in the producer gas [3].

4.2 Biomass availability and logistics

The collection, storage, and transport of biomass need to be carefully evaluated as biomass have a poor volumetric energy density, and transporting it at a cost of high energy density fuel (fossil fuels) may not be worth the benefit. Even in the practical scenarios, processing, handling, and transporting biomass from production sites to conversion plants may contribute 20 to 50% of the total costs of bioenergy production [3, 88].

The wide range of power production scale of fluidized bed gasification ($1MW - 1000MW$) presents the flexibility of exploiting biomass in via fluidized bed gasification in local small-scale plants based on the local resource. Yet, factors are affecting the reliability of locally available biomass such as weather conditions, agricultural practices, regulations, and competing uses for local biomass [78]. As a result, it requires critical assessment and forecasting of technically and financially available biomass quantities on an annual basis with a well-structured and proven methodology.

Densification techniques, such as baling, pelleting, briquet, and pyrolysis/torrefaction, help mitigate logistics costs associated with biomass transportation, storage, and handling, but the role of densification within the overall biomass-to-biofuel supply chain context is not yet well understood [89]. Densification by mechanical method increases the energy density of the biomass while torrefaction reduces the H/C and O/C ratio with the remaining 75–95% of total biomass energy

Property	Wood Pellets	Torrefied biomass Pellets	Coal (bituminous)
Moisture content (%)	7–10	1–5	5–10
Mass density (kg/m^3)	600–650	750–850	800–1000
LHV (MJ/kg)	16.2	19–22	>25
CV (MWh/t)	4.5	5.2–6.2	7
Energy density (MWh/m^3)	3	4.2–5	5.6–7
Bulk density (GJ/m^3)	7.5–10.4	15–18.7	18.4–23.8
Nature	Hygroscopic	Hydrophobic	Hydrophobic
Grindability	Poor	Good	Good
Biological degradation	Yes	No	No

Adapted [6].

Table 4.
Comparison of properties of Wood Pellets, Torrefied Biomass Pellets, and Coal.

after removing moisture, hemicellulose, and partially cellulose [29]. CO and H_2 production apparently increases with torrefaction [26, 90]. It results in near-complete degradation of its hemicellulose content while maximizing the mass and energy yield of the solid product [91]. It increases the calorific value, energy density, grindability and makes the biomass more hydrophobic so that it can be more suitable for storage [92]. **Table 4** compares the characteristics of torrefied biomass against wood pellets and coal. Such feedstock upgrading techniques are getting increasing attention, yet have their associated investment and operational costs.

4.3 Lack of adequate policy frameworks and strategies

Policy and regulatory barriers can significantly hinder the deployment of biomass waste for decentralized heat and energy generation. For the case of Europe, although there are some at the national and industry levels, there are currently no legally binding sustainability criteria for biomass at the EU level [2]. Several countries offer varying incentives and frameworks for biomass utilization, thereby creating disparities in policies and regulations [93]. These barriers come in the form of legal, regulatory, institutional, and economic constraints. Government regulatory policies may lack clarity, and inconsistent or overly restrictive policies can impede progress in this field.

For instance, a lack of uniformity in the requirements for emissions monitoring can make it much more difficult and expensive for project developers to advance [94]. Additionally, unclear feedstock requirements can make it challenging for suppliers to comply, which could prevent the facility from operating at full capacity or necessitate significant changes in the facility's design. Furthermore, it is less profitable than fossil-based power due to the lack of financial incentives and subsidies for decentralized heat and energy production facilities. Another significant obstacle is the unequal playing field between renewable and non-renewable energy sources. Potential investors may be deterred from investing as a result of these regulatory restrictions that may make it more difficult to implement decentralized heat and energy generation facilities using biomass wastes.

In conclusion, there is a need for supportive policies and clear regulatory frameworks to encourage the deployment of decentralized heat and energy generation facilities that use biomass waste.

5. Techno economics

The ability of fluidized bed gasification technology to effectively transform a variety of feedstocks into a clean and renewable form of energy makes it a desirable option for distributed heat and power generation. To determine a biomass conversion plant's economic viability, techno-economic considerations must be taken into account. In addition to end product application and feedstock availability, challenges such as cost-competitiveness and sustainability constrain bioenergy development [95]. The cost of fluidized bed gasification varies depending on factors like the region of the world, the type of feedstock, the cost of supplies for conversion processes, the size of the production, and the length of the production process.

Numerous academic institutions and international organizations recognize biomass as one of the most cost-effective renewable energy sources for power production [3]. Various biomass power technologies are mature and have production costs that are competitive with electricity generation rates in the OECD, particularly with low-cost agricultural or forestry waste sources like wood pellets, according to a recent report by the International Renewable Energy Agency (IRENA) [96]. Due to their low capital and operating expenses, small-scale gasification technologies have been shown to be both technically and economically viable for the production of heat. Industries that need process heat and have access to biomass resources should use these systems [3]. Additionally, by combining electrification with readily available biomass fuels from the local area, small-scale gasification systems have the potential to promote rural development [3].

Diego et al. [3], stated that the levelized cost of commercial bioenergy for electricity or combined heat and power ranges from US $cents_{2005}$ 3.5 to $25/kWh$ (USD_{2005} $10to50/GJ$) for liquid and gaseous biofuels and roughly USD_{2005} 2 to $48/GJ$ for electricity or combined heat and power (CHP) systems larger than about $2MW$. The estimated price for domestic or district heating systems is between USD_{2005} 2 and $77/GJ$, with feedstock costs between USD_{2005} 0 and $20/GJ$ [3]. Appropriate gasifier systems with internal combustion engines can produce $1kWh$ of electricity from $1.1 - 1.5kg$ wood, $0.7 - 1.3kg$ charcoal, or $1.8 - 3.6kg$ rice husks [3].

A Cost Benefit Analysis framework can be used to evaluate the feasibility of fluidized bed gasification projects from an economic perspective. This framework aids in the technical and economic analysis of the bioprocess's domain data, including capital and operational costs, simulation, equipment installation costs, and project profitability [97, 98]. Truong et al. [99], studied the effect of plant size on the total capital investment (TCI), total production cost (TPC), and specific capital cost for a fluidized bed gasification plant with a gas engine, gas turbine, and combined gas and steam turbines. TCI and TPC increase linearly between $10 - 140\$M$ and between $1 - 12.5\$M/year$, respectively for plant size from 10 to $550t/d$. The gasification system with a gas engine has the lowest TCP while the gasification system with gas and steam turbines has the highest TCP, with a narrow difference between them. The SCC on the contrary, has a very steep decline in plant size between 10 and $50t/d$, a medium decline between 50 and $150t/d$, a slight decline between 150 and $300t/d$, and almost no change beyond $300t/d$. Also, the plant with steam and gas turbines has the least SCC ($/(kWh/yr)).

State-of-the-art tools like Aspen Plus can be utilized to make TEA based on thermodynamic modeling, and energy and/or exergy analysis. In simulating and optimizing the biorefinery process flow, the size of the equipment and utilities can be calculated. Based on variables like equipment cost and material, raw material and utility costs, operating time, and product yield, the operational cost and investment are quantified. The discounted cash flow (DCF) method additionally aids in evaluating the project's profits, excises, internal rate of return (IRR), net present value (NPV), remuneration period, and lowest selling price [88]. It is also important to take into account the financial incentives if applicable. Antonio et al. [11] showed that the NPV is about four times when the financial incentives are considered in the techno-economic analysis of a fluid bed gasifier and combined gas-steam cycle.

In order to account for uncertainties in the TEA of biorefineries and to take into account societal sustainability factors like employment, the Monte Carlo simulation can also be used. Financial metrics like revenue, lowest selling price, and profitability—which are frequently expressed are typically used to deliver the TEA's results [88]. **Table 5** summarizes some of the TEA conducted on fluidized bed biomass gasification plants.

6. Future prospects

With the increasing global focus on climate change abatement and transition towards renewable energy sources, biomass gasification via fluidized bed gasification provides an opportunity to reduce greenhouse gas emissions while generating energy. Several research studies have outlined the future prospects of fluidized bed gasification technology.

According to the IPCC 2012 special report [1], advanced biomass integrated gasification combined-cycle (IGCC) power plants are among the technologies that are at a pre-commercial stage. With higher biomass to the power conversion efficiency of about 35–40%, the IGCC is a very competitive and attractive technology compared to conventional ones, especially in small-scale decentralized heat and power plants. For countries with existing huge small-scale decentralized plants, converting the conventional technology with a relatively small modification is considered a huge opportunity. Jesper et al. [102] showed the prospects of decentralized heat and power generation via gasification in the Case of Denmark where there are many existing small-scale heat and power plants with versatile technological options including FICFB (fast Internally Circulating Fluidized Bed) and LT-CFB (Low temperature circulating fluidized bed). The FICFB is designed for the cogeneration of electricity and bio-SNG and the LT-CFB is designed for the cogeneration of power, heat, and bio-fertilizer, with overall estimated efficiencies of 97 and 90% respectively.

The future prospect of fluidized bed gasification lies in leveraging key instruments such as advances in research and development, following proven methodologies in techno-economic feasibility and sustainability, and standardized and consistent policies and strategies at the regional, national, and global levels as depicted in **Figure 4**.

Gasification plant	Techno-economic analysis	Refs
Forest biomass blends gasification for small-scale power production facilities (1000KW Units) in the Azores (Portugal)—validated against 250$kWth$ quasi-industrial biomass gasifier in a pilot-scale fluidized bed reactor	• NPV: 486$k€$2020 • IRR: 17.44% • PBP: 7.4$years$	[100]
Fluidized bed gasification-based polygeneration from SCB (sugar cane bagasse) in Brazil—annual operating hours 8000—representative biomass price 25$€/t$ biomass	• PBP: 5.4 (DFB) and 7.6 (CFB) years • Plant operating time: 20 years • TCI: $k€$ (309, 427 & 398, 712 for BFB and CFB respectively) • SNPC: $k€_{2009}/kWh$ (0.09 & 0.12 for BFB and CFB respectively)	[101]
Fluidized bed gasification of wood chips with gas engine—various capacities: 15, 50, 150, 300, 500t/d	• TCI $\$M$: 11.5, 27.4, 44.9, 78.8, 127.3 with respect to capacities • SCC, $\$/(kWh/yr)$): 1.9, 1.3, 0.7, 0.6, 0.6 with respect to capacities • ASR $M\$/yr$: 1.3, 4.2, 12.7, 25.3, 42.3 with respect to capacities • PBP, yr.: –, 11.9, 5.4, 4.5, 4.3 with respect to capacities • ROI, %: –, 2.3, 11.0, 14.1, 15.2 with respect to capacities • DCFROR, %: –, 4.2, 15.4, 18.8, 19.9 with respect to capacities	[99]
Fluidized bed gasification of wood chips with gas turbine—various capacities: 15, 50, 150, 300, 500t/d	• TCI $\$M$: 18.4, 30.1, 49.6, 78.7, 124.2 with respect to capacities • SCC, $\$/(kWh/yr)$: 4.7, 1.9, 0.9, 0.6, 0.6 with respect to capacities • ASR, $M\$/yr$: 1.0, 3.7, 12.1, 25.5, 43.9 with respect to capacities—PBP, yr: –, 19.1, 6.7, 4.7, 4.1 with respect to capacities • ROI, %: –, −0.4, 8.0, 13.6, 16.1 with respect to capacities • DCFROR, %: –, −1.0, 11.9, 18.2, 20.8 with respect to capacities	[99]
Fluidized bed gasification of wood chips with gas and steam turbines—Various capacities: 15, 50, 150, 300, 500t/d	• TCI $\$M$: 22.2, 34.1, 58.4, 94.6, 143.5 with respect to capacities • SCC, $\$/(kWh/yr)$: 3.7, 1.3, 0.6, 0.5, 0.4 with respect to capacities • ASR, $M\$/yr$: 1.2, 4.7, 15.5, 32.8, 57.0 with respect to capacities—PBP, yr: -, 18.3, 5.8, 4.1, 3.5 with respect to capacities • ROI, %: -, −0.2, 10.0, 16.3, 20.0 with respect to capacities • DCFROR, %: -, −0.5, 14.3, 21.0, 24.8 with respect to capacities	[99]

DFB, Dual Fluidized Bed; CFB, Circulating Fluidized Bed; NPV, Net Present Value; IRR, Internal Rate of Return; PBP, Pay Back Period; TCI, Total Investment Cost; SNPC, Specific Net Production Cost; SCC, Specific Capital Cost; ASR, Annual Sale Revenue; ROI, Return on Investment; DCFROR, Discount Cash Flow Rate of Return.

Table 5.
Techno economic analysis conducted on different gasification plant designs.

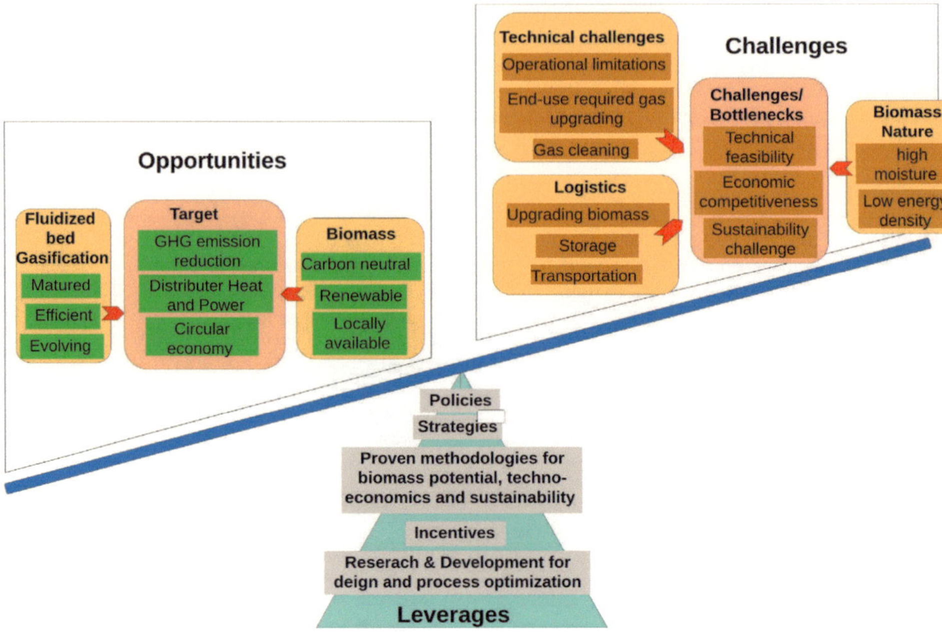

Figure 4.
Leveraging comprehensive instruments for opportunities outweigh challenges.

7. Conclusion

In conclusion, this study has demonstrated the significant promise of fluidized bed gasification technology in converting biomass wastes into viable energy sources for distributed heat and power production, thereby contributing to climate change mitigation efforts. By effectively harnessing waste materials, this technology offers a compelling means of reducing greenhouse gas emissions.

However, the widespread adoption of fluidized bed gasification necessitates the resolution of various challenges related to technology, supply chains, and policies. The optimization of the gasification process, targeting enhanced system performance, gas yield, and quality, emerges as a primary technical challenge. Additionally, managing the heterogeneity of feedstock compositions and developing efficient approaches for treating byproducts like tar and particulates pose substantial technical and economic obstacles.

Nonetheless, the overall conversion of biomass waste into energy through fluidized bed gasification presents a promising avenue for climate change mitigation, sustainable energy provision, and the promotion of circular economy principles. To fully capitalize on the benefits of this technology, it is imperative to establish well-defined policies. These policies should be designed to promote renewable energy sources and facilitate the widespread adoption of fluidized bed gasification technology.

By effectively optimizing the implementation of fluidized bed gasification in biomass-to-heat and energy conversion pathways, significant contributions can be made towards sustainable energy provision and addressing the pressing issue of climate change. Future research and development efforts should continue to focus on overcoming technical challenges, refining supply chains, and formulating comprehensive policies to unlock the full potential of fluidized bed gasification in biomass waste utilization.

Acknowledgements

The authors would like to give our heartfelt thanks to the CRECK team members of Politecnico di Milano University, Italy. We also acknowledge the grant support from KFW through EXiST Project: Excellence in Science and Technology being implemented at Jimma Institute of Technology, Ethiopia (KFW Project No. 51235).

Conflict of interest

The authors declare no conflict of interest.

Author details

Henok Mekonnen Mossissa[1,2*], Million M. Afessa[1,2], Derese T. Nega[1,2,3], Muhammad Ahsan Amjed[2,4] and A. Venkata Ramayya[1]

1 Faculty of Mechanical Engineering, Jimma Institute of Technology, Ethiopia

2 Department of Chemistry, Materials and Chemical Engineering "Giulio Natta" (DCMC), Politecnico di Milano University, Polimi, Italy

3 Pacific Water Research Centre, Simon Fraser University, Canada

4 Department of Energy "Giulio Natta" (DCMC), Politecnico di Milano University, Polimi, Italy

*Address all correspondence to: henatan99@gmail.com

References

[1] Edenhofer O, Pichs-Madruga R, editors. Renewable energy sources and climate change mitigation: Summary for policymakers and technical summary. Intergovernmental Panel on Climate Change (IPCC). 2012. Available from: https://archive.ipcc.ch/pdf/special-reports/srren/SRREN_FD_SPM_final.pdf

[2] Bourguignon D. Biomass for electricity and heating: Opportunities and challenges. EPRS—European Parliamentary Research Service. PE 568.329. 2015 Availabe from: https://www.europarl.europa.eu/RegData/etudes/BRIE/2015/568329/EPRS_BRI(2015)568329_EN.pdf

[3] Masera D. Renewable energy forinclusive and sustainable industrial development: The case of biomass gasification. United Nations Industrial Development Organization. 2014. Available from: https://www.unido.org/sites/default/files/2014-10/Gasification_FINAL_0.pdf

[4] Gallucci F, Palma A, Vincenti B, Carnevale M, Paris E, Ancona V, et al. Fluidized bed gasification of biomass from plant assisted bioremediation (pabr): Lab-scale assessment of the effect of different catalytic bed material on emissions. Fuel. 2022;**322**:124214

[5] Lakshana PV, Nair G, Agrawal K. An overview of sustainable approaches for bioenergy production from agro-industrial wastes. Energy Nexus. 2022;**6**:100086

[6] Bharadwaj K, Henbagaraj S, Sharma PK, Sharma AK, Ghodke PK, Chen W-H. Biomass torrefaction: An overview of process and technology assessment based on global readiness level. Fuel. 2022;**324**:124663

[7] Kumar A M, Sindhu R, Sirohi R, Kumar V, Ahluwalia V, Binod P, et al. Agricultural waste biorefinery development towards circular bioeconomy. Renewable and Sustainable Energy Reviews. 2022;**158**:112122

[8] Siedlecki M, de Jong W, Verkooijen AHM. Fluidized bed gasification as a mature and reliable technology for the production of bio-syngas and applied in the production of liquid transportation fuels—A review. Energies. 2011;**4**:389-434

[9] Ltd. Qingdao Kexin New Energy Technology Co. Industrijski gasifikator biomase. 2017. Available from: https://ba.biomasspowerchina.com/biomass-gasifier/industrial-biomass-gasifier.html

[10] Ruiz JA, Juárez MC, Morales MP, Muñoz P, Mendívil MA. Biomass gasification for electricity generation: Review of current technology barriers. Renewable and Sustainable Energy Reviews. 2013;**18**:174-183

[11] Caputo AC, Palumbo M, Pelagagge PM, Scacchia F. Economics of biomass energy utilization in combustion and gasification plants: Effects of logistic variables. Biomass and Bioenergy. 2005;**28**:35-51

[12] Wang C, Zhang M, Han Z, Bai D, et al. Pilot verification of a two-stage fluidized bed gasifier with a downer pyrolyzer using oxygen-rich air. Fuel. 2022;**307**:121816

[13] Gómez-Barea A, Leckner B, Perales AV, Nilsson S, Cano DF. Improving the performance of fluidized bed biomass/waste gasifiers for distributed electricity: A new three-stage

gasification system. Applied Thermal Engineering. 2013;**50**:1453-1462

[14] Lalji N S, Panwar NL, Gupta T, Meena KK. Bubbling fluidized bed gasification of biomass: A review on the effect of selected operational parameters. Biointerface Research in Applied Chemistry. 2023;**13**(5):474

[15] Guan G, Fushimi C, Tsutsumi A, Ishizuka M, Matsuda S, Hatano H, et al. High-density circulating fluidized bed gasifier for advanced igcc/igfc—Advantages and challenges. Particuology. 2010;**8**:602-606

[16] Liu Z, Peng W, Motahari-Nezhad M, Shahraki S, Beheshti M. Circulating fluidized bed gasification of biomass for flexible end-use of syngas: A micro and nanoscale study for production of bio-methanol. Journal of Cleaner Production. 2016;**129**:249-255

[17] Pfeifer C, Puchner B, Hofbauer H. Comparison of dual fluidized bed steam gasification of biomass with and without selective transport of CO_2. Chemical Engineering Science. 2009;**64**:5073-5083

[18] Rauch R, Hofbauer H. Bolh'ar-Nordenkampf; M. Biomass chp plant gussing—using gasification for power generation. In: Proceedings of the Second RCETCE Conference. Phuket, Thailand; 2003. pp. 567-572

[19] Guan H, Fan X, Zhao B, Yang L, Sun R. Application and discussion of dual fluidized bed reactor in biomass energy utilization. IOP Conference Series: Earth and Environmental Science. 2018;**108**:052019

[20] Verbong GPJ, Kirkels AF. Biomass gasification: Still promising? A 30-year global overview. Renewable and Sustainable Energy Reviews. 2011;**15**: 471-481

[21] Sheth PN, Patra TK. Biomass gasification models for downdraft gasifier: A state-of-the-art review. Renewable and Sustainable Energy Reviews. 2015;**50**:583-593

[22] Susastriawan AAP, Saptoadi H. Small-scale downdraft gasifiers for biomass gasification: A review. Renewable and Sustainable Energy Reviews. 2017;**76**:989-1003

[23] Hanchate N, Ramani S, Mathpati CS, Dalvi VH. Biomass gasification using dual fluidized bed gasification systems: A review. Journal of Cleaner Production. 2021;**280**:123148

[24] Heyne S, Thunman H, Harvey S. Exergy-based comparison of indirect and direct biomass gasification technologies. Biomass Conversion and Biorefinery. 2013;**2013**:2190-6815

[25] Smith C, Evans G. 5.11 - biomass to liquids technology. Biomass Conversion and Biorefinery. 2012;**5**:155-204

[26] Dutta A, Acharya B, Sule I. Ia review on advances of torrefaction technologies for biomass processing. Biomass Conversion and Biorefinery. 2012;**2**: 349-369

[27] Jankes GG et al. Biomass gasification with chp production: A review of state of the art technology and near future perspectives. Thermal Science. 2012;**16**: 115-130

[28] Larsson A, Kuba M, Vilches TB, Seemann M, Hofbauer H, Thunman H. Steam gasification of biomass—Typical gas quality and operational strategies derived from industrial-scale plants. Fuel Processing Technology. 2021;**212**:106609

[29] Karmakar MK, Chandra P, Chatterjee P. Circulating fluidized bed

gasification: Status, challenges and prospects in Indian perspective. Indian Journal of Science and Technology. 2016; **9**:48

[30] Al-Ansari T, AlNouss A, McKay G. A comparison of steam and oxygen fed biomass gasification through a techno-economic-environmental study. Energy Conversion and Management. 2020; **208**(112612):115-130

[31] Zabaniotou A, Damartzis T, Michailos S. Energetic assessment of a combined heat and power integrated biomass gasification–internal combustion engine system by using aspen plus. Fuel Processing Technology. 2012;**95**:37-44

[32] González-Vázquez MP, García R, Gil MV, Pevida C, Rubiera F. Comparison of the gasification performance of multiple biomass types in a bubbling fluidized bed. Energy Conversion and Management. 2018;**176**: 309-323

[33] Alauddin ZABZ, Lahijani P, Mohammadi M, Mohamed AR. Gasification of lignocellulosic biomass in fluidized beds for renewable energy development: A review. Renewable and Sustainable Energy Reviews. 2010;**14**: 2852-2862

[34] Moilanen A, Nasrullah M, Kurkela E. The effect of biomass feedstock type and process parameters on achieving the total carbon conversion in the large scale fluidized bed gasification of biomass. Environmental Progress in Sustainable Energy. 2009;**28**:355-359

[35] Zainal ZA, Pooya Lahijani A. Gasification of palm empty fruit bunch in a bubbling fluidized bed: A performance and agglomeration study. Bioresource Technology. 2011;**102**: 2068-2076

[36] Rauch R, Hofbauer H, Bosch K, Siefer I, Aichernig C, et al. Steam gasification of biomass at chp plant in güssing e status of the demonstration plant. In: Proceedings of 2nd World Conference on Biomass for Energy, Industry and Climate Protection. Rome; 2004. pp. 1687-1690

[37] Paisley MA, Overend RP. The sylvagas process from future energy resources: A commercialization success. In: Proceedings of 12th European Conference on Biomass for Energy, Industry and Climate Protection. Amsterdam; 2002. pp. 975-978

[38] Narvaez I, Orio A, Aznar M, Corella J. Biomass gasification with air in an atmospheric bubbling fluidized bed. Effect of six operational variables on the quality of the produced raw gas. Industrial and Engineering Chemistry Research. 1996;**35**:2110-2120

[39] Nemmour A, Inayat A, Janajreh I, Ghenai C. New performance correlations of municipal solid waste gasification for sustainable syngas fuel production. Biomass Conversion and Biorefinery. 2022;**12**:4271-4289

[40] Yuhong Q, Feng J, Li WY. Formation of tar and its characterization during air–steam gasification of sawdust in a fluidized bed reactor. Fuel. 2010; **89**(7):1344-1347

[41] Sajid M, Raheem A, Ullah N, Asim M, Rehman MSU, Ali N. Gasification of municipal solid waste: Progress, challenges, and prospects. Renewable and Sustainable Energy Reviews. 2022;**168**:112815

[42] Leckner B, Gómez-Barea A. Gasification of biomass and waste. Handbook of Combustion. 2009;**4**: 365-397

[43] Lu Y, Jin H, Guo L, Zhang X, Cao C, Guo X. Hydrogen production by biomass gasification in supercritical water with a fluidized bed reactor. International Journal of Hydrogen Energy. 2008;**33**: 6066-6075

[44] Asadullah M, Miyazawa T, Ito SI, Kunimori K, Tomishige K. Catalyst performance of rh/ceo2/sio2 in the pyrogasification of biomass. Energy & Fuels. 2003;**17**(4):842-849

[45] Corella J, Herguido J, Gonzalez-Saiz J, Alday FJ, Rodriguez-Trujillo JL. Fluidized bed steam gasification of biomass with dolomite and with a commercial FCC catalyst. Research in Thermochemical Biomass Conversion. 1988;**1988**:754-765

[46] Ji X, Bai J, Xu M, Tian Y, Zhou X, Lin S. Syngas production from air-steam gasification of biomass with natural catalysts. Science Total Environment. 2018;**645**:518-523

[47] Thornley P, Adams P. Greenhouse Gas Balances of Bioenergy Systems. London, UK: Academic Press; 2018

[48] Singh J. Management of the agricultural biomass on decentralized basis for producing sustainable power in India. Journal of Cleaner Production. 2017;**142**:3985-4000

[49] Patricia Pérez-Rodríguez C, Ríos LA, González CSD, Montaña A, García-Marroquín C. Harnessing residual biomass as a renewable energy source in Colombia: A potential gasification scenario. European Conference on Renewable Energy Systems. 2022; **14**(19):12537-12537

[50] Kumar A, Eskridge K, Jones DD, Hanna MA. Steam–air fluidized bed gasification of distillers grains: Effects of steam to biomass ratio, equivalence ratio and gasification temperature. Bioresource Technology. 2009;**100**: 2062-2068

[51] Skoulou V, Swiderski A, Yang W, Zabaniotou A. Process characteristics and products of olive kernel high-temperature steam gasification (htsg). Bioresource Technology. 2009;**100**: 2444-2451

[52] Ioannidou O, Zabaniotou A. Agricultural residues as precursors for activated carbon production—A review. Renewable and Sustainable Energy Reviews. 2007;**11**:1966-2005

[53] Skoulou V, Zabaniotou A, Ioannidou O. Rapeseed residues utilization for energy and 2nd generation biofuels. Fuel. 2008;**87**:1492-1502

[54] Lv P, Xiong Z, Chang J, Wu C, Chen Y, Zhu J. An experimental study on biomass air–steam gasification in a fluidized bed. Bioresource Technology. 2004;**95**:95-101

[55] Maglinao AL, Capareda SC, Nam H. Fluidized bed gasification of high tonnage sorghum, cotton gin trash and beef cattle manure: Evaluation of synthesis gas production. Energy Conversion and Management. 2015;**105**: 578-587

[56] Cao Y, Wang Y, Riley J, Pan W. A novel biomass air gasification process for producing tar-free higher heating value fuel gas. Fuel Process Technology. 2006; **87**:343-353

[57] Arvelakis S, Gehrmann H, Beckmann M, Koukios E. Agglomeration problems during fluidized bed gasification of olive-oil residue: Evaluation of fractionation and leaching as pre-treatments. Fuel. 2003;**82**: 1261-1270

[58] Vélez J, Chejne F, Valdés C, Emery E, Londöo C. Co-gasification of colombian coal and biomass in fluidized bed: An experimental study. Fuel. 2009; **88**:424-430

[59] Li K, Zhang R, Bi J. Experimental study on syngas production by co-gasification of coal and biomass in a fluidized bed. International Journal of Hydrogen Energy. 2009;**35**:2722-2726

[60] Miccio F, Piriou B, Ruoppolo G, Chirone R. Biomass gasification in a catalytic fluidized reactor with beds of different materials. Chemical Engineering Journal. 2009;**154**:369-374

[61] Murakami T, Xu G, Suda T, Matsuzawa Y, Tani H, Fujimori T. Some process fundamentals of biomass gasification in dual fluidized bed. Fuel. 2007;**86**:244-255

[62] Oesch P, Leppamaki E, Stahlberg P. Sampling and characterization of high-molecular-weight polyaromatic tar compounds formed in the pressurized fluidized-bed gasification of biomass. Fuel. 1996;**75**:1406-1412

[63] Baruah DC, Hiloidhari M, Das D. Bioenergy potential from crop residue biomass in India. Renewable and Sustainable Energy Reviews. 2014;**32**: 504-512

[64] Ministry of New and Government of India Renewable Energy. Mnre-undp-gef biomass power project. Biopower India - A quarterly magazine on biomass energy, 2014

[65] Ram M, Mondal MK. Biomass gasification: A step toward cleaner fuel and chemicals. Biofuels and Bioenergy. 2022;**2022**:253-276

[66] UNFCCC. The paris agreement. 2015

[67] Kikas T, Cahyanti MN, Doddapaneni TRKC. Biomass torrefaction: An overview on process parameters, economic and environmental aspects and recent advancements. Bioresource Technology. 2020;**301**:12273

[68] Alfonso D, Perpiñá C, Pérez-Navarro A, Peñalvo E, Vargas C, Cárdenas R. Methodology for optimization of distributed biomass resources evaluation, management and final energy use. Biomass and Bioenergy. 2009;**33**:1070-1079

[69] Littlechild S, Vaidya K. Energy Strategies for the UK. Abingdon, Oxon, United Kingdom: Routledge; 2019

[70] Bacovsky D. et al. Iea bioenergy countries' report: Bioenergy policies and status of implementation. 2016

[71] Raturi AK. Renewables 2016 Global Status Report. 2016

[72] Daly H, Walton MA, Cozzi L, Yong LG, Adam Z, Bouckaert S, et al. Energy Access Outlook 2017: From Poverty to Prosperity (World Energy Outlook Special Report). Intenational Energy Agency; 2017. Available from: https://iea.blob.core.windows.net/assets/9a67c2fc-b605-4994-8eb5-29a0ac219499/WEO2017SpecialReport_EnergyAccessOutlook.pdf

[73] Kaza SY, Lisa C, Bhada-Tata P, van Woerden F. Publication: What a Waste 2.0: A Global Snapshot of Solid Waste Management to 2050. OPEN KNOWLEDGE REPOSITORY BETA—World Bank. 2018

[74] Helena C, Andre F, Jose M, Göran B, et al. Chapter 2 – Bioenergy: Renewable Energy Sources and Climate Change Mitigation Special Report of the Intergovernmental Panel on Climate

Change. Cambridge, United Kingdom: Cambridge University Press; 2011. pp. 209-332. Available from: https://www.cambridge.org/core/books/abs/renewable-energy-sources-and-climate-change-mitigation/bioenergy/ACD7914E3E50CFB05DA2A828CA49E25D

[75] Dasappa S. Potential of biomass energy for electricity generation in sub-saharan africa. Energy for Sustainable Development. 2011;**15**:203-213

[76] Quaschning V. Renewable Energy and Climate Change. Hoboken, NJ, USA: John Wiley Sons, Ltd.,; 2019

[77] Ralph EHS, Mabee W, Saddler JN, Taylor M. An overview of second generation biofuel technologies. Bioresource Technology. 2010;**101**: 1570-1580

[78] Fytili D, Zabaniotou A. Circular economy synergistic opportunities of decentralized thermochemical systems for bioenergy and biochar production fueled with agro-industrial wastes with environmental sustainability and social acceptance: A review. Current Sustainable/Renewable Energy Reports. 2018;**5**:150-155

[79] Muerza V, Urciuoli L, Habas SZ. Enabling the circular economy of bio-supply chains employing integrated biomass logistics centers - a multi-stage approach integrating supply and production activities. Journal of Cleaner Production. 2023;**384**:135628

[80] Commission to European Parliament. Communication from the commission to the european parliament, the council, the european economic and social committee and the committee of the regions. European Commission. 2020. Available from: https://eur-lex.europa.eu/legal-content/EN/TXT/?uri=CELEX%3A52020DC0066

[81] Husgafvel R, Linkosalmi L, Hughes M, Kanerva J, Dahl O. Forest sector circular economy development in Finland: A regional study on sustainability driven competitive advantage and an assessment of the potential for cascading recovered solid wood. Journal of Cleaner Production. 2018;**181**:483-497

[82] Ahmed A, Ge T, Peng J, Yan W-C, Tee BT, You S. Assessment of the renewable energy generation towards net-zero energy buildings: A review. Energy Buildings. 2022;**256**:111755

[83] Antoine D, Nachtigall D, Venmans F. The joint impact of the european union emissions trading system on carbon emissions and economic performance. 2018

[84] Directive (EU) 2018/2001 of the European Parliament and of the Council of 11 December 2018 on the promotion of the use of energy from renewable sources (recast). Official Journal of the European Union, L 328/82, 21 December 2018

[85] Arena U, Zaccariello L, Mastellone ML. Fluidized bed gasification of waste-derived fuels. Waste Management. 2010;**30**:1212-1219

[86] Devi L, Ptasinski KJ, Janssen FJJG. A review of the primary measures for tar elimination in biomass gasification processes. Biomass and Bioenergy. 2003; **24**:125-140

[87] Jignesh P, Jay M, Mishra PG. Improving the properties of producer gas using high temperature gasification of rice husk in a pilot scale fluidized bed gasifier (fbg). Renewable Energy. 2019; **130**:943-951

[88] Goswami L, Kayalvizhi R, Dikshit PK, Sherpa KC, Roy S, Kushwaha A, et al. A critical review on prospects of bio-refinery products from second and third generation biomasses. Chemical Engineering Journal. 2022;**448**: 137677

[89] Nibal TA, Heier Stamm JL. Optimization of lignocellulosic biomass-to-biofuel supply chains with densification: Literature review. Biomass and Bioenergy. 2020;**144**:105888

[90] Commandre JM, Couhert C, Salvador S. Impact of torrefaction on syngas production from wood. Fuel. 2009;**88**(11):2286-2290

[91] Basu P. Biomass Gasification, Pyrolysis and Torrefaction: Practical Design and Theory. London, UK: Academic Press; 2018

[92] Kartal UÖF. Investigation of an integrated circulating fluidized bed gasifier/steam turbine/proton exchange membrane (pem) fuel cell system for torrefied biomass and modeling with artificial intelligence approach. Energy Conversion and Management. 2022;**263**: 115718

[93] Daneshmandi M, Sahebi H, Ashayeri J. The incorporated environmental policies and regulations into bioenergy supply chain management: A literature review. Science of the Total Environment. 2022;**820**:153202

[94] Wylie L, Sutton-Grier AE, Moore A. Keys to successful blue carbon projects: Lessons learned from global case studies. Marine Policy. 2016;**65**:76-84

[95] Ali B M, Mehranzamir K, Ahmed J, Nafea M, Nabipour-Afrouzi H, Wooi CL, et al. Techno-economic analysis of direct combustion and gasification systems for off-grid energy supply: A case for organic rankine cycle and dual fluidized-bed. IET Renewable Power Generation. 2021;**15**:2596-2614

[96] Taylor M, Ralon P, Al-Zoghoul S, Epp B, Jochum M. Renewable power generation costs in 2020. International Renewable Energy Agency. 2020. Available from: https://www.irena.org/-/media/Files/IRENA/Agency/Publication/2021/Jun/IRENA_Power_Generation_Costs_2020.pdf

[97] Piok K, Rakshit SK, Kosamia NM, Samavi M. Perspectives for scale-up of biorefineries using biochemical conversion pathways: Technology status, techno-economic, and sustainable approaches. Fuel. 2022;**324**:124532

[98] Brown RC, Brown TR. Techno-economics of advanced biofuels pathways. RSC Advances. 2013;**3**(17):5758-5764

[99] Xuan D T, Lim Y, Yeo H, Lee U, Choi Y, Song J. Techno-economic analysis of power plant via circulating fluidized-bed gasification from woodchips. Energy. 2014;**70**:547-560

[100] Cardoso CS, Silva V, Eusébio D, Lima Azevedo I, Tarelho LAC. Techno-economic analysis of forest biomass blends gasification for small-scale power production facilities in the Azores. Fuel. 2020;**279**:118552

[101] Silva O P, Maier S, Dietrich R-U, Mariano AP, Filho RM, Posada J. Comparative techno-economic and exergetic analysis of circulating and dual bed biomass gasification systems. Frontiers in Chemical Engineering. 2021;**3**: 727068

[102] Ahrenfeldt J, Thomsen TP, Henriksen U, Clausen LR. Biomass gasification cogeneration – A review of state of the art technology and near future perspectives. Applied Thermal Engineering. 2013;**50**(2):1407-1417

Chapter 12

Optimizability of Biogenic Hydrogen Production

Natascha Eggers, Celia Kirsch, Fabian Giebner and Torsten Birth

Abstract

The biogenic hydrogen production by dark fermentative digestion of biomass shows excellent potential for solving the recent energy difficulties. To augment the efficiency of this process, the biochemical degradation needs to be understood better. The physical optimum (PhO) method was found to be particularly suitable to illustrate the maximum potential of the process. It is based on a theoretical, ideal reference state. For this purpose, the dark fermentation was modeled mathematically using Petri nets, a modeling language based on graph theory. The Petri net model was developed based on the biochemical reactions system of anaerobic digestion. All degradative steps relevant to hydrogen production were included. Then the modeled metabolic rates were analyzed to be used as a reference base in evaluating the process's efficiency. This ideal reference state, the anaerobic digestion's PhO, was compared with experimental data by the PhO factor. This factor summarizes the efficiency of the whole process.

Keywords: dark fermentation, efficiency assessment, physical optimum, modeling, Petri net

1. Introduction

Hydrogen technologies, combined with sector coupling, can decarbonize sectors that cannot be electrified (e.g., the chemical industry, steel industry, etc.) [1].

In 2030, hydrogen will account for 0.5% of the electricity mix, according to studies by DNV. In 2050, DNV expects it to be 5%. However, according to the Paris Agreement, at least 15% hydrogen is required in the electricity mix [1]. Yet, expanding renewable energy and electrolysis technologies at the necessary pace will be nearly impossible. Both renewables and electrolysis-ready water are critical resources that could lead to supply insecurities [2].

For this reason, we need new technologies for the production of hydrogen in the course of decarbonization.

1.1 Biological hydrogen as an alternative to electrolysis

In 2021 bioenergy covered about 55 % of global renewable energy supply (6 % of total energy supply) which makes it the largest source of renewable energy [3]. According to Net Zero Emissions by 2050 the use of bioenergy will increase rapidly by 2030 [4].

The conversion of biomass can take place *via* three routes: thermo-chemical (pyrolysis, gasification), physicochemical (e.g., chemical conversion), or biochemical (anaerobic fermentation, alcoholic fermentation). All three approaches can produce gaseous and liquid fuels, with the first two having high energy requirements. Biochemical transformation, on the other hand, proceeds essentially in a self-determined manner and with a lower external energy requirement. [5, 6]

A general advantage of biomass conversion is the sensible use of biological waste products (e.g., from agriculture) [6]. A significant disadvantage is the current inefficiency of the hydrogen production method [7]. The resulting biogas contains only small amounts of hydrogen since, in most cases, hydrogen is directly decomposed to form methane [8].

This effect can be reduced by so-called dark fermentation, an anaerobic fermentation without light. Methanogenesis is suppressed using bacterial cultures in which few hydrogen-consuming microorganisms (MOs) are present [8]. The process's main advantages are the simple reactor configuration and the ability to produce hydrogen continuously regardless of ambient conditions.

1.2 The efficiency of hydrogen *via* dark fermentation

Dark fermentation has not yet been investigated on a large scale at the current stage. Laboratory studies suggest substantial production variations depending on inoculum, substrate, and substrate pretreatment. Values between 27 L H_2/kg TVS (Total Volatile Solids) [9] and 125 L H_2/kg TVS [10] were measured for continuously operated biogas plants. This corresponds to a maximum of one-third of converted initial biomass [11].

1.3 Challenges in optimizing biological processes

Efficiency assessment is essential, especially for sector coupling processes, but it is also difficult. Depending on the (sub)process of the overall process under consideration, there are various methods and key figures for efficiency evaluation. This makes a holistic, objective comparison of the impossible processes. However, this is indispensable if the optimizability of the overall process is to be determined [12]. While the efficiency of purely technical processes can at least be described thoroughly, biological (sub)processes are even more difficult to describe due to their complexity. Biological processes can hardly ever be considered completely isolated since they usually interact with many other processes and are thus dependent on them. On the one hand, this makes it challenging to draw balance boundaries. On the other hand, not all interaction partners are known, so a complete consideration of the isolated process is impossible. Thus, a comprehensive efficiency assessment of these processes is (almost) impossible, even if the state of knowledge yields new information.

Therefore, this chapter develops a model to describe dark fermentation's physical optimum (PhO). Based on this model, the efficiency potentials of the process can be determined, and recommendations for improvement can be derived. An in-depth understanding of the process is required for model development. For this reason, the anaerobic degradation of biomass is described first. The dark fermentation process is then mapped using Petri nets (PNs) to create an idealized comparison process for evaluating existing processes.

2. Theoretical basics of dark fermentation

In biogas plants, biomass is converted into biogas by bacterial decomposition. Primarily methane (60% [13]) and carbon dioxide (35% [13]), but also hydrogen is produced. The biochemical conversion of biomass to biogas is divided into four reaction stages: hydrolysis, acidogenesis, acetogenesis, and acetylation [5, 13].

Dark fermentation, the anaerobic degradation of biomass without light, is the term for hydrogen formation. Dark fermentation is thus identical to anaerobic digestion up to hydrogen formation (acetogenesis). Only methanogenesis is inhibited in dark fermentation [14]. In an actual biogas plant, the degradation process does not simply stop after hydrogen formation. Instead, the resulting H_2 is further converted to methane with CO_2. Accordingly, this step must be eliminated to increase the hydrogen yield.

2.1 Degradation step 1: Hydrolysis

During hydrolysis, the complex biomass is enzymatically split into its monomers. This process divides complex carbohydrates into monosaccharides, primarily glucose, by amylases (Eq. 1 [5, 15]).

$$(C_6H_{10}O_5)_n + nH_2O \rightarrow nC_6H_{12}O_6 \tag{1}$$

Proteins are cleaved by extracellular proteases into shorter peptide chains (di- and tripeptides) and amino acids (AA) (**Figure 1** [15]). Their degradability depends on their AA sequence, hydrophobicity, and the milieu conditions (especially acidity).

The peptides are subsequently cleaved intracellularly into AA. These are used for protein synthesis and energy production of the MO [16]. Lipids are cleaved by lipases into fatty acids and glycerol (**Figure 2** [17, 18]). This degradation is done by anaerobic bacteria such as *Bacteroides*, Clostridia, or Bifidobacteria [19, 20].

2.2 Degradation step 2: Acidogenesis

The monomers are converted into even simpler components in the second degradation step. Here, AA and monosaccharides are metabolized, and long-chain saturated

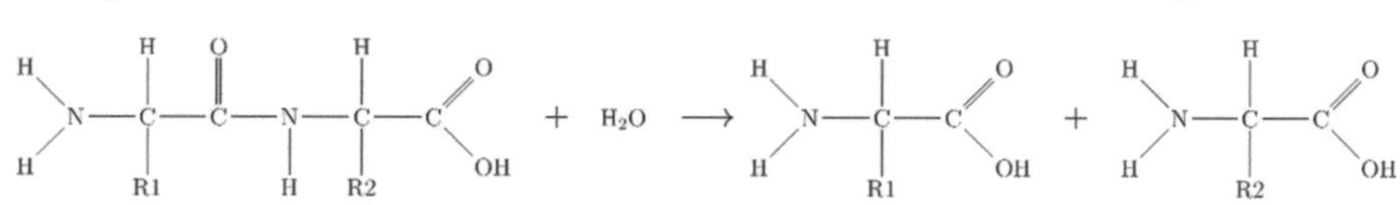

Figure 1.
Protein hydrolysis reaction scheme [15].

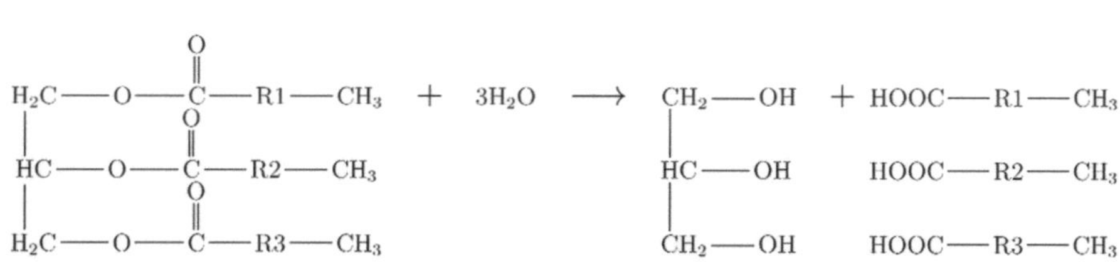

Figure 2.
Reaction scheme of lipid hydrolysis.

fatty acids (LCFA, e.g. C4 to C18 by *Syntrophomonas* spp. [21]) are converted by β-oxidation [17, 22] by facultative and anaerobic acidogenic bacteria such as *Bacteroides*, *Clostridium*, *Butyribacterium*, *Propionibacterium*, *Klebsiella*, *Escherichia*, *Syntrophomonas,* and *Thermoanaerobacterium* to various organic acids (OAs) such as propionic or butyric acid, but mainly to acetic acid (**Figure 3**) [13, 20, 24, 25]. Carbon dioxide, water, and hydrogen are also formed as waste products [13].

The glycerol is converted in the glycolysis process at which hydrogen is produced [17]. Protein-rich substrates yield exceptionally high levels of butyric and valeric acids [13]. These are released as the carbon residue of the AA after deamination or transamination (loss of the amino group). The amino group is separated as an ammonia molecule at the end of the metabolic pathway. It is, therefore, the main product of protein catabolism [16].

2.3 Degradation step 3: Acetogenesis

In the third degradation step, the metabolic products of acidogenesis, mainly propionic and butyric acids, but also ethanol (Eq. 2 [19]) and lactate (eq. 3 [19]), are further degraded by anaerobic oxidation.

$$H_3C - CH_2 - OH + H_2O \rightarrow H_3C - COOH + H_2 \tag{2}$$

$$H_3C - HCOH - COOH + H_2 \rightarrow H_3C - CH_2 - COOH + H_2O \tag{3}$$

Butyric and other longer OAs are degraded according to the same principle during β oxidation (Eq. 3 [26]). Propionic acid is metabolized to acetic acid (Eq. 4) [13, 19].

$$H_3C - CH_2 - COOH + 3H_2O \rightarrow H_3C - COOH + CO_2 + H_2O + 3H_2 \tag{4}$$

However, the butyrate degradation to acetate and H2 is, under standard conditions, an endergonic reaction [27]. For this reaction to occur, hydrogen elimination is

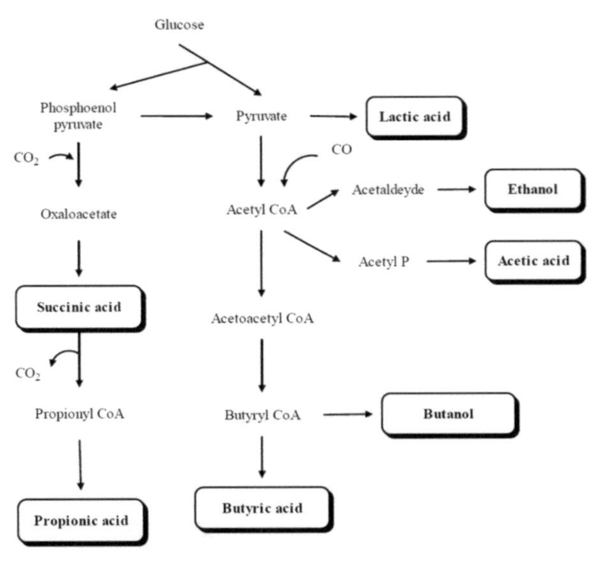

Figure 3.
Various fermentation reactions with glucose as a reactant and various organic acids as products [23].

strictly necessary. By forming so-called syntrophic communities consisting of hydrogen producers (e.g., Clostridium spp.) and hydrogen consumers (e.g., Methanoculleus spp.), the metabolization of butyrate to methane can be achieved despite being thermodynamically unfavorable.

Similarly, lactate degradation can be performed by several members of the genus Clostridium, resulting in the formation of acetate, propionate, and carbon dioxide. In sulfate-limited and hydrogen-consuming milieus, Desulvovibrio vulgaris is furthermore able to generate H2, besides acetate and CO2, from lactate, making it a desirable species for dark fermentation [28]. Additionally, lactate seems to be cometabolized by syntrophic acetate oxidizers such as Syntrophaceticus schinkii [29].

2.4 Degradation step 4: Methanogenesis

In the last fermentation step, methane is formed. There are two possible metabolic pathways.

On the one hand, there are hydrogenotrophic methanogenic MO, such as *Methanoculleus*, which produces methane from hydrogen and carbon dioxide (Eq. 5 [30]). On the other hand, there are acetoclastic methane producers, such as Methanosaeta, which form methane based on acetate (Eq. 6 [20, 30]. The hydrogenotrophic methane formation is energetically preferred.

$$4H_2 + CO_2 \rightarrow CH_4 + CO_2 \tag{5}$$

$$H_3C - COOH \rightarrow CH_4 + CO_2 \tag{6}$$

This changes as soon as the critical hydrogen partial pressure is below a certain threshold. A peculiarity is that the transforming MO is now exclusively Archeans, in contrast to the previous bacteria [19]. These exhibit two temperature optima: at 32–42°C in the mesophilic [20] and at 48–55°C in the thermophilic range [20], as well as a pH optimum at pH 7 to 7.5 [13, 20]. If the pH value falls below a critical threshold, e.g., to pH 6.5, short OAs and acetate accumulate because methanogenesis is inhibited, further lowering the pH [20].

Archaea activity shows a stronger inhibition by lower pH values than bacteria in the previous degradation steps [13].

2.5 Inhibition of methanogenesis in favor of hydrogen production

Methanogenesis is the final step in the anaerobic degradation of biomass. In this step, methane is formed from H_2 and CO_2 on the one hand (hydrogenotrophic) and from acetate on the other (acetoclastic). Since hydrogenotrophic methane formation is H_2-consuming, this step must be inhibited for a higher hydrogen yield and the H_2-producing reactions must be enabled [20, 30].

Hydrogen partial pressure is the most crucial factor influencing H_2 formation and consumption. According to Ahlert, this must be below 50,66 Pa [19] so that H_2 can be formed. According to Cazier, a value of 101,33 Pa [31] is sufficient to inhibit acetogenesis. A higher value inhibits production [13, 24]. At the same time, methanogenesis is inhibited by a low H_2 partial pressure (below 6,48 Pa [19]). Then, acetoclastic methanogenesis is favored instead of hydrogenotrophic methanogenesis, so no H_2 is consumed but acetate.

Another possibility to inhibit hydrogenotrophic methanogenesis is the removal of CO_2 from the reactor air [31]. However, this does not solve the problem of H_2 formation inhibition when the H_2 partial pressure becomes too high, so this method should be used as a supplement, but not alone.

Two other important influencing factors are temperature and pH. Methanogenic MO is archaea, which are more susceptible to stress than bacteria. They can better withstand temperature fluctuations and survive somewhat lower pH values, whereas methanogenic archaea can withstand higher pH values. Methanogenic archaea are strongly inhibited at a pH below seven, already in the slightly acidic range [13, 19, 20]. Therefore, to inhibit methanogenesis, a slightly acidic pH value such as 6.5 would be advantageous for inhibiting methanogenesis.

The low pH must not be caused by LCFA enrichment, as this inhibits hydrolysis and acidogenesis in addition to methanogenesis, which is essential for H_2 formation. Methanogenic MO is the first to be inhibited by LCFA accumulation, starting at 1.0–2.9 kg COD/m^3 biomass [32]. Acidogenic MO is inhibited from 2.1–7.9 kg COD/m^3 biomass [32] and hydrolytic ones only from 2.6–9.4 kg COD/m^3 biomass [32]. Thus, if the concentration of LCFA is between 1.0–2.1 kgCOD/m^3 biomass, methanogenesis should be inhibited, not the other two degradation steps.

3. Development of an efficiency evaluation method for dark fermentation hydrogen production

To develop a model for evaluating the efficiency of dark fermentation, the system must first be defined and limited as a process. This process is to be worked out as a PN. The results of the modeled biomass conversion will then be compared with experimental data, and the efficiency of the process will be summarized based on an efficiency factor.

3.1 Definition of the system under consideration

Before a system can be considered mathematically, its accounting limits must first be defined. The system of dark fermentation in a biogas plant must be represented mathematically. In particular, changes in the process flow leading to hydrogen production are relevant. The balance space must contain all desired processes and structures [33]. This work focuses solely on the biodegradation of biomass without technically required secondary components (**Figure 4**).

The input is the already prepared biomass, and the output is the end products of the reaction pathways during fermentation, mainly hydrogen and methane formation. The material balance is formed based on the effluents described. Here, all material flows have to be considered, which influence the considered target substances, in this case, the hydrogen formation [33].

Figure 5 shows the transit-adjusted distribution of the chemical reactions considered. It is a highly simplified representation of the relevant reactants and products. This figure focuses on classifying the products and intermediates to the degradation steps of anaerobic digestion. It should be noted that the degradation steps are not strictly sequential. To simplify matters, overlaps have not been considered in this figure.

Each reaction is divided into "contributes positively" (green), "negatively" (red), "not directly" (blue), or "not at all" (black) to hydrogen formation. In the red

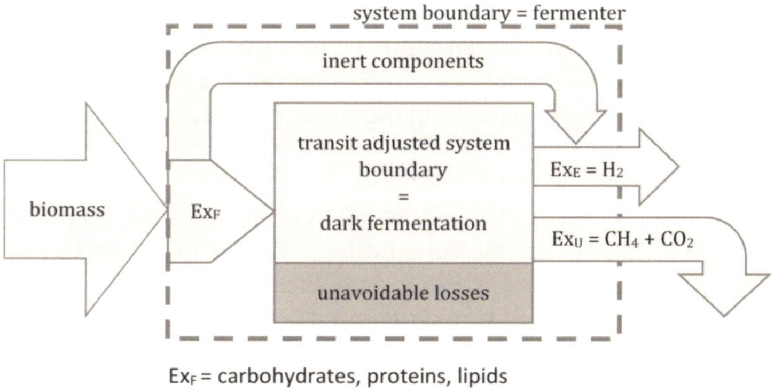

Figure 4.
Schematic illustration of the accounting process.

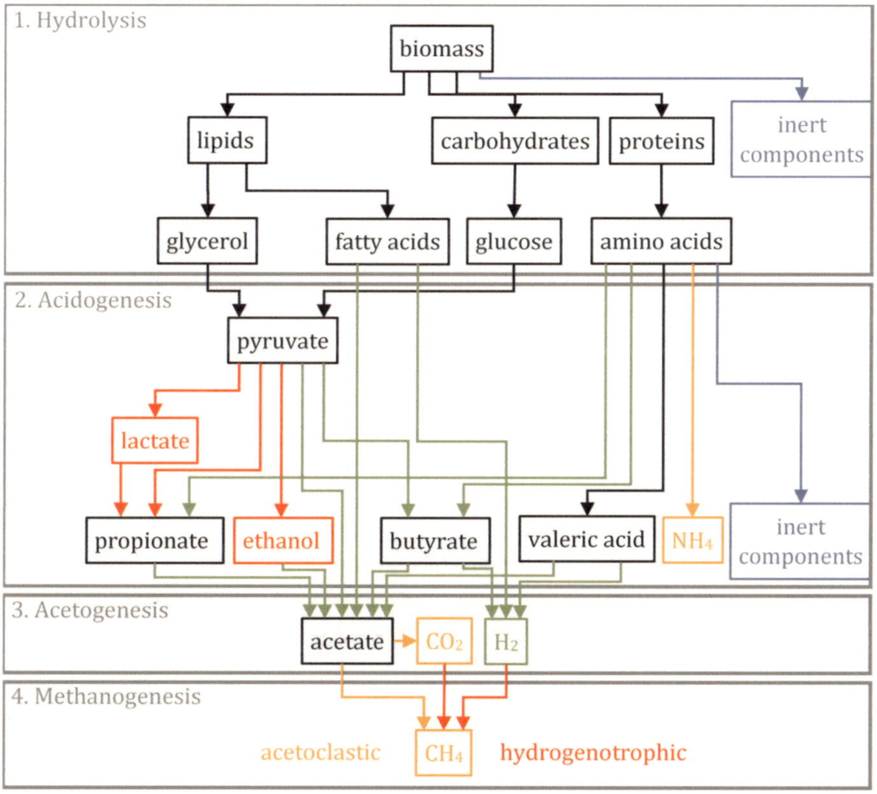

Figure 5.
Schematic illustration of the four steps of biomass degradation to methane —Hydrolysis, acidogenesis, acetogenesis, and methanogenesis—Their intermediates and their interactions (transit-adjusted balance area; green: Contributes positively to H_2 formation; red: Contributes negatively to H_2 formation; blue: Does not contribute directly to H_2 formation; and black: Does not contribute at all to H_2 formation).

reactions, H_2 is consumed, whereas H_2 is formed in the green reactions. Undesired by-products, which are formed independently of the desired H_2, but do not negatively influence the yield of the main product, are shown in orange.

There are two transit streams: the inert biomass parts, e.g., lignin, and the inert AA. These are hardly or not at all degraded. The desired main product is H_2; the undesired by-products are CH_4, CO_2, and NH_4. Ammonia is formed during AA degradation. Carbon dioxide is produced during sugar fermentation and acetoclastic methanogenesis.

It is striking that the sugar fermentation shows red and green reactions, while the AA fermentation and fat degradation show only green reactions. In particular, the lactate and ethanol pathways are H_2-consuming, whereas in the other reaction pathways, H_2 is produced during the further metabolism of OAs.

By definition, methanogenesis is no longer part of dark fermentation. Since it is the primary H_2 consumer, it is nevertheless considered.

3.2 Development of a Petri net

PNs are mathematical structures based on graph theory. They function as a formal modeling language with strict syntax [34]. PNs are directed bipartite multigraphs. They are constructed from two types of nodes (bipartite): places and transitions. Places describe states, which can be, e.g., molecular species and physical parameters like temperature. They are graphically represented as circles. Transitions are state changes and represent, e.g., chemical reactions, changes of location, or an interaction. They are graphically represented as squares. So-called tokens mark places. The terminology used for the developed model is shown in **Figure 6**.

With PNs, causal processes can be represented particularly well, although they can also run independently. Thus, PNs are well suited for describing complex biochemical systems, such as gene regulation or metabolism. The exact syntax of PNs prevents mathematically nonsensical models, and the graphical representation allows an intuitive application that hardly requires mathematical or computer science knowledge.

The balance boundary of the considered process was drawn in the last section. The boundary encompasses all anaerobic digestion, including methanogenesis. The PN was developed based on **Figure 5**, which summarizes the results of the literature review on biomass degradation. Accordingly, the structure of the PN (**Figure** 7), created in SNOOPY (version 1.22), is similar to that of the schematic illustration of the degradation process (see **Figure 5**).

The different components were treated as plates and the chemical reactions as transitions. At the arrows, there are numerical values, which reflect stoichiometry. In the model, a reaction will only occur if all necessary components are available in sufficient quantity and if the reaction rates conditional probability of a reaction is high. Once these conditions are fulfilled, the reaction takes place until the conditions are no longer fulfilled. For example, 4 H_2 molecules and 1 CO_2 molecule are required to hydrogenotrophically form 1 CH_4 molecule (see methanogenesis at the bottom of **Figure** 7).

3.3 Calculation of transition rates

The transition rates correspond to the stochastic probability in a stochastic PN that a transition switches. This corresponds to the reaction rate of the chemical reaction.

Initially, the biomass is added to the digester in grams per liter. The following values are specific conversion values of maize silage (see **Table 1**).

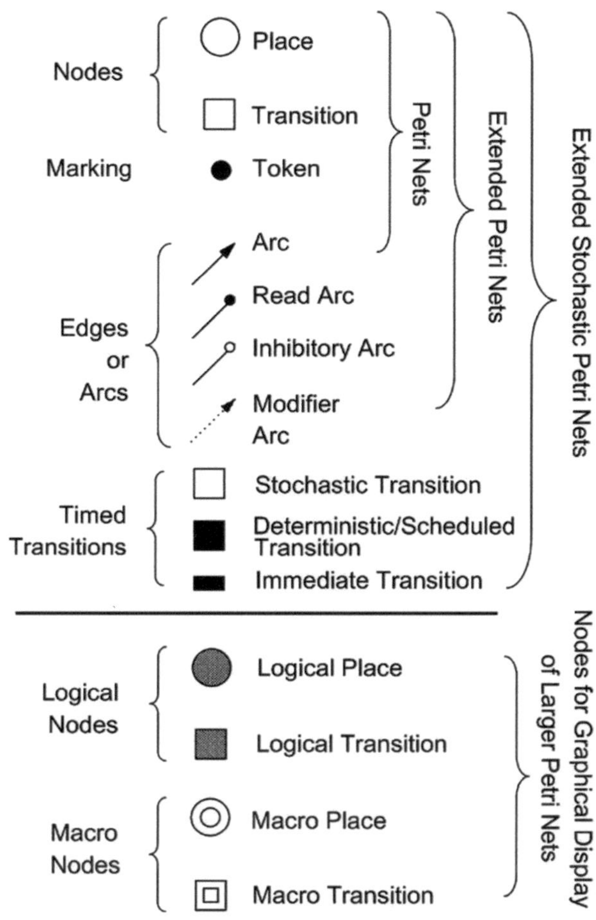

Figure 6.
Graphical representation of the elements that can be used in a Petri net [35].

Maize silage was chosen as a substrate for simplicity since the data are available. The transition rates have to be adjusted for each new substrate. First, each substrate contains a specific ratio of carbohydrates, proteins, lipids, and inert substances (those that are not degraded and thus do not contribute to the biogas yield).

In this context, a test report on maize silage from 2021 (commissioned by the Fraunhofer IFF) was used and converted into relative shares of the total dry matter (DM). As an example, this calculation is carried out in Eq. 8. Digestible carbohydrates can be found among nitrogen-free extractives. The proteins are noted as "crude proteins." The inert substances are composed of crude ash and crude fiber. The lipids are indicated as "crude fats."

$$protein\ content = \frac{raw\ protein\ DM}{total\ DM} = \frac{10,4\%}{92,7\%} = 10,36\%DM \tag{7}$$

Up to this point, all values are in g DM/L. One unit corresponds to one gram DM per liter of reactor content. These must be converted into several molecules. The calculation is shown for glucose as an example in Eq. 9.

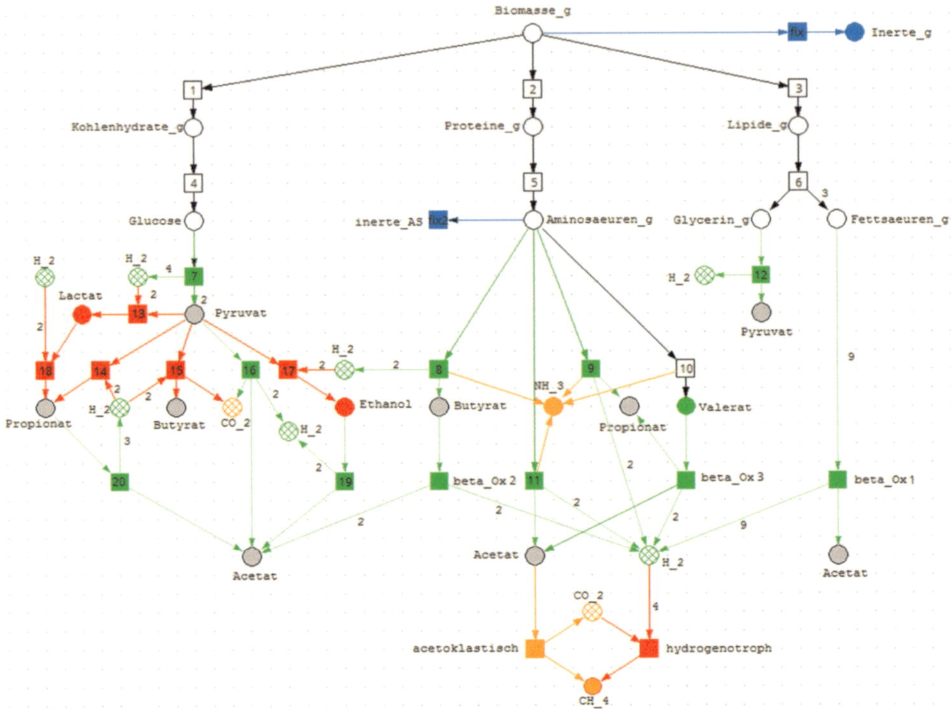

Figure 7.
Petri net model of the four steps of biomass degradation to methane-hydrolysis, acidogenesis, acetogenesis, methanogenesis, their intermediates, and their interactions.

$$N_{Glucose} = \frac{m_{Glucose}}{M_{Glucose}} * N_A = \frac{1g}{180{,}906\,\frac{g}{mol}} * 6{,}002 * 10^{23}\,\frac{1}{mol} = 3,3 * 10^{21} \quad (8)$$

$m_{Glucose}$=Mass of glucose, [g].
$M_{Glucose}$=Molecular weight of glucose, [g/mol].
N_A=Avogadro constant, [1/mol].
$N_{Glucoce}$=Number of molecules per gram of glucose, [−].

Thus, 1 g glucose corresponds to $3*10^{21}$ glucose molecules. Finally, the reactor volume must be considered for all molecule numbers to know the total number of molecules in the fermenter.

The other conversion factors in **Table 1** were calculated analogously based on average molecular weights.

3.4 Derivation of the differential equations

The finished PN can be derived with the calculated reaction rates by SNOOPY to a differential equation system. The corresponding differential equations are formed according to the following scheme:

$$\frac{dX}{dt} = \sum_i p_i * c_{in,m} - \sum_j{}' t - 1 \quad (9)$$

Conversion coefficient	Description	Value	Literature
p1	Carbohydrate content	0.546	(74)
p2	Protein content	0.1036	(74)
p3	Lipid content	0.0183	(74)
pfix	Inert content	0.3311	(74)
p4	Glucose factor	$3*10^{21}$	(75)
p5	Protein hydrolysis	1	—
p6	Lipid hydrolysis	$1.64*10^{20}$	(73,75)
p7	Glucose ➔ pyruvate	1	—
pfix2	Inert amino acids	$0.2607*4581*10^{22}$	(72,75)
p8	Butyric acid amino acids	$0.0073*2019*10^{22}$	(72,75)
p9	Propionic acid amino acids	$0.0871*2208*10^{22}$	(72,75)
p10	Valeric acid amino acids	$0.1718*1977*10^{22}$	(72,75)
p11	Acetic acid amino acids	$0.4421*2272*10^{22}$	(72,75)
p12	Glycerol ➔ pyruvate	1	—
p13	Pyruvate ➔ lactate	1	—
p14	Pyruvate ➔ propionate	1	—
p15	Pyruvate ➔ butyrate	1	—
p16	Pyruvate ➔ acetate	1	—
p17	Pyruvate ➔ ethanol	1	—
p18	Lactate ➔ propionate	1	—
p19	Ethanol ➔ acetate	1	—
p20	Propionate ➔ acetate	1	—
betaOx1	β-oxidation of fatty acids	8.76	(73)
betaOx2	β-oxidation of butyric acid	1	—
betaOx3	β-oxidation of valeric acid	1	—
hydrogen	Hydrogenotrophic methanogenesis	1	—
acetogen	Acetoclastic methanogenesis	1	—

Table 1.
Data used when creating the Petri net.

$\frac{dX}{dt}$=Concentration of component X at time t, [−].
p_i=Stoichiometric coefficient of reaction i, [−].
$c_{in,m}$=Concentration of input component m, [−].

SNOOPY created the complete differential equation system of the anaerobic degradation. With the help of this chemical equation system, the efficiency analysis of the degradation can be performed.

The results of the modeling of the concentration course in the reactor for the time of the total of 23 reaction steps are shown in **Figure 8**. By modeling each individual reaction step, it was possible to represent the idealized concentration curve of all reactants, products, and intermediates.

4. Efficiency assessment of laboratory data

This paper describes the development of a model characterizing an idealized reference process of dark fermentation. Based on the deviation between experimental data and the idealized reference process, the efficiency of the actual process can be quantified.

By comparing the real and the simulated concentration curves, the efficiency of the process can be mapped. At the same time, each concentration present in the reactor provides information about biodegradation *via* the model.

MicroPro GmbH experimentally obtained reaction rates of the components H_2 and CO_2 in a corn silage fermentation. Other components could not be measured for technical reasons.

Since the real-time process does not relate to the reaction steps in a linear matter, data points have been adopted to the respective reaction steps. At this point, there is no accurate way to measure the progress of the reaction in terms of reaction steps. The experimental data has, therefore, been adjusted to the simulated data based on H_2 and CH_4 formation.

Figure 9 compares the experimental data with the calculated reaction conversions.

4.1 Performance indicator for the biological reaction

The deviation between the experimental and simulated data can be represented by the PhO factor defined by eq. 11. D. Volta first described the PhO factor, which compares an actual process and an idealized reference process, the so-called PhO [36].

$$F_{PhO,turnover} = \frac{Modeled\ concentration}{Experimental\ concentration} \leq 1 \quad (10)$$

Figure 10 shows the course of the PhO factor over the respective reaction steps. As expected, the idealized reference process was not achieved under laboratory conditions. However, it is noticeable that the efficiency of the process fluctuates strongly over the course of the process.

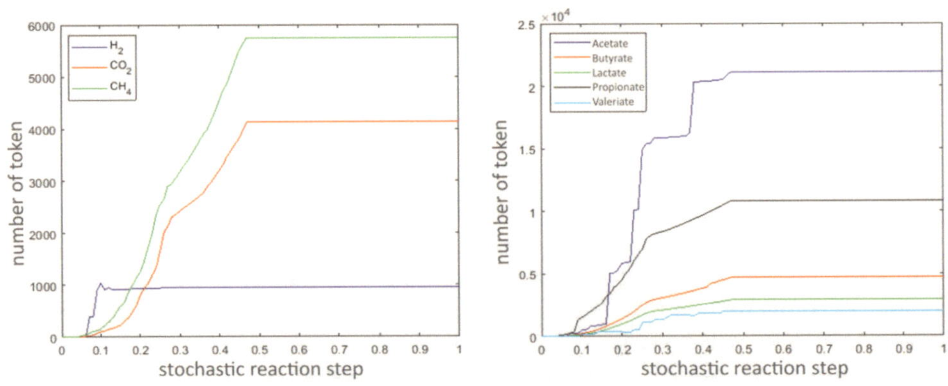

Figure 8.
Substance conversions of various (intermediate) products during anaerobic degradation determined by the Petri net. Left: Resulting biogases CH_4. CO_2 and H_2; right: Organic acids.

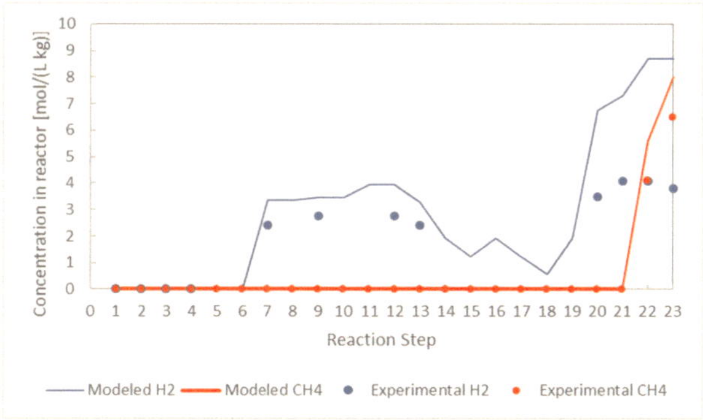

Figure 9.
Comparison of the model-based and experimental reaction turnovers of hydrogen H_2 (blue) and methane CH_4 (red). The concentration is in Mol per kilogram of corn silage per liter of fermenter contents.

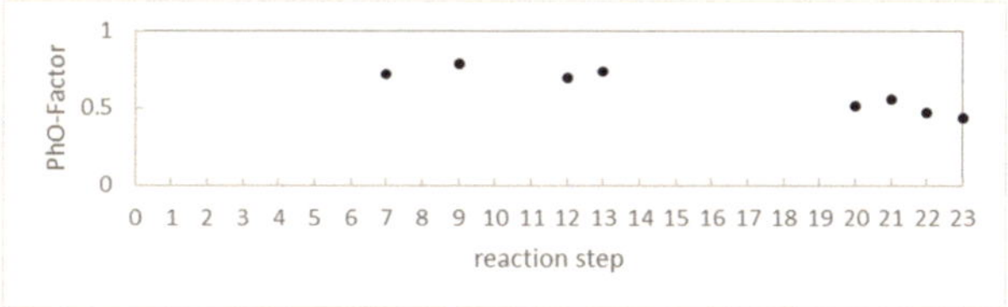

Figure 10.
PhO turnover factor through comparison of experimental and modeled data for H_2 concentration in the reactor.

During the reaction, the PhO factor is a maximum of 0.79 (reaction step 9) and a minimum of 0.44 (reaction step 23). This suggests an optimization potential of the biological process of 21–56%. However, to be able to make a more precise statement as to whether this optimization potential is technically achievable, more detailed investigations must be carried out.

4.2 Discussion and outlook

A model was developed to evaluate the efficiency of the metabolism of dark fermentation on a laboratory scale, simulating the degradation and formation of all reactants, products, and intermediates.

The developed model is based on a PN, which uses defined rates and conditions for the course of a reaction for each defined step. A change in these rates and conditions directly influences the results of the PN and thus shifts the modeled optimum of the reaction. While the requirements for a reaction step are already very well described in the literature, the reaction rates can only be determined by approximating experimental data. The practical reaction rates used for the model were taken from various sources (cf. **Table 1**).

Since it was not technically possible to measure the concentration curves over time in the reactor, a more concrete validation had to be omitted at this point that the final product composition of the simulated process agrees with literature data and values collected in the experiment implies that most reaction rates were chosen realistically. Which reaction rates were defined too inaccurately at the current time can only be found by parameter variation and behavioral analysis of the model under comparison with experimental data.

Thus, the hydrogen and methane production rates were mapped using pressure differences and converted into concentrations for evaluation. Therefore, the pure hydrogen increase could only be measured up to reaction step 21. From then on, a comparison had to be made with combination experiments to estimate the concentration of H_2 and CH_4, separately. In these combination experiments, reactor contents were fed unchanged to methanation after dark fermentation.

Since the concentrations of the reactants, intermediates, and products cannot be monitored technically during the reaction, the temporal assignment of the experimental data to the respective reaction steps was based solely on the calculated H_2 concentrations in the reactor. Further experiments must prove the reproducibility of this assignment.

Another reason for possible inaccuracies in the model is that a change in pH was not considered in the modeling. However, these effects were deliberately neglected since the process is intended to represent an idealized reference process to identify optimization potential by comparison with an actual process.

As part of the HyPerFerment II project, the construction of a demonstration plant for dark fermentation is planned. Here, data can be collected for the first time on a small industrial scale and under nonlaboratory conditions. Further experimental data must be collected to better adapt the model of the idealized reference process of dark fermentation to actual conditions.

5. Conclusion

An idealized reference model was needed for an accurate efficiency evaluation of dark fermentation. A PN was chosen, which was modeled based on the thermodynamics of the relevant reaction pathways. A total of 27 reactions and 21 components were considered to calculate the reactor's idealized concentration of H_2 and CH_4. Comparing the results of the PN and the experimental data showed a qualitative agreement of the graphs, which support the model's suitability as an ideal reference state.

The model was validated using initial experimental data at a laboratory scale. Divergences between the expected and the observed reaction turnover were found. These are mainly due to experimental uncertainties. The authors plan to validate the model based on further experimental data, which will be obtained during the operation of a newly constructed demonstration plant.

Based on the existing experimental data, a substantial fluctuation in the efficiency becomes visible during the reaction. The efficiency evaluation based on the developed model suggests an optimization potential of 21–56%. Since the model is a purely theoretical idealized reference process, it can be assumed that this optimization potential cannot be achieved technically. However, the assessment method gives a better idea of where the biological process can be optimized due to the detailed analysis of separate reaction steps. By monitoring the concentrations of the reactants, intermediates, and end products in the reactor, the model can be used to identify which process steps need to be optimized in detail.

Acknowledgements

We want to extend our heartfelt appreciation to our project partners of MicroPro GmbH and Streicher Anlagenbau GmbH & Co. KG for their valuable contribution of expertise and experimental data, which greatly enhanced the quality and significance of this research. Without their support and collaboration, this study would not have been possible. We are grateful for their dedication and commitment to advancing scientific knowledge in this field.

Conflict of interest

The authors declare no conflict of interest.

Nomenclature

AA	Amino acids
DM	Dry matter
LCFA	long-chained fatty acids
MOs	Microorganisms
OA	Organic acid
PhO	Physical optimum
PN	Petri net
TVS	Total volatile solids
$c_{in,m}$	Concentration of input component m, [−]
$\frac{dX}{dt}$	Concentration of component X at time t, [−]
$m_{Glucose}$	Mass of glucose, [g]
$M_{Glucose}$	Molecular weight of glucose, [g/mol]
N_A	Avogadro constant, [1/mol]
$N_{Glucoce}$	Number of molecules per gram of glucose, [−]
p_i	Stoichiometric coefficient of reaction i, [−]

Author details

Natascha Eggers[1*], Celia Kirsch[2], Fabian Giebner[3] and Torsten Birth[4]

1 Fraunhofer Institute for Factory Operation and Automation IFF, Magdeburg, Germany

2 IOI Oleochemical, Wittenberge, Germany

3 MicroPro GmbH, Gommern, Germany

4 Hamburg University of Applied Sciences, Hamburg, Germany

*Address all correspondence to: natascha.eggers@iff.fraunhofer.de

References

[1] DNV. Hydrogen at Risk of Beeing the Great Missed Opportunity of the Energy Transition: Hydrogen Forecast to 2050 [cited 2023 March 16] Available from: https://www.dnv.com/news/hydrogen-at-risk-of-being-the-great-missed-opportunity-of-the-energy-transition-226628.

[2] Birth T, Jentsch S, Hayen S, Scheffler M. Wasser als kritische Ressource für die Wasserstofferzeugung. gwf-wa. 2022; **162**(09):73-88. DOI: 10.17560/gwfwa.v162i09.2574

[3] Hodgson D, Bains P, Morrhouse J. Bioenergy - Energy systems overview. Paris: International Energy Agency; 2022 Available from: https://www.iea.org/reports/bioenergy; last checked: 05. June 2023

[4] Gül T, Cozzi L, Havlik P. What does net-zero emissions by 2050 mean for bioenergy and land use? Paris: International Energy Agency; 2021 Available from: (https://www.iea.org/articles/what-does-net-zero-emissions-by-2050-mean-for-bioenergy-and-land-use; last checked: 05. June 2023)

[5] Kaltschmitt M, Hartmann H, Hofbauer H, editors. Energie aus Biomasse: Grundlagen, Techniken und Verfahren. 3., aktual. Aufl. 2016 ed. Berlin, Heidelberg: Springer Berlin Heidelberg; 2016

[6] Götz M, Lefebvre J, Mörs F, et al. Renewable power-to-gas: A technological and economic review. Renewable Energy. 2016;**85**:1371-1390. DOI: 10.1016/j.renene.2015.07.066

[7] Schenuit C, Heuke R, Paschke J. Potentialatlas Power to Gas: Klimaschutz umsetzen, erneuerbare Energien integrieren, regionale Wertschöpfungskette ermöglichen. Berlin: dena (Deutsche Energie-Agentur); 2016

[8] Liao Q, editor. Bioreactors for Microbial Biomass and Energy Conversion. Singapore: Springer; 2018

[9] Li S-L, Kuo S-C, Lin J-S, Lee Z-K, Wang Y-H, Cheng S-S. Process performance evaluation of intermittent–continuous stirred tank reactor for anaerobic hydrogen fermentation with kitchen waste. International Journal of Hydrogen Energy. 2008;**33**(5):1522-1531. DOI: 10.1016/j.ijhydene.2007.09.049

[10] Shin H-S, Youn J-H. Conversion of food waste into hydrogen by thermophilic acidogenesis. Biodegradation. 2005;**16**(1):33-44. DOI: 10.1007/s10531-004-0377-9

[11] Cavinato C, Gottardo M, Micolucci F, Bolzonella D, Pavan P. Ammonia concentration and ph control in pilot scale two-phase anaerobic digestion of food waste for hydrogen production: Focus on start-up. Chemical Engineering Transactions. 2016;**49**:151-156. DOI: 10.3303/CET1649026

[12] Eggers N, Birth T. Effizienzbewertung biomassebasierter PtX-Systeme. In: Herlitzius T, editor. Effizienzbewertung biomassebasierter PtX-Systeme. Wiesbaden: Springer Fachmedien Wiesbaden; 2021. pp. 15-26

[13] Gerber M. Ganzheitliche stoffliche und energetische Modellierung des Biogasbildungsprozesses. Doctoralthesis. Bochum, Germany: Ruhr-Universität Bochum, Universitätsbibliothek; 2010

[14] Khan MA, Ngo HH, Guo WS, et al. Optimization of process parameters for

production of volatile fatty acid, biohydrogen and methane from anaerobic digestion. Bioresource Technology. 2016;**219**:738-748. DOI: 10.1016/j.biortech.2016.08.073

[15] Cypionka H. Grundlagen der Mikrobiologie. 4. Aufl. 2010. Berlin, Heidelberg: Springer Berlin Heidelberg; 2010

[16] Gresner N, Südekum KH, Höltershinken M. Umsetzungen von Stickstoffverbindungen des Futters im Pandsen. Übersetzer Nierernährung. 2014;**42**:27-80

[17] Harmjanz F. Biochemie - Zelle, Enzyme, Praktische Biochemie. 1. Aufl. 2021. Berlin, Heidelberg: Springer Berlin Heidelberg; 2021

[18] Fromenty B. Inhibition of mitochondrial fatty acid oxidation in drug-induced hepatic steatosis. Liver Research. 2019;**3**(3–4):157-169. DOI: 10.1016/j.livres.2019.06.001

[19] Ahlert S. Propionsäure in NawaRO-Biogasanlagen. Dissertation. Mainz, Germany: Biologie, Johannes Gutenberg-Universität; 2015

[20] Rojas Reiner CJ. Simulation und Modellierung des anaeroben Prozesses der Biogaserzeugung mit verschiedenen Substraten. Hamburg, Germany: TUHH Universitätsbibliothek; 2014

[21] Schink B, Muñoz R. The family Syntrophomonadaceae. In: Rosenberg E, DeLong EF, Lory S, Stackebrandt E, Thompson F, editors. The Prokaryotes. 4th ed. Berlin: Springer Reference; 2014. pp. 371-379

[22] Kim H, Moon S, Abug A, Choi S-C, Zhang R, Oh Y-S. Effect of fermentation conditions on biohydrogen production from lipid-rich food material. International Journal of Hydrogen Energy. 2012;**37**(20):15062-15069. DOI: 10.1016/j.ijhydene.2012.07.104

[23] Lyberatos G, Antonopoulou G, Koutrouli E, Kalfas H, Gavala H, Skiadas I. Gaseous biofuels production from sweet sorghum and olive pulp. In: Gaseous Biofuels Production from Sweet Sorghum and Olive Pulp. 2006.

[24] O-Thong S. Microbial population optimization for control and improvement of dark hydrogen fermentation. In: Jozala AF, editor. Fermentation Processes. Erscheinungsort nicht ermittelbar. London, UK, London, UK: IntechOpen; 2017

[25] Harirchi S, Wainaina S, Sar T, et al. Microbiological insights into anaerobic digestion for biogas, hydrogen or volatile fatty acids (VFAs): A review. Bioengineered. 2022;**13**(3):6521-6557. DOI: 10.1080/21655979.2022.2035986

[26] Batstone DJ, Pind PF, Angelidaki I. Kinetics of thermophilic, anaerobic oxidation of straight and branched chain butyrate and valerate. Biotechnology and Bioengineering. 2003;**84**(2):195-204. DOI: 10.1002/bit.10753

[27] Thauer RK, Jungermann K, Decker K. Energy conservation in chemotrophic anaerobic bacteria. Bacteriological Reviews. 1977;**41**(1):100-180. DOI: 10.1128/br.41.1.100-180.1977

[28] Zellner G, Neudörfer F, Diekmann H. Degradation of lactate by an anaerobic mixed culture in a fluidized-bed reactor. Water Research. 1994;**28** (6):1337-1340. DOI: 10.1016/0043-1354 (94)90299-2

[29] Moestedt J, Müller B, Nagavara Nagaraj Y, Schnürer A. Acetate and lactate

production during two-stage anaerobic digestion of food waste driven by lactobacillus and Aeriscardovia. Frontiers in Energy Research. 2020;**8**. DOI: 10.3389/fenrg.2020.00105. Available from: https://www.frontiersin.org/articles/10.3389/fenrg.2020.00105/full

[30] Schieder D, Gronauer A, Lebuhn M, et al. Prozessmodell Biogas. Freising, Germany: Arbeitsgemeinschaft Landtechnik und landwirtschaftliches Bauwesen in Bayern e.V.; 2010

[31] Cazier EA, Trably E, Steyer JP, Escudie R. Biomass hydrolysis inhibition at high hydrogen partial pressure in solid-state anaerobic digestion. Bioresource Technology. 2015;**190**:106-113. DOI: 10.1016/j.biortech.2015.04.055

[32] Ma J, Zhao QB, Laurens LLM, et al. Mechanism, kinetics and microbiology of inhibition caused by long-chain fatty acids in anaerobic digestion of algal biomass. Biotechnology for Biofuels. 2015;**8**(141). DOI: 10.1186/s13068-015-0322-z. Available from: https://biotechnologyforbiofuels.biomedcentral.com/articles/10.1186/s13068-015-0322-z

[33] Keichel C. Methode der grenzwertorientierten Bewertung. Dissertation. Clausthal, Germany: Technische Universität Clausthal; 2017

[34] Petri CA. Kommunikation mit Automaten. Dissertation. Bonn, Germany: Universität Bonn; 1962

[35] Marwan W, Rohr C, Heiner M. Petri nets in Snoopy: A unifying framework for the graphical display, computational modelling, and simulation of bacterial regulatory networks. Methods in Molecular Biology. 2012;**804**:409-437. DOI: 10.1007/978-1-61779-361-5_21

[36] Volta D. Das physikalische Optimum als Basis von Systematiken zur Steigerung der Energie- und Stoffeffizienz von Produktionsprozessen. Clausthal, Germany: Universitätsbibliothek der TU Clausthal; 2014

Chapter 13

Cellulases from Mycelial fungi *Penicillium verruculosum* as a Real Alternative to *Trichoderma* Enzymes in Industrial Hydrolysis of Cellulosic Biomass

Olga G. Korotkova, Alexandra M. Rozhkova, Ekaterina A. Rubtsova and Arkady P. Sinitsyn

Abstract

Abstract The possibility of using the recipient strain *Penicillium verruculosum* B1-537 (ΔniaD) as a producer of laboratory and industrial enzymes was considered. The advantage of this strain is its ability to secrete a basic cellulase complex consisting of cellobiohydrolases, endoglucanases, and β-glucosidase, which exceeds in its hydrolytic ability the enzyme complex of *Hypocrea* (*Trichoderma*) strains. Using the expression system, the basic complex of cellulases of the recipient strain *Piptochaetium verruculosum* B1-537 (ΔniaD) was supplemented with new (booster) enzymes that are necessary to increase its hydrolytic activity. Enzyme preparations adapted to the processing of various types of renewable plant biomass were obtained.

Keywords: *Penicillium verruculosum*, *Trichoderma*, cellulase complex, cellulose bioconversion, renewable plant biomass, industrial enzymes

1. Introduction

Currently, the attention of scientists is focused on the depletion of reserves of fossil energy sources. Rapid population growth provides high energy needs for fossil resources such as coal, natural gas, and crude oil [1]. The burning of fossil fuels poses a threat to the environment due to the emission of greenhouse gases, which, in turn, lead to global climate change. As for renewable energy sources, fuels and products derived from cellulosic waste do not harm the environment and can be considered as an alternative to fossil energy sources. Plant waste not only supports economic development but also creates an ecologically friendly environment for the production of energy and biochemicals [2–10].

Biomass cellulose is a plant polysaccharide and is an almost inexhaustible source of renewable raw materials that can be converted enzymatically into glucose. In turn, glucose is a raw material for microbiological processes of obtaining liquid and gaseous fuels (ethanol, butanol, etc.), organic and amino acids, feed protein, and many other useful products of microbiological synthesis [11]. Coniferous and deciduous wood and its waste are of particular interest [12].

An important stage of bioconversion of cellulose-containing biomass, which prevents its commercial use, is the enzymatic conversion of cellulose into glucose. Natural wood and other lignocellulose materials are resistant to enzymes due to the crystalline structure of cellulose and the presence of lignin and hemicellulose-protecting cellulose fibers [12]. The reactivity of natural crystalline cellulose (for example, cotton) or lignocellulose (wood of various species, grass straw) during the enzymatic conversion is low, which is accompanied by extremely low yields of glucose and other sugars. Effective enzymatic hydrolysis of cellulose-containing biomass requires its pretreatment to increase reactivity by destroying the crystal structure of cellulose and completely or partially removing the lignin. As methods of pretreatment, gamma irradiation, mechanical grinding in a ball mill, treatment with mineral acids (sulfuric, phosphoric), cadoxene, and alkali delignification are used [12].

The influence of such factors on the efficiency of enzymatic conversion as changes in the degree of crystallinity of cellulose (determined by X-ray diffraction), the availability of cellulose surface to enzyme molecules (measured by such methods as protein adsorption, as well as by thermal desorption of nitrogen-hydrogen mixture), the size of cellulose particles (by dispersion analysis using optical microscopy), and the degree of polymerization of cellulose (determined by the viscosity of cellulose solutions in cadoxene) was investigated. Based on the results of these studies, the influence of these parameters on the efficiency of enzymatic cellulose conversion was quantified—it linearly depends on a decrease in the degree of crystallinity and an increase in the surface area available to the enzyme molecules, but it does not depend much on the geometric size of cellulose particles and its degree of polymerization [13].

Deep cleavage of crystalline cellulose to soluble sugars (glucose) is carried out by a complex of cellulolytic enzymes, including endoglucanases (EG), cellobiohydrolase (CBH), and β-glucosidase (BGL). The activity of the components of this complex and their interaction determine the action of the enzymes on cellulose-containing substrates [14].

Many research groups are looking for new effective producers of cellulases and hemicellulases. Research is being conducted to improve existing strains of microorganisms in order to increase the production of various cellulases and reduce their cost. In this chapter, we compare the hydrolytic potential of multi-enzyme cellulose complexes produced by different strains of fungi and evaluate the role of different enzymes in the hydrolysis of pretreated cellulose-containing substrates.

2. Enzymatic complex for deep destruction of cellulose-containing biomass

Enzymatic hydrolysis allows the destruction of plant raw materials to monosaccharides without significant energy costs and anthropogenic impact on the environment. Since plant biomass is a complex substrate that includes polysaccharides of various compositions, a complex of enzymes of various specificities is necessary for its deep destruction, which carries out cooperative hydrolysis of its constituent components. In industrial biotechnology, various microscopic fungi are widely used as producers of such enzyme complexes. They are the main source of commercial cellulase preparations produced on an industrial scale in different countries of the world. For a long time, it was believed that cellulolytic fungi belonging to the genus *Trichoderma* are the undisputed leaders in the secretion of the most active cellulases [15–19]. Commercial cellulase preparations based on mutant strains of *Trichoderma reesei* (synonym of *Hypocrea jecorina*) are most common on the enzyme market. Therefore, it is not surprising that most of the research and developments on the bioconversion of lignocellulose raw materials (primarily for the production of second-generation bioethanol) were based on the use of *T. reesei* cellulases. However, in recent years, more and more publications demonstrating that there are alternatives to *T. reesei*-based enzyme preparations have appeared [16–18, 20–22]. In particular, it is related to cellulases produced by fungi from the genera *Penicillium* and *Myceliophthora*. Modern developments aimed at obtaining cellulose preparations of a new generation are characterized by using methods of protein and genetic engineering; recombinant (modified) enzymes with improved properties from various microorganisms are included in the composition of these preparations.

The leading role of mutant or recombinant strains of fungi of the genus *Trichoderma* (*T. reesei, T. viride, T. longibrachiatum*) is explained, firstly, by their high secretory ability, and, secondly, by the variety of enzymes produced with different substrate specificity [23–26]. Therefore, it is not surprising that biocatalysts (enzyme preparations) of cellulases based on *Trichoderma* fungi are produced in many countries by leading manufacturers of industrial enzymes, in particular, Novozymes (Denmark), Genencor and Danisco Division (USA), Iogen (Canada), PrimAlko (Finland), Röhm Gmbh (Germany), EnMex (Mexico), Meiji Seika Kaisha Ltd. and Shin Nihon Chemical Co. (Japan), etc. At the same time, the search for new producers of enzymes suitable for the hydrolysis of plant raw materials, as well as an increase in the overall activity and balance in the component composition of already known enzyme complexes, is still an actual task of modern biotechnology.

Strains of fungi of the genera *Penicillium*, *Acremonium*, *Chrysosporium*, *Chaetomium*, and *Humicola* can become a worthy alternative to strains of the genus *Trichoderma* since the enzyme complexes produced by them are not inferior in hydrolytic ability and sometimes are superior to those obtained with the help of known strains of *Trichoderma* [27–31].

Promising producers of highly active cellulase complexes are some species of fungi from the genus *Penicillium*. Cellulases from *Penicillium*, as a rule, surpass *T. reesei* enzymes in the rate of hydrolysis and glucose yield from various cellulose-containing substrates at the same dosage of protein concentration or cellulase activity (**Table 1**). Almost all the authors note a relatively high level of beta-glucosidase activity in enzyme preparations based on penicillin. This property is one of the main advantages of fungi of the genus *Penicillium* over *Trichoderma*. As a consequence, the gain of penicillin enzymes in glucose output sometimes reaches a 5-fold size (**Table 1**) [17, 18, 32–34]. Only after the addition of an excess of exogenous BGL, *T. reesei* preparations are able to provide comparable glucose yields. However, in some cases, even with the addition of BGL cellulase from *T. reesei* could not compete with cellulase complexes from *Penicillium* [17, 18, 20]. For example, with the same cellulase activity on filter paper in the reaction mixture (10 U/g of substrate), the preparation of *T. reesei* RUT-C30, enhanced with the addition of aspergilli BGL (6 U/g), was inferior in depth of hydrolysis of microcrystalline cellulose (MCC) after 48 hours to the preparation of *P. pulvirrolum* F-2220 (the difference

Cellulase producer	Substrate	Difference, times*
Penicillium sp.	Abies alba pretreated by steam explosion or organosolve	1.6–3.6 [17, 18]
Penicillium sp.	Poplar wood pretreated by organosolve	1.4–2.0 [17, 18]
Penicillium sp.	Red maple wood pretreated by organosolve	1.5–2.1 [17, 18]
P. brasilianum	Spruce wood pretreated by steam explosion	2.1 [18]
Panicum echinulatum	Bleached eucalyptus kraft pulp	1.1 [17, 18]
P. funiculosum	Crushed corn cobs	1.1–1.5 [18]
P. occitanis	Stipa tenacissima	1.8 [18]
P. oxalicum	Corn cobs alkali pretreated	2.0–2.2 [32]
P. oxalicum	Microcrystalline cellulose	3.6 [33]
P. pinophilum	Spruce wood pretreated by steam explosion	3.1 [17, 18]
Piptochaetium verruculosum	Microcrystalline cellulose	1.3–1.7 [18]
P. verruculosum	Cellulose from eucalyptus	1.3–3.9 [17, 18]
P. verruculosum	Coniferous wood pretreated by organosolve	1.3–5.1 [17, 18]
P. verruculosum	Wheat straw alkali pretreated	1.6 [34]

**The ratio of product concentrations (usually glucose) obtained by hydrolysis of cellulose-containing substrates under the action of enzyme preparations based on fungi Penicillium and T. reesei.*

Table 1.
Comparison of the hydrolytic ability of cellulase complexes from fungi of the genus Penicillium and T. reesei.

was 25%), whereas in the hydrolysis of pretreated spruce wood and wheat straw, both preparations showed comparable results [20].

Sequencing and annotation of the genomes of *P. decumbens* 114-2, *P. funiculosum* NCIM 1228, and *Piptochaetium verruculosum* TS63-9 have shown that these species of fungi differ in a richer set of enzymes that catalyze the cleavage of components of lignocellulose materials compared with *T. reesei* [35–37]. It especially concerns cellulases with a cellulose-binding module (CSM) and hemicellulases. Analysis of *P. decumbens* 114-2 secretome obtained by cultivation on a medium with wheat bran showed that it was richer in carbohydrates than that obtained on a medium with glucose [35]. In the secretome of *P. funiculosum* NCIM 1228, 113 enzymes acting on carbohydrates were detected by quantitative mass spectrometry, of which 92 were glycoside hydrolases [37]. Apparently, the genomes and secretomes, which are characterized by a high content of glycoside hydrolases and their accompanying enzymes, are also characteristics of other species of fungi of the genus *Penicillium*. Another reason for the high efficiency of cellulase complexes based on *Penicillium* fungi is the high specific activity of their key components, namely CBH I and CBH II, compared to the corresponding *T. reesei* enzymes. The difference can reach 2–2.5 times. In particular, this was demonstrated in the example of CBH *P. funiculosum*, *P. pulvirrolum*, *P. verruculosum*, and *P. canescens* [18–20, 38, 39]. One of the reasons for such high specific activity in the case of CBH I and CBH II of *P. verruculosum* is the optimal distribution of N-bound glycans on the surface of the catalytic domain of these enzymes [40, 41].

One of the properties that negatively affect the activity of cellulases during the hydrolysis of lignocellulose substrates is their unproductive adsorption on lignin [42–48]. During the hydrolysis of MCC, enzymes from *P. verruculosum* are significantly less inhibited by two types of lignin artificially added to the reaction system than cellulases from five different *T. reesei* preparations. At the same time, the lignin regenerated from the solution in the organosolve caused a stronger inhibition (by 11–84%) than the residual lignin obtained by exhaustive enzymatic hydrolysis of pretreated wood (8–58% inhibition). Similar results demonstrating a lower negative effect of lignin on *P. verruculosum* cellulase compared to *T. reesei* enzymes were obtained in the work [34].

In the example of enzymes from *P. oxalicum*, it was shown that the degree of unproductive adsorption on lignin decreases in the raw of CBH – xylanase – EG – BGL, and mainly electrostatic interactions are responsible for the binding of xylanase and EG to lignin, whereas, in the case of CBH and BGL, hydrophobic interactions have a decisive influence. CBM in CBH plays a noticeable role in binding to the lignin [48]. The changes in the properties of lignin, caused by the pretreatment of lignocellulose raw materials, had a noticeable effect on the adsorption of *P. oxalicum* enzymes, as in the work [46]. Thus, as a result of pretreatment with hot water, lignin from corn waste got an increased ability to bind proteins [48].

Despite the high hydrolytic potential, commercial cellulase preparations from penicillins have not yet become widespread, although they are produced on an industrial scale by the company Adisseo in France (Rovabio ® Excel based on *P. funiculosum*) [36] and in China (enzyme preparations based on *P. decumbens*) [35]. Perhaps this is due to the fact that the existing strains of *Penicillium* are still inferior to the best industrial strains of *T. reesei* in extracellular protein secretion [17, 18]. One of the reasons may be the insufficiently aggressive marketing policy of the producers of these enzymes.

2.1 Recombinant strains of *Penicillium verruculosum*—Highly effective producers of cellulases

It is possible to increase the effectiveness of complex cellulase preparations both by improving the properties of individual enzymes and by optimizing the component composition of preparations. Protein engineering allows to change in the properties of already known individual enzymes, significantly increasing their activity and operational stability [49–51]. Since the composition and structure of various types of cellulose-containing raw materials can vary quite a lot depending on the source and type of pretreatment, the component composition of the enzyme preparations used for their hydrolysis should correspond to the composition of the raw materials.

The selection of the composition optimal for processing a particular type of raw material preparation is carried out by comparing the hydrolytic ability of different mixtures of individual enzymes [31, 52]. Genetic engineering of cellulase producers with a subsequent screening of the obtained mutants allows the creation of new highly effective strains of microorganisms secreting a complex of enzymes of optimal composition. At the same time, classical nondirectional mutagenesis also remains an effective tool for obtaining such producers.

The wild strain of *P. verruculosum* WA30 was isolated about 40 years ago. He was identified as a producer of a cellulase complex with high activity in relation to MCC. The productivity of the wild strain was low and was about 7 g/l of secreted protein in the culture fluid [53]. On the basis of the wild strain WA30 by the method of mutagenesis using nitrosoguanidine and ultraviolet irradiation with subsequent selection *P. verruculosum* strains were obtained. The protein concentration in the culture fluid reached 30–40 g/l [54, 55]. The synthesis of extracellular cellulolytic enzymes of *P. verruculosum* is induced by cellulose and gentiobiose, and in most cases, it is subject to catabolic repression by glucose. The peculiarity of the obtained *P. verruculosum* mutants with a high level of extracellular cellulase secretion is that, during the mutagenesis, catabolic repression was reduced. Changes in pH during fermentation can influence the composition of the secreted cellulase complex. Some side activities, such as xylanase and amylase, become higher at higher pH values.

The enzyme complex secreted by *P. verruculosum* contains more than 20 enzymes differing in biochemical and catalytic properties. The main cellulases secreted by the control highly productive strains B1-221-151 (contracted B151) and B1-537 are CBH I (Cel7A), CBH II (Cel6A), EG I (Cel7B), EG IIa (Cel5A), EG IIb (Cel5B), EG III (Cel12A), and BGL (Cel3A). The content of CBH I is about 35%, CBH II is 21–33%, the content of these EGases varies from 2 to 5%, and the percentage of BGL is 3–4% [38, 56, 57]. In addition to cellulases, the enzyme complex also includes other carbohydrases—xyloglucanases, xylanases, α-galactosidase, and glucoamylase (a total of 18–30% of the total protein). The molecular weights of enzymes vary from 19 to 120 kDa and isoelectric points from 2.0 to 5.8. For most of these cellulases, the pH-optima of activity is in the pH range of 4.0–5.0 [38]. As a result of the limited proteolysis of the linker *P. verruculosum* cellulases, having CBM (CBH I, CBH II, EG I), as a rule, is represented by high-molecular (full-size) and low-molecular forms, the latter being the catalytic domain of a specific enzyme without CBM [38].

As mentioned above, CBH I and CBH II of *P. verruculosum*, when acting on crystalline cellulose, are significantly superior in hydrolytic ability to both CBH from *T. reesei* (of which CBH I is actually the "gold standard" in the study of properties of the cellulases [58]); at the same time, even the low molecular weight form of CBH II from *P. verruculosum* is comparable in effectiveness to the full-size CBH I from *T. reesei* [18]. This, together with a higher level of BGL activity, leads to a higher potential of *P. verruculosum*-based enzyme preparations compared to traditional commercial *T. reesei* preparations [17, 18, 34].

2.2 Hydrolytic potential of *P. verruculosum* cellulase complex in comparison with industrial analogs

One of the important criteria for comparing the effectiveness of enzyme preparations intended for the biodegradation of cellulose-containing raw materials is their specific activity on a number of substrates—soluble and insoluble polysaccharides (FP, MCC, CMC, xylan), synthetic substrates (pNPG, pNPX), and oligosaccharides (cellobiose). These activities on the example of the enzyme preparations obtained with the help of fungi of the genus *Trichoderma* (all the preparations Accelerase and Cellic CTec) and *Penicillium* as producers are shown in **Table 2**.

The presented data make it possible to characterize the individual components of the enzyme complex and their balance in the composition of the preparation, which mainly determines the effectiveness of their action in the hydrolysis of a substrate of complex composition, which is plant biomass.

Activity on filter paper characterizes the overall activity of enzyme preparations in relation to insoluble cellulose [59]. On average, it was 0.6–1.2 U/mg of protein, while the commercial preparation Accelerase1000 was characterized by the increased activity of 1.7 U/mg, and Accelerase XY had extremely low activity on filter paper.

MCC activity illustrates the ability of enzyme preparations to hydrolyze highly ordered crystalline zones of cellulose (crystallites).

Preparations of Cellic CTec-1 and Cellic CTec-2 had the same MCC activity (0.3 U/mg protein); preparations of Accelerase 1500 and DUET, as well as *P. verruculosum*

Preparation	FP	MCC	CMC	pNPG	Cellobiose	Xylan	pNPX
	50°C, pH 5.0	40°C, pH 5.0	50°C, pH 5.0	40°C, pH 5.0	40°C, pH 5.0	50°C, pH 5.0	40°C, pH 5.0
Accelerase 1000	1.7	1.2	12.3	3.6	2.7	2.4	<0.01
Accelerase 1500	1.2	0.8	10.5	3.8	2.3	1.7	<0.01
Accelerase XY	<0.1	0.1	0.8	0.6	0.4	92.5	0.3
Accelerase DUET	1.1	0.8	7.9	3.1	2.2	14.0	1.3
Cellic CTec-1	0.7	0.3	10.9	2.8	2.4	1.5	<0.01
Cellic CTec-2	0.6	0.3	15.6	5.5	4.0	28.2	<0.01
Piptochaetium verruculosum B151	0.9	0.7	18.3	1.7	0.7	30.3	<0.01
P. verruculosum F10	0.2	1.1	5.9	45.5	60.2	0.8	<0.01

Table 2.
Specific activities (units/mg of protein) of enzyme preparations obtained with the help of fungi of the genus Trichoderma *and* Penicillium *as producers.*

B151, were characterized by activity of 0.7–0.8 U/mg of protein; while Accelerase XY corresponded to a low level of MCC activity (0.1 U/mg of protein), and Accelerase 1000 had the highest level (1.2 U/mg of protein) compared to other drugs.

The activity of preparations on CMC demonstrates their ability to hydrolyze less ordered, amorphous zones of cellulose [60, 61]. The hydrolysis of these regions of the substrate is carried out mainly by endoglucanases, which hydrolyze internal β-1,4-glucosidic bonds remote from the ends of the cellulose polymer chain, with the formation of fragments of the polymer substrate chain and cellooligosaccharides, which is accompanied by a significant decrease in the degree of substrate polymerization and, as a result, a decrease in its viscosity (which is especially important in the first stages of the bioconversion process). Reducing the viscosity of the reaction mixture is a very significant factor in the industrial implementation of enzymatic hydrolysis processes since increased viscosity can lead to a significant decrease in the efficiency of some production stages (for example, due to a decrease in heat transfer rates, mixing, mass transfer, etc.). The considered commercial enzyme preparations, on average, were characterized by CMC activity of 8–12 U/mg of protein, while Accelerase XY corresponded with a low level of activity (0.8 U/mg of protein), and the *P. verruculosum* B151 preparation had the highest value (18, 3 U/mg protein) compared to other preparations.

In general, the specific activities of the *P. verruculosum* B151 preparation in relation to various substrates, which characterize the effectiveness of individual components of multienzyme complexes, are comparable with similar activities of commercial *Trichoderma* preparations. However, the specific activities given in **Table 2** were determined from the initial rates of hydrolysis of the corresponding substrates, and their comparison is not sufficient to compare the hydrolytic activity of enzyme preparations, since during long-term hydrolysis of biomass, the processes of enzyme inactivation, their inhibition by lignin (for due to unproductive adsorption of enzymes on it), as well as inhibition of enzymes by the reaction products.

For a more detailed consideration, **Table 3** shows the data characterizing the results of testing enzyme preparations during long-term (exhaustive) hydrolysis of biomass of various types [62]. Pretreated corn stalks, bagasse, and coniferous and deciduous wood were used as substrates. The selected substrates differ in composition, including the content of cellulose and hemicelluloses (xylans), as well as the content of lignin. The content of xylans decreased in the series: corn stalks, bagasse, aspen wood, and pine wood. In addition, MCC was used, which made it possible to evaluate the effectiveness of cellulase enzyme preparations in relation to a lignin-free substrate [63].

The hydrolytic efficiency of enzyme preparations based on fungi of the genus *Trichoderma* and *Penicillium* was evaluated by the formation of reducing sugars (RS) at various dosages of enzyme preparations (**Table 3**).

The enzyme preparations Accelerase 1000, Accelerate 1500, and Accelerate DUET had approximately the same efficiency of hydrolysis (RS yield) of pretreated corn stalks and bagasse, shredded pine, and aspen wood. The effectiveness of these preparations can be taken as an "average level".

Accelerase XY, which is essentially a xylanase preparation, turned out to be the least effective among those studied for the hydrolysis of pretreated corn stalks, shredded pine and aspen wood, and MCC but proved to be competitive in the hydrolysis of pretreated bagasse.

Cellic CTec-1 and Cellic CTec-2 preparations had approximately equal activity values for FP and MCC, which turned out to be less than the corresponding values for Accelerase 1000, Accelerase 1500, and Accelerase DUET preparations (**Table 2**). At the same time, Cellic CTec-2 was characterized by increased activities in relation to CMC, pNPG, and cellobiose, compared to Cellic CTec-1, as well as Accelerase 1000,

Preparation	RS yield (mg/ml)				
	Pretreated with steam explosion		Grind wood		MCC
	Corn stalks	Bagasse	Pine	Aspen	
Accelerase 1000	37.8-38.0-38.5	19.6-25.8-40.0	22.4-25.3-30.6	21.5-34.3-41.3	26.0-52.2-76.8
Accelerase 1500	33.9-36.5-38.9	17.6-24.4-34.6	19.9-24.5-29.1	19.3-35.2-40.9	27.0-50.6-77.3
Accelerase XY	18.2-18.3-18.3	18.7-19.5-32.8	3.6-4.9-8.6	11.8-13.9-17.3	2.9-4.8-5.8
Accelerase DUET	34.3-34.8-36.9	20.3-31.7-43.0	20.4-24.5-29.5	25.9–41.5-49.6	25.3-53.5-75.6
Cellic CTec-1	23.3-26.3-26.3	12.0-21.8-32.0	12.1-14.7-22.6	9.2-18.9-23.0	16.2-30.2-60.8
Cellic CTec-2	48.8–53.2-54.4	28.5-37.0-50.5	24.5-29.2-31.6	27.6–40.7-48.4	35.5-73.7-84.4
Piptochaetium verruculosum B151 + F10	48.1-51.0-52.2	32.2-34.5-48.6	27.3-29.0-32.4	29.8-44.9-52.3	32.5-58.0-75.1

Table 3.
Yields of RS as a result of 48-hour hydrolysis of biomass with the enzyme preparations obtained using fungi of the genus Trichoderma and Penicillium. The dosage of enzyme preparations is 2, 5, and 10 mg of the preparation protein per 1 g of dry matter of the cellulose-containing substrate, 50°C, pH 5.0. The substrate concentration was 100 mg/ml on a dry matter basis.

Accelerase 1500, and Accelerase DUET. In addition, Cellic Ctec-2 also had high xylanase activity. Such an advantage of Cellic CTec-2 in relation to Cellic CTec-1 and other studied commercial preparations, revealed in a comparison of their activities in terms of initial hydrolysis rates, was also observed during long-term biomass hydrolysis. Cellic CTec-2 has proven to be the most effective for hydrolyzing pretreated corn stalks and bagasse, ground pine wood, and MCC. The overall effectiveness of the Cellic CTec-1 preparation in long-term biomass hydrolysis was lower than the efficiencies of the Accelerase1000, Accelerase1500, and Accelerase DUET preparations, despite comparable values of the main corresponding activities (**Table 2**). Thus, among the studied commercial preparations, Cellic CTec-2 was the most effective for the hydrolysis of the considered biomass samples.

According to the data given in **Table 2**, the *P. verruculosum* B151 enzyme preparation had increased activities for FP, MCC, CMC, and xylan compared to the commercial Cellic CTec-2 preparation. At the same time, the BGL activity of the *P. verruculosum* B151 preparation was lower; therefore, the combined preparation *P. verruculosum* B151 + *P. verruculosum* F10 in a ratio of 9:1 for the protein was used in the hydrolysis experiments. The combination of enzyme preparations B151 + F10 exceeded the effectiveness of the preparation Cellic CTec-2 with prolonged hydrolysis of crushed aspen wood, both preparations hydrolyzed pine wood in the same way; with the hydrolysis of pretreated corn stalks and bagasse, the effectiveness was slightly lower than the result of Cellic CTec-2, but nevertheless higher than the effectiveness of the other investigated commercial preparations.

As can be seen from **Table 3**, the increase in the depth of hydrolysis of pretreated corn stalks (48 hours of hydrolysis) with an increase in the dosage of the most effective enzyme preparations (Cellic CTec-2 and B151 + F10) turns out to be very insignificant, which indicates the achievement of the maximum degree of enzymatic conversion of this substrate and its high reactivity (and high efficiency of the pretreatment processing of this type of biomass). Pretreated bagasse and crushed aspen wood were characterized by a greater increase in the yield of RS than crushed pine wood with an increase in the dosage of enzyme preparations, which is explained by the increased content of lignin in pine wood compared to other substrates under consideration. Also in pine, unlike bagasse, there are pitch and hardly hydrolyzable galactomannan. For the hydrolysis of bagasse, which contains readily available polysaccharides (arabinoxylan and xylan), the presence of xylanase, which is present in penicillin preparations, is important.

The maximum values of the RS increase also allow us to characterize and compare the effectiveness of the studied enzyme preparations conveniently analyzed by the results of saccharification of the MCC model substrate. It was shown that the degree of hydrolysis of MCC by all enzyme preparations except Accelerase XY turned out to be significantly higher than the maximum degree of hydrolysis of various types of biomass provided by them. At the same time, the maximum increase in the RS yield per unit mass of the consumed enzyme preparation corresponded to the Cellic CTec-2 and *P. verruculosum* B151 + F10 enzyme preparations, which demonstrated the highest efficiency in the hydrolysis of various types of biomass.

The effectiveness of the action of enzyme complex that performs the bioconversion of plant biomass is its most important characteristic, which determines the feasibility of its use in biotechnology and the resulting economic benefit. This characteristic, in turn, generalizes such important properties of the preparation as specific activities for a number of substrates and the limiting degrees of conversion of plant materials during long-term hydrolysis.

A comparative study of commercial enzyme preparations based on *Trichoderma* and an enzyme preparation based on *P. verruculosum* indicates that it is not inferior to commercial *Trichoderma* biocatalysts or surpasses them in hydrolytic activity in the processes of saccharification of various types of plant materials. This allows us to state that biocatalysts obtained on the basis of *P. verruculosum* producer strains can be a competitive alternative when scaling up biotechnological processes of bioconversion of renewable plant materials.

2.3 Methods for increasing the efficiency of bioconversion of plant materials under the action of an enzyme complex produced by *P. verruculosum*

An auxotrophic mutant B1-537 (ΔniaD) was obtained for genetic engineering manipulations based on the highly productive strain *P. verruculosum* B151. It has a defective nitrate reductase gene, which makes it possible to use it as a basis for obtaining recombinant producers of homologous and heterologous proteins by selecting transformants on the medium with nitrate [55–57, 64].

The expression of the gene of the target enzyme can be carried out under the control of various "promoter-terminator" systems selected by the researcher. This may be a proprietary system inherent in the target protein gene, or a promoter and terminator of another gene. To express various genes in the strain B1-537 (ΔniaD), a promoter and a transcription terminator of the inducible CBH I gene *P. verruculosum* (*cbh1*) [55–57, 65, 66], which is a major component of the cellulase complex of the fungus *P. verruculosum*, were used. The CBH I expression system makes it possible to obtain enzyme preparations with a high content of the target protein. For example, on the basis of strain B1-537 (ΔniaD) using the promoter of the *cbh1* gene, a recombinant strain F10 was created—a superproducer of highly active BGL *A. niger*. The enzyme preparations based on this strain contain up to 80% of the heterologous enzyme from the total secreted protein [56, 62, 63]. The strain F10 contained a large number of copies of the aspergillus BGL gene under the *cbh 1* promoter, whereas the expression of *P. verruculosum*'s own enzymes (and primarily CBH I) was largely suppressed. However, in many other cases, the use of the *cbh1* gene promoter made it possible to obtain recombinant *P. verruculosum* strains capable of secreting a largely preserved cellulase complex of this fungus together with homologously or heterologously expressed additional enzymes of various specificity to enhance the action of cellulases during hydrolysis of various types of cellulose-containing raw materials (or for use in other biotechnological processes), examples are given in **Table 4**.

One of the approaches to obtain recombinant strains is so-called "fusion construct", consisting, for example, of sequentially connected genes encoding CBH I, EG II *P. verruculosum*, and BGL *A. niger*, expressed under the promoter and terminator cbh1 [65]. The use of such a construct made it possible to obtain an enzyme preparation enriched with the three most important components for cellulose hydrolysis (**Table 4**). The use of this preparation together with the initial preparation P. *verruculosum* B151 led to an increase in the yield of sugars during the hydrolysis of pretreated pine and aspen wood by up to 70% [65].

Another example, as described in the previous chapter, is the composite preparation of *P. verruculosum*, which was obtained on the basis of strains B151 and F10 and which, during hydrolysis of various types of lignocellulose raw materials, provided a sugar yield comparable to that demonstrated by modern commercial enzyme preparations based on a new generation of mutant strains of *T. reesei* (Accelerase 1500

Cloned genes	Expressed enzymes	Content in the preparation, %*
xylA P. canescens	Xylanase A	18 (0) [55]
manB T. reesei	Mannanase B	45–70 (0) [55]
egl1 T. reesei	EG I	3 (0) [55]
eglIV T. reesei	LPMO (EG IV)	15 (0) [55]
cbhII T. reesei	CBH II	2 (0) [55]
xyl3 T. reesei	Xylanase III	17 (0) [55]
bgl1 A. niger	BGL I	21 (0) [56]
cbh1 P. verruculosum	CBH I	66 (20) [65]
cbh1 P. verruculosum + egII P. verruculosum	CBH I + EG II	56 + 19 (20 + 3) [65]
cbh1 P. verruculosum + egII P. verruculosum + bgl1 A. niger	CBH I + EG II + BGL I	37 + 34 + 12 (20 + 3 + 0) [65]
eglIV T. reesei	LPMO (EG IV)	24 (0) [65]
eglIV T. reesei + bgl1 A. niger	LPMO (EG IV) + BGL I	1 + 18 (0 + 0) [57]
cbh1 P. verruculosum + egII P. verruculosum + bgl1 A. niger + eglIV T. reesei	CBH I + EG II + BGL I + LPMO (EG IV)	24 + 35 + 10 + 1 (20 + 3 + 0 + 0) [57]

**The content of enzymes in % of the total protein content in the preparation obtained using the corresponding recombinant strain of P. verruculosum is given. In parentheses the content of the expressed enzymes in the control preparation P. verruculosum B151 is indicated.*

Table 4.
Using the promoter and terminator of the cbh1 gene to produce recombinant strains and enzyme preparations of Piptochaetium verruculosum with homologously or heterologously expressed enzymes.

and DUET, Cellic CTec1 and CTec2) [62]. In most cases, the composite preparation of *P. verruculosum* was superior to the indicated preparations of *T. reesei* in terms of effectiveness.

To obtain the strains with a moderate level of expression of target recombinant enzymes alternative expression systems can be used. This type includes, for example, a constitutive expression system based on the histone gene promoter hist4.1 [64]. Its use makes it possible to increase the production of the target protein without significantly changing the composition of the main secreted enzyme complex. Using an expression system based on a histone promoter, the B1_PrHist enzyme preparation was obtained with heterologous *A. niger* BGL, the content of which was about 13% of the total secreted protein. This preparation provided a higher yield of glucose (by 10–21%) during the hydrolysis of MCC and crushed aspen wood compared to the control preparation *P. verruculosum* B151. Together with the heterologous *A. niger* BGL under the control of a histone promoter, the expression of *P. verruculosum*'s own CBH I was increased, as a result, the enzyme preparation obtained on the basis of the recombinant strain demonstrated a 29–100% gain in sugar yield compared to the control preparation [57].

Thus, both of the genetic engineering approaches discussed above, in which the target protein is expressed either under the control of a strong inducible (*cbh1*) promoter or under the control of a weaker constitutive (*hist4.1*) promoter, can significantly increase the hydrolytic ability of secreted cellulase complexes. The choice of one or another approach largely depends on the specific application of the final enzyme preparation in a particular biotechnological process. In the case of enzymatic hydrolysis of cellulose, it depends on the type of plant material and the method of its pretreatment.

Cellulases are an example of enzymes that are extremely demanded in practice and are used in Europe, the USA, and a number of other countries to implement the ideas of bioeconomics, implying the replacement of non-renewable fossil raw materials and chemical processes of its processing with alternative biotechnological processes with integrated use of renewable raw materials and various wastes. In particular, with the use of cellulases in a number of countries (USA, Italy, Brazil), the concept of biofactory (biorefinery) has been practically implemented, consisting in the realization of large-scale processes of enzymatic conversion of renewable lignocellulose plant biomass (agricultural waste) into sugars, followed by the production of alcohols (second-generation biofuels) from them [67]. Other biotechnological processes based on the use of lignocellulose raw materials have been tested and are close to practical implementation in order to obtain solvents, organic and amino acids, biologically active substances, monomers for the synthesis of various polymers, etc. In addition, cellulases are traditionally used in the textile industry, pulp and paper industry, and various branches of the food industry, such as brewing, alcohol, and bakery, as well as feed additives in poultry and animal husbandry [68].

3. Conclusions

Cellulolytic enzymes are found in various living systems, but the main destroyers of cellulose are microorganisms, primarily microscopic aerobic fungi, as well as various types of aerobic and anaerobic bacteria. Fungi from the genus *Trichoderma* are the most widely used industrial producers of cellulases in different countries, from which intensive research on cellulases began.

Using the methods of mutagenesis, selection, and genetic engineering, highly productive *T. reesei* strains capable of secreting up to 100 g/l of extracellular protein were obtained. Some species of fungi from the genus *Penicillium* (in particular *P. verruculosum*) have great prospects as an alternative to *T. reesei*, producing cellulase complexes whose enzymes have significantly higher specific activity compared to *T. reesei* cellulases. Highly productive mutant strains of *P.verruculosum* capable of secreting up to 60 g/l of extracellular protein were obtained. Based on the *P. verruculosum* strain, a host-vector system has been created that allows the expression of homologous and heterologous genes and the production of enzyme preparations, the qualitative and quantitative composition of which is appropriately optimized in terms of ensuring its maximum activity in relation to various types of cellulose-containing plant raw materials.

Abbreviations

EG	endoglucanase
CBH	cellobiohydrolase
BGL	β-glucosidase
LPMO	lytic polysaccharide monooxygenase
CBM	cellulose binding domain

MCC	microcrystalline cellulose
CMC	carboxymethyl cellulose
FP	filter paper
RS	reducing sugars
pNPG	*p*-nitrophenyl-β-D-glucoside
pNPX	*p*-nitrophenyl-β-D-xyloside

Author details

Olga G. Korotkova[1], Alexandra M. Rozhkova[1*], Ekaterina A. Rubtsova[1] and Arkady P. Sinitsyn[2]

1 Federal State Institution "Federal Research Centre "Fundamentals of Biotechnology" of the Russian Academy of Sciences", Moscow, Russian Federation

2 Lomonosov Moscow State University, Moscow, Russian Federation

*Address all correspondence to: amrojkova@yahoo.com

References

[1] Robak K, Balcerek M. Review of second generation bioethanol production from residual biomass. Food Technology and Biotechnology. 2018;**56**(2):174-187. DOI: 10.17113/ftb.56.02.18.5428

[2] Srivastava R, Sarangi P, Bhatia L. Convertion of methane to methanol: Technologies and future challenges. Biomass Convertion and Biorefinery. 2022;**12**:1851-1875. DOI: 10.1007/s13399-024-01872-5

[3] Niphadkar S, Bagade P, Ahmed S. Bioethanol production: Insight into past, present and future perspectives. Biofuels. 2018;**9**(2):229-238. DOI: 10.1080/17597269.2017.1334338

[4] Ahmed B, Aboudi K, Tyagi V, Alvarez-Gallego C, Fernandez-Guelfo L, Romero Garcia L, et al. Improvement of anaerobic digestion of lignocellulosic biomass by hydrother-mal pretreatment. Applied Sciences. 2019;**9**(18):3853. DOI: 10.3390/app9183853

[5] Ali S, Nugent B, Mullins E, Doohan F. Fungal-mediated consolidated bioprocessing: The potential of fusarium oxysporum for the lignocellulosic ethanol industry. AMB Express. 2016;**6**:1-13. DOI: 10.1186/s13568-016-0185-0

[6] Sarangi P, Singh T, Singh N, Shadangi K, Srivastava R, Singh A, et al. Sustainable utilization of pineapple wastes for production of bioenergy, bio-chemicals and value-added products: A review. Bioresource Technology. 2022;**351**:127085. DOI: 10.1016/j.biortech.2022.127085

[7] Schuster B, Chinn M. Consolidated bioprocessing of ligno-cellulosic feedstocks for ethanol fuel production. BioEnergy Research. 2012;**6**(2):416-435. DOI: 10.1007/s12155-012-9278-z

[8] Sarangi P, Nayak M. Agro-waste for second generation biofuels. In: Shadangi KP, editor. Wiley-Scrivener, Scrivener. Liquid Biofuels: Fundamentals, Characterization, and Applications. Beverly: Publishing LLC; 2021. pp. 697-706. DOI: 10.1002/9781119793038.ch20

[9] Pandey A, Kumar B. Utilization of agricultural and industrial waste as replacement of cement in pavement quality concrete: A review. Environmental Science and Pollution Research. 2022;**29**:24504-24546. DOI: 10.1007/s11356-021-18189-5

[10] Ocak S, Acar S. Biofuels from wastes in Marmara region, Turkey: Potentials and constraints. Environmental Science and Pollution Research. 2021;**28**:66026-66042. DOI: 10.21203/rs.3.rs-245086/v1

[11] Kurabi A, Berlin A, Gilkes N, Kilburn D, Bura R, Robinson J, et al. Enzymatic hydrolysis of steam-exploded and ethanol organosolv-pretreatd Douglas-fir by novel and covvertial fungal cellulases. Applied Biochemistry and Biotechnology. 2005;**121**:219-230. DOI: 10.1007/978-1-59259-991-2_19

[12] Berlin A, Gilkes N, Kilburn D, Bura R, Robinson J, Markov A, et al. Evaluation of novel fungal cellulase preparations for ability to hydrolase softwood substrates - evidence for the role of accessory enzymes. Enzyme and Microbial Technology. 2005;**37**(2):175-184. DOI: 10.1016/j.enzmictec.2005.01.039

[13] Sinitsyn A, Gusakov A, Vlasenko E. Effect of structural and physic-chemical

features of cellulosic substrates on the efficiency of enzymatic hydrolysis. Applied Biochemistry and Biotechnology. 1991;**30**:43-59. DOI: 10.1007/BF02922023

[14] Kastel'yanos O, Sinitsyn A, Vlasenko E. The investigation of properties of the cellulase system from Penicillium sp. Applied Biochemistry and Microbiology. 1994;**30**(6):799-811

[15] Payne C, Knott B, Mayes H, Hansson H, Himmel M, Sandgren M, et al. Fungal cellulases. Chemical Reviews. 2015;**115**(3):1308-1448. DOI: 10.1021/cr500351c

[16] Gusakov A. Cellulases and hemicellulases in the 21st century race for cellulosic ethanol. Biofuels. 2013;**4**:567-569. DOI: 10.4155/bfs.13.55

[17] Gusakov A. Alternatives to Trichoderma reesei in biofuel production. Trends in Biotechnology. 2011;**29**(9):419-425. DOI: 10.1016/j.tibtech.2011.04.004

[18] Gusakov A, Sinitsyn A. Cellulases from Penicillium species for producing fuels from biomass. Biofuels. 2012;**3**:463-477. DOI: 10.4155/bfs.12.41

[19] Gupta V, Steindorff A, Paula R, March-Aigner A, March R, Silva R. The post-genomic era of Trichoderma reesei: what's next? Trends in Biotechnology. 2016;**34**:970-982. DOI: 10.1016/j.tibtech.2016.06.003

[20] Marjamaa K, Toth K, Bromann P, Szakacs G, Kruus K. Novel Penicillium cellulases for total hydrolysis of lignocellulotics. Enzyme and Microbial Technology. 2013;**52**:358-369. DOI: 10.1016/j.enzmictec.2013.03.003

[21] Maeda R, Barcelos C, Anna L, Pereira N. Cellulase production by Penicillium funiculosum and its application in the hydrolysis of sugarcane bagasse for for second generation ethanol production by fed batch operation. Journal of Biotechnology. 2013;**163**:38-44. DOI: 10.1016/j.jbiotec.2012.10.014

[22] Mäkelä M, Mansouri S, Wiebenga A, Rytioja J, Vries R, Hilden K. Penicillium subrubescens is a promising alternative for aspergillus Niger in enzymatic plant biomass saccharafication. New Biotechnology. 2016;**33**(6):834-841. DOI: 10.1016/j.nbt.2016.07.014

[23] Merino S, Cherry J. Progress and challenges in enzyme development for biomass utilization. Advances in Biochemical Engineering/Biotechnology. 2007;**108**:95-120. DOI: 10.1007/10_2007_066

[24] Margeot A, Hahn-Hagerdal B, Edlund M, Slade R, Monot F. New improvements for lignocellulosic ethanol. Current Opinion in Biotechnology. 2009;**20**(3):372-380. DOI: 10.1016/j.copbio.2009.05.009

[25] Nieves R, Ehrman C, Adney W, Elander R, Himmel M. Survey and analysis of commercial cellulase preparations suitable for biomass conversion to ethanol. World Journal of Microbiology and Biotechnology. 1997;**14**:301-304. DOI: 10.1023/A:1008871205580

[26] Kubicek C, Mikus M, Schuster A, Schmoll M, Seiboth B. Metabolic engineering strategies for the improvement of cellulase production by Hypocrea jecorina. Biotechnology for Biofuels. 2009;**2**(1):19-33. DOI: 10.1186/1754-6834-2-19

[27] Skomarovsky A, Gusakov A, Okunev O, Solov'eva I, Bubnova T, Kondrat'eva E, et al. Studies of hydrolytic activity of enzyme preparation of Penicillium and Trichoderma fungi.

Applied Biochemistry and Microbiology. 2005;**41**(2):182-184. DOI: 10.1007/s10438-005-0032-6

[28] Martins L, Kolling D, Camassola M, Dillon A, Ramos L. Comparison of Penicillium echinulatum and Trichoderma reesei cellulases in relation to their activity against various cellulosic substrates. Bioresource Technology. 2008;**99**(5):1417-1424. DOI: 10.1016/j.biortech.2007.01.060

[29] Ikeda Y, Hayashi H, Okuda N, Park E. Efficient cellulase production by the filamentous fungus Acremonium cellulolyticus. Biotechnology Progress. 2007;**23**(2):333-338. DOI: 10.1021/bp060201s

[30] Fujii T, Fang X, Inoue H, Murakami K, Sawayama S. Enzymatic hydrolyzing performance of Acremonium cellulolyticus and Trichoderma reesei against three lignocellulosic materials. Biotechnology for Biofuels. 2009;**2**(1):24-32. DOI: 10.1186/1754-6834-2-24

[31] Gusakov A, Salanovich T, Antonov A, Ustinov B, Okunev O, Burlingame R, et al. Design of highly efficient cellulase mixtures for enzymatic hydrolysis of cellulose. Biotechnology and Bioengineering. 2007;**97950**:1028-1038. DOI: 10.1002/bit.21329

[32] Liao H, Fan X, Mei X, Wei Z, Raza W, Shen Q, et al. Production and characterization of cellulotic enzyme from Penicillium oxalicum GZ-2 and its application in lignocelluloses saccharafication. Biomass and Bioenergy. 2015;**74**:122-134. DOI: 10.1016/j.biombioe.2015.01.016

[33] Zhao X, Wang W, Tong B. A newly isolated Penicillium oxalicum 16 cellulase with high efficient synergism and high tolerance of monosaccharide. Applied Biochemistry and Biotechnology. 2016;**178**:173-183. DOI: 10.1007/s12010-015-1866-x

[34] Steffien D, Aubel I, Bertau M. Enzymatic hydrolysis of pre-treated lignocellulose with Penicillium verruculosum cellulases. Journal of Molecular Catalysis B: Enzymatic. 2014;**103**:29-35. DOI: 10.1016/j.molcatb.2013.11.004

[35] Liu G, Zhang L, Wei X, Zou G, Qin Y, Ma L, et al. Genomic and secretomic analyses reveal unique features of the lignocellulolytic enzyme system of Penicillium decumbens. PLoS One. 2013;**8**(2):e55185. DOI: 10.1371/journal.pone.0055185

[36] Ogunmolu F, Kaur I, Gupta M, Bashir Z, Pasari N, Yazdani S. Proteomics insights into the biomass hydrolysis potentials of a hypercellulotic fungus Penicillium funiculosum. Journal of Proteome Research. 2015;**14**(10):4342-4358. DOI: 10.1021/acs.jproteome.5b00542

[37] Hu L, Taujale R, Liu F, Song J, Yin Q, Zhang Y, et al. Draft genome sequence of Talaromyces verruculosus ("Penicillium verruculosum") strain TS63-9, a fungus with great potential for industrial production of polysaccharide-degrading enzymes. Journal of Biotechnology. 2016;**219**:5-6. DOI: 10.1016/j.jbiotec.2015.12.017

[38] Morozova V, Gusakov A, Sinitsyn A, Andrianov R, Pravilnikov A, Osipov D. Cellulases of Penicillium verruculosum. Biotechnology Journal. 2010;**5**(8):871-880. DOI: 10.1002/biot.201000050

[39] Volkov P, Rozhkova A, Gusakov A, Sinitsyn A. Homologous cloning, purification and characterization of highly active cellobiohydrolase I (cel7A) from

Penicillium canescens. Protein Expression and Purification. 2014;**103**:1-7. DOI: 10.1016/j.pep.2014.08.011

[40] Gusakov A, Dotsenko A, Rozhkova A, Sinitsyn A. N-linked glycans are an important component of the processive machinery of cellobiohydrolases. Biochimie. 2017;**132**:102-108. DOI: 10.1016/j.biochi.2016.11.004

[41] Dotsenko A, Gusakov A, Rozhkova A, c Volkov P. N-linked glycosylation of recombinant cellobiohydrolase I (cel7A) from Penicillium verruculosum and its effect on the enzyme activity. Biotechnology and Bioengineering. 2016;**113**:283-291. DOI: 10.1002/bit.25812

[42] Donohoe B, Resch M. Mechanisms employed by cellulase systems to gain access through the complex architecture of lignocellulosic substrates. Current Opinion in Chemical Biology. 2015;**29**:100-107. DOI: 10.1016/j.cbpa.2015.08.014

[43] Rahikainen J, Martin-Sampedro R, Heikkinen H, Rovio S, Marjamaa K, Tamminen T, et al. Inhibitory effect of lignin during cellulose bioconversion: The effect of lignin chemistry on non-productive enzyme adsorption. Bioresource Technology. 2013;**133**:270-278. DOI: 10.1016/j.biortech.2013.01.075

[44] Zeng Y, Zhao S, Yang S, Ding S. Lignin plays a negative role in the biochemical pocess for producing lignocellulotic biofuels. Current Opinion in Biotechnology. 2014;**27**:38-45. DOI: 10.1016/j.copbio.2013.09.008

[45] Liu H, Sun J, Leu S-Y. Toward a fundamental understanding of cellulase-lignin interactions in the whole slurry enzymatic saccharification process. Biofuels, Bioproducts and Biorefining. 2016;**10**(5):648-655. DOI: 10.1002/bbb.1670

[46] Berlin A, Balakshin N, Gilkes N, Kadla J, Maximenko M, Kubo S, et al. Inhibition of cellulase, xylanase and beta-glucosidase activities by softwood lignin preparations. Journal of Biotechnology. 2006;**125**(2):198-209. DOI: 10.1016/j.jbiotec.2006.02.021

[47] Ko J, Ximenes E, Kim Y, Ladisch M. Adsorption of enzyme onto lignins of liquid hot water pretreated hardwoods. Biotechnology and Bioengineering. 2015;**112**(3):447-456. DOI: 10.1002/bit.25359

[48] Lu X, Zheng X, Li X, Zhao J. Adsorption and mechanism of cellulase enzymes onto lignin isolated from corn Stover pretreated with liquid hot water. Biotechnology of Biofuels. 2016;**9**:118. DOI: 10.1186/s13068-016-0531-0

[49] Schülein M. Protein engineering of cellulases. Biochimica et Biophysica Acta. 2000;**1543**(2):239-252. DOI: 10.1016/s0167-4838(00)00247-8

[50] Bommarius A, Sohn M, Kang Y, Lee J, Realff M. Protein engineering of cellulases. Current Opinion in Biotechnology. 2014;**29**:139-145. DOI: 10.1016/j.copbio.2014.04.007

[51] Gusakov A. Recent advances in selection and engineering of biomass degrading enzymes. Jacobs Journal of Enzymology and Enzyme Engineering. 2014;**1**:1

[52] Billard H, Faraj A, Ferreira N, Menir S, Heiss-Blanquet S. Optimization of a synthetic mixture composed of major Trichoderma reesei enzymes for the hydrolysis of steam-exploded wheat straw. Biotechnology for Biofuels. 2012;**5**:9. DOI: 10.1186/1754-6834-5-9

[53] Szakacs G, Reczey K, Hemady P, Dobozi M. Pénicillium verruculosum WA30, a new source of cellulase. Applied Microbiology and Biotechnology. 1981;**11**:120-124. DOI: 10.1007/BF00518054

[54] Solov'eva I, Okunev O, Velk'ov V, Koshelev A, Bubnova T, Kondrat'eva E, et al. The selection and properties of Penicillium verruculosum mutants with enhanced production of cellulases and xylanases. Microbiology. 2005;**74**:141-146. DOI: 10.1007/s11021-005-0043-6

[55] Sinitsyn A, Osipov D, Rozhkova A, Bushina E, Dotsenko G, Sinitsyna O, et al. Production of highly effective enzyme complexes of cellulases and hemicellulases on the basis of *Penicillium verruculosum* strain for hydrolysis of plant raw materials. Biothechnology. 2013;**29**(5):40-53. DOI: 10.1134/S0003683814080055

[56] Dotsenko G, Gusakov A, Rozhkova A, Korotkova O, Sinitsyn A. Heterologous β-glucosidase in a fungal Cellulase system: Comparison of different methods for development of multienzyme cocktails. Process Biochemistry. 2015;**50**:1258-1263. DOI: 10.1016/j.procbio.2015.05.008

[57] Sinitsyn A, Korotkova O, Sinitsyna O, Rozhkova A, Dotsenko G, Proskurina O, et al. Optimization of the composition of the cellulase enzyme complex Penicillium verruculosum: Increasing the hydrolytic capacity using genetic engineering methods. Catalysis in Industry. 2015;**15**(6):78-83. DOI: 10.18412/1816-0387-2015-6-78-83

[58] Gusakov AV. Comment on "revealing nature's cellulase diversity: The digestion mechanism of Caldicellulosiruptor bascii CelA". Science. 2014;**344**(6184):578. DOI: 10.1126/science.1251248

[59] Ghose T. Measurement of cellulase activities. Pure and Applied Chemistry. 1987;**59**(2):257-268. DOI: 10.1351/PAC198759020257

[60] Wood T, McCrae S, Bhat M. The mechanism of fungal cellulase action. Synergism between components of Penicillium pinophilum cellulase in solubilizing hydrogen bond-ordered cellulose. Biochemical Journal. 1989;**260**(1):37-43. DOI: 10.1042/bj2600037

[61] Garcia E, Johnston D, Whitaker J, Shoemaker S. Assessment of endo-1,4-β-D-glucanase activity by a rapid colorimetric assay using disodium 2,2′-bicinchoninate. Journal of Food Biochemistry. 1993;**17**:135-145. DOI: 10.1111/J.1745-4514.1993.TB00463.X

[62] Chekushina A, Dotsenko G, Sinitsyn A. Comparison of the efficiency of plant materials bioconversion processes using biocatalysts based on enzyme preparations Trichoderma and Penicillium verruculosum. Catalysis in Industry. 2013;**5**:98-104. DOI: 10.1134/S2070050413010042

[63] Chekushina A, Dotsenko G, Kondratieva E, Sinitsyn A. Enzyme preparations of Penicillium verruculosum for bioconversion of plant raw materials is an alternative to commercial preparations obtained using Trichoderma fungi species. Biotechnology. 2013;**29**(3):69-80

[64] Korotkova O, Rozhkova A, Matys V, Koshelev A, Okunev O, Nemashkalov V, et al. Receiving of complex biocatalysts based on enzyme preparations from the recombinant fungus Penicillium verruculosum and their application in the hydrolysis of waste from the wood-processing and agricultural industries. Catalysis in Industry. 2011;**5**:61-68

[65] Proskurina O, Korotkova O, Rozhkova A, Matys V, Koshelev A, Okunev O, et al. Application of "fusion" technology to create highly efficient biocatalysts based on recombinant strains of the fungus Penicillium verruculosum for the convertion of cellulose-containing biomass. Catalysis in Industry. 2013;**5**:65-73

[66] Proskurina O, Korotkova O, Rozhkova A, Kondrat'eva E, Sinitsyn A, Matys V, et al. Use of endoglucanase IV from Trichoderma reesei to enhance the hydrolytic activity of a cellulase complex from the fungus Penicillium verruculosum. Applied Biochemistry and Microbiology. 2015;**51**(6):592-599. DOI: 10.1134/S0003683815060125

[67] Singh N, Singhania R, Nigam P, Dong C-D, Patel A, Puri M. Global status of lignocellulosic biorefinery: Challenges and perspectives. Bioresource Technology. 2022;**344**(Pt B):126415. DOI: 10.1016/j.biortech.2021.126415

[68] Fatma S, Hmeed A, Noman M, Ahmed T, Shahid M, Tariq M, et al. Lignocellulosic biomass: A sustainable bioenergy source for the future. Protein and Peptide Letters. 2018;**25**(2):148-163. DOI: 10.2174/0929866525666180122144504

Chapter 14

Application of the Six Sigma DMAIC Methodology to the Gasification Process

José Antonio Mayoral Chavando, Valter Silva, João Sousa Cardoso, Daniela Eusébio and Luís A.C. Tarelho

Abstract

Despite the advantages of gasification over combustion, some elements remain to improve. Fortunately, it is not necessary to reinvent the wheel to improve efficiency and quality because there are already methodologies that have been proven successful with other processes, like the Six Sigma DMAIC methodology. Therefore, this chapter explores the synergies between gasification and Six Sigma DMAIC to improve gas quality and hydrogen production, using RDF and wood as feedstock. Furthermore, the blends and equivalence ratio influence the produced gas is explored.

Keywords: Six Sigma, DMAIC, refuse-derived fuels, gasification, biomass

1. Introduction

Climate change is an urgent problem of our time since it threatens the equilibrium of our planet and, with it, the livelihood of billions of people and species [1]. Our reliance on fossil fuels, urbanization, population growth, and the increase in municipal solid waste (MSW) have influenced climate change, prompting us to reconsider how we produce and consume energy [2]. Global gross final energy consumption was 370EJ in 2017, with oil accounting for 38.7%, coal 20.3%, natural gas 21.1%, nuclear 2.1%, and renewables 17.8% (13% biomass, 3% hydro, 0.9% wind, 0.7% solar and 0.23% geothermal). The gross final energy consumption in continents in 2017 in renewable energy was: Africa 54.5%, Americas 16.0%, Asia 15.9%, Europe 12.7%, Oceania 11.9%, and the world 17.8% [3]. These figures demonstrate the dominance of fossil fuels over renewables. Therefore, substantial work remains to be done to shift the balance toward renewable energy and prevent climate change's effects.

Intergovernmental organizations and policymakers are the cornerstones of combating climate change, looking for cost-effective change by shifting toward non-conventional energy sources, for example, the EU emissions trading system (EU ETS) [4]. In this regard, some countries have formally stated a deadline to stop all coal burning. For example, the UK has set a deadline of 2025 to phase out coal use and, to expedite the process, has imposed a carbon tax of £18 per ton of carbon dioxide equivalent [5, 6]. It is worth noting that the UK remains in the EU ETS until December 31, 2020, aligning with the withdrawal agreement [7]. Likewise, Netherlands and Italy plan to phase out coal burning by 2030 and 2025 [6]. Other countries, such as Portugal, have already phased out coal use. Indeed, it completed the project 2 years ahead of schedule (from 2023 to 2021), and it now intends to use the coal-burning facilities to generate green hydrogen [8, 9].

All efforts to accelerate the phase-out of coal are also favorable from an economic point of view since the cost of emitting greenhouse gases continues to rise as policymakers increase their efforts to curb pollution-induced climate change. Carbon Pulse predicts that EU carbon prices will triple by 2030, reaching €90 [10]. Other forecasts place the price of CO_2 equivalent at between €32 and €65 per ton by 2030 [11]. **Figure 1** illustrates the upward trend in carbon emission futures prices over time.

In light of rising carbon prices, renewable energy sources appear viable for meeting global energy demand while reducing the reliance on fossil fuels in the energy sector [14]. As companies are phasing out coal burning, they turn to biomass combined with other feedstocks like MSW, reducing carbon footprint. The technology to pass from coal to biomass is already mature [6]. Indeed, Valmet upgrades old units based on bubbling fluidized bed (BFB) or circulating fluidized bed (CFB) technology or converting existing grate, oil, or pulverized coal boilers to BFB. The latest Valmet solution is biomass gasification, which partially replaces fossil fuel with biomass and RDF on a large scale, providing fuel flexibility and decreasing CO_2 emissions economically [15]. Gasification is a thermochemical conversion of carbonaceous materials into a combustible gas through partial oxidation and oxidizing agents, namely air, vapor, oxygen, or carbon dioxide [1, 16]. Gasification has numerous advantages over combustion, such as larger molecules being completely broken down into syngas. Gasification has an oxygen-deficient atmosphere. Thus, it prevents the formation of furans and dioxins since their formation demands enough oxygen. Another advantage is that the resulting syngas can produce energy or chemicals like ammonia [17].

Despite the advantages of gasification over combustion, some elements remain to improve, particularly analyzing biomass blends with other feedstocks (co-gasification) [18]. Co-gasification offers additional advantages. For example, RDF can

Figure 1.
Carbon emissions futures price (euros/ton) [12, 13].

blend materials with little added value. It also reduces CO_2 emissions by avoiding the extraction of new fossil fuels. Furthermore, blends with RDF improve the overall feedstock's LHV since RDF has a higher LHV. It also increases the CH_4 and C_2H_4 concentrations, decreasing CO concentration, which may be related to the interaction between the thermal cracking of the plastic and the catalytic ashes contained in RDF [19]. Finally, blends of RDF with biomass dilute some negative features of the RDF char, like high ash and chlorine contents, allowing its energetic valorization in existing gasification facilities [20].

Another improvement element is business process optimization, an integrated activity to make business processes manageable, reaching the best asset utilization and performance through measurable factors like efficiency and quality [21]. Fortunately, it is unnecessary to reinvent the wheel to improve efficiency and quality since many methodologies and tools for business process optimization have been designed and proved successful, for example, Six Sigma DMAIC [22] and the design of experiments. DMAIC is an essential part of the Six Sigma methodology that can be executed independently as a quality improvement method. Define, Measure, Analyze, Improve, and Control [23].

Six Sigma DMAIC methodology improves the bottom line of a product, service, or process by reducing waste and resources and increasing customer satisfaction. Although many believe Six Sigma aims to reach Six Sigma levels of quality, the truth is that Six Sigma and DMAIC aim to improve profitability. Therefore, efficiency and quality are excellent value by-products of its correct implementation [23]. According to Mikel Harry (the creator of Six Sigma), six areas drive its implementation [23]: (1) basic organizational capabilities, (2) industrial process variations, (3) business process variation, (4) engineering/design process, and documentation, (5) quality of specifications, and (6) supplier capabilities.

Even though the Six Sigma DMAIC methodology has proven successful in improving a process, it remains a considerable gap between gasification and Six Sigma DMAIC in the literature because little information or null is available. In this regard, the objective of this chapter is to explore the synergies between gasification and Six Sigma DMAIC by (1) Analyzing the Six Sigma DMAIC, history, and achievements, (2) proposing an integrated Six Sigma DMAIC framework for continuous incremental improvement and optimization of co-gasification, enhancing efficiencies of the overall process, and (3) analyzing the effects of blending biomass with refuse-derived fuels (RFD).

To achieve that, a set of experimental co-gasification runs was performed, changing blending percentages and equivalence ratio (ER) to maximize energetic efficiency based on Low Heating Value (LHV) and gas composition.

2. Six Sigma DMAIC methodology

Six Sigma has been used, tasted, and adapted to different industries and businesses from 1985 until now, optimizing processes and improving profitability. **Figure 2** [24–33] shows a historical background of the evolution and the use of Six Sigma in different fields, among them mechanical design, electrical design, manufacturing, value creation, environmental sustainability, education, etc.

A quantum leap in Six Sigma occurred in 2000 when Mikel Harry published the book *The Breakthrough Management Strategy*. That book provides a strategy called The Breakthrough Management Strategy that gathers the experiences of 15 years to reach Six Sigma Quality through a highly efficient method. In other words, Six Sigma is the

Figure 2.
Gasification history.

Land of Oz, and the Breakthrough Strategy is the Yellow Brick Road. This strategy is based on eight phases: Recognize, Define, Measure, Analyze, Improve, Control, Standardize, and Integrate (RDMAICSI) [23]. The five core phases are called DMAIC, which may implement as a standalone method [34].

The Define phase aims to understand the why of the project and what it is intended to reach. In this phase, the objectives and scope must be defined [35–37]. In the Measure phase, the applicable measurement systems and tools focusing on data collection and reporting are reviewed to identify the opportunity for improvement and the baseline performance. Critical variables are measured and collected in this phase [35, 38, 39].

The Analyze phase provides statistical methods and tools to isolate critical information that will expose the number of defective products. Here, practical business problems are shifted into statistical problems, and it is glimpsed the cause of the problems and possible solutions [35, 38]. Next, the Improve digs into the key variables that cause the problem. It may also encompass the tool Design for Six Sigma (DFSS) to guarantee a complete understanding of the problem or customer's requirements and expectations before design completion or selection of the optimum solution [3, 6, 7]. Finally, the Control phase sustains the Six Sigma initiative through continuous monitoring to avoid falling into the same problem [3, 6, 7].

The application of Six Sigma DMAIC has been quite extensive since it has proven successful in guiding companies to reduce mistakes in day-to-day operations, focusing on eliminating or reducing lapses in quality at the earliest possible time of occurrence by implementing quality control programs to detect and correct commercial, industrial, and design faults [23]. In addition, the correct implementation of Six Sigma DMAIC results in an economic benefit. For example, Motorola's savings was $15 billion over 11 years, General Electric's savings of 2 billion, Honeywell's Savings of $1.2 billion, Texas Instruments' savings of $600 million, Johnson & Johnson's savings of $ 500 million, among many more [40].

Companies have adopted the Six Sigma DMAIC methodology to improve their processes and margins. Before Six Sigma, improvements in quality programs or process improvements usually had no evident impact on a company's net income. Organizations that cannot track the effect of quality improvements on profitability do not identify what must be changed to increase their profit margins. Thus, implementing this methodology for gasification might bridge the gap between Six Sigma and gasification to the continuous incremental improvement and optimization of gasification, enhancing efficiencies of the overall process. **Table 1** shows examples of applying the Six Sigma DMAIC methodology in some processes and equipment, namely boilers, heat exchangers, ovens, compressors, cooling towers, etc.

Study	Results	Ref
Enhancing Effectiveness of Shell and Tube Heat Exchanger through Six Sigma DMAIC Phases	• Reduce the thermal energy in exhaust flue gas, significantly impacting the furnace's efficiency. • The sigma level was improved from 1.34 to 2.01. • The monetary savings was achieved by about Rs. 0.34 million per year.	[41]
Defect analysis and lean six sigma implementation experience in an automotive assembly line	• drastic reduction of unproductive activities expending 19 min work time, and a 37.2% defect ratio	[42]
Lean Six Sigma in the Energy Service Sector: A Case Study	• The company significantly improved the actualization rate from 2.6 to 20%, outperforming the 10% target in just 3 months	[43]
Improve the extrusion process in tire production using the Six Sigma methodology	• decrease of 0.89% on the indicator of work-off generated by the production system, resulting in annual savings of over 165 thousand Euros.	[44]
A systematic approach to industrial oven optimization for energy saving.	• Annual gas saving of 1,658,000 kWh (29%)	[45]

Study	Results	Ref
Improved Boiler Sootblowing	• Improve boiler efficiency by 1.2% by reducing average stack temperature to 50 F. • −0.86/0.58 (1.44 Sigma improvement) • $26,000/year fuel savings	[46]
Improve 110 psig Compressed Air System	• Reduce plant compressed air demand by 10% • 1.54/1.98 (0.44 Sigma improvement) • $140,000/year increase in recurring revenue	[46]
Improved Micronizer Steam Condensate Heat Recovery	• Reduce the steam required to heat the wash water through increased condensate recovery. • −3.2/0.25 (3.5 Sigma improvement) $577,000/year energy savings	[46]
Reduce Cooling Tower Water Header Pressure	• Reduce CTW header pressure from 68 psig (average) to 62 psig (average) • $133,000/year electrical energy savings	[46]

Table 1.
Application of the Six Sigma DMAIC methodology in processes and equipment.

3. Materials and method

3.1 Characterization of the feedstock

Three feedstocks were used (1) RDF pellets, (2) Pine Chips, and (3) pine pellets. **Table 2** contains the main aspects of the characterization of the feedstock.

3.2 Gasification pilot-scale infrastructure

The experiments were performed in a Pilot-scale Bubbling Fluidized Bed Reactor at the University of Aveiro. **Figure 3** is the P&ID of the process where each part of the instrumentation and equipment are indicated. This drawing will be helpful later in identifying potential improvements.

3.3 Methodology

To determine the best conditions, four parameters are considered gas lower heating value (LHV), specific dry gas production (Ygas), cold gas efficiency (CGE), and carbon conversion efficiency (CCE).

$$LHV_{Gas} = \frac{\sum y_{iG} * m_i * LHV_i}{m_G} \quad (1)$$

$$Y_{gas} = \frac{V_G}{m_F} \quad (2)$$

$$CGE[\%] = \frac{V_G * LHV_G}{m_F * LHV_F} * 100 \quad (3)$$

$$CCE[\%] = \frac{V_G * \frac{P_G}{R * T_G} * M_C * \sum_i \in c, i * y_i}{m_F * W_{CF}} * 100 \quad (4)$$

Material	RDF pellets	Pine chips	Pine pellets
Proximate analysis (wt.%, wet basis)			
Moisture	4.3	11.0	4.6
Volatile matter	75.2	77.9	78.5
Fixed carbon	7.1	10.8	16.6
Ash	13.4	0.3	0.3
Ultimate analysis (wt.%, dry basis)			
Ash	13.4	0.3	0.3
C	54.0	46.4	47.5
H	7.4	6.6	6.2
N	0.5	0.2	0.1
S	nd	nd	nd
O (by difference)	24.1	46.5	45.9
Ash composition (mg/kg dry basis)			
Ca	29,000	540	600
Al	20,000	22	96
Si	18,000	<200	<200
S	<6000	<6000	<6000
Fe	3100	29	73
Na	1400	280	280
Mg	950	190	280
Cl–	710	10	1500
K	680	410	590
Cu	380	<3	3
P	370	33	48
Ti	200	<3	4
Ba	190	<3	<3
Sr	180	3	5
Zn	180	5	7
Pb	42	<3	<3
Ni	34	6	<3
Cr	21	<3	<3
V	19	<3	<3
Sn	9	<1	<1
Co	8	<1	<1
Lower heating value (MJ/kg) (dry basis)	24.8	18.8	18.0
Bulk Density	864	577	911

Table 2.
Feedstock characterization.

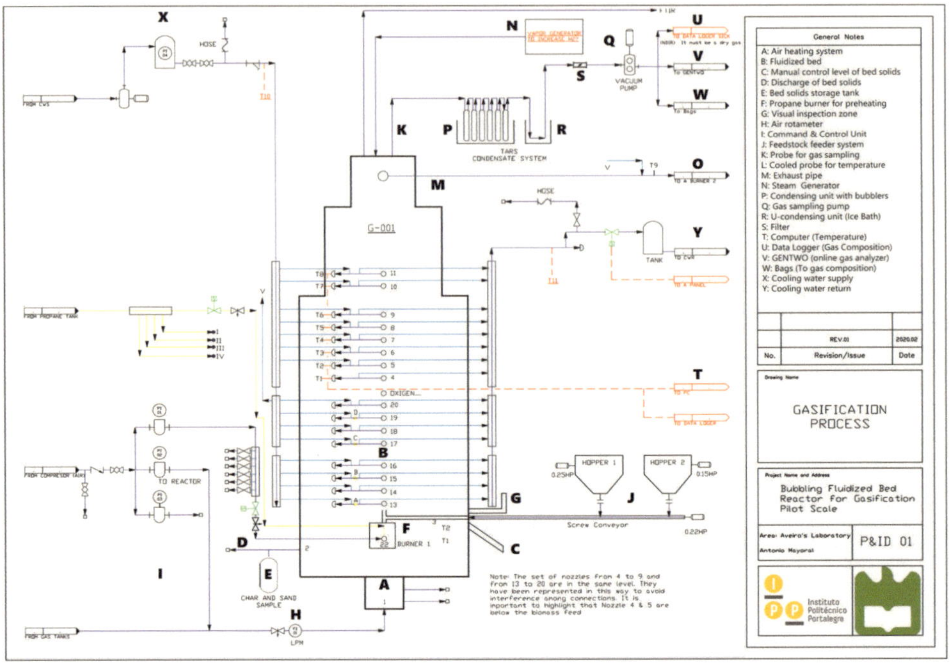

Figure 3.
Bubbling fluidized bed reactor P&ID.

4. Six Sigma DMAIC methodology applied to gasification

Six Sigma DMAIC is a systematic methodology with phases, steps, and tools. **Figure 4** shows the main steps when applying the Six Sigma DMAIC methodology. It also shows the activities and tools to use and the expected outputs; this chapter will follow some of the steps using some of these tools.

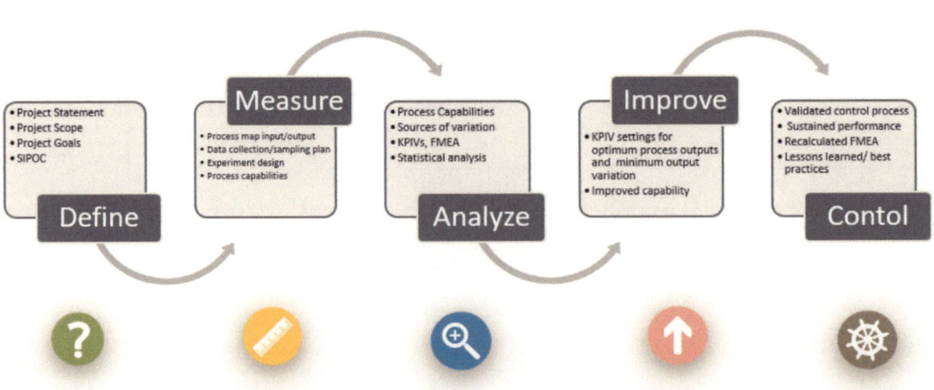

Figure 4.
DMAIC's main aspects.

4.1 Define

4.1.1 Problem statement

Six Sigma DMAIC's success must be explored and replicated in academic gasification studies. Unfortunately, no meaningful works in this scope are available.

4.1.2 Project scope

The project scope is to use the Six Sigma DMAIC methodology to find two optimal process conditions (1) RDF/Wood Blending and (2) Equivalence Ratio by running co-gasification experiments at 785°C and atmospheric pressure, using different RDF/Wood blending and different Equivalence Ratio, and parallelly look for improvements in the co-gasification process of the pilot-scale bubbling fluidized bed reactor of the University of Aveiro to reduce variance in the syngas composition and COPO for future studies.

4.1.3 Project goals

- The primary objective of this chapter is to explore the synergies between gasification and Six Sigma DMAIC to:
    - Propose an integrated Six Sigma DMAIC framework for continuous incremental improvement and co-gasification optimization, enhancing overall efficiency.
- Analyze the effects of blending biomass with refuse-derived fuels (RFD).

4.1.4 SIPOC (supplier, input, process, output, customer)

It is a tool that summarizes the inputs and outputs of one or more processes in table form (see **Figure 5**).

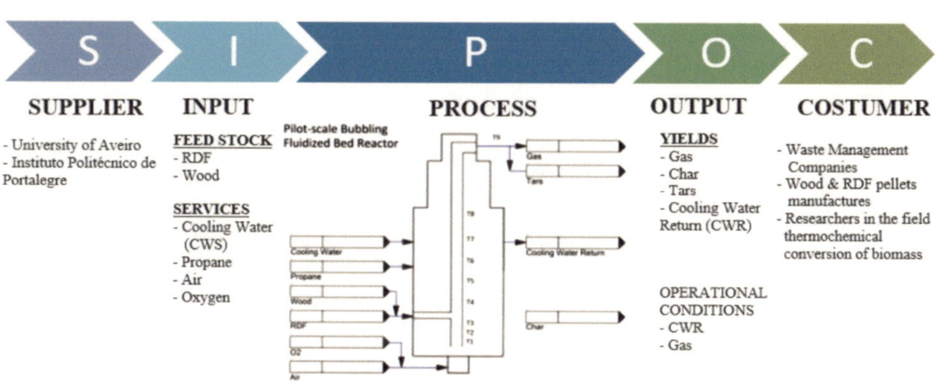

Figure 5.
SIPOC diagram.

4.1.5 Business impact

The exploration and implementation of Sigma DMAIC in Gasification can make it attractive to investors since this methodology has improved many processes, increasing profit margins. Furthermore, this can catalyze the utilization of agroforestry residues and MSW to produce energy, leading to economic and environmental benefits.

4.2 Measure

The measure phase delivers a detailed process map, presented in **Figure 3** as a piping and instrumentation diagram (P&ID), presenting input/output variables, sampling points, and other process details.

Detailed process map: The P&ID shows the parts of the process, highlighting the following process Input/output variables:

Input Variables

- Air/O_2 flow rate: This flow is measured by a flowmeter (**Figure 3H**).
- Biomass flow rate: The flowrate of biomass is calculated by the dimensions of the screw feeder and the rpm (**Figure 3J**).
- Cooling water supply temperature: Temperature is measured by a thermocouple T10 (**Figure 3**).

Output Variables

- Reactor temperature: The temperature of the reactor is measured in 8 parts; they can also be visualized in the SIPOC diagram, where T1 & T2 are the temperatures of Air/O_2 before going inside the reactor, T3 is the temperature before biomass goes inside the reactor and, T4, T5, T6, T7, T8 are the temperatures along the reactor.
- Exhaust pipe temperature: It is tagged as T9 and is the temperature of the syngas.
- Cooling water temperature: The cooling water supply temperature is T10, and the temperature of the cooling water return is T11.
- Syngas composition: Syngas from gasification contains CO, CO_2, CH_4, H_2, and N_2, mostly.

Data collection/sampling plan: The syngas is collected and analyzed in the areas named U, V, and W.

Design of experiments: The first part of the project is to identify the optimal conditions of two process variables (1) a Mixture of RDF and wood and (2) Equivalence ratios. Syngas' quality typically determines the optimal conditions considering five indicators: carbon conversion, H_2/CO ratio, CH_4/H_2 ratio, gas yield, and gasification efficiency [47]. **Table 3** shows the DoE to determine the best RDF/Wood mixture and ER.

TAG	G-CG reference	Biomass Type	%wt RDF	ER
1	PC100: ER023	Pine Chips	0	0.21
2	PC100: ER031	Pine Chips	0	0.31
3	PC90 - RDF10: ER022	Pine Chips	10	0.22
4	PC90 - RDF10: ER025	Pine Chips	10	0.25
5	PC90 - RDF10: ER030	Pine Chips	10	0.30
6	PC80 - RDF20: ER022	Pine Chips	20	0.22
7	PC80 - RDF20: ER025	Pine Chips	20	0.25
8	PC80 - RDF20: ER031	Pine Chips	20	0.31
9	PC50 - RDF50: ER032	Pine Chips	50	0.32
10	PP100: ER022	Wood pellets	0	0.22
11	PP100: ER030	Wood pellets	0	0.30
12	PP90 - RDF10: ER022	Wood pellets	10	0.22
13	PP90 - RDF10: ER031	Wood pellets	10	0.31
14	PP80 - RDF20: ER022	Wood pellets	20	0.22
15	PP80 - RDF20: ER031	Wood pellets	20	0.31
16	PP50 - RDF50: ER021	Wood pellets	50	0.21
17	PP50 - RDF50: ER030	Wood pellets	50	0.30
18	RDF100: ER023	—	100	0.23
19	RDF100: ER027	—	100	0.27

Table 3.
DoE for best mixture and equivalence ratio.

4.3 Analyze

The analysis phase delivers process setup baselines, capability analysis, and identifying sources of variation. So, the first step is to calculate the process's baseline Sigma to understand how well it performs and how much work will be required to reach Six Sigma quality.

4.3.1 Experimental results

This section presents the gasifier's operating conditions, temperature profiles over time, gas composition (CO, CO_2, CH_4, and C_2H4) profile, and the average gas composition. Finally, the dry gas LHV and efficiency parameters (Ygas, CGE, and CCE) are exhibited. The results are presented in **Table 4** and are discussed below.

4.3.2 Operational conditions

The pilot-scale gasifier of the University of Aveiro operates under the auto-thermal regime. Therefore, an external heating supply was not necessary. The average temperature was 785°C, sustained by the feedstock's ash fusibility

TAG	Bed Tem [°C]	%wt RDF	ER	H_2	N_2	CH_4	CO	CO_2	C_2H4	C_2H_6	C_3H_8	SUM	Molar Weight (Dry) [kg/kmol]	Q [NL Dry gas/min]	LHV [MJ/Nm3]	Ygás [Nm^3 dry gas/kg feedstock db]	CGE [%]	CCE [%]	std CO	std CO_2	std CH4	std C2H4	Std H2	Std N2	Std C2H6	Std C3H8
1	803	0	0.21	6.5	53.0	5.3	18.6	15.9	2.2	0.2708	0.0470	101.8	28.7	298.9	6.2	1.52	52.9	78.4	0.7	0.5	0.2	0.1	0.3	2.0	0.0	0.0
2	806	0	0.31	5.7	59.8	3.9	13.7	16.6	1.5	0.0794	0.0152	101.3	29.1	264.5	4.7	1.77	46.3	76.7	0.9	1.4	0.3	0.2	0.5	5.3	0.0	0.0
3	803	10	0.22	5.8	57.2	4.6	15.9	15.4	2.1	0.3098	0.0449	101.4	28.8	277.0	5.6	1.35	41.0	63.1	0.9	0.2	0.4	0.2	0.4	3.6	0.0	0.0
4	804	10	0.25	5.3	58.1	4.4	15.5	15.5	2.2	0.1492	0.0551	101.3	28.9	272.4	5.4	1.56	46.0	71.3	0.5	0.1	0.2	0.2	0.2	2.2	0.0	0.0
5	807	10	0.30	4.0	63.5	3.8	13.2	16.4	1.8	0.0875	0.0167	102.8	29.9	249.2	4.5	1.67	41.0	70.7	0.4	0.1	0.2	0.1	0.2	2.8	0.0	0.0
6	785	20	0.22	5.9	54.4	4.7	16.4	15.7	2.7	0.3981	0.0394	100.3	28.5	291.1	6.0	1.47	46.6	70.8	0.4	0.1	0.2	0.2	0.2	2.1	0.0	0.0
7	794	20	0.25	5.7	56.9	4.3	15.1	15.6	2.5	0.2270	0.0602	100.4	28.6	278.3	5.6	1.62	47.8	74.0	0.6	0.1	0.3	0.3	0.3	3.2	0.0	0.0
8	811	20	0.31	4.9	61.6	3.8	12.6	15.8	2.1	0.0934	0.0325	100.9	29.0	256.9	4.7	1.85	45.9	75.6	0.4	0.1	0.2	0.2	0.2	2.8	0.0	0.0
9	819	50	0.32	4.8	62.1	4.0	13.8	15.4	2.0	0.1263	0.0467	102.3	29.4	254.9	4.9	2.10	49.9	84.1	0.4	0.1	0.1	0.1	0.1	1.9	0.0	0.0
10	791	0	0.22	7.1	56.8	4.5	15.5	15.4	1.8	0.1637	0.0231	101.3	28.4	278.4	5.4	1.36	41.2	60.4	0.2	0.1	0.1	0.0	0.1	0.7	0.0	0.0
11	829	0	0.30	5.1	65.4	3.0	11.4	15.6	1.2	0.0532	0.0000	101.6	29.3	241.8	3.7	1.56	32.6	56.8	0.3	0.3	0.3	0.2	0.4	5.7	0.0	0.0
12	797	10	0.22	6.5	59.3	4.1	14.2	15.7	2.1	0.2507	0.0416	102.2	29.0	267.0	5.2	1.30	36.7	56.0	0.6	0.3	0.2	0.1	0.3	2.3	0.0	0.0
13	816	10	0.31	6.7	60.3	3.6	13.9	16.0	1.6	0.1115	0.0184	102.2	29.0	262.3	4.7	1.85	47.0	75.8	0.8	0.6	0.1	0.1	0.3	2.6	0.0	0.0
14	801	20	0.22	6.9	58.6	4.4	13.9	15.5	2.3	0.1618	0.0256	101.8	28.7	270.3	5.5	1.37	39.5	58.6	0.3	0.1	0.1	0.1	0.1	1.1	0.0	0.0
15	806	20	0.31	5.3	63.2	3.5	11.4	15.8	2.1	0.1308	0.0250	101.4	29.1	250.4	4.5	1.79	42.6	69.1	1.7	1.8	0.7	0.6	1.0	11.5	0.0	0.0
16	812	50	0.21	5.4	56.4	4.7	11.7	16.4	3.7	0.2394	0.0542	98.6	28.3	281.2	5.9	1.52	43.9	65.9	0.6	0.6	0.2	0.1	0.2	2.1	0.0	0.0
17	818	50	0.30	5.8	61.7	4.1	11.6	15.0	2.8	0.1577	0.0303	101.2	28.8	256.4	5.2	2.01	50.9	77.8	0.4	0.1	0.1	0.1	0.2	1.7	0.0	0.0
18	818	100	0.23	5.2	64.6	5.1	6.9	15.2	4.3	0.1451	0.0951	101.5	28.9	245.6	5.8	1.70	42.3	61.3	0.3	0.6	0.3	0.4	0.3	3.7	0.0	0.0
19	793	100	0.27	4.8	64.6	5.6	7.3	14.7	5.0	0.1961	0.0357	102.2	29.1	245.4	6.4	1.95	53.5	73.7	0.5	0.5	0.5	0.4	0.3	4.3	0.0	0.0

Table 4.
Experimental results.

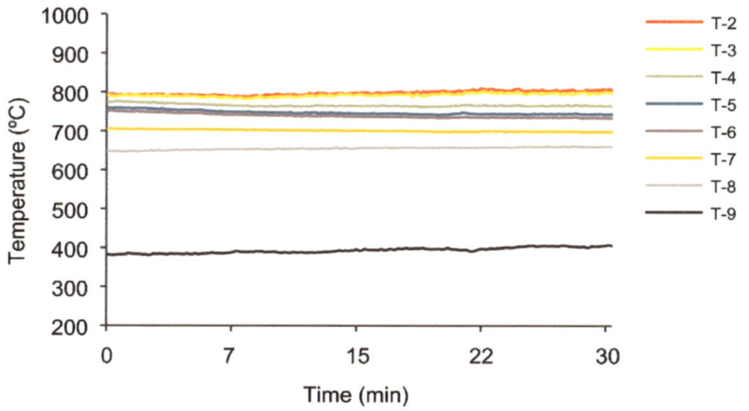

Figure 6.
Temperature profile over time.

temperatures (>1000°C). **Figure 6** depicts the temperature profiles over time (see also Bubbling Fluidized Bed Reactor P&ID).

The temperature profiles for the different experiments showed similar behavior. Yet, the experiments performed with pine chips had higher temperature fluctuations than those performed with pine pellets. Pine chip particle size is heterogeneous. At the same time, wood pellets have a more homogeneous particle size, which can justify the temperature fluctuations.

4.3.3 Gas composition

Adding RDF to the fuel mixes significantly reduced the CO content in the produced gas for comparable ER. The phenomenon may be related to the methanation reactions described below.

$$2CO + 2H_2 \rightarrow CH_4 + CO_2 \quad (5)$$

$$CO + 3H_2 \rightarrow CH_4 + H_2O \quad (6)$$

However, there is a discrepancy between experiment PC90 - RDF10: ER022 and experiment PC80 - RDF20: ER022. Both experiments were performed at ER = 0.22, having 10 and 20% of RDF, respectively. Therefore, a lower CO concentration in the mixture of 20% RDF was expected (See **Table 4**).

On the other hand, If ER increases, then CO decreases even more. However, there is a discrepancy between experiments PC80 - RDF20: ER031 and PC50 - RDF50: ER032, which were run at 20% of RDF and ER = 0.31 for the first one and 50% of RDF and ER=0.31 for the second one. The results show that the CO concentration was 12.6 and 13.6%, respectively. These discrepancies may be related to a wrong ER since a higher ER means more nitrogen. Never less. In these examples, the nitrogen concentration is lower in blends with higher RDF.

Generally, **Figure 7** illustrates the effect of the RDF weight % in the fuel mixture on the composition of the produced gas. The gasification of pine chips with 0.23 ER produced the highest CO concentration (18.6 vol%, experiment reference PC100: ER0.23), whereas RDF with 0.23 ER produced the lowest CO concentration (6.9 vol%, experiment reference RDF100: ER0.23). Increasing the amount of RDF in the

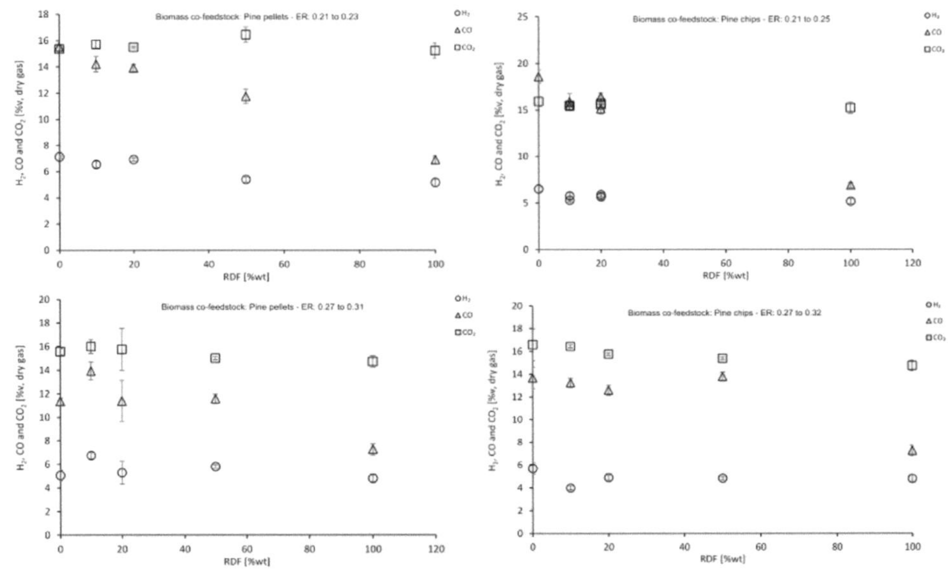

Figure 7.
Influence of the RDF weight percentage on the gas composition (H_2, CO, and CO_2).

feedstock combination from 10 to 20%, 20 to 50%, and 50 to 100% resulted in CO reductions of 6.3, 1.5, and 42.0%, respectively. In contrast, increasing the RDF weight percentage from 0 to 10% in the fuel combination resulted in an average CO increase of 5.5%.

The effect of adding RDF on CH_4 and C_2H_4 results in a higher composition as wt%. of RDF increases. For the gasification of RDF with ER 0.27 (experiment reference RDF100: ER 0.27), the maximum CH_4 and C_2H_4 concentrations were 5.6 and 5%, respectively. On the one hand, this may be rationalized by the thermal breaking of polymers in RDF pellets, which yields light hydrocarbons. On the other hand, the increased quantity of ashes rich in alkali and alkali earth metals (e.g., calcium, sodium, magnesium, potassium) found in RDF pellets compared to biomass (**Table 2**) may stimulate a catalytic effect that also results in the synthesis of light hydrocarbons. However, **Figure 8** shows a higher amount of CH_4 for blends of pine chips with no RDF, which can be an error derivate from a wrong ER measured or wrong feedstock measure derivate from the particle size and shape since this problem is seen to be reduced with pine pellets.

4.3.4 LHV, Ygas, CGE, and CCE

As shown in **Table 2**, although RDF contains a considerable amount of ash, it also contains more carbon and hydrogen than pine chips and pellets. Therefore, mixtures with more RDF will need more air in a given ER than those with less RDF in the same ER. For example, if it is assumed that carbon of 100 g of samples of RDF, pine chips, and pine pellets will burn completely (ER = 1), then 18.84% more oxygen will be required for RDF than for pine chips. At the same time, 13.68% more oxygen will be needed for RDF than for pine pellets (see **Table 5**). This situation indicates that blends with more RDF will deliver higher Ygas values than those with less RDF.

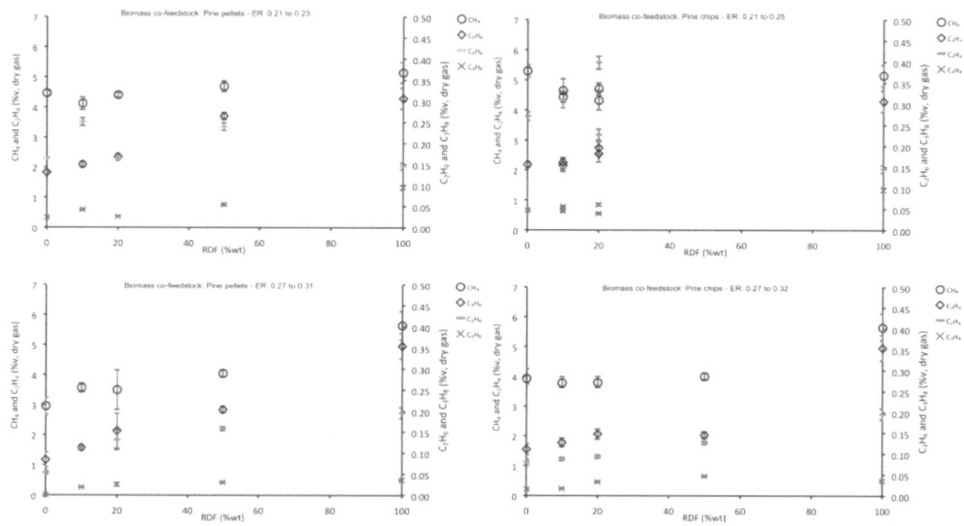

Figure 8.
Influence of the RDF weight percentage on the gas composition (CH_4, C_2H_4, C_2H_6, and C_3H_8).

Feedstock	C(g)	O2(g)
RDF	54	143.863
Pine Pellets	45.4	120.951
Pine chips	47.5	126.546

Table 5.
Required oxygen for complete combustion of 100 g of sample of each feedstock.

This analysis indicates that blends with more RDF will deliver higher Ygas values than those with less RDF. However, **Figure 9** is not entirely aligned with this analysis, which may indicate a potential error in the ER or the fed feedstock amount.

On the other hand, the LHV of the generated gas improved with increasing ER (from 5.8 to 6.4 MJ/Nm^3), mainly due to an increase in CH_4 and C_2H_4. This behavior is not expected in operations involving biomass gasification. It may be due to the higher ER promoting the thermal breaking of the organic molecules in the plastic fractions of RDF. Thus, this effect increases in blends with higher RDF amounts. However, this effect is unclear, so it may be an error in ER or the fed feedstock.

Figure 10 depicts the influence of RDF wt.% on CGE and CCE. Adding RDF to the fuel mixture has no appreciable effect on the CGE. However, there is a slight tendency for CGE to grow when the RDF weight % rises. The RDF gasification with an ER of 0.27 yielded the highest CGE value (53.5%). (Experiment reference RDF100: ER0.27). The lowest CGE value (32.6%) was reported for the gasification of pine pellets with an ER of 0.30. (Experiment reference PP100: ER0.30).

Gas Composition: Higher RDF wt.% increases CH_4 and C_2H_4 and reduces CO concentration. This effect might be due to the thermal cracking of the plastic polymers in the RDF pellets and the catalytic effect promoted by the ashes (alkali and alkali earth metals). In contrast, no significant trends were observed for the variation of H_2.

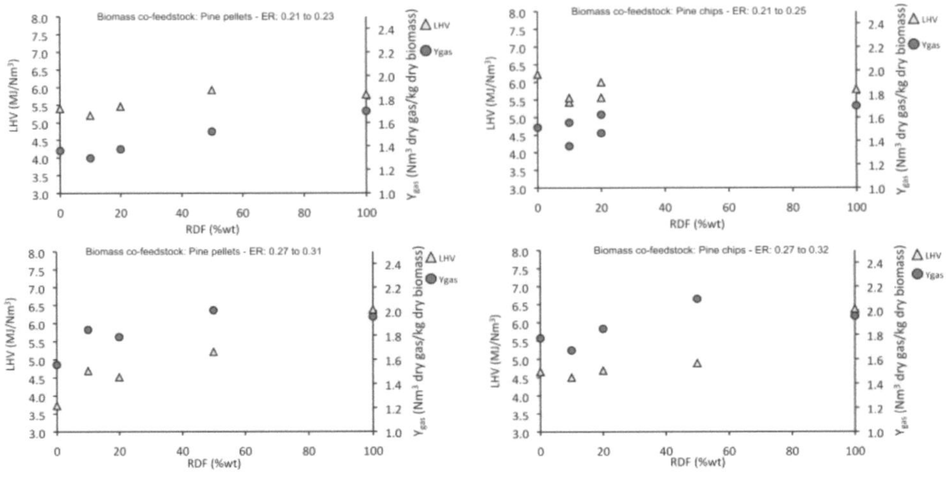

Figure 9.
Influence of the RDF weight percentage on the LHV and Ygas.

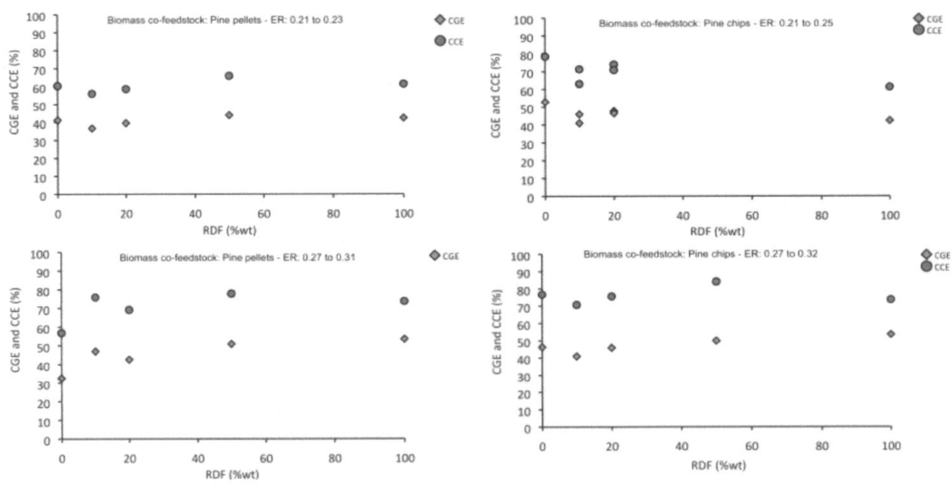

Figure 10.
Influence of the RDF weight percentage on the CGE and CCE.

- LHV: Higher RDF wt.% increases LHV because of the increasing CH_4 and C_2H_4 concentration.

- Ygas: Higher RDF weight percentage also led to slightly higher Ygas values. This result might be concealed by involuntary changes in the ER, which has a prominent effect on the Ygas.

- CGE and CCE: Higher RDF wt.% seems optimistic, although this is unclear due to conflicting effects.

In conclusion, the following tendencies are noticed, and numbers outside the trend may sometimes suggest a measurement mistake, causing variance in the process.

4.3.5 Source of variation

The variance in the Aveiro's Pilot-scale Bubbling Fluidized Bed Reactor for gasification can be visualized by monitoring the temperature and syngas composition:

1. Temperature: It is a crucial process variable that modifies the syngas composition, and its variance might result from the following improvement opportunities.

   Cooling system: Gasification is partial oxidation when an excess char is produced. It is also oxidized and produces CO_2, H_2O, and heat, increasing the reactor's temperature. The current equipment has a complicated cooling system comprising 16 small pipes along and around the reactor, supplying and returning cooling water (**Figure 2**). When the temperature increases, the operator introduces a few centimeters of some 16 pipes. This kind of technology depends entirely on the operator's expertise. Thus, the temperature variance of the reactor will change from operator to operator. Furthermore, some 16 small pipes are inaccessible to the operator, clearly a poor engineering design.

   Equivalence ratio: To produce an oxidation reaction, it is necessary to have oxygen, so the temperature will also depend on the equivalence ratio. A single flowmeter indicates oxygen and airflow, so the chosen equivalence ratio will be inaccurate. Furthermore, flowmeters are just indicators. Thus, the flow control is performed by partially opening or closing a valve. Therefore, the flowmeter measurement adds an error to the flow ratio because an operator performs this opening and closing of the valve. Hence, the precision depends on how well an operator's eyesight is.

   Refuse-derived Fuel Composition: The composition of RDF is given by moisture, volatile matter, fixed carbon, elementary composition, impurities, and ashes, and those elements might impact the temperature by promoting specific exothermic reactions.

2. Syngas Composition

   Equivalence Ratio: The equivalence ratio corresponds to the ratio between the oxygen content in the oxidant supply required for complete stoichiometric combustion. Usually, ER is between 0.2 and 0.4. ER < 0.2 results in incomplete gasification, excessive char formation, and low calorific value of the product gas. Whereas ER > 0.4 results in excessive formation of CO_2 and H_2O, rather than CO and H_2, it also decreases the calorific value of the gas.

   Refused Derived Fuel: The composition of RDF is given by moisture, volatile matter and fixed carbon, elementary composition, impurities, and ashes, and those elements might impact the gas composition by promoting the production of products with low calorific value.

   The following fishbone diagram (**Figure 11**) helps to understand the source of the variance.

   The Analyze phase offers statistical methodologies and tools for isolating vital data that will reveal the number of defective products. Business issues are transformed into statistical problems, exposing their causes and potential

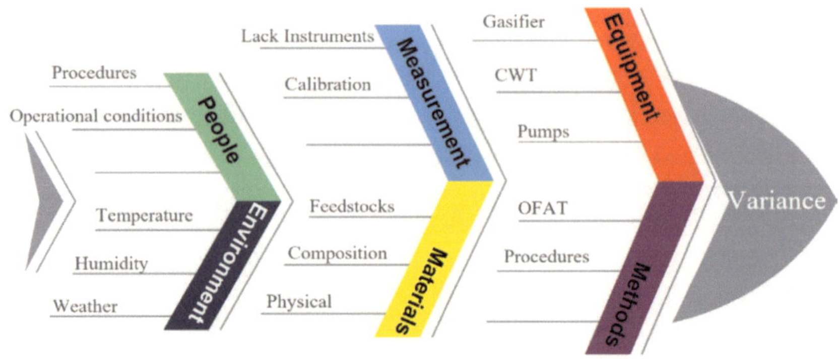

Figure 11.
Fishbone diagram.

solutions. The process settings, experimental findings, efficiency metrics, and gas composition standard deviation are detailed in **Table 3**. In contrast, **Table 6** presents an FMEA that hints at alternative solutions to the source of variance.

4.4 Improve

To look for improvements in a process, it is necessary first to understand the engineering/design process. Unfortunately, the current process lacks this documentation, making it hard to identify a flaw in the engineering design or if the operation is out of design operating conditions. However, the problems and variability caused by the lack and understanding of this documentation continuously show up, which is a wake-up call to adopt a process methodology like Six Sigma. That is why the Engineering/design process and documentation are some of the drivers of Six Sigma. (1) Key technology and process description, (2) General mass balance, (3) General energy balance, (4) Thermal Rating, (5) Process flowsheets, (6) Piping and (7) Instrument Diagrams, (8) Definition and sizing of significant equipment resulting in the process specifications, (9) Definition of control and safety devices, (10) Mechanical data sheets of the leading equipment, (11) HAZOP.

Poor industrial process capabilities often result in high COPO (rework, scrap, field failure).

Piping clogging: Gasification also produces tars, a combination of char and oils. When the gas temperature decreases, tars condensate and clog the exhaust pipe (**Figure 2**, section M). This problem can be caused by poor heating in the exhaust pipe, inappropriate feedstock flow for the facilities, or the wrong size of pipes and equipment. This results in high COPO (rework, scrap, field failure).

4.4.1 Brain-writing

Temperature control: (1) Replace the cooling temperature system with an internal serpentine can help to provide temperature homogeneity. In addition, this serpentine can be sectioned along the reactor body to provide better temperature profile control. **Figure 12** depicts the internal serpentine with a cooling water supply nozzle and a cooling water supply return.

Process Step/ Input	Potential Failure Mode	Potential Failure Effects	Severity (1–10)	Potential Causes	OCCURRENCE (1–10)	Current Controls	DETECTION (1–10)	RPN	Action Recommended
What is the process step or feature under investigation?	**In what ways could the step or feature go wrong?**	**What is the impact on the customer if this failure is not prevented or corrected?**		**What causes the step or feature to go wrong? (How could it occur?)**		**What controls exist that either prevent or detect the failure?**			**What are the recommended actions for reducing the occurrence of the cause?**
Biomass Size	(a)Clog the feeding system if too big	Rework, loss of time and resources	10	Inappropriate selection of particle size	1	Pre-processing with sieves	1	10	Establish a particle size range
	(b) Deficient Heat transfer	Increase residence time	7	Inappropriate selection of particle size	1	Pre-processing with sieves	1	7	Establish a particle size range
Biomass Shape	(a)Clog the feeding system	Loss of time and resources	10	No pre-processing of feedstock	1	Pre-processing by chopping	1	10	Establish a particle size shape range
	(b) Increase equipment size	Increase the cost of equipment	10	No pre-processing of feedstock	1	Pre-processing by chopping	1	10	Establish a particle size shape range
Biomass Flow Rate	(a)Clog the feeding system	Loss of time and resources	10	Inappropriate equipment design	2	Screw calculation	8	160	Select a proper Flow Indicator Transmitter
	(b) Reactions	Increase residence timeand cost	7	Inappropriate Flow rate selection	2	Screw calculation	8	112	Select a proper Flow Indicator Transmitter
Equivalence Ratio (air feed)	(a) Incomplete gasification	Excessive char formation and low calorific value of the product gas	10	Inaccurate RE calculation	5	Flowmeters	8	400	Select a proper Flow Indicator Transmitter
	(b) Excessive formation of CO_2 and H_2O, rather than CO and H2	Decreases the calorific value of the gas	10	Flow meters are not calibrated or were incorrectly chosen	5	Flowmeters	8	400	Select a proper Flow Indicator Transmitter

Process Step/ Input	Potential Failure Mode	Potential Failure Effects	Severity (1–10)	Potential Causes	OCCURRENCE (1–10)	Current Controls	DETECTION (1–10)	RPN	Action Recommended
What is the process step or feature under investigation?	**In what ways could the step or feature go wrong?**	**What is the impact on the customer if this failure is not prevented or corrected?**		**What causes the step or feature to go wrong? (How could it occur?)**		**What controls exist that either prevent or detect the failure?**			**What are the recommended actions for reducing the occurrence of the cause?**
Gasification Agent	(a) Excessive production of tar	Lower calorific value and clogged exhaust pipe because of Tars production	7	Flow meters are not calibrated or were incorrectly chosen	5	Flowmeters	8	280	Select a proper Flow Indicator Transmitter
Temperature	(a) Too high	Operational problems	10	Char and combustible gases combustion More air than what should be	5	Thermocouples	8	400	Redesign and automate the cooling system
	(b) Too low	No gasification reactions	10	Feed combustion gas to the chamber	5	Thermocouples	8	400	Redesign and automate the cooling system
Ash Content	(a) Too High	The catalytic effect, changing gas composition	10	RDF with a higher number of ashes	1	No control	1	10	Establish an ash content range
Moisture	(a) Too High	Waste of energy and gas composition changes	10	Improper storage of biomass	1	No control	1	10	Establish an ash content range

Table 6.
FMEA.

Figure 12.
Serpentine.

(2) Furthermore, the cooling water flow can be controlled by a control valve (CV) linked to a Temperature Indicator Transmitter (TIT) to partially open or close the CV. Furthermore, a Flow Indicator Transmitter (FIT) can indicate the actual cooling water flow. **Figure 13** depicts the control system.

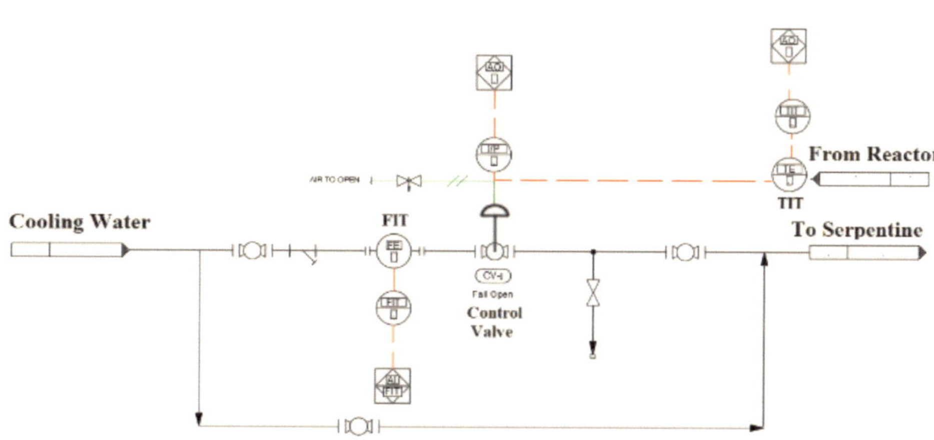

Figure 13.
Temperature control system.

Airflow system: The airflow is controlled by partially opening or closing a valve based on the flowmeter indication. Therefore, an error is added to the flow ratio since an operator performs the opening and closing of the valve, so the precision depends on how well an operator's eyesight is. The proposed solutions are: (1) Automate the airflow system through a control valve and two flow indicator transmitters. The control system aims to link the air's FIT with the biomass FIT, so the control valve will open or close to allow an airflow based on the amount of fed biomass and the established ER. **Figure 14** depicts the control system.

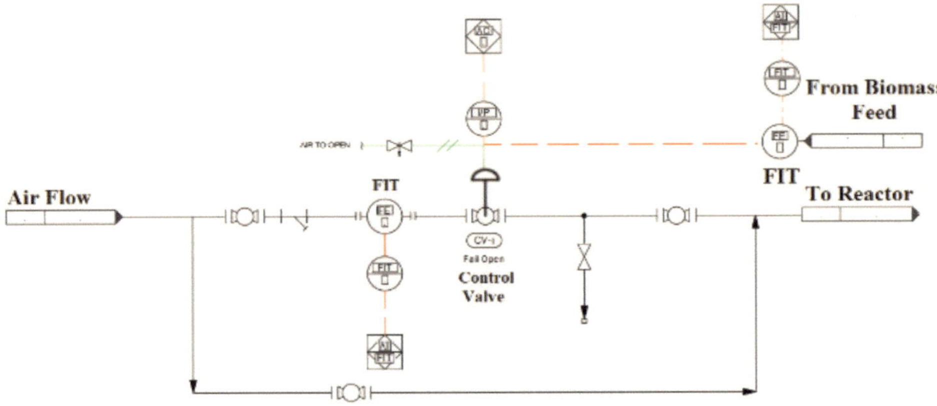

Figure 14.
Air flow control system.

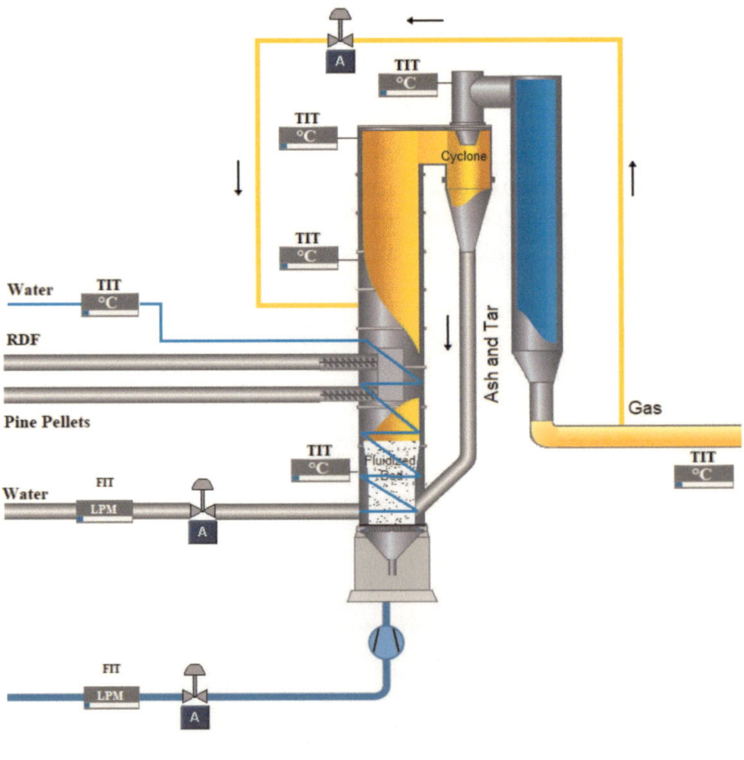

Figure 15.
PLC gasification process.

Biomass flow system: Several previous actions are necessary to control biomass flow, such as pretreating the biomass by establishing a homogeneous particle diameter range. This consideration is essential so that the feeder screw feeds the biomass homogeneously. (2) Another option is to change the mechanical feeding system to a pneumatic one, requiring biomass pretreatment and screening.

4.5 Control

The control phase sustains the Six Sigma DMAIC initiative through continuous monitoring to avoid the same problem [48]. So, a PLC can be installed to monitor and control the process as described in the improvement section. The PLC allows the operator to interact with the process without opening or closing valves to a particular flow, adding errors to the process. Furthermore, the PLC can save data by analyzing the process and promoting continuous improvement. **Figure 15** shows a typical PLC interface adapted to the gasification process.

5. Conclusion

In this chapter, the Six Sigma DMAIC methodology is used to identify causes of variance in the gasification process and suggest opportunities for improvement. In terms of the Equivalence Ratio (ER), the developed Failure Modes and Effects Analysis (FMEA) has a high-Risk Priority Number (RPN). This high figure is due to the method of feeding biomass, which significantly impacts airflow. In addition, the temperature has a high RPN value as well. This situation is due to the cooling mechanism of this device and the ER employed, which may be inaccurate. Finally, some ways are proposed to improve the air and biomass feeding and process cooling systems.

On the other hand, it demonstrated the process's stability and the synergy between RDF and biomass, resulting in enhanced gasification products. Furthermore, no slag, agglomeration, or defluidization phenomena were observed. Again, implementing the DMAIC methodology helped identify the source of variance and ways to enhance the overall process. Therefore, it has the potential to strengthen the gasification process, promoting the economic viability and environmental benefits of future and existing gasification plants.

Acknowledgements

The authors would like to thank FCT for the grant SFRH/BD/146155/2019, contract CEECIND/00641/2018, and the projects SAICTALT/39486/2018 and PTDC/EME-REN/4124/2021. CESAM thanks the FCT and MCTES for UIDP/50017/2020 + UIDB/50017/2020 + LA/P/0094/2020, through national funds.

Author details

José Antonio Mayoral Chavando[1,2], Valter Silva[1,2*], João Sousa Cardoso[1,3], Daniela Eusébio[1] and Luís A.C. Tarelho[2]

1 Polytechnic Institute of Portalegre, Portalegre, Portugal

2 Centre for Environmental and Marine Studies (CESAM), Department of Environment and Planning, University of Aveiro, Aveiro, Portugal

3 Instituto Superior Técnico, Universidade de Lisboa, Lisboa, Portugal

*Address all correspondence to: valter.silva@ipportalegre.pt

References

[1] Silva VBRE, Cardoso J. Overview of biomass gasification modeling: Detailed analysis and case study. In: Computational Fluid Dynamics Applied to Waste-to-Energy-Processes. Oxford, UK: Elsevier; 2020. pp. 123-149. DOI: 10.1016/b978-0-12-817540-8.00004-2

[2] Cardoso J, Silva V, Eusébio D. Techno-economic analysis of a biomass gasification power plant dealing with forestry residues blends for electricity production in Portugal. Journal of Cleaner Production. 2019;**212**: 741-753. DOI: 10.1016/j.jclepro.2018.12.054

[3] World Bioenergy Association (WBA). Global Bioenergy Statistics. 2019. Available from: https://www.worldbioenergy.org/global-bioenergy-statistics/

[4] Cardoso J, Silva V, Eusébio D, Brito P, Hall MJ, Tarelho L. Comparative scaling analysis of two different sized pilot-scale fluidized bed reactors operating with biomass substrates. Energy. 2018;**151**: 520-535. DOI: 10.1016/j.energy.2018.03.090

[5] Gov UK. Carbon Emissions Tax Consultation. London, UK: HM Revenue & Customs; 2020

[6] Power, The Shift from Coal to Biomass Is on in Europe. May 2018

[7] Gov UK, Carbon Emissions Tax Consultation. Jul 2020

[8] EnergyWatch, Portugal expedites coal plant phase-out to next year. Jul. 2020

[9] ClimateHomeNews. Portugal ends coal burning two years ahead of schedule. Jul. 2020

[10] Carbon Pulse, Analyst team sees EU carbon prices near €90 by 2030 as industrials forced to step up abatement. Oct. 2020

[11] ArgusMedia, EU ETS price €32-65/t under 2030 scenarios. Sep. 2020

[12] Bloomberg. EU carbon permits hits record 50 Euros on tighter pollution rules. May 04 2021. Available from: https://www.bloomberg.com/news/articles/2021-05-04/carbon-permits-hit-record-50-euros-on-tighter-pollution-rules [accessed Jun. 29, 2021]

[13] Investing.com. Carbon emissions futures price. Jun. 29, 2021. Available from: https://uk.investing.com/commodities/carbon-emissions [accessed Jun. 29, 2021]

[14] Dudek T. The impacts of the energy potential of Forest biomass on the local market: An example of south-eastern Poland. Energies (Basel). 2020;**13**(18):4985. DOI: 10.3390/en13184985

[15] Valmet, Biomass to energy solutions for high-efficient and sustainable power generation

[16] Daniell J, Köpke M, Simpson SD. Commercial biomass syngas fermentation. Energies. 2012;**5**(12) MDPI AG:5372-5417. DOI: 10.3390/en5125372

[17] GSTC. Gasification vs. incineration

[18] Alam MT et al. Co-gasification of treated solid recovered fuel residue by using minerals bed and biomass waste blends. Energies (Basel). 2020;**13**(8):2081. DOI: 10.3390/en13082081

[19] Pio DT, Tarelho LAC, Tavares AMA, Matos MAA, Silva V. Co-gasification of refused derived fuel and biomass in a pilot-scale bubbling fluidized bed reactor. 2020;**206**. DOI: 10.1016/j.enconman.2020.112476

[20] Nobre C, Longo A, Vilarinho C, Gonçalves M. Gasification of pellets produced from blends of biomass wastes and refuse derived fuel chars. Renewable Energy. 2020;**154**:1294-1303. DOI: 10.1016/j.renene.2020.03.077

[21] Poe WA, Mokhatab S. Process optimization. In: Modeling, Control, and Optimization of Natural Gas Processing Plants. Oxford, UK: Elsevier; 2017. pp. 173-213. DOI: 10.1016/B978-0-12-802961-9.00004-8

[22] Lei G, Zhu J, Guo Y, Liu C, Ma B. A review of design optimization methods for electrical machines. Energies. 2017; **10**(12):1962. MDPI AG. DOI: 10.3390/en10121962

[23] Mikel Harry PD, Schroeder R. Six Sigma the Breakthrough Management Strategy Revolutionizing the World's Top Corporations. New York, USA: Doubleday; 2000

[24] Chakrabarty A, Tan KC. The current state of six sigma application in services. Managing Service Quality. 2007;**17**(2): 194-208. DOI: 10.1108/09604520710735191

[25] Virender Narula SG, Narula V, Grover S. Six Sigma: Literature review implications for future research. International Journal of Industrial Engineering & Production Research. 2015;**26**(1):13-26. DOI: 10.1016/j.ijproman.2010.01.006

[26] Maurus S. Lean Six Sigma in Versicherungen. In: Handbuch Versicherungs Marketing. Berlin Heidelberg: Springer; 2019. pp. 505-531. DOI: 10.1007/978-3-662-57755-4_29

[27] Mikel J. Harry Harry's contributions to Six Sigma

[28] Gibbons PM, Kennedy C, Burgess S, Godfrey P. The development of a value improvement model for repetitive processes (VIM): Combining lean, Six Sigma and systems thinking. International Journal of Lean Six Sigma. 2012;**3**(4):315-338. DOI: 10.1108/20401461211284770

[29] Sreeram TR, Thondiyath A. Combining lean and Six Sigma in the context of systems engineering design. International Journal of Lean Six Sigma. 2015;**6**(4):290-312. DOI: 10.1108/IJLSS-07-2014-0022

[30] Garza-Reyes JA, Kumar V, Chen FF, Wang YC. Seeing green: Achieving environmental sustainability through lean and Six Sigma. International Journal of Lean Six Sigma. 2017;**8**(1):2-6. DOI: 10.1108/IJLSS-01-2017-0005

[31] da Silva FF, Filser LD, Juliani F, de Oliveira OJ. Where to direct research in lean Six Sigma?: Bibliometric analysis, scientific gaps and trends on literature. International Journal of Lean Six Sigma. 2018;**9**(3):324-350. DOI: 10.1108/IJLSS-05-2017-0052

[32] Li N, Laux CM, Antony J. How to use lean Six Sigma methodology to improve service process in higher education: A case study. International Journal of Lean Six Sigma. 2019;**10**(4):883-908. DOI: 10.1108/IJLSS-11-2018-0133

[33] Petrenko Y, Denisov I, Koshebayeva G, Biryukov V. Energy efficiency of Kazakhstan enterprises: Unexpected findings. Energies (Basel). 2020;**13**(5):1055. DOI: 10.3390/en13051055

[34] Li P, Jiang P, Zhang G. An enhanced DMAIC method for feature-driven continuous quality improvement for multi-stage machining processes in one-of-a-kind and small-batch production. IEEE Access. 2019;7:32492-32503. DOI: 10.1109/ACCESS.2019.2900461

[35] Roger B. Lean Six Sigma Business Transformation for Dummies. New Jersey, USA: John Wiley & Sons, Ltd.; 2014

[36] Hanafi NA, Tan OK, Goh CG. Categorization of lean six sigma tools and techniques based on DMAIC framework. Journal of Information System and Technology Management. 2019;**4**(13):1-12. DOI: 10.35631/JISTM.413001

[37] Noori B, Latifi M. Development of Six Sigma methodology to improve grinding processes: A change management approach. International Journal of Lean Six Sigma. 2018;**9**(1): 50-63. DOI: 10.1108/IJLSS-11-2016-0074

[38] Praveen G. Chapter Two. Six Sigma —An Overview. in Six Sigma Business Scorecard Ensuring Performance for Profit. New York, USA: McGraw-Hill, Ed.; 2004. pp. 17-39

[39] B. Carreira, Lean Six Sigma That Works a Powerful Action Plan for Dramatically Improving Quality, Increasing Speed, and Reducing Waste. NY, USA: AMACOM Publisher; 2006.

[40] Wasage C. Implementation of Six Sigma projects in fortune 500 companies. Journal of Modern Accounting and Auditing. 2016;**12**(4): 208-216. DOI: 10.17265/1548-6583/2016.04.002

[41] Srinivasan K, Muthu S, Devadasan SR, Sugumaran C. Enhancing effectiveness of shell and tube heat exchanger through six sigma DMAIC phases. Procedia Engineering. 2014;**97**: 2064-2071. DOI: 10.1016/j.proeng.2014.12.449

[42] Krishna Priya S, Jayakumar V, Suresh Kumar S. Defect analysis and lean six sigma implementation experience in an automotive assembly line. Materials Today: Proceedings. 2020;**22**:948-958. DOI: 10.1016/j.matpr.2019.11.139

[43] Bloj M-D, Moica S, Veres C. Lean Six Sigma in the energy service sector: A case study. Procedia Manufacturing. 2020;**46**:352-358. DOI: 10.1016/j.promfg.2020.03.051

[44] Costa T, Silva FJG, Ferreira LP. Improve the extrusion process in tire production using Six Sigma methodology. Procedia Manufacturing. 2017;**13**:1104-1111. DOI: 10.1016/j.promfg.2017.09.171

[45] Pask F, Sadhukhan J, Lake P, McKenna S, Perez EB, Yang A. Systematic approach to industrial oven optimisation for energy saving. Applied Thermal Engineering. 2014;**71**(1):72-77. DOI: 10.1016/j.applthermaleng.2014.06.013

[46] Kane J, Dupont EI. Using Six Sigma to drive energy efficiency improvements at DuPont

[47] Silva V et al. Multi-stage optimization in a pilot scale gasification plant. International Journal of Hydrogen Energy. 2017;**42**(37):23878-23890. DOI: 10.1016/j.ijhydene.2017.04.261

[48] C. Sarah and S. S. Academy, The Black Belt Memory Jogger, Second. Massachusetts, USA: Goal/QPC; 2016

Chapter 15

Dialing Back the Doomsday Clock with Circular Bioeconomy

Sammy N. Aso

Abstract

Present day status of planet earth is perilous. In January 2023, the "Doomsday Clock" (a serving global indicator for worldwide catastrophe) crept up to 90 seconds before midnight. According to the bulletin of atomic scientists, the Doomsday Clock aims to designate humanity's closeness to annihilation; with midnight being the instance of ignition and thus, the point of no return from Armageddon. Because 90 seconds is the closest the clock has ever been to midnight, the year 2023 is therefore, planet earth's nearest to Armageddon. But why is planet earth perilously close to extinction? The bulletin of atomic scientists cited threats from War; Disease; Climate change; and Disruptive technologies as major contributors. In the context of climate change mitigation, this chapter attempts to present contributions of the circular bioeconomy paradigm that could help humanity to dial back the Doomsday Clock. Anaerobic digestion (AD), integrated regenerative agriculture (IRA), controlled ecological life support system (CELSS), bioregenerative life support system (BLSS), note by note cuisine (NNC), circularity, and molecular pharming are some of the solutions isolated.

Keywords: aerosols, anaerobic digestion, biofertilizer, biofuels, biogas, circular bioeconomy, climate change, digestate, doomsday clock, ecological intensification, extreme weather events, greenhouse gases, particulate matter, regenerative agriculture, regolith, renewable energy, space food system, sustainability

1. Introduction

The "Doomsday Clock" is a construct of the bulletin of atomic scientists that serves as a barometer for humanity's nearness to irreparable destruction. Midnight hour on the clock is designated to be the instance of ignition and thus, the point of no return from disaster. On 24th January 2023, the Doomsday Clock got to 90 seconds before midnight. This was the closest the clock had ever been to humanity's insertion into the extinction process. Analyses by the bulletin of atomic scientists indicated that the threats of war (biological, chemical, nuclear); disease (zoonotic epidemics and pandemics); climate change (weather extremes such as radical storms and heat waves); and disruptive technologies (cyber space, drones, artificial intelligence, social media) are major contributors. These contributors were also indicted in the Stanford existential risk initiative (SERI) study [1]. Luckily, several of these factors are amenable to anthropogenic interventions. But are humans doing enough to prevent the progression of Doomsday Clock to the point of no return (aka Armageddon)? Would humanity mobilize all that it may take (cooperation, technology, political resolve and courage, and any other resource needed) to constrain anthropogenic polluting emissions, attenuate the exceedance of planetary boundaries, and forestall irreversible perturbation of the earth's system? This chapter proffers mitigation strategies for climate change. Specifically, climate change attenuating applications of biomass resources as indicators of the circular bioeconomy model that could help humanity to dial back the Doomsday Clock are presented.

2. Signatures of climate change

Climate change may manifest in weather extremes: floods, torrential rainfalls, radical storms, cold spells, heatwaves, severe droughts, and wildfires. These events could cost life; destroy residential and commercial buildings, public utilities, and infrastructures; lead to contamination of natural resources (e.g., air, soil, and water bodies); disrupt numerous functions including schools; social events; air, land, and sea transportation.

2.1 Storms, rainfalls, and floods

Drastic storms like cyclones, hurricanes, tornados, and typhoons may encompass blizzards, hail, rain, snow, lightning, thunder, and strong winds. As shown in **Figure 1**, the outcome could be devastating for communities in technologically advanced and developing nations alike.

- In 2016, a flood impacted households in Sri Lanka. A case study of extreme weather events revealed that about 55% of households surveyed in a rural community were exposed to the 2016 flood event [6].

- In 2020, land use influences and heavy rains caused Lake Victoria to rise by >1 m. Trigged floods submerged jetties in Uganda [7]. The rains, the worst in 19 years, affected most of sub-Saharan Africa above 10°S [8].

Figure 1.
Effects of storms, rains, and floods on planet earth. Photo sources. (A): Joe Raedle/Getty Images: Residents of Houston evacuate their homes after the area was flooded from Hurricane Harvey, – Yahoo Image Search Results (Accessed 1st June 2023); (B): [2]; (C): [3]; (D and E): [4]; and (F): [5].

- The July 2021 flood in Europe impacted schools, public utilities, and the health system. Bridges, motorways, railway lines, clinics, hospitals, and pharmacies were damaged and over 220 people died [2].

- In 2022, heavy rains caused flooding in West Africa. Nigeria and Niger in June; Chad in July; Cameroon in August; and Benin in September. Over 4.6 million people were affected with at least 829 killed [9].

- In 2022, unsafe storms from the Indian Ocean visited Southeastern Africa. Winds, rains, and floods impacted Madagascar, Mozambique, and Malawi. Deaths and infrastructure destruction were recorded [5, 10].

- In 2022, extraordinary heavy rainfall hit vulnerable communities in Northeast Brazilian states, generating flash floods and landslides. Over 25,000 persons were displaced and at least 133 fatalities were reported [11].

- In June 2022, extreme monsoon rainfall unleashed floods and landslides in Pakistan. As of 9th September 2022, reported impacts included over 33 million people affected, and 1396 deaths [12, 13].

- In December 2022/January-February 2023, atmospheric rivers and storms hit California State in the USA. Ensued ruins of infrastructures and communities were reported with several photos (**Figures 1D** and **E** [4]).

2.2 Cold and heat extremes

Cold and heat extremes disrupt energy facilities, food supplies, economic activities, and endanger humans, plants, and animals. Impacts of non-optimal temperatures have been reported [14, 15]; and are stressed by age, comorbidities of health condition, gender, race, socioeconomic status, and pharmaceuticals. For example, medications that impair thermoregulation or depress sweat gland functions have been implicated [16].

2.3 Severe droughts

Drought is a key constraint affecting food supply, especially in areas dependent on rain-fed agriculture. Water stress impedes water bodies, food, and forage crops. Livestock grazing/production, lakes, and ponds become devastated (**Figure 2**), exacerbating the economic status and welfare of subsistence farmers and households.

2.4 Excessive wildfires

Figure 3 presents sites that are becoming rather common on planet earth. From country to country, wildfires have been occurring with alacrity. Brazil, Greece, Indonesia, Italy, Portugal, and Spain are among the victims in recent years. Wildfires destroy human life, and rare plant and animal species; and emit heat energy, smoke, carbon monoxide (CO), and other pollutants with adverse environmental and health effects. The wildfires of 2019–2020 in Australia aggravated health issues, caused 417 deaths, and emitted about 715 Tg CO_2 [19, 20]. In February 2023, the searing heat wave in the Parana-La Plata Basin, Chile, caused flames to spread, killed over 20 persons, injured over 1000 others, and burnt about 270,000 hectares of land [21].

Figure 2.
Impacts of drought on planet earth. The remains of dead livestock and a donkey (left), and cracked earth in an area once under the water of a Lake. Source: [3].

Figure 3.
Excessive wildfires cause destruction and pollution on planet earth. (A): Raging wildfires. Source- Munich Re; https://www.munichre.com/natcatservice (Accessed 26th February 2023). (B): Wildfires smoke aerosol over Zama City, Alberta, Canada. Source: Alberta Wildfire/Handout via Reuters [17]. (C): Smoke pollution over the Pacific Ocean. Source: [18].

In early June 2023, various news channels were reporting the fallout of wildfires from Canada (**Figure 3B**) that affected several states in the USA; including New Jersey, New York, Pennsylvania, and Washington DC; the nation's capital. Some air flights were interrupted, schools closed, Broadway shows canceled, and residents were urged to stay indoors, and/or wear masks [NPR news bulletin of 8th June 2023]. By mid-August 2023, the Maui Hawaii wildfires in the USA had traumatized families, erased homes, devastated infrastructures, killed over 100 persons, and hundreds more reported missing [CBS and BBC news bulletins of 16th August 2023].

3. Who (or what) is to be blamed?

Who or what causes breaches of planetary boundaries that trigger climate change? Some authors have blamed natural phenomena. Solar proton events inject nitrogen oxides that destroy ozone in the stratosphere [22]. In 2010, natural aerosols affected the global energy balance more than anthropogenic aerosols [23]. In 2021, the amount of sulfur dioxide (SO_2) emitted into the troposphere by the La Palma volcanic eruption ($\approx$ 1.8 Tg), exceeded the total anthropogenic SO_2 emitted from the 27 European Union countries in 2019 [24]. Thundercloud coronas may be a major source of greenhouse gases [25]. Other authors assert anthropogenic activities are culpable [9, 26–28]. A classic example that illustrates this dichotomy surrounds the extremely hot summer of 2010 in Russia. One group of scientists concluded that the intense heatwave was mainly due to nature [29]. Another group, with 80% probability blamed human influences [30]. The same event with different determinations that may be both right? [31]. Yet other researchers surmise that natural and human activities are both contributors, with the human inputs perhaps more pronounced in the Anthropocene epoch [8, 32, 33]. Irrespective of the force (human or natural), extreme events may be key indicators of climate change, and environmental responses to its perturbations, such as the effects on glaciers [34, 35]. It also appears that climate change could be precipitated and driven by greenhouse gases and aerosols. A question may be asked, what are the anthropogenic emanators of these drivers, and what strategies might be deployed to mitigate them? If the natural contributors to climate change cannot be controlled, it is imperative or rather obligatory that human-made causes be attenuated.

4. Causes of climate change

Climate change seems to be generally forced by greenhouse gases and aerosols. If so, then, what are these so-called greenhouse gases and aerosols that pollute the environment and influence the climate?

4.1 Greenhouse gases

To maintain earth's radiation budget, the radiant energy coming from the sun must be in a fair balance with the radiation leaving the earth. If the emission from earth falls short, energy would accumulate and force the earth's temperature to rise. There are gases in the atmosphere that absorb and prevent the radiation emitted by the earth from escaping into outer space. The energy trapped by these gases is radiated back to earth, causing the earth's temperature to warm up. Gases with this ability are called greenhouse gases (GHGs). GHGs include Carbon dioxide (CO_2), Methane (CH_4), Nitrous oxide (N_2O), and Halocarbons. GHGs do occur naturally. But a group of halocarbons that contain fluorine, known as Fluorinated gases (F-gases), are mostly human-generated. F-gases include sulfur hexafluoride (SF_6) used as an insulating gas, and for the production of aluminum and magnesium; hydrofluorocarbons (HFCs) used as blowing agents and refrigerants; and perfluorocarbons (PFCs) used in the electronics industry for semiconductor manufacture. F-gases are potent GHGs that contribute significantly to climate change [36, 37]. **Table 1** highlights the characters of some GHGs.

S/N	GHG	Concentration (mol/mol)	Lifetime (Years)	Radiative forcing (W/m^2)	Global warming potential (100 years)*
1	Carbon dioxide (CO_2)	$365–409 \times 10^{-6}$	≈100	1.66	1
2	Methane (CH_4)	$1.7–1.857 \times 10^{-6}$	10–12.4	0.48	25
3	Nitrous oxide (N_2O)	$319–331 \times 10^{-9}$	114–131	0.16	298–320
4	Chlorofluorocarbon (CFC-12)	538×10^{-12}	102	0.17	8500
5	Sulfur hexafluoride (SF_6)	$5.6–9.6 \times 10^{-12}$	3200	0.0029	22,800–24,900
6	Perfluoromethane (CF_4)	$74–79 \times 10^{-12}$	50,000	0.0034	6300–7390

*Source: Compiled from [27, 36, 38–40]. *Relative to CO_2.*

Table 1.
Characteristics of some greenhouse gases.

S/N	Aerosol	Diameter (μm)	Concentration (μg m^{-3})	Lifetime (Days)	Radiative forcing (W/m^2)
1	Black carbon (BC)	2.5	0.10–0.63	1.8–15	0.002–5.335
2	Organic carbon (OC)/ Particulate organic matter (POM)	2.5	(3.0 ± 0.6) to (191.2 ± 104.3)	4.3–11	(−0.31) to (−0.09)
3	Nitrate (NO_3^-)	2.5–192	0.72–84.24	0.8–25	(−0.17) to (−0.03)
4	Sulfate (SO_4^{2-})	8.0–30.6	0.1–50.0	2.3–5.2	(−1.5) to (−0.96)
5	Dust	0.01–100	0.3–180	1.3–7	(−0.56) to (+0.1)
6	Ammonium (NH_4^+)	2.5–192	0.56–43.68	4–6	(−0.11) to (+0.21)

Source: Assembled from [18, 40, 43–57].

Table 2.
Potential characteristics of aerosol pollutants.

4.2 Aerosols

Air quality may be eroded by particulates. Particles of public health attention are defined by an aerodynamic diameter in the range of ≤10 μm (PM_{10}). Fragments with diameters less than 2.5 μm ($PM_{2.5}$) are identified as fine particles. Aerosol is the technical parlance used to denote airborne liquid and solid droplets and particles. Aerosols are ever-present (inside/outside homes, buildings, atmosphere); could be of human or natural origin; primary (when emitted directly from pollution sources) or secondary (when made from precursor pollutants); and could exhibit greenhouse gas properties or function as reflectors. Aerosols include sulfates (SO_4^{2-}), nitrates (NO_3^-), black carbon (BC), and a plethora of other compounds, and chemical species [41, 42]. Examples of natural aerosols are desert dust, sea salt particles, meteoritic fragments, as well as ash and sulfates from volcanic eruptions. Aerosols affect atmospheric and ocean conditions including cloud droplets, circulation, snow covers, precipitation, temperature, and heat transfer; thereby influencing planetary energy and water budgets [18, 26]; and the references in these citations. Those aerosols that absorb radiation from the sun (e.g., BC, mineral dust) could fulfill the effects of GHGs and warm the earth, while those that reflect and scatter radiant energy (e.g., SO_4^{2-}, NO_3^-) may impart cooling effect. Collectively and on average, aerosols appear to have a net cooling impact [18]. In 2007, the solar direct radiative effect (DRE) for aerosols estimated from satellite remote sensing studies was reported to be −5.4 ± 0.9 W/m^2 [40]. The possible properties of some aerosols are presented in **Table 2**, and their pollution impacts exhibited in **Figure 3**.

5. Anthropogenic sources of the breaches and exceedances of planetary boundaries

The major anthropogenic GHGs breaching planetary boundaries and perturbing the global climate system are CO_2, CH_4, N_2O, and F-gases. Their percentage repartitions to the global GHG budget are estimated at 72–76% for CO_2, 16–19% for CH_4, 6% for N_2O, and 2–3% for F-gases [58, 59]. The major aerosols include BC, organic carbon

(OC), SO_4^{2-}, and NO_3^-. BC and OC are primary aerosols because they are emitted directly from pollution sources, while SO_4^{2-} and NO_3^- are secondary aerosols given that they are generated from precursor pollutants. The F-gases are totally anthropogenic, and their industrial applications are highlighted in Section 4.1. Anthropogenic sources of the other GHGs, primary aerosols, and precursor pollutants are intricately tied to humanity's methods of exploitation of fossil and biomass energy resources. Their three main sources are:

1. Fossil fuel: For example, coal, crude oil (also known as oil or petroleum), and natural gas. Emissions arise from their production, combustion, and use, including flaring.

2. Biofuel: Such as wood, charcoal, dung, bioethanol, and biodiesel. In their production, combustion, and use, encompassing domestic and industrial applications.

3. Open biomass burning: Such as for shifting cultivation, combustion of agricultural waste, land use change, and permanent deforestation.

Global annual GHG emissions increased by about 57% between 1990 and 2018, from 32.7 to 51.2 Gt CO_2 eq, with the increment dominated by CO_2 emissions from fossil fuels. Anthropogenic CO_2 emissions declined from ≈ 37.7 Gt in 2018 to ≈ 36 Gt in 2020 [37]. Fossil energy CO_2 rose to 40.0 ± 2.9 $GtCO_2$ in 2021 [60], and in 2022 was reported to be about 36.8 Gt [61]. Anthropogenic share of global BC aerosol inventory in 2010 was reported as 66%. Those for precursor gases were 64%, 68%, and 83% respectively for nitrogen oxides ($NOx = NO + NO_2$), ammonia (NH_3), and sulfur oxides (SOx) [23]. The global percentage contribution of GHG emissions by the economic sector in year 2010 is shown in **Table 3**.

6. Analysis of agriculture connections with exceedances of planetary boundaries

Table 3 indicates that agriculture is a major source of GHGs with a repartition of 24%. This share does not tell the whole story because GHGs do not all have equal impact on global warming potential (GWP). As **Table 1** shows, on a one-hundred-year time horizon, CH_4 would exert 25 times the GWP of CO_2; and N_2O, 298 to 320 times. Hence impact of a sector on climate change is related not only to the ratio of GHG but also to the proportion and potency of constituent GHGs. Also, **Table 3** did not disclose any information on the other climate forcer; aerosols. We will now take a closer look at agricultural connections with GHGs and aerosols.

6.1 Agriculture connections with GHGs

CO_2 captures the lion's share of anthropogenic GHGs, accounting for ≈ 74% of the global ≈ 51.2 Gt CO_2 eq GHG emitted in 2018 [37]. Agriculture's input to global CO_2 inventory is minimal, at 6–8% [62, 63]. In the USA in 2018, agriculture added just 0.63% of the over 5.4 Gt CO_2 inventoried [36]. On non-CO_2 GHGs, agriculture emits virtually no F-gases but is a huge emitter of CH_4 and N_2O. Estimates for CH_4 range from 31% [58] to 45% [64]; and for N_2O, from 66% [65] to 84% [62]. In 2005,

S/N	Economic sector	Share of GHG (%)
1	Energy: • combustion of coal, natural gas, and oil, for heat and electricity generation (25%) • fuel extraction, processing, refining, and conveyance (10%)	35
2	Agriculture: encompassing emissions from crop cultivation, livestock production, manure management, land use, land-use change, and forestry (e.g., deforestation, afforestation, population pressure), etc.	24
3	Industry: includes fossil fuels combustion at onsite facilities for industrial operations (e.g., chemical, metallurgical, and mineral transformation processes)	21
4	Transport: includes emissions generated by fuels combusted by mobile constructs as in aviation, marine (shipping, including boats in inland waterways), and terrestrial (e.g., rail, road, off-road/non-road) conveyances	14
5	Buildings: includes emissions from small-scale immobile facilities such as residential homes, and private and public offices.	6

Source: Ref. [59].

Table 3.
Year 2010 distribution of global GHG emission by economic sector.

agriculture emitted almost 54% of the non-CO_2 GHGs; 0% of the F-gases; about 45% of the CH_4; and over 82% of the N_2O (Author's derivation, based on data in **Table 3** of Ref. [66]). Agricultural CH_4 sources are livestock rumination (enteric fermentation), rice cultivation, manure management, and combustion of agricultural residues. Sources of N_2O are the burning of agricultural residues, synthetic fertilizer, manure, and soil management (nitrogen fixing, leaching/run-off, *etc.*).

6.2 Agriculture connections with aerosols

There are two major ways that agriculture contributes to aerosols and their effects. One is by direct generation of primary aerosols. The second mode is indirectly, by generation of precursor gases. Precursors can undergo chemical reactions that produce secondary aerosols and other GHGs. Incomplete combustion of agricultural biomass, and fossil fuels by agricultural machinery for example, liberates primary aerosols (BC, OC); precursor gases (CO, SO_2, NOx); as well as non-methane volatile organic compounds (NMVOCs) [42]. Precursors may also arise via manure management (e.g., NOx), fertilizer use (e.g., NH_3), and pesticide applications. In the aquatic ecosystem, pesticides were blamable for 81%, 87%, and 96% of the detected exceedances of the acute risk threshold (ART) related to fish, invertebrates, and algae, respectively [67].

6.3 Attribution assessment

Studies of tropical cyclones Ana and Batsirai that devastated Southeastern Africa deduced that human acts are at least in part liable for climate change. [9, 10, 28]. So did a study of the desert locust (*Schistocerca gregaria*) outbreak that impacted Ethiopia, Kenya, and Somalia [68, 69]. Anthropogenic GHGs are linked to energy and agricultural industries (**Table 3**). Agriculture is also a big aerosols emitter [48, 63, 70]. About 20% of the 3.3 million premature mortalities incurred in the world in 2010 from ozone and $PM_{2.5}$ pollution were attributed to agriculture [71]. A good amount of agriculture's GHG and aerosol burden emanates from fertilizer manufacture and use. The Haber-Bosch fertilizer process via ammonia synthesis is a gargantuan fossil energy consumer and GHG emitter [72]. In the USA corn (maize) production system, for instance, baseline GHG emission is ≈ 2469 kg CO_2 eq per ha, and nitrogen fertilizer alone (manufacture and field losses) accounts for about 56% [73]. In livestock production, pig feed accounted for 79% of fossil fuel consumption and 88% of GHG emissions. The values for chicken feed were respectively 84% and 91%. In these studies, corn and soybean were basic inputs, and inorganic fertilizer was used in their cultivation [74]. In view of these values and recalling that nitrogen fertilizer and manure contribute 92% of all N_2O attributable to agriculture in the USA [75], and because fertilizer is required for the cultivation of maize and rapeseed, used for bioethanol and biodiesel creation respectively, a Nobel Laureate invested some computations. The results motivated the Nobel Laureate to caution that due to the larger GWP of N_2O relative to CO_2, and contingent upon the uptake efficiency of nitrogen fertilizer, the production of bioethanol from maize, and biodiesel from rapeseed, may lead to the emission of enough N_2O that wipes out the benefits of using biofuels to replace CO_2 emitting fossil fuels [76]. So, how may humanity manage and mitigate GHG emissions, to constrain climate change?

7. Circular bioeconomy to the rescue

The linear economic model operates on the norm of produce, use, and discard. Circular bioeconomy emulates nature's edict of circularity (Chapter 6 in [77]; Fig. 2 in [78]; Fig. 1 in [79]; [80]; Fig. 5 in [81]; [82]). In the circular model, there is no waste. "Waste" from one process is "resource" and "input" for another. Circular bioeconomy was described as embracing wasteless design, and management tactics that recycle so-called wastes into useful goods [83, 84]. One study estimated the menace imposed by linear methods on nature's worth at over one-half of global GDP (≈ US$44 trillion). The study also indicated that more than US$5.5 trillion in business opportunities and 163 million jobs could be generated by 2030 with the circular bioeconomy approach [84].

7.1 Approaches to achieving circular bioeconomy

Circular bioeconomy requires input from all stakeholders, including those that risk their well-being to put out wildfires and rescue flood victims. CO_2 is a major GHG tied to energy use, and agriculture is a major GHG and aerosol generator, via inorganic fertilizer use. Thus, decarbonization of energy systems and another fertilizer source are necessary. Approaches advocated here obligate renewable energy and the concept of integrated regenerative agriculture (IRA). Renewable energy [85], encompasses a broad gamut of energy carrier types, technologies, and names such as bioenergy [86], biofuels [87], biogas [88], biomass [89], biomethane [90], hydro [91], solar [92], wind [93], and several others. The work presented here highlights anaerobic digestion (AD) as the purveyor of renewable energy. This is because AD is perhaps the only current technology capable of addressing both issues of energy systems decarbonization, and substitution of ammonia-based fertilizer. In addition, AD will simultaneously address another issue threatening planet earth: waste management.

7.2 Anaerobic digestion (AD) technology

AD is a biochemical process that decomposes organic matter into flammable gas and nutritious watery sludge. The process occurs in a near oxygen-free environment with natural microorganisms. AD was probably recognized in a scientific sense about 392 years ago in 1630, when Jan Baptist van Helmont (1580–1644), observed that decomposing organic matter produced combustible gas; and during 1804–1808 John Dalton and Humphrey Davy established that the combustibility of the gas was due to the presence of methane [94, 95]. Today, AD is carried out within artificial environments named digesters, which are designed to optimize the process [96–98]. The combustible gas is called biogas, and the watery sludge, digestate. Biogas is composed of methane (40–75%), carbon dioxide (25–40%), and trace contaminants ($\approx$ 0.1–3%). Digestate is rich in plant growth macronutrients; with their quantities and micronutrients contents related to the quality of the feedstock [98, 99]. Digestate attributes, AD's technical feasibilities, types of feedstocks utilized (virtually all organic matter), limitations, advantages, benefits, applications, and much more have been presented in the literature [83, 96, 98, 100–102]. AD may be now considered a mature technology that could help humanity decarbonize industries, supplant inorganic fertilizers, and address waste management shortcomings.

7.3 Integrated regenerative agriculture (IRA)

IRA system combines food, feed, bioenergy, and livestock production in the farm space. IRA practices minimize the fuel/energy versus food mindset competition. The "this or that" image of pitting food production against energy production could be abrogated, thereby promoting synergy and coexistence of fuel, feed, and food cropping; enhancing carbon sequestration, and fostering sustainable ecological intensification. As of 2018, it was estimated that about 163 million farms in 100 countries, engaged 453 million hectares of agricultural land in some form of IRA [103]. The origin/history, definitions, lessons, modifications, and various aspects of IRA have been explored in published literature [104–107]; tried, and/or implemented in several countries such as Colombia [108]; Italy [109]; Egypt [110]; Bangladesh [111]; USA [112]; Finland [113]; Chile and Vietnam [114]. Advocates argue that IRA would also enhance biodiversity, create wealth, support rural communities, improve the environment, increase equity, and mitigate injustice [84, 103, 115–118]. Perhaps these reasons could invoke concern globally to justify a more stringent climate policy [119].

8. Implications of circular bioeconomy for the space food system

In low gravity, e.g., onboard spacecraft, and celestial body (Lunar-, Asteroid-) surfaces, providing food is more taxing than on earth. Humans operating in space face the dynamics of microgravity, radiation, and restricted activity [120–122]. In near-earth missions lasting days (Apollo, Space shuttle, Shenzhou), weeks (Skylab, Salyut), or even months (Mir, International space station, Tiangong space station), food could be carried at liftoff and resupplied from earth. Even in these cases, the space food system is mediated by biological, engineering, environmental, operational, and psychosocial factors. The specific constraints such as safety (biological); mass/volume (engineering); ergonomics (operational); and others have been discussed [122–128]; and detailed for the Apollo food system [129]. Longer durations and interplanetary destinations make food supply from earth onerous for various reasons. On cost for example, depending on the destination orbit, payload launch cost may range from $2719/kg for Falcon 9 launch destined to low earth orbit (LEO), to $28964.52/kg for Delta IV heavy launch destined to geosynchronous transfer orbit (GTO) [130]. Nevertheless, the performance of a launch system is governed by its mass fraction (ratio of the mass of propellants to the total mass of the launch system). The larger the mass fraction, the farther the destination and/or longer the duration. A larger rocket, such as the SpaceX Starship under development, would be able to carry more food, go farther, and last longer than the Saturn V rocket used for the Apollo moon program. But for a given launch system, the larger the food component of the payload, the nearer the destination and/or shorter the mission duration. With no infinite mass fraction launch system, onboard and in-situ food production become vital for successful long-time deep space missions, and extraterrestrial body habitations. To achieve this, space farming would need to benefit from circular bioeconomy technologies and concepts such as AD [131–133]; IRA [116, 134]; circularity [77–82, 135]; bioregenerative life support system (BLSS) [136, 137]; controlled ecological life support system (CELSS) [138–140]; and note by note cuisine (NNC) [141, 142]. For greater plant utility and optimization, molecular pharming [143, 144] could be applied as well.

8.1 Sample study efforts on earth and in space

Research efforts have been expended both on earth and in space to ascertain the possibility of cultivating plant and animal food stocks in the space environment. Some examples are stated below.

In Apollo quarantine tests of moon soil (regolith), a wide variety of biological species from the animal and plant kingdoms were utilized. Researchers [145–147] concluded that no microbe or harmful agents were found in the tested Apollo Luna regolith; some benefits could be associated with lunar soil cultivation; and that Lunar soil material is a potential source of nutrients for many plants. But experiments with Apollo 11, 12, and 17 Luna regolith reported challenging performance for *Arabidopsis thaliana* (Arabidopsis). Luna regolith seedlings demonstrated normal stems and cotyledon development. However, roots were found stunted after day 6, and the plants showed stress morphologies indicative of ionic stresses [148].

Lettuce (*Lactuca sativa* L. cv. Dasusheng) was grown onboard Tiangong II Spacelab [149]. Germination rate was 37% for space and 78% for ground control. Plant morphology and number of leaves were similar but produced biomass fresh weight differed notably (space-grown lettuce was 40% lower). In nutrition, magnesium ($\approx$ 33%), iron ($\approx$ 35%), and calcium ($\approx$ 45%) were lower, while potassium (by about 10%) was higher in space-grown samples. In food safety, pathogens such as *Staphylococcus aureus*, and *Salmonella* were not detected in both groups. But ground control lettuce presented a higher net colony count. Also, space lettuce grew taller, and had more deep green appearance without transitions (**Figure 4**).

Parabolic flights were used to conduct microgravity research. An interesting discovery was made when thick-toed geckos were flown. Apart from a few rodents that could move by clinging to their container gratings, virtually all objects, mammals (including humans), amphibians, and reptiles float when exposed to the space environment without restraints [122]. However, when the thick-toed geckos were flown in parabolic flights, it was reported that geckos had the ability to remain attached to smooth surfaces. This peculiar capacity apparently allows the geckos to keep normal activities and behavior during weightlessness [150].

Figure 4.
Lettuce plant growths and appearances. In space (left), and on earth. Source: [149].

Future studies: Planet Mars' regolith & rocks are now being gathered by the perseverance rover. These would be studied in detail when delivered to Earth following a future sample return mission [NASA Science: Available @ Video Gallery: Perseverance Rover - NASA Mars; Accessed 12th August 2023]. I venture to speculate that the Martian regolith would be analyzed for agricultural services once in earth's laboratories.

9. Dialing back the doomsday clock

The bulletin of atomic scientists and SERI study cited war, disease, climate change, and disruptive technology as the proverbial Swords of Damocles hanging over planet earth [1, 151]. Due to these threats, the doomsday clock crept 90 seconds to midnight (the ignition time and point of no return from Armageddon) on 24th January 2023. Thus, mitigation of any of the threats is tantamount to dialing back the doomsday clock.

9.1 Contributions of circular bioeconomy to climate change mitigation

Fossil fuels and inorganic fertilizers are sources of GHGs and aerosols implicated as agents of climate change. Their substitutions with benign alternatives are needed. Below, are some of the advantages, attributes, benefits, and quantified contributions of the circular bioeconomy paradigm to climate change mitigation.

9.1.1 Production of renewable energy

Biomasses are renewable energy creation inputs, and AD is used for biofuel (biogas, biohydrogen, biomethane, *etc.*) production [152–154]. Work with watermelon waste for bioethanol and biomethane production reported biogas yield at 581 ± 11 mLN/gVS [155]. Biogas may be used as is or converted to higher-quality biomethane [156, 157]. One study subjected cassava peeling residue (CPR) derived from 1000 kg of cassava root to AD. Output biogas provided sufficient energy to process the 1000 kg root into flour, gari, and starch [158]. Biogas canceled the global warming potential of fossil energy otherwise required for the production of these products, hence contributing to climate change mitigation.

9.1.2 Production of biofertilizer

As noted in Section 6.3, fertilizer via ammonia synthesis is a huge fossil energy consumer and GHG emitter. One ton of oil, 108 tons of water, and 7 tons of CO_2 are linked to one ton of synthetic fertilizer [159]. Inorganic fertilizer volatilizes ammonia, a known precursor [70], that enables nitrate and sulfate aerosols. Studies indicate that nitrates will be a major anthropogenic climate forcer in the future because of inorganic fertilizers in agriculture [160, 161]. Substitution with digestate biofertilizer generated from AD operations indicated viability. In cassava root production, a liquid fraction of CPR digestate could supplant up to 28% nitrogen, 11% phosphorus, and 38% potassium sourced from chemical fertilizers for cassava root production [83]. Digestate biofertilizer, therefore, seems to be a climate change mitigator.

9.1.3 Minimization of CO_2 footprint

Circular bioeconomy could decarbonize energy systems and reduce CO_2 footprint. Substituting Haber-Bosch fertilizer with digestate for example avoids the 7 tons of CO_2 emitted for each ton of fertilizer produced. Studies replaced fossil fuel with biogas in the transport sector. Biogas caused only 21–36% of the GHG emissions of fossil-based fuels [162], and biomethane reduced CO_2 footprint by 49–84% [163]. Emissions linked to AD production of renewable energy and biofertilizer were lower than those for production of fossil energy and mineral fertilizer [164]. A study of combined heat and power (CHP) production found that pressurized AD systems achieved a direct CO_2 footprint of about 13 $kgCO_2/MW\ h_f$. This footprint is quite low when compared to the 700 $kgCO_2/MW\ h_{el}$ direct CO_2 footprint of conventional CHP systems [165]. In the analysis of sorghum cultivation as an energy crop, inorganic fertilizer accounted for 51% of GHG emissions; digestate decreased carbon footprint by 11%. Digestate was then recommended for sustainable energy crop production and GHG mitigation [166]. Furthermore, a study from Switzerland quantified their national biomass availability and confirmed its potential for biofertilizer provision and climate change mitigation. The authors concluded that by 2050, circular bioeconomy application of AD would save significant amounts of mineral fertilizer (over 10 kt/a), and GHG emissions (38 kt/a of CO_2eq) [167].

9.1.4 Ozone (O_3) and waste management

Some controls bear more benefits: mitigate climate change, and upgrade agriculture and health. Ozone control with methane reduction embodies such impacts [52, 168]. Tropospheric ozone is a GHG and air pollutant. Ozone degrades crop output and human health [71, 169]. Because methane is an ozone precursor {via oxidation reaction: $CH_4 + 4O_2 \rightarrow CO + H_2 + H_2O + 2O_3$ [170]}, CH_4 reduction would minimize ozone and ozone's allied ill outcomes. About 17% reduction of global methane emission per year is estimated to reduce ozone by $\approx$ 1 ppb, and radiative forcing by $\approx 0.12\ W/m^2$ [168]. Avoided global warming of such measures in 2050 is $\approx 0.28 \pm 0.10°C$ [52]. Methane does arise from animal manure, municipal wastes, and landfills. Circular bioeconomy using AD as waste manager converts organic matter (energy crops/double-cropped biomasses, farm residues, animal manure, food processing wastes, space mission wastes, wastewater treatment plant effluent, organic fraction of municipal solid waste, and virtually any organic waste: whether previously sent to landfills or not), into valuable energy carriers (e.g., biogas and digestate). AD code would eliminate these ozone precursor sources, and their effects as environment-polluting wastes. As methane itself is a GHG of a global spatial scale, its reduction reduces its GHG effect; and combined with ozone & waste management, the benefits scale the entire world -all: human health, agriculture, ecosystems, countries, etc.

9.1.5 Monetization, cost savings, and energy justice

Circular bioeconomy could contribute to climate change mitigation by appealing to the morality, justice inkling, pocketbooks, and bottom lines of stakeholders. Profit margins and pecuniary savings might incentivize entrepreneurs/businesses to adopt circular bioeconomic tools. Studies monetized the benefits of climate mitigation via gains associated with O_3 management [52, 168, 171]; and cost savings inherent in the application of digestate biofertilizer [102]. Monetized benefits of O_3 reduction through NOx mitigation were estimated at $1875 per metric ton of NO_2, and that for NMVOCs at $1100 per metric ton NMVOC [171]. For agriculture, forestry, and non-mortality human health, the estimate of global annual monetized benefits of O_3 mitigation through 59 M ton CH_4 reduction included $7.8 billion avoided damages, and $1.7 billion net cost savings at marginal cost of < $81 per ton CH_4 [168]. When human health mortality, crops, and climate were evaluated, monetized benefits of CH_4 reduction as an ozone control measure were estimated at $700 to $5000/ metric ton CH_4. This value was much higher than the usual marginal costs (< $250/ metric ton CH_4) required to implement abatement solutions [52]. Likewise, a study analyzed the economic implications of supplanting the inorganic fertilizer required for cassava root production with liquid fraction digestate. Results showed that about 25% of cost incurred in chemical fertilizer purchase could be saved with liquid fraction digestate. The monetized benefits were valued at $0.141 billion for the year 2019 global cassava root output. However, this valuation was conservative as the monetization did not include external costs of inorganic fertilizers such as air pollution, eutrophication, GHGs, and the contamination of potable water supply reserves [102]. Related to profit motive is the ethical and philosophical inclinations of humans to do good albeit for self-preservation. Perhaps the moral allures of global energy access and security would prod policymakers to address issues of equity and justice in energy availability, accessibility, distribution, and affordability. Anaerobic digestion and biogas energy have been recognized to have the potential to achieve more sustainability and energy justice in society [100, 172].

10. Conclusions and perspectives

There is a palpable sense climate change is likely due to GHGs and aerosols. There is debate though, as to who or what is more blamable for their emissions. This controversy was illustrated with attributions to the hot summer of 2010 in Russia. A group of authors attributed the cause to natural events [29], while another attributed it to human actions [30]. Yet a third group argued that both sides may be correct, subject to how the data is checked [31]. So, what should citizens/ people/ the public believe? There are reports that nature continues to contribute climate change agents [22–25, 173–175], and human activities are also doing the same [9, 26–28, 176–180]. Since at present, there is no human ability to bar natural contributors of these agents (e.g., stellar outbursts, volcanic detonations), it becomes obligatory that the human-made agents must be restricted, especially in the current Anthropocene era. This author suggests human talents and ingenuity be expended on constraining anthropogenic sources and creating mitigating alternatives.

Remarkably, climate change is no respecter of borders, be it local, national, regional, or international; and not on land, at sea, in air, or in space. Climate change does not respect wars and nations at war. Heatwave of 2010 affected countries in

Europe. Helsinki in Finland recorded a daily mean temperature of 26.1°C; Kiev in Ukraine, 25°C at night, and Moscow in Russia reached a daytime temperature of 38.2°C [181]. Climate change is killing humanity; humanity should stop killing humanity. From Ukraine to Sudan, all wars on planet earth must stop. Countries should cooperate to fight common enemies of humanity (including climate change, hunger, disease, inadequate- health facilities, potable water and electricity supply, etc.); not fight each other. The industrially advanced nations that contributed most of the emissions breaching planetary boundaries, also have the resources and ability to adapt and manage the fallouts. The onus is on the technological economies to use the usual 20/20 hindsight to share answers and assist, encourage, and guide developing countries. Doing so would enable developing nations to avert the mistakes advanced countries made in their developmental trajectories.

This chapter identified ways circular bioeconomy minimize exceedances of planetary boundaries, mitigate climate change, and dial back the doomsday clock. Approaches of IRA and AD were discussed. AD is a wise way to recycle resources of land, nutrients (e.g., N, P, K), carbon, water, and sunlight [83, 101, 102, 108, 109, 115, 153, 158, 167, 182, 183]. AD converts organic matter into valuable energy carriers (e.g., biogas, biomethane, electricity, and biofertilizer) for beneficial domestic and industrial applications. In particular, the solutions of renewable bioenergy and biofertilizer provision, attenuation of carbon dioxide footprint, ozone and waste management, as well as monetization, cost savings, and energy justice were presented. These solutions are anchored to IRA in general, but more specifically to AD; a mature, cost-effective, and environment-friendly technology. Both approaches, however, could be scaled to the industrial competence of virtually any country on earth. To strengthen the robustness and sustainability of these approaches/ solutions, three additional answers are pinpointed:

No till crop management. This method was applied to maize (*Zea mays*, L.) farming and offered better economic, environmental, and ecosystem benefits as compared to other practices tested [184]. No till was also found to be advantageous in phosphorus accumulation, availability, and overall soil ecosystem function [185].

Reduction of food waste. Near 1.3 to 1.8 billion metric tons of food ≈ 30% of global food output is wasted annually [186, 187]. Because 24 to 30% of GHGs emanate from agriculture [59, 187], up to 10% of GHGs may be attributable to wasted food, besides emissions inherent in food decaying in trash cans, landfills, and dumpsites. Consequently, reduction of food waste would contribute to climate change mitigation.

Reduction of local pollution. Nations may attend to local pollutants like those generated by the transport sector. This is because local pollutants respond to the environmental Kuznets curve (EKC); **Figure 5**. EKC hypothesis applies an inverted U-shaped relationship to pollution and economic prosperity. Income increases with pollution until a peak regarded as the turning point is reached. As **Figure 5** shows, after the peak threshold, further rise in income results in decrease in pollution indicators [188–191].

The chapter also highlighted the implications of circular bioeconomy for the space food system. Along with AD and IRA; CELSS, BLSS, NNC, circularity, and molecular pharming were identified as strong links to sustainable space enterprise food system. Despite the challenges of microgravity, decades of research efforts have demonstrated the possibility and capability of growing plants in microgravity environments.

Food is an indispensable requirement for life on earth and in space. Its provision in terms of production, processing, consumption, and disposal must be carefully

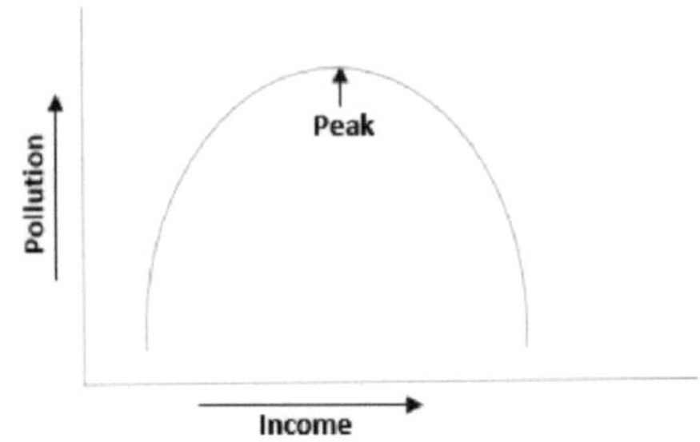

Figure 5.
A generalized schematic of the environmental Kuznets curve (EKC) hypothesis.

managed to minimize the footprints of resources consumption and external costs of GHGs and aerosols on the one hand. On the other hand, the provisioning should maximize solutions that mitigate climate change. AD in the context of circular bioeconomy appears to embody viable climate mitigation solutions. Recycling with AD judiciously utilizes biomass resources to generate climate mitigation solutions (e.g., bioenergy, biofertilizer); and simultaneously reducing the waste management menace. Evidently, mitigating climate change with a circular bioeconomy would enable dialing back the doomsday clock for the betterment of humanity. Studies that engage in quantitative determination of the contributions of these solutions (including their scientific foundations, environmental strengths, and financial viabilities) would be endeavors deployed in the appropriate path.

Author details

Sammy N. Aso
Food Engineering Laboratory, Rivers State University, Port Harcourt, Nigeria

*Address all correspondence to: sammyasso@yahoo.com

† Present address: Brookings, SD USA

References

[1] Stanford University. Stanford cascading risk study. 2023. Available from: https://seri.stanford.edu/research/stanford-cascading-risk-study [Accessed: July 24, 2023]

[2] World Weather Attribution. Heavy rainfall which led to severe flooding in western Europe made more likely by climate change. 2021. Available from: https://www.worldweatherattribution.org/heavy-rainfall-which-led-to-severe-flooding-in-western-europe-made-more-likely-by-climate-change/ [Accessed: February 27, 2023]

[3] O'Malley I. Confirmed: Global Floods, Droughts Worsening with Warming. Sunnyvale, CA, USA: Associated Press, via Yahoo! News; 2023 [Accessed: March 13, 2023]

[4] Antczak J, Dazio S. Storms end Southern California Water Restrictions for 7M. Sunnyvale, CA, USA: Associated Press, via Yahoo! News; 2023 [Accessed: March 16, 2023]

[5] World Weather Attribution. Climate change increased rainfall associated with tropical cyclones hitting highly vulnerable communities in Madagascar, Mozambique & Malawi. 2022. Available from: https://www.worldweatherattribution.org/climate-change-increased-rainfall-associated-with-tropical-cyclones-hitting-highly-vulnerable-communities-in-madagascar-mozambique-malawi/ [Accessed: 4th February 2023]

[6] De Silva MMGT, Kawasaki A. Socioeconomic vulnerability to disaster risk: A case study of flood and drought impact in a rural Sri Lankan community. Ecological Economics. 2018;**152**:131-140

[7] Mafaranga H. Heavy rains, human activity, and rising waters at Lake Victoria. Eos. 2020;**101**. DOI: 10.1029/2020EO146582

[8] Rodell M, Li B. Changing intensity of hydroclimatic extreme events revealed by GRACE and GRACE-FO. Nature Water. 2023;**1**:241-248. DOI: 10.1038/s44221-023-00040-5

[9] Zachariah M et al. Climate Change Exacerbated Heavy Rainfall Leading to Large Scale Flooding in Highly Vulnerable Communities in West Africa. Oxford, England, UK: World Weather Attribution Scientific Report; 2022

[10] Otto FEL et al. Climate Change Increased Rainfall Associated With Tropical Cyclones Hitting Highly Vulnerable Communities in Madagascar, Mozambique & Malawi. Oxford, England, UK: World Weather Attribution Scientific Report; 2022 [Accessed: February 4, 2023]

[11] Zachariah M et al. Climate Change Increased Heavy Rainfall, Hitting Vulnerable Communities in Eastern Northeast Brazil. Oxford, England, UK: World Weather Attribution Scientific Report; 2022

[12] Khan MA. NDMA Monsoon 2022 Daily Situation Report No 088. National Disaster Management Authority (NDMA). Islamabad, Pakistan: Government of Pakistan; 2022 [Accessed: February 19, 2023]

[13] Otto FEL et al. Climate change increased extreme monsoon rainfall, flooding highly vulnerable communities in Pakistan. Environmental Research Climate. 2023;**2**(2):025001

[14] Díaz J et al. Comparison of the effects of extreme temperatures on daily mortality in Madrid (Spain),

by age group: The need for a cold wave prevention plan. Environmental Research. 2015;**143**:186-191

[15] Chen TH, Li X, Zhao J, Zhang K. Impacts of cold weather on all-cause and cause-specific mortality in Texas, 1990-2011. Environmental Pollution. 2017;**225**:244-251. DOI: 10.1016/j.envpol.2017.03.022

[16] Ellis F. Heat wave deaths and drugs affecting temperature regulation. British Medical Journal. 1976;**2**(6033):474. DOI: 10.1136%2Fbmj.2.6033.474

[17] Wilson C. Midwest air quality plummets as Canada wildfire smoke moves in. Residents of Chicago woke up Tuesday to the worst air quality in the world. Yahoo News. Available from: https://www.yahoo.com/news/midwest-air-quality-canada-wildfire-smoke-chicago-wisconsin-grand-rapids-172026191.html [Accessed: June 27, 2023]

[18] Chin M, Kahn RA, Schwartz SE. Atmospheric aerosol properties and climate impacts. US Climate Change Science Program Report. Washington DC: NASA; 2009. p. 128

[19] Arriagada NB et al. Unprecedented smoke-related health burden associated with the 2019-2020 bushfires in eastern Australia. The Medical Journal of Australia. 2020;**213**(6):282-283

[20] van der Velde IR et al. Vast CO_2 release from Australian fires in 2019-2020 constrained by satellite. Nature. 2021;**597**(7876):366-369. DOI: 10.1038/s41586-021-03712-y

[21] European Space Agency. Chile battles raging wildfires. ESA 2023. Available from: https://www.esa.int/ESA_Multimedia/Images/2023/02/Chile_battles_raging_wildfires [Accessed: February 7, 2023]

[22] Heath DF, Krueger AJ, Crutzen PJ. Solar proton event: Influence on stratospheric ozone. Science. 1977;**197**(4306):886-889. Available from: https://www.jstor.org/stable/1744475

[23] Heald CL et al. Contrasting the direct radiative effect and direct radiative forcing of aerosols. Atmospheric Chemistry and Physics. 2014;**14**:5513-5527

[24] Milford C et al. Impact of the 2021 La Palma volcanic eruption on air quality: Insights from a multidisciplinary approach. The Science of the Total Environment. 2023;**869**:161652

[25] Neubert T, Gordillo-Vázquez FJ, Huntrieser H. Optical observations of thunderstorms from the International Space Station: Recent results and perspectives. Npj Microgravity. 2023;**9**:12

[26] Bond TC et al. Bounding the role of black carbon in the climate system: A scientific assessment. Journal of Geophysical Research-Atmospheres. 2013;**118**(11):5380-5552

[27] Hartmann DL et al. Observations: Atmosphere and Surface. Chapter 2. In: Stocker TF et al., editors. Climate Change 2013: The Physical Science Basis. Contribution of Working Group I to the Fifth Assessment Report of the Intergovernmental Panel on Climate Change. Cambridge, UK and New York, USA: Cambridge University Press; 2013. pp. 159-254

[28] Pinto I et al. Climate Change Exacerbated Rainfall Causing Devastating Flooding in Eastern South Africa. Oxford, England, UK: World Weather Attribution Scientific Report; 2022 [Accessed: February 4, 2023]

[29] Dole R et al. Was there a basis for anticipating the 2010 Russian

heat wave? Geophysical Research Letters. 2011;**38**(6):L06702. DOI: 10.1029/2010GL046582

[30] Rahmstorf S, Coumou D. Increase of extreme events in a warming world. Proceedings of the National Academy of Sciences. 2011;**108**(44):17905-17909

[31] Otto FEL et al. Reconciling two approaches to attribution of the 2010 Russian heat wave. Geophysical Research Letters. 2012;**39**(4):L04702. DOI: 10.1029/2011GL050422

[32] Ehlers E, Krafft T, editors. Earth system science in the Anthropocene. Berlin Heidelberg: Springer-Verlag; 2006. p. 273

[33] Bellouin N et al. Bounding global aerosol radiative forcing of climate change. Reviews of Geophysics. 2020;**58**(1):e2019RG000660. DOI: 10.1029/2019RG000660

[34] Barry RG. The status of research on glaciers and global glacier recession: A review. Progress in Physical Geography, Earth and Environment. 2006;**30**(3):285-306

[35] Zemp M et al. Historically unprecedented global glacier decline in the early 21st century. Journal of Glaciology. 2015;**61**(228):745-762. DOI: 10.3189/2015JoG15J017

[36] United States Environmental Protection Agency. Inventory of US greenhouse gas emissions and sinks: 1990-2018. Washington, DC: US EPA; 2020. p. 730

[37] Crippa M et al. GHG emissions of all world countries - 2021 Report. EUR 30831 EN. Luxembourg: Publications Office of the European Union. 2021. DOI: 10.2760/173513

[38] Houghton JT et al., editors. Climate change 1994: Radiative forcing of climate change and an evaluation of the IPCC IS92 Emission Scenarios. Cambridge: Cambridge University Press; 1995. p. 339

[39] Jacob DJ. Introduction to Atmospheric Chemistry. Princeton: Princeton University Press; 1999. p. 267. Available from: http://acmg.seas.harvard.edu/publications/jacobbook/index.html

[40] Forster P et al. Changes in atmospheric constituents and in radiative forcing. Chapter 2. In: Solomon S et al., editors. Climate Change 2007. The Physical Science Basis. Contribution of Working Group I to the Fourth Assessment Report of the Intergovernmental Panel on Climate Change. Cambridge: Cambridge University Press; 2007. pp. 129-234

[41] Jacobson MZ. Global direct radiative forcing due to multicomponent anthropogenic and natural aerosols. Journal of Geophysical Research-Atmospheres. 2001;**106**(D2):1551-1568

[42] Jacobson MZ. The short-term cooling but long-term global warming due to biomass burning. Journal of Climate. 2004;**17**(15):2909-2926

[43] Roberts JM, Fajer RW. UV absorption cross sections of organic nitrates of potential atmospheric importance and estimation of atmospheric lifetimes. Environmental Science & Technology. 1989;**23**(8):945-951

[44] Boucher O, Lohmann U. The sulfate-CCN-cloud albedo effect: A sensitivity study with two general circulation models. Tellus. 1995;**47B**:281-300. DOI: 10.3402/tellusb.v47i3.16048

[45] Adams PJ, Seinfeld JH, Koch DM. Global concentrations of tropospheric sulfate, nitrate, and ammonium aerosol simulated in a general circulation model. Journal of

Geophysical Research-Atmospheres. 1999;**104**(D11):13791-13823

[46] Tindale NW, Pease PP. Aerosols over the Arabian Sea: Atmospheric transport pathways and concentrations of dust and sea salt. Deep Sea Research Part II. 1999;**46**(8-9):1577-1595

[47] Uno I et al. Analysis of surface black carbon distributions during ACE-Asia using a regional-scale aerosol model. Journal of Geophysical Research-Atmospheres. 2003;**108**(D23):8636

[48] Vignati E et al. Sources of uncertainties in modeling black carbon at the global scale. Atmospheric Chemistry and Physics. 2010;**10**(6):2595-2611. DOI: 10.5194/acp-10-2595-2010

[49] Schulz M et al. Radiative forcing by aerosols as derived from the AeroCom present-day and pre-industrial simulations. Atmospheric Chemistry and Physics. 2006;**6**(12):5225-5246

[50] Textor C et al. Analysis and quantification of the diversities of aerosol life cycles within AeroCom. Atmospheric Chemistry and Physics. 2006;**6**(7):1777-1813

[51] Cape JN, Coyle M, Dumitrean P. The atmospheric lifetime of black carbon. Atmospheric Environment. 2012;**59**:256-263. DOI: 10.1016/j.atmosenv.2012.05.030

[52] Shindell D et al. Simultaneously mitigating near-term climate change and improving human health and food security. Science. 2012;**335**(6065):183-189. DOI: 10.1126/science.1210026

[53] Zhao PS et al. Characteristics of concentrations and chemical compositions for $PM_{2.5}$ in the region of Beijing, Tianjin, and Hebei, China. Atmospheric Chemistry and Physics. 2013;**13**(9):4631-4644

[54] Guo S et al. Elucidating severe urban haze formation in China. Proceedings of the National Academy of Sciences. 2014;**111**(49):17373-17378. DOI: 10.1073/pnas.1419604111

[55] Wang H et al. Using an explicit emission tagging method in global modeling of source-receptor relationships for black carbon in the Arctic: Variations, sources, and transport pathways. Journal of Geophysical Research-Atmospheres. 2014;**119**(22):12888-12909

[56] Wang G et al. Source apportionment and seasonal variation of $PM_{2.5}$ carbonaceous aerosol in the Beijing-Tianjin-Hebei Region of China. Environmental Monitoring and Assessment. 2015;**187**:143

[57] Gen M, Liang Z, Zhang R, Mabato BRG, Chan CK. Particulate nitrate photolysis in the atmosphere. Environmental Science, Atmospheres. 2022;**2**(2):111-127. DOI: 10.1039/D1EA00087J

[58] Olivier JGJ, Peters JAHW. Trends in Global CO_2 and GHG Emissions–2020 Report. PBL Report 2020. Available from: https://www.pbl.nl/en/publications/trends-in-global-co2-and-total-greenhouse-gas-emissions-2020-report [Accessed: 24th July 2023]

[59] United States Environmental Protection Agency. Washington, D.C. USA: Global Greenhouse Gas Emissions Data. Updated on February 25, 2022. USEPA; 2022 [Accessed: February 7, 2023]

[60] Friedlingstein P et al. Global carbon budget 2022. Earth System Science Data. 2022;**14**:4811-4900

[61] International Energy Agency. CO_2 Emissions in 2022. Paris, France: IEA; 2023 [Accessed: July 24, 2023]

[62] Olivier JGJ et al. Recent trends in global greenhouse gas emissions: Regional trends 1970-2000 and spatial distribution of key sources in 2000. Environmental Sciences. 2005;**2**(2-3):81-99

[63] Lelieveld J et al. The Indian Ocean Experiment: Widespread Air Pollution from South and Southeast Asia. Chapter 9 in: Paul J. Crutzen: A pioneer on atmospheric chemistry and climate change in the Anthropocene. Crutzen PJ, Brauch HG. (Editors). SpringerNature Nobel Laureates. 2016;**50**:197-209

[64] Isaksen IS et al. Atmospheric composition change: Climate–chemistry interactions. Atmospheric Environment. 2009;**43**(33):5138-5192. DOI: 10.1016/j.atmosenv.2009.08.003

[65] Davidson EA, Kanter D. Inventories and scenarios of nitrous oxide emissions. Environmental Research Letters. 2014;**9**(10):105012. DOI: 10.1088/1748-9326/9/10/105012

[66] U.S. Environmental Protection Agency. Summary Report: Global Anthropogenic Non-CO_2 Greenhouse Gas Emissions: 1990-2030. Washington, DC: Office of Atmospheric Programs, US EPA; 2012. p. 22

[67] Malaj E et al. Organic chemicals jeopardize freshwater ecosystems health on the continental scale. Proceedings of the National Academy of Sciences. 2014;**111**:9549-9554

[68] Kennedy M. Why are swarms of locusts wreaking havoc in East Africa? NPR. 2020. Available from: https://n.pr/2zkJciW

[69] Salih AA, Baraibar M, Mwangi KK, Artan G. Climate change and locust outbreak in East Africa. Nature Climate Change. 2020;**10**(7):584-585. DOI: 10.1038/s41558-020-0835-8

[70] Domingo NG et al. Air quality–related health damages of food. Proceedings of the National Academy of Sciences. 2021;**118**(20):e2013637118. DOI: 10.1073/pnas.2013637118

[71] Lelieveld J et al. The contribution of outdoor air pollution sources to premature mortality on a global scale. Nature. 2015;**525**(7569):367-371. DOI: 10.1038/nature15371

[72] Schiffer ZJ, Manthiram K. Electrification and decarbonization of the chemical industry. Joule. 2017;**1**(1):10-14. DOI: 10.1016/j.joule.2017.07.008

[73] Northrup DL et al. Novel technologies for emission reduction complement conservation agriculture to achieve negative emissions from row-crop production. Proceedings of the National Academy of Sciences. 2021;**118**(28):e2022666118. DOI: 10.1073/pnas.2022666118

[74] Benavides PT, Cai H, Wang M, Bajjalieh N. Life-cycle analysis of soybean meal, distiller-dried grains with solubles, and synthetic amino acid-based animal feeds for swine and poultry production. Animal Feed Science and Technology. 2020;**268**:114607

[75] Decock C. Mitigating nitrous oxide emissions from corn cropping systems in the midwestern U.S.: Potential and data gaps. Environmental Science & Technology. 2014;**48**(8):4247-4256

[76] Crutzen PJ, Mosier AR, Smith KA, Winiwarter W. N_2O Release from Agro-biofuel Production Negates Global Warming Reduction by Replacing Fossil Fuels. Chapter 12 in: Paul J. Crutzen: A pioneer on atmospheric chemistry and climate change in the Anthropocene. Crutzen PJ, Brauch HG. Editors. SpringerNature Nobel Laureates. 2016;**50**:227-238

[77] Berry W. Chapter 6: The use of energy. In: The Unsettling of America: Culture and Agriculture. 1986 edition Sierra Club Books Softcover. San Francisco: Sierra Club Books; 1977. pp. 81-95

[78] Galston AW. Photosynthesis as a basis for life support on Earth and in space. BioScience. 1992;**42**(7):490-493. (Figure 2, pp. 492). DOI: 10.2307/1311878

[79] Ferl R, Wheeler R, Levine HG, Paul AL. Plants in space. Current Opinion in Plant Biology. 2002;**5**(3):258-263 (Figure 1, pp. 259). DOI: 10.1016/S1369-5266(02)00254-6

[80] Wheeler RM. Plants for human life support in space: From Myers to Mars. Gravitational and Space Biology. 2010;**23**(2):25-36

[81] Haeuplik-Meusburger S, Paterson C, Schubert D, Zabel P. Greenhouses and their humanizing synergies. Acta Astronautica. 2014;**96**:138-150 (Fig. 5, pp. 143)

[82] Jones J, Verma B, Basso B, Mohtar R, Matlock M. Transforming food and agriculture to circular systems: A perspective for 2050. Resource Magazine. 2021;**28**(2):7-9

[83] Aso SN. Mitigation of external costs of inorganic fertilizers with liquid fraction digestate. Biomass Conversion and Biorefinery. 2022;**2022**. DOI: 10.1007/s13399-022-02497-y

[84] Palahi M. Why the world needs a 'circular bioeconomy' –for jobs, biodiversity and prosperity. In: The Jobs Reset Summit. Cologny, Switzerland: World Economic Forum; 2020. [Accessed: March 1, 2023]

[85] Renewable Energy Policy Network for the 21st Century (REN21). Renewables 2019 Global Status Report. Paris: REN21 Secretariat; 2019. p. 335

[86] International Energy Agency (IEA). Technology Roadmap: Delivering Sustainable Bioenergy. Paris: OECD/IEA; 2017. p. 89

[87] Eurobsev'er. Biofuels Barometer. Paris, France: Eurobsev'er; 2015. p. 16 [Accessed: January 5, 2016]

[88] World Biogas Association -UK. Global Potential of Biogas. London: WBA; 2019. p. 56

[89] World Bioenergy Association -Sweden. Global Bioenergy Statistics 2021. Stockholm: WBA; 2021. p. 50

[90] World Bioenergy Association -Sweden. Biomethane vision document: A 5 point plan to scale up biomethane globally. Stockholm: WBA; 2023. p. 7 [Accessed January 31, 2023]

[91] International Hydropower Association. 2023 World Hydropower Outlook: Opportunities to Advance Net Zero. London: IHA; 2023. p. 71

[92] EurObserv'ER. Photovoltaic Barometer. Paris: European Commission; 2022. p. 12

[93] EurObserv'ER. Wind energy Barometer. Paris: European Commission; 2022. p. 14

[94] Abbasi T, Tauseef SM, Abbasi SA. A brief history of anaerobic digestion and "biogas". Chapter 2. In: Biogas Energy. New York, NY: Springer Briefs in Environmental Science; 2012. pp. 11-23

[95] Grando RL et al. Technology overview of biogas production in anaerobic digestion plants: A European evaluation of research and development. Renewable and Sustainable Energy Reviews. 2017;**80**:44-53

[96] Al Seadi T, Lukehurst C. Quality management of digestate from biogas

plants used as fertilizer. IEA Bioenergy. 2012;**2012**:4-36

[97] Aso SN. Synergistic Enzymatic Hydrolysis of Cassava starch and Anaerobic Digestion of Cassava Waste. PhD Dissertation. Gainesville, FL, USA: Univ FL, Dept Ag Bio Eng; 2013. Available from: https://search.proquest.com/docview/1547360007/fulltextPDF/29A82F37217B4D11PQ/1?accountid=28594

[98] Aso SN. Digestate: The coproduct of biofuel production in a circular economy, and new results for cassava peeling residue digestate. Chapter 13. In: Taner T, Tiwari A, Ustun TS, editors. Renewable Energy – Technologies and Applications. London: IntechOpen; 2021. pp. 199-225. DOI: 10.5772/intechopen.91340

[99] Lukehurst CT, Frost P, Al Seadi T. Utilisation of digestate from biogas plants as biofertiliser. IEA Bioenergy. 2010;**2010**:1-22

[100] Lettinga G. Digestion and degradation, air for life. Water Science and Technology. 2001;**44**(8):157-176

[101] Aso SN, Teixeira AA, Achinewhu SC. Cassava residues could provide sustainable bioenergy for cassava producing nations. Chapter 13. In: Waisundara VY, editor. Cassava. Rijeka, Croatia: InTechOpen; 2018. pp. 219-240. DOI: 10.5772/intechopen.72166

[102] Aso SN, Achinewhu SC, Iwe MO. Global fertilizer contributions from specific biogas coproduct. Chapter 5. In: Abomohra AE, Salama E, editors. Biogas – Basics, Integrated Approaches, and Case Studies. London: IntechOpen; 2022. pp. 71-92. DOI: 10.5772/intechopen.101543

[103] Pretty J. Intensification for redesigned and sustainable agricultural systems. Science. 2018;**362**:eaav0294

[104] Al Mamun S, Nasrat F, Debi MR. Integrated farming system: Prospects in Bangladesh. Journal of Environmental Science and Natural Resources. 2011;**4**(2):127-136

[105] Zhu T, Curtis J, Clancy M. Promoting agricultural biogas and biomethane production: Lessons from cross-country studies. Renewable and Sustainable Energy Reviews. 2019;**114**:109332

[106] Schreefel L, Schulte RPO, De Boer IJM, Schrijver AP, Van Zanten HHE. Regenerative agriculture – the soil is the base. Global Food Security. 2020;**26**:100404

[107] O'Donoghue T, Minasny B, McBratney A. Regenerative agriculture and its potential to improve farmscape function. Sustainability. 2022;**14**(10):5815. DOI: 10.3390/su14105815

[108] Preston TR. Future strategies for livestock production in tropical third world countries. Ambio. 1990;**19**(8):390-393. Available from: https://www.jstor.org/stable/4313747

[109] Dale BE et al. Biogasdoneright™: an innovative new system is commercialized in Italy. Biofuels, Bioproducts and Biorefining. 2016;**10**(4):341-345. DOI: 10.1002/bbb.1671

[110] Soliman NF. Aquaculture in Egypt under Changing Climate: Challenges and Opportunities. Alexandria, Egypt: Alexandria Research Center for Adaptation to Climate Change (ARCA); 2017

[111] Shefat SHT. Integrated Aqua-Farming in Bangladesh: SWOT Analysis. Acta Scientific Agriculture. 2018;**2**:112-118

[112] LaCanne CE, Lundgren JG. Regenerative agriculture: Merging farming and natural resource conservation profitably. PeerJ. 2018;**6**:e4428. DOI: 10.7717/peerj.4428

[113] Koppelmäki K, Helenius J, Schulte RP. Nested circularity in food systems: A Nordic case study on connecting biomass, nutrient and energy flows from field scale to continent. Resources, Conservation and Recycling. 2021;**164**:105218. DOI: 10.1016/j.resconrec.2020.105218

[114] Nestlé. Nestlé global virtual event: Implementing regenerative agriculture at scale Webinar. Ian Welsh, Moderator. 2023. Available from: https://www.youtube.com/watch?v=4xtrdXzgiho [Accessed: April 27, 2023]

[115] Valli L et al. Greenhouse gas emissions of electricity and biomethane produced using the Biogasdoneright™ system: Four case studies from Italy. Biofuels, Bioproducts and Biorefining. 2017;**11**(5):847-860

[116] Meyer BL. Farming in space for future space settlements: Not just glob and salad! In: Space Farming. Washington, D.C. USA: National Space Society Space Forum; 28th October 2021. Available from: https://youtu.be/s2oNVOhfS-E

[117] Lal R. Regenerative agriculture for food and climate. Journal of Soil and Water Conservation. 2020;**75**(5):123A-124A. DOI: 10.2489/jswc.2020.0620A

[118] Schulte LA et al. Meeting global challenges with regenerative agriculture producing food and energy. Nature Sustainability. 2002;**5**(5):384-388. DOI: 10.1038/s41893-021-00827-y

[119] Tol RS. Targets for global climate policy: An overview. Journal of Economic Dynamics and Control. 2013;**37**(5):911-928. DOI: 10.1016/j.jedc.2013.01.001

[120] Bourland CT, Fohey MF, Kloeris VL, Rapp RM. Designing a food system for Space Station Freedom. Food Technology. 1989;**1989**:76-81

[121] Evert MF, Glaus-Late KD, Hill SD, Bourland CT. Food provisioning considerations for long duration space missions and planet surface habitation. In: 22nd International Conference on Environmental Systems; 13-16 July 1992. Seattle, Washington: SAE, Warrendale; 1992. Paper No. 921415

[122] Aso SN. Requirements of Food for the Space Environment. Washington, DC: US Copyrights Office. 2001; Reg. #: TXu 1 -846-985

[123] Finkelstein B, Taylor AA. Food, nutrition and the space traveler. The American Journal of Clinical Nutrition. 1960;**8**(6):793-800. DOI: 10.1093/ajcn/8.6.793

[124] Klicka MV, Smith MC. Food for US manned space flight. US army Natick Research and Development Laboratories. US Army: Natick; 1982 Report No. NATICK/TR – 82/019

[125] NASA. Food for Space Flight. NASA Facts. NASA Headquarters. Washington, D.C. USA; 1982. Publication No. NF-133/6-82

[126] CARRAR. Nutremsi. Final report of the Study on Food Management for EMSI. ESA Headquarters. Paris, France; 1990; ESA Contract No. 8697/89/F/BZ/(SC).

[127] Sandal GM, Leon GR, Palinkas L. Human challenges in polar and space environments. Environmental Science and Biotechnology. 2006;**5**:281-296

[128] Odeh R, Guy CL. Gardening for therapeutic people-plant interactions during long-duration space missions. Open Agriculture. 2017;**2**(1):1-13. DOI: 10.1515/opag-2017-0001

[129] Smith MC et al. Apollo food technology. Chapter 1. In: Section VI

in: Biomedical Results of Apollo (NASA SP-368). Washington, DC: Scientific and Technical Information Office, NASA; 1975. pp. 437-468

[130] Helms CC. A survey of launch services 2016-2020. AIAA Propulsion and Energy 2020 Forum. AIAA. 2020. p. 21. DOI: 10.2514/6.2020-3532

[131] Strayer RF, Finger BW, Alazraki MP. Evaluation of an anaerobic digestion system for processing CELSS crop residues for resource recovery. Advances in Space Research. 1997;**20**(10):2009-2015

[132] Chynoweth DP, Owens JM, Teixeira AA, Pullammanappallil P, Luniya SS. Anaerobic digestion of space mission wastes. Water Science and Technology. 2006;**53**(8):177-185

[133] Benvenuti AJ, Drecksler SR, Gupta SS, Menezes AA. Design of Anaerobic Digestion Systems for Closed Loop Space Biomanufacturing. Emmaus, PA USA: 2020 International Conference on Environmental Systems; 2020

[134] Monje O, Stutte GW, Goins GD, Porterfield DM, Bingham GE. Farming in space: Environmental and biophysical concerns. Advances in Space Research. 2003;**31**(1):151-167

[135] Di Bartolo A, Infurna G, Dintcheva NT. A review of bioplastics and their adoption in the circular economy. Polymers. 2021;**13**(8):1229 (Figure 2, pp. 19 of 26)

[136] Knie M et al. Approaches to assess the suitability of Zooplankton for bioregenerative life support systems. Chapter 11. In: Russomano T, Rehnberg L, editors. Into Space – A Journey of How Humans Adapt and Live in Microgravity. London: IntechOpen; 2018. pp. 171-207

[137] Verbeelen T, Leys N, Ganigué R, Mastroleo F. Development of nitrogen recycling strategies for bioregenerative life support systems in space. Frontiers in Microbiology. 2021;**12**:700810

[138] McElroy RD. CELSS and regenerative life support for manned missions to MARS. NASA. Marshall Space Flight Center Manned Mars Missions. Working Group Papers. 1986;**1**:1-4

[139] Olson RL, Oleson MW, Slavin TJ. CELSS for advanced manned mission. HortScience. 1988;**23**(2):275-286. DOI: 10.21273/HORTSCI.23.2.275

[140] Mitchell CA. Bioregenerative life-support systems. The American Journal of Clinical Nutrition. 1994;**60**(5):820S-824S. DOI: 10.1093/ajcn/60.5.820S

[141] This H. Molecular gastronomy is a scientific discipline, and note by note cuisine is the next culinary trend. Flavor. 2013;**2**:1-8. Available from: http://www.flavourjournal.com/content/2/1/1

[142] Burke R, Danaher P. Note by note: A new revolution in cooking. 2016. Available from: https://arrow.tudublin.ie/cgi/viewcontent.cgi?article=1060&context=dgs [Accessed April 16, 2023]

[143] Brown CS, Sederoff HW, Davies E, Ferl RJ, Stankovic B. Plan(t)s for space exploration. Chapter 9. In: Gilroy S, Masson PH, editors. Plant Tropisms. Oxford, UK: Blackwell Publishing; 2008. pp. 183-195

[144] McNulty MJ et al. Molecular pharming to support human life on the moon, mars, and beyond. Critical Reviews in Biotechnology. 2021;**41**(6):849-864

[145] Walkinshaw CH, Sweet HC, Venketeswaran S, Horne WH. Results of Apollo 11 and 12 quarantine studies

on plants. Bioscience. 1970;**20**(24):1297-1302. DOI: 10.2307/1295365

[146] Walkinshaw CH, Johnson PH. Analysis of vegetable seedlings grown in contact with Apollo 14 Lunar Surface Fines. HortScience. 1971;**6**(6):532-535. DOI: 10.21273/HORTSCI.6.6.532

[147] Taylor GR, Mieszkuc BJ, Simmonds RC, Walkinshaw CH. Quarantine testing and biocharacterization of lunar materials. Chapter 2, Section V. In: Biomedical Results of Apollo (NASA SP-368). Washington, DC: Scientific and Technical Information Office, NASA; 1975. pp. 425-434

[148] Paul AL, Elardo SM, Ferl R. Plants grown in Apollo lunar regolith present stress-associated transcriptomes that inform prospects for lunar exploration. Communications Biology. 2022;**5**(1):382

[149] Shen Y, Guo S, Zhao P, Wang L, Wang X, Li J, et al. Research on lettuce growth technology onboard Chinese Tiangong II Spacelab. Acta Astronautica. 2018;**144**:97-102

[150] Gulimova V et al. Reptiles in space missions: Results and perspectives. International Journal of Molecular Sciences. 2019;**20**(12):3019. DOI: 10.3390/ijms20123019

[151] Mecklin J. A time of unprecedented danger: It is 90 seconds to midnight. 2023 Doomsday Clock Statement. Bulletin of the Atomic Scientists. 2023. pp. 1-9. [Accessed: January 28, 2023]

[152] Zhu H et al. Biohydrogen production by anaerobic co-digestion of municipal food waste and sewage sludges. International Journal of Hydrogen Energy. 2008;**33**(14):3651-3659

[153] Pan SY, Tsai CY, Liu CW, Wang SW, Kim H, Fan C. Anaerobic co-digestion of agricultural wastes toward circular bioeconomy. Iscience. 2021;**24**(7):102704

[154] Wang W, Xie L, Luo G, Zhou Q, Lu Q. Optimization of biohydrogen and methane recovery within a cassava ethanol wastewater/waste integrated management system. Bioresource Technology. 2012;**120**:165-172. DOI: 10.1016/j.biortech.2012.06.048

[155] Scapini T et al. Bioethanol and biomethane production from watermelon waste: A circular economy strategy. Biomass and Bioenergy. 2023;**170**:106719

[156] Baccioli A et al. Small scale bio-LNG plant: Comparison of different biogas upgrading techniques. Applied Energy. 2018;**217**:328-335. DOI: 10.1016/j.apenergy.2018.02.149

[157] Seong MS et al. Optimization of pilot-scale 3-stage membrane process using asymmetric polysulfone hollow fiber membranes for production of high-purity CH_4 and CO_2 from crude biogas. Chemical Engineering Journal. 2020;**384**:123342. DOI: 10.1016/j.cej.2019.123342

[158] Aso SN, Pullammanappallil PC, Teixeira AA, Welt BA. Biogasification of cassava residue for on-site biofuel generation for food production with potential cost minimization, health and environmental safety dividends. Environmental Progress & Sustainable Energy. 2019;**2019**:ep.13138. DOI: 10.1002/ep.13138

[159] European Biogas Association (EBA). Digestate factsheet: The value of organic fertilizers for Europe's economy, society and environment. Brussels: EBA; 2015. p. 4

[160] Bauer SE et al. Nitrate aerosols today and in 2030: Importance relative to other aerosol species and tropospheric ozone. Atmospheric Chemistry and Physics Discussions. 2007;**7**(2):5553-5593

[161] Hauglustaine DA, Balkanski Y, Schulz M. A global model simulation of present and future nitrate aerosols and their direct radiative forcing of climate. Atmospheric Chemistry and Physics. 2014;**14**(20):11031-11063

[162] Tuomisto HL, Helenius J. Comparison of energy and greenhouse gas balances of biogas with other transport biofuel options based on domestic agricultural biomass in Finland. Agricultural and Food Science. 2008;**17**(3):240-251. Available from: https://orgprints.org/id/eprint/15981/

[163] Uusitalo V et al. Carbon footprint of selected biomass to biogas production chains and GHG reduction potential in transportation use. Renewable Energy. 2014;**66**:90-98

[164] Timonen K et al. LCA of anaerobic digestion: Emission allocation for energy and digestate. Journal of Cleaner Production. 2019;**235**:1567-1579. DOI: 10.1016/j.jclepro.2019.06.085

[165] Budzianowski WM, Postawa K. Renewable energy from biogas with reduced carbon dioxide footprint: Implications of applying different plant configurations and operating pressures. Renewable and Sustainability Energy Reviews. 2017;**68**:852-868

[166] Głąb L, Sowiński J. Sustainable production of sweet sorghum as a bioenergy crop using biosolids taking into account greenhouse gas emissions. Sustainability. 2019;**11**(11):3033

[167] Burg V et al. Agricultural biogas plants as a hub to foster circular economy and bioenergy: An assessment using substance and energy flow analysis. Resources, Conservation and Recycling. 2023;**190**:106770

[168] West JJ, Fiore AM. Management of tropospheric ozone by reducing methane emissions. Environmental Science & Technology. 2005;**39**(13):4685-4691

[169] Van Dingenen R et al. The global impact of ozone on agricultural crop yields under current and future air quality legislation. Atmospheric Environment. 2009;**43**(3):604-618

[170] Crutzen PJ, Birks JW. The atmosphere after a nuclear war: Twilight at noon. Chapter 5. In: Paul J. Crutzen: A Pioneer on Atmospheric Chemistry and Climate Change in the Anthropocene. Crutzen PJ, Brauch HG. Eds. Cham, Switzerland: Springer Nature Nobel Laureates; 2016. pp. 125-152

[171] Rabl A, Eyre N. An estimate of regional and global O_3 damage from precursor NOx and VOC emissions. Environment International. 1998;**24**(8):835-850

[172] Sovacool BK, Dworkin MH. Energy justice: Conceptual insights and practical applications. Applied Energy. 2015;**142**:435-444. DOI: 10.1016/j.apenergy.2015.01.002

[173] Prospero JM. Atmospheric dust studies on Barbados. Bulletin of the American Meteorological Society. 1968;**49**(6):645-652. DOI: 10.1175/1520-0477-49.6.645

[174] Schmidt A et al. Satellite detection, long-range transport, and air quality impacts of volcanic sulfur dioxide from the 2014-2015 flood lava eruption at Bárðarbunga (Iceland). Journal of Geophysical Research-Atmospheres. 2015;**120**(18):9739-9757. DOI: 10.1002/2015JD023638

[175] Bègue N et al. Statistical analysis of the long-range transport of the 2015 Calbuco volcanic plume from ground-based and space-borne observations. Annales Geophysicae. 2020;**38**(2):395-420

[176] Crutzen PJ, Stoermer EF. The "Anthropocene". International Geosphere-Biosphere Programme (IGBP) Newsletter. 2000;**41**:17-18

[177] Karnieli A et al. Temporal trend in anthropogenic sulfur aerosol transport from central and eastern Europe to Israel. Journal of Geophysical Research-Atmospheres. 2009;**114**:D10

[178] Lewis SC, Karoly DJ. The role of anthropogenic forcing in the record 2013 Australia-wide annual and spring temperatures. [In "Explaining Extremes of 2013 from a Climate Perspective"]. Bulletin of the American Meteorological Society. 2014;**95**(9):S31-S34

[179] Marzeion B et al. Attribution of global glacier mass loss to anthropogenic and natural causes. Science. 2014;**345**(6199):919-921

[180] Waters CN et al. The Anthropocene is functionally and stratigraphically distinct from the Holocene. Science. 2016;**351**(6299):aad2622. DOI: 10.1126/science.aad2622

[181] Barriopedro D et al. The hot summer of 2010: Redrawing the temperature record map of Europe. Science. 2011;**332**:220-224. DOI: 10.1126/science.1201224

[182] Foresti E. Perspectives on anaerobic treatment in developing countries. Water Science and Technology. 2001;**44**(8):141-148

[183] Lema JM, Omil F. Anaerobic treatment: A key technology for a sustainable management of wastes in Europe. Water Science and Technology. 2001;**44**(8):133-140

[184] Basso B et al. Economic and environmental evaluation of site-specific tillage in a maize crop in NE Italy. European Journal of Agronomy. 2011;**35**(2):83-92

[185] Rodrigues M et al. Legacy phosphorus and no tillage agriculture in tropical oxisols of the Brazilian savanna. Science of the Total Environment. 2016;**542**(Part B):1050-1061

[186] Gustavsson J et al. Global food losses and food waste—Extent, causes and prevention. Rome, Italy: FAO; 2011. p. 29. Available from: http://www.fao.org/3/mb060e/mb060e.pdf

[187] Oakes K. How cutting your food waste can help the climate. BBC Futures. 2020

[188] Grossman GM, Krueger AB. Economic growth and the environment. The Quarterly Journal of Economics. 1995;**110**(2):353-377. DOI: 10.2307/2118443

[189] Cole MA, Rayner AJ, Bates JM. The environmental Kuznets curve: An empirical analysis. Environment and Development Economics. 1997;**2**(4):401-416

[190] Dinda S. Environmental Kuznets curve hypothesis: A survey. Ecological Economics. 2004;**49**(4):431-455

[191] Al-Mulali U, Ozturk I. The investigation of environmental Kuznets curve hypothesis in the advanced economies: The role of energy prices. Renewable and Sustainable Energy Reviews. 2016;**54**:1622-1631

Chapter 16

Production of Biochar from Oilseed Residue (Deoiled Cakes): State-of-the-Art

Mattaparthi Lakshmi Durga, Lalita Pal and Aseeya Wahid

Abstract

Even today, the generation of chemicals and energy is still reliant on fossil-based resources in industrialized countries. Biomass could be a valuable renewable energy source that could reduce dependence on fossil fuels as well as provide a significant reduction of carbon dioxide and greenhouse gas emissions. In this scenario, residue from natural oil extraction units is uplifted to produce biofuels as replacement of fossil fuels. In the process of bio-refinery, well established technologies were presented. Those are thermochemical treatment (pyrolysis, liquefaction, gasification, etc.), anaerobic digestion, catalysis, etc. Especially, importance is given to pyrolysis as it is the feasible technique to utilize residue and to produce wealthy products. The role of intrinsic bio-polymers in quantity of final pyrolytic products was discussed. Major process parameters were critically elucidated, however, the investigation of advanced pyrolysis technologies requires further research.

Keywords: pyrolysis, biochar, oilseed residue, bio-polymers, temperature

1. Introduction

The future energy demand is projected to rise further in accordance with ever-growing population. In addition to this, global energy consumption relies on non-renewable sources (like coal, petroleum, and natural gas) and those depleting day by day [1]. Moreover, the growth in greenhouse gas emissions from fossil fuel usage generates major menace to the climate. Therefore, it is imperative to search for environmentally and sustainable resources and utilize them to their full potential in light of biofuels. Biomass, as a renewable energy source, can be widely viewed as a promising alternative, because of its numerous applications, especially biofuels and bio-fertilizers [2]. There are several bio-refinery platforms that will be capable of processing biomass feed-stocks, which are the key intermediates between biomass feed-stocks and their value-added products. The bio-refineries usually add value to biomass supply chains through the production of various bio-based products. Selection of biomass has numerous advantages such as (a) abundant availability of precursor material; (b) environmentally friendliness; and (c) biodegradability and sustainability. In a typical bio-refinery, by using appropriate technologies, biomass by-products can be processed to generate various high-valued fuels. The bioenergy

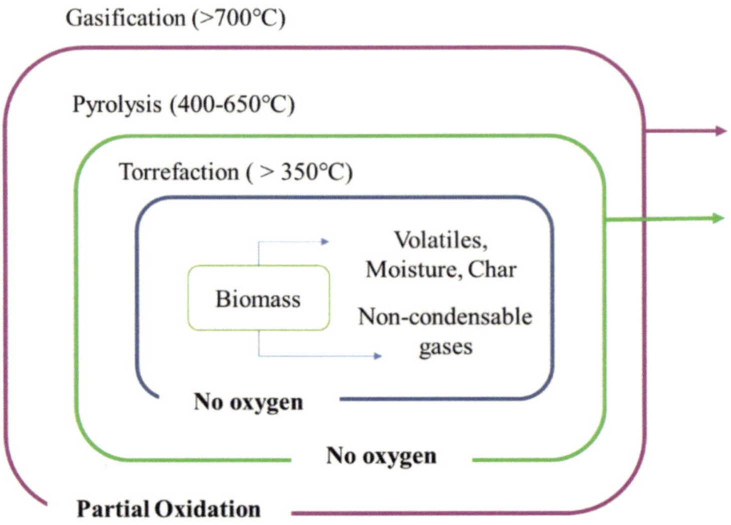

Figure 1.
Comparison of common thermochemical conversion techniques.

sector has many major platforms, for example, syngas from gasification, fuel and fertilizer from anaerobic digestion, sugars from starch, extraction of cellulose, hemicellulose and lignin from biomass, solid carbon and bio-oil from pyrolysis, etc. A comparative ideology of thermochemical conversion technologies is illustrated in **Figure 1**.

Much research has been conducted on converting industrial and agricultural waste into valuable products [3, 4]. Still, this research topic is emerging because of its abundant scope for future energy demands. However, agricultural residue is prominent in energy conversion policies. Primarily, rice straw, rice husk, wheat straw, sugar cane bagasse, bamboo, areca nut shells, corn cob, coconut shells, etc., were used to produce energy fuels [5–9]. Apart from that, a vast range of oilseeds can be grown in India due to its diverse agro-ecological conditions and these all are coming under two major categories, such as edible oil seeds and non-edible oil seeds. The oil seeds are used for extraction of either edible oils or other purposes. In general, extraction of kg oil seeds can produce 250–350 g of oil [7], however, 650 g of residue seed cake is also obtained in the process of oil extraction [8].

Consequently, almost 65% of residue will available at the end of the extraction process. Some portion of extracted residue from edible seeds such as soybean, ground nut, flax seeds, sunflower, and etc., can be used as feed material for animal and the remaining unused. But, the residue from non-edible seeds contains toxic compounds, which limit their utilization as edible material. Hence, the oilseed residue (OSR) especially from non-edible seeds can be a potential material to produce biofuels [9], instead of unused. To convert oil seed residue (OSR) into valuable biofuels, thermochemical or bio-chemical conversion techniques can be used [10].

Thermochemical conversion involves thermal cracking of chemical organic bonds of biomass (OSR) into solids, gases, or liquid form [10]. Possible thermochemical conversion processes are torrefaction, pyrolysis, gasification, and liquefaction. The interest has been increasing day by day toward thermochemical conversion, because it is user friendly and sustainable too. However, in bio-chemical conversion, specified bacteria (yeast) may be used to digest or convert biomass into different biofuels. Bio-chemical conversion includes anaerobic digestion, alcoholic fermentation, transesterification, and so on. Compared to bio-chemical conversion methods, thermochemical process has superior qualities. The final product obtained from thermochemical process has higher carbon content, lower volatile matter, and reduced contaminants. Also, it is more energy-efficient method and can mitigate the release of harmful emissions and pollutants. Hence, the subsequent categories in thermochemical conversion technique are discussed as follows.

2. Value addition technologies

2.1 Pyrolysis

In pyrolysis, the organic material and volatiles present in the biomass undergo an irreversible thermochemical decomposition reaction to produce biofuels. Pyrolysis technologies can be classified as slow, flash, and fast [1]. The fast heating rate and short vapor residence times result in rapid exit of the volatile hydrocarbon vapors from the reactor and condensation into bio-oil, whereas slow pyrolysis results in greater biochar production due to increasing of carbonization of biomass due to slow heating rate and longer vapor residence time [11]. However, slow pyrolysis is the prominent technique than others to produce biochar. The remaining pyrolysis techniques enable the production of bio-oil and gases. A flow chart of pyrolysis is present in **Figure 2**.

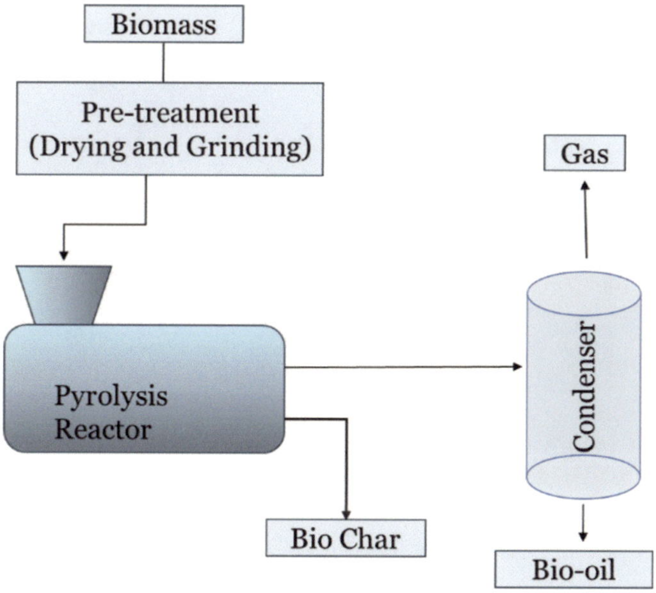

Figure 2.
A simple flow chart of pyrolysis process.

2.2 Torrefaction

Torrefaction is the one of the clean thermal treatments of biomass. Further, torrefaction actively removes the moisture content from biomass, however, similarly improves the efficiency of subsequent thermochemical conversion methods [12]. Biomass that has been torrefied has a number of superior quality over raw biomass, including higher grindability, lower hydrophobicity, higher heating value, and pelletization properties [13]. In the process of torrefaction, several factors influence their performance, such as, torrefaction temperature, holding time, heating rate, environment, moisture content, particle size [14]. **Figure 3** illustrates general flow chart of torrefaction process.

2.3 Gasification

Gasification is another of the thermochemical conversion method, used to produce gaseous material from OSR [11]. Gasification flow diagram can be observed in **Figure 4**. Though, the produced gaseous material can be used for heat, electricity, methane and hydrogen generation [15]. However, gasification not only produces gaseous products but also produces bio-chemical liquids. In general, gasification occurs at high temperature value, tar content in the gas is very low, and the cleaning process is very simple [16].

2.4 Liquefaction

Other than torrefaction, pyrolysis, gasification, an alternative thermochemical process for converting biomass to biofuel is liquefaction [17]. In order to form bio-crude oil, biomass must be hydrogenated and thermally disintegrated at high

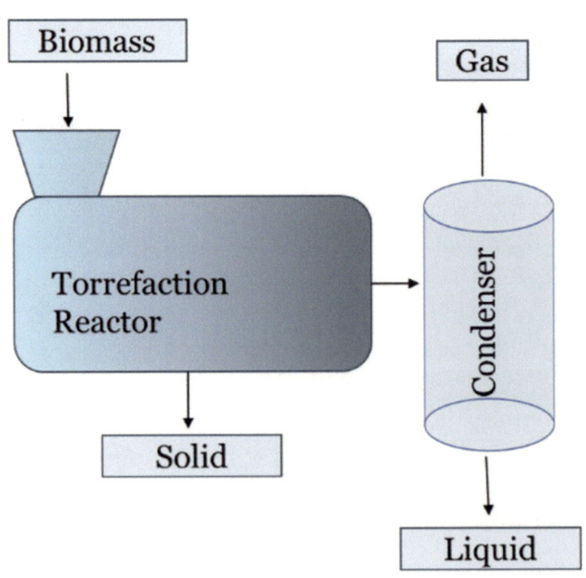

Figure 3.
Illustration of torrefaction process.

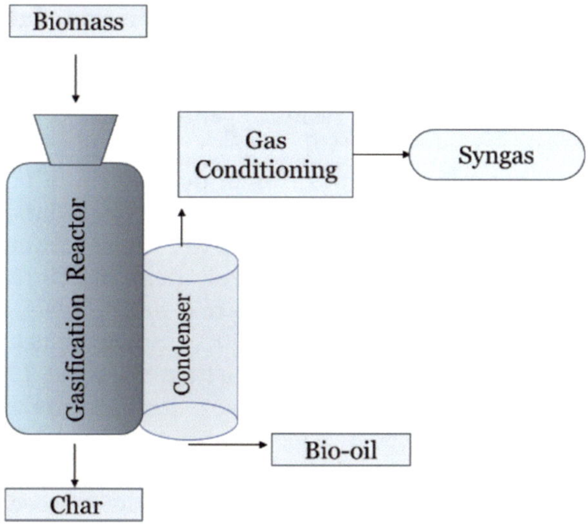

Figure 4.
Process flow diagram of gasification.

pressures. In the process of hydrothermal liquefaction, solid waste contains lots of moisture and it is transformed into bio-crude oil in the presence of some catalysts [18]. The temperature and pressure ranges for liquefaction generally fall within the ranges of 240–380°C and 5–30 MPa, respectively [11]. In **Figure 5**, a view of liquefaction process is reported.

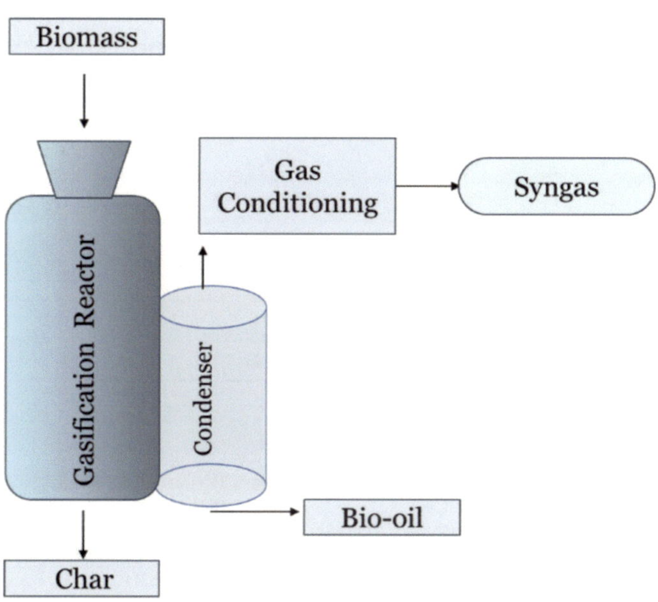

Figure 5.
Simple flow diagram of liquefaction.

3. Role of bio-polymeric compounds in characteristics of the oil seed formed

Lignocellulosic materials (oilseed residue) generally consist of several intrinsic bio-polymeric components [19]. Among others, lignin, cellulose, and hemicellulose play a vital role in composition of pyrolytic products. When the lignocellulosic material undergoes heat treatment, the degradation trends of each bio-polymer decompose differently [20]. Moreover, the percentage of total carbon also depends on the degradation of these bio-polymers. Briefly, hemicellulose is susceptible to lower temperatures of torrefaction or pyrolysis, average range of the temperature for hemicellulose is 150–300°C. However, cellulose also gets volatized at slighter high temperatures than the hemicellulose. Lignin is a complex long chain bio-polymer and it starts to decompose partially at the temperature of cellulose and hemicellulose, however, it gets decomposed at higher temperature i.e., > 500°C [21].

Table 1 reports that the liquid production from safflower seed cake (residue) is more, when the average percentages of cellulose and hemicellulose are 40% and 16%, respectively [30]. However, in another study, the authors suggested that if the mustard oilseed residue contains an average percentage of cellulose and hemicellulose at 52.21% and 10.85%, respectively, the residue is potentially suitable for bio-fuel production [8]. In the same way, a comparative study of African star apple seed (ASA) and silk cotton seed residue (SCS), bio-oil production is more in ASA because of its high lignin content at optimum temperature of 400°C [23]. Kinetic studies of Jatropha de-oiled cake residue were conducted to examine the utilization of Jatropha residue for bio-fuel production [25]. Withal, bio-polymeric compositions in OSR affect the yield of outcome of pyrolysis. The further studies of variations with heat treatment could have great scope in future [31, 32]. In addition, proximate and ultimate analysis also plays a key role in pyrolysis. Here, elucidation of their role and importance is not mentioned, as they are mentioned in many literatures [30, 33].

Residue	Lignin (%)	Cellulose (%)	Hemicellulose (%)	Reference
Safflower seed press cake	26.7	40	16	[22]
Mustard oil residue	15.27	52.21	10.85	[8]
African star apple seed cake	10.0	16.21	14.0	[23]
Silk cotton seed cake	4.0	19.14	19.33	
Raspberry seed cake	19.3	42.1	16.9	[24]
Jatropha de-oiled cake	24.9	53.5	16.6	[25]
Pongammia pinnata de-oiled cake	21.87	21.34	49.66	[26]
Black cumin seed cake	26.73	37.14	10.44	[27]
Raspberry Seed Cake	19.3	42.1	16.9	[28]
Radish seed cake	5.62	3.89	4.15	[29]

Table 1.
Bio-polymeric composition of oilseed residues.

4. Effect of process parameters on pyrolytic products

Specifically, pyrolysis reactor selection is the principal factor to decide the quality and quantity of final products [34]. From the studies, it is observed that fixed bed reactor (slow pyrolysis) is mostly used to produce biofuels as it is simple in operation and maintenance as well easy in construction. However, advanced pyrolytic techniques such as plasma pyrolysis, microwave pyrolysis, ultrasonic pyrolysis, solar-based pyrolysis are being explored and they will likely emerge as significant concepts in the future.

From tabulated studies, it is observed that biochar yield increases at slow pyrolysis rate, i.e., approximately 350–400°C at 5°C/min for 30 minutes of holding time. Whereas, bio-oil and bio-gas occur at little higher temperatures and higher heating rates (>500°C for 20°C/min). The above phenomenon clarifies that slower heating rate provides enough time to settlement of volatiles during thermal treatment, but not in higher heating rates [35]. Similarly, pyrolysis of safflower seed cake occurs at 400–600°C at 50°C interval and at 10, 30, and 50°C/min. The authors claimed that with increasing heating rates, bio-oil yield increases, but with increase in temperature from 400 to 600°C, yield of bio-oil decreases. Because, at higher temperatures, secondary reactions occur and lead to an increase in the content of gases rather than bio-oil. However, biochar yield decreases with increment in both heating rate and temperature [22]. In **Table 2**, the effect of different thermal process parameters and their influence on pyrolytic output are listed.

OSR	Reactor	Temperature (°C)/Heating rate (°C/min)	Time (min)	Pyrolysis products			References
				Solid (%)	Liquid (%)	Gas (%)	
Flaxseed cake	Fixed bed	350–650/5, 20	30	44.6	43.3	12.1	[35]
Safflower seed cake	Fixed bed	400–600/10, 30, 50	30	34.2	27.5	18.9	[22]
Sesame seed cake	Semi-Batch type	350–700/25	—	29	58.5	12.4	[36]
Neem seed cake	--do--	350–700/25	—	51.1	40.2	8.5	
Mustard seed cake	--do--	350–700/25	—	29.9	53.2	16.7	
Mahua seed cake	--do--	350–600/25		33	41.36	25.64	[37]
Sunflower seed cake	Heniz retort	550/7	—	—	34.8	—	[38]
Mahua seed cake	Furnace	400–600	—	26.43	30.9	—	[39]
Karanja seed cake	—	800/5, 10, 20	—	18.11	—	—	[40]

Table 2.
Process parameters of pyrolysis process.

5. Challenges and limitations

The conversion and utilization of oilseed residue could provide better scope for future economy. However, it encounters with several challenges and difficulties while processing the residue and those are discussed below.

5.1 Logistical challenges

Collecting oilseed waste from diverse sources presents a logistical challenge. Establishing an efficient collection system requires collaboration with local bodies and cooperatives. Optimizing biomass logistics, including transportation from collection points to processing facilities, is crucial to reduce costs and energy inputs.

5.2 Oilseed variability

OSR from different plant species exhibits significant variability in oil content due to various genetic, environmental, and agricultural factors. Understanding and managing this variability are crucial for ensuring a consistent and reliable feedstock supply. Plant breeding and selection, monitoring of crop growth, proper soil management, and suitable irrigation strategies can help mitigate the impact of environmental factors on oil yield [11].

5.3 Non-edible seed residue

The development of non-edible crops as potential feedstock for biodiesel presents unique challenges toward achieving self-reliance in energy security. These challenges include the forest origin of non-edible crops, making harvesting, collection, and transportation difficult. Additionally, lower fuel economy, seasonal availability, improper marketing channels, and high FFAs and moisture content in non-edible oils require pre-treatment before the transesterification process. The lack of post-harvest technologies further affects the oil quality of non-edible crops [41].

5.4 Seasonal nature and annual variability

The seasonal nature of oilseed cultivation results in fluctuations in feedstock supply, which can affect continuous production processes. Addressing these fluctuations requires effective planning and coordination of agricultural activities. Despite the challenges, non-edible crops offer opportunities for cultivating harsh and unused lands in developing countries, such as seashores, riverbanks, deserts, and wastelands. These crops could maximize land utilization for crop production. Implementation of advanced technology, particularly waterless extraction technology like supercritical fluids, offers potential solutions to avoid FFAs and moisture content in non-edible oil [42].

5.5 Efficient storage and self-decomposition

Proper storage of oilseed waste is essential to maintain its quality and energy potential. Prolonged on-site storage, however, can lead to self-decomposition and the release of harmful gases, posing environmental and safety risks. Developing appropriate storage techniques is critical to preserve the feedstock's quality [43].

5.6 Environmental impact and carbon cycle

A life cycle assessment (LCA) of the entire bioenergy production process is crucial to understand the environmental implications and carbon cycle. This assessment helps in identifying potential areas for improvement in terms of sustainability [44].

5.7 Acceptance and demand

Market and community acceptance of biochar and biofuels derived from oilseed waste are vital for sustainable production. Promoting sustainable agriculture practices and establishing standards can enhance demand and acceptability in the local market.

Overcoming the challenges in producing biochar and biofuels from oilseed waste requires collaborative efforts from researchers, policymakers, and industry stakeholders. By addressing these obstacles, we can unlock the full potential of oilseed waste as a valuable feedstock for bioenergy production, contributing to a more sustainable and environmentally friendly future. However, careful consideration is necessary to balance the use of agricultural crops and edible oils for biodiesel production to avoid potential conflicts between "food vs. fuel" [41, 45].

6. Conclusion

Different technologies such as pyrolysis, liquefaction, gasification, torrefaction are reported to convert oilseed residues as high-valued products. Withal, in terms of "waste-to-energy" technology, pyrolysis is considered one of the most cost-effective and environmentally benign techniques. In order to reduce GHG emissions, pollution, global warming, and fossil fuel dependence, bio-fuel products can be used as drop-in alternatives to fossil fuels or blended with fossil fuels. An idea on importance of bio-polymer compositions in preparation of pyrolytic products is reported. The different process parameters and their role in production process are reported. However, this article highlighted the selection of pyrolysis method and bio-polymer compositions in preparation of pyrolytic products. It critically explained the difference between process parameters depending on final product requirement. Yet, the research needs to establish toward the adaption of new methods of pyrolysis such as plasma, ultrasonic, microwave-assisted pyrolysis etc.

Acknowledgements

The authors are sincerely thankful to Director (ICAR-CIAE, Bhopal); Dean (IARI, New Delhi) for their support and encouragement.

Author contribution

All authors contributed to the study conception and design. Material preparation, initial draft preparation, conceptualization were performed by Mattaparthi Lakshmi Durga, Aseeya Wahid. The first draft of the manuscript was written by Mattaparthi Lakshmi Durga, Aseeya Wahid, Lalita.

Conflict of interest

The authors declare that they have no known competing financial interests or personal relationships that could have appeared to influence the work reported in this paper.

Author details

Mattaparthi Lakshmi Durga*, Lalita Pal and Aseeya Wahid
ICAR- Central Institute of Agricultural Engineering, Bhopal, India

*Address all correspondence to: lakshmidurga1225@gmail.com

References

[1] Rajpoot L, Tagade A, Deshpande G, Verma K, Geed SR, Patle DS, et al. An overview of pyrolysis of de-oiled cakes for the production of biochar, bio-oil, and pyro-gas: Current status, challenges, and future perspective. Bioresources Technology Reports. 2022;**19**:101205. DOI: 10.1016/j.biteb.2022.101205

[2] Gangil S, Bhargav VK. Influence of torrefaction on intrinsic bioconstituents of cotton stalk. Energy. 2018;**142**:1066-1073. DOI: 10.1016/j.energy.2017.10.128

[3] Chandrasekaran A, Ramachandran S, Subbiah S. Determination of kinetic parameters in the pyrolysis operation and thermal behavior of *Prosopis juliflora* using thermogravimetric analysis. Bioresource Technology. 2017;**233**:413-422. DOI: 10.1016/j.biortech.2017.02.119

[4] Gangil S. Positive transformations in intrinsic bioconstituents due to briquetting of soybean crop residues, energy. Sustainable Development. 2015;**27**:112-119. DOI: 10.1016/j.esd.2015.05.005

[5] Durga ML, Gangil S, Bhargav VK. Thermal influx induced biopolymeric transitions in paddy straw. Renewable Energy. 2022;**199**:1024-1032. DOI: 10.1016/j.renene.2022.09.054

[6] Song M, Jin B, Xiao R, Yang L, Wu Y, Zhong Z, et al. The comparison of two activation techniques to prepare activated carbon from corn cob. Biomass and Bioenergy. 2013;**48**:250-256. DOI: 10.1016/j.biombioe.2012.11.007

[7] Volli V, Singh RK, Murugan S. The use of mustard cake pyrolytic oil blends as fuel in a diesel engine. Waste and Biomass Valorization. 2014;**5**:661-668. DOI: 10.1007/s12649-013-9270-7

[8] Singh RK, Patil T, Pandey D, Sawarkar AN. Bioresource technology pyrolysis of mustard oil residue: A kinetic and thermodynamic study. Bioresource Technology. 2021;**339**:125631. DOI: 10.1016/j.biortech.2021.125631

[9] Mirpoor SF, Giosafatto CVL, Porta R. Biorefining of seed oil cakes as industrial co-streams for production of innovative bioplastics. A review. Trends in Food Science and Technology. 2021;**109**:259-270. DOI: 10.1016/j.tifs.2021.01.014

[10] Lee SY, Sankaran R, Chew KW, Tan CH, Krishnamoorthy R. Energy waste to bioenergy: A review on the recent conversion technologies. BMC. 2019;**2019**:1-22

[11] Jha S, Nanda S, Acharya B, Dalai AK. A review of thermochemical conversion of waste biomass to biofuels. Energies. 2022;**15**:1-23. DOI: 10.3390/en15176352

[12] Chen WH, Kuo PC. Torrefaction and co-torrefaction characterization of hemicellulose, cellulose and lignin as well as torrefaction of some basic constituents in biomass. Energy. 2011;**36**:803-811. DOI: 10.1016/j.energy.2010.12.036

[13] Tumuluru JS, Ghiasi B, Soelberg NR, Sokhansanj S. Biomass Torrefaction process, product properties, reactor types, and moving bed reactor design concepts. Frontier in Energy Research. 2021;**9**:1-20. DOI: 10.3389/fenrg.2021.728140

[14] Gan YY, Ong HC, Ling TC, Chen WH, Chong CT. Torrefaction of de-oiled Jatropha seed kernel biomass for solid fuel production. Energy. 2019;**170**:367-374. DOI: 10.1016/j.energy.2018.12.026

[15] Centi G, Perathoner S. Chemistry and energy beyond fossil fuels. A perspective view on the role of syngas from waste sources. Catalysis Today. 2020;**342**:4-12. DOI: 10.1016/j.cattod.2019.04.003

[16] Makwana JP, Pandey J, Mishra G. Improving the properties of producer gas using high temperature gasification of rice husk in a pilot scale fluidized bed gasifier (FBG). Renewable Energy. 2019;**130**:943-951. DOI: 10.1016/j.renene.2018.07.011

[17] Gollakota ARK, Kishore N, Gu S. A review on hydrothermal liquefaction of biomass. Renewable and Sustainable Energy Reviews. 2018;**81**:1378-1392. DOI: 10.1016/j.rser.2017.05.178

[18] Zhang S, Yang X, Zhang H, Chu C, Zheng K, Ju M, et al. Liquefaction of biomass and upgrading of bio-oil: A review. Molecules. 2019;**24**:1-30. DOI: 10.3390/molecules24122250

[19] Mandal S, Haydary J, Gangil S, Husar J, Jena PC, Bhattacharya TK. Inferences from thermogravimetric analysis of pine needles and its chars from a pilot-scale screw reactor. Chemical Papers. 2020;**74**:689-698. DOI: 10.1007/s11696-019-00998-1

[20] Gangil S. Distinct splitting of polymeric cellulosic signals in cashew shell: A TG-diagnosis. Cellulose. 2014;**21**:2913-2924. DOI: 10.1007/s10570-014-0309-0

[21] Amalina F, Razak ASA, Krishnan S, Sulaiman H, Zularisam AW, Nasrullah M. Biochar production techniques utilizing biomass waste-derived materials and environmental applications – A review. Journal of Hazardous Materials Advance. 2022;**7**:100134. DOI: 10.1016/j.hazadv.2022.100134

[22] Şensöz S, Angın D. Pyrolysis of safflower (*Charthamus tinctorius* L.) seed press cake: Part 1. The effects of pyrolysis parameters on the product yields. Bioresource Technology. 2008;**99**(13):5492-5497. DOI: 10.1016/j.biortech.2007.10.046

[23] Abdullahi M, Biswas B, Kumar J, Bhaskar T. Slow pyrolysis of Defatted Seeds Cakes of African star apple and silk cotton for production of bio-oil. Renewable Energy. 2020;**146**:1710-1716. DOI: 10.1016/j.renene.2019.07.145

[24] Smets K, Schreurs S, Carleer R, Yperman J. Valorization of raspberry seed cake by flash and slow pyrolysis: Product yield and characterization of the liquid and solid fraction. Journal of Analytical and Applied Pyrolysis. 2014;**107**:289-297. DOI: 10.1016/j.jaap.2014.03.014

[25] Sharma R, Sheth PN, Gujrathi AM. Kinetic modeling and simulation: Pyrolysis of Jatropha residue de-oiled cake. Renewable Energy. 2016;**86**:554-562. DOI: 10.1016/j.renene.2015.08.066

[26] Kumar D, Pant KK. Production and characterization of biocrude and biochar obtained from non-edible de-oiled seed cakes hydrothermal conversion. Journal of Analytical and Applied Pyrolysis. 2015;**115**:77-86. DOI: 10.1016/j.jaap.2015.06.014

[27] Şen N, Kar Y. Pyrolysis of black cumin seed cake in a fixed-bed reactor. Biomass and Bioenergy. 2011;**35**:4297-4304. DOI: 10.1016/j.biombioe.2011.07.019

[28] Schedler M, Ortner-Pichler A, Reitbauer E, Mahringer G, Landschützer C. Interdisciplinary system simulation of a tracked compost turner. Procedia Manufacturing. 2020;**51**:1005-1013. DOI: 10.1016/j.promfg.2020.10.141

[29] Silvestre WP, Pauletti GF, Baldasso C. Industrial crops & products Fodder

radish (*Raphanus sativus* L.) seed cake as a feedstock for pyrolysis. Industrial Crops and Products. 2020;**154**:112689. DOI: 10.1016/j.indcrop.2020.112689

[30] Amalina F, Syukor A, Razak A, Krishnan S, Sulaiman H, Zularisam AW, et al. Advanced techniques in the production of biochar from lignocellulosic biomass and environmental applications. Cleaning Materials. 2022;**6**:100137. DOI: 10.1016/j.clema.2022.100137

[31] Thers H, Djomo SN, Elsgaard L, Knudsen MT. Science of the Total environment biochar potentially mitigates greenhouse gas emissions from cultivation of oilseed rape for biodiesel. Science Total Environment. 2019;**671**:180-188. DOI: 10.1016/j.scitotenv.2019.03.257

[32] Myounghoon M, Park WK, Lee SY, Hwang KR, Lee S, Kim MS, et al. Utilization of whole microalgal biomass for advanced biofuel and biorefinery applications. Renewable Sustainale Energy. 2022;**160**:112269. DOI: 10.1016/j.rser.2022.112269

[33] Komandur J, Vinu R, Mohanty K. Pyrolysis kinetics and pyrolysate composition analysis of *Mesua ferrea* L: A non-edible oilseed towards the production of sustainable renewable fuel. Biotechnology. 2022;**351**:126987

[34] Jahirul MI, Rasul MG, Chowdhury AA, Ashwath N. Biofuels production through biomass pyrolysis: A technological review. Energies. 2012;**5**:4952-5001. DOI: 10.3390/en5124952

[35] Sun Y, Li C, Li Q, Zhang S, Xu L, Gholizadeh M, et al. Pyrolysis of flaxseed residue: Exploration of characteristics of the biochar and bio-oil products. Journal of the Energy Institute. 2021;**97**:1-12. DOI: 10.1016/j.joei.2021.03.020

[36] Volli V, Singh RK. Production of bio-oil from de-oiled cakes by thermal pyrolysis. Fuel. 2012;**96**:579-585. DOI: 10.1016/j.fuel.2012.01.016

[37] Volli V, Singh RK. Production of bio-oil from mahua de-oiled cake by thermal pyrolysis. Journal of Renewable and Sustainable Energy. 2012;**4**(1). DOI: 10.1063/1.3676074

[38] Yorgun SAİT, Şensöz S, Koçkar ÖM. Characterization of the pyrolysis oil produced in the slow pyrolysis of sunflower-extracted bagasse. Biomass and Bioenergy. 2001;**20**(2):141-148. DOI: 10.1016/S0961-9534(00)00064-7

[39] Mulimani HV, Navindgi MC. Production of solid fuel biochar from de-oiled seed cake by pyrolysis. Journal of Mechanical and Civil Engineering. 2017;**14**:57-61. DOI: 10.9790/1684-1402035761

[40] Muktham R, Ball AS, Bhargava SK, Bankupalli S. Study of thermal behavior of deoiled karanja seed cake biomass: thermogravimetric analysis and pyrolysis kinetics. Energy Science & Engineering. 2016;**4**(1):86-95. DOI: 10.1002/ese3.109

[41] Shaah MAH, Hossain MS, Allafi FAS, Alsaedi A, Ismail N, Ab Kadir MO, Ahmad MI. A review on non-edible oil as a potential feedstock for biodiesel: physicochemical properties and production technologies. RSC advances. 2021;**11**(40):25018-25037. DOI: 10.1039/D1RA04311K

[42] Demirbas A. Competitive liquid biofuels from biomass. Applied Energy. 2011;**88**:17-28. DOI: 10.1016/j.apenergy.2010.07.016

[43] Antizar-Ladislao B, Turrion-Gomez JL. Second-generation biofuels and local bioenergy systems. Biofuels, Bioproducts and Biorefining:

Innovation for a Sustainable Economy. 2008;**2**(5):455-469. DOI: 10.1002/bbb.97

[44] Demirbas A. Political, economic and environmental impacts of biofuels: A review. Applied Energy. 2009;**86**:S108-S117. DOI: 10.1016/j.apenergy.2009.04.036

[45] Cerón Ferrusca M, Romero R, Martínez SL, Ramírez-Serrano A, Natividad R. Biodiesel Production from Waste Cooking Oil: A Perspective on Catalytic Processes. Processes. 2023;**11**(7):1952. DOI: 10.3390/pr11071952

Chapter 17

Using LmPy-GC/MSMS to Molecular Characterization of Organic Components in Heterogeneous Mixtures: Instrumental Development

João Graciano Mendonça Filho, Noelia del Valle Franco Rondón, Luiz Guilherme Costa dos Santos, Ana Cristina de Oliveira Macedo, Milton Cézar da Silva, Thiago dos Santos Barbosa, Tais Freitas, Jaqueline Torres de Souza and Antonio Donizeti de Oliveira

Abstract

Pyrolysis techniques use thermal energy to degrade large molecules into smaller and volatile molecules, which can be analyzed by gas chromatography to determine the composition or structure of the initial sample. A laser system coupled to a GC/MSMS, for analysis of these species, gave rise to the technique called Laser MicroPyrolysis-GC/MSMS (LmPy-GC/MSMS). The laser micropyrolyzer is a hyphenated system with a laser source, an optical microscope, and a GC/MSMS technique. In this hyphenated technique, a collimated, short-duration laser beam with a large amount of energy can be focused on isolated components, or a macromolecule inside a heterogeneous matrix, promoting the thermal cleavage and producing smaller components, which are finally analyzed and characterized by GC/MSMS. The LmPy-GC/MSMS has been proven to be able to supply important molecular data on organic fossils, coals, source rocks, etc., improving the knowledge and the understanding of the compositions and opening a horizon in the study of several heterogeneous materials.

Keywords: laser micropyrolysis, organic matter, instrumental development, gas chromatography, mass spectrometry

1. Introduction

The pyrolysis technique has been extensively used for many years as an analytical technique in which large molecules are degraded into smaller volatile species using solely thermal energy under an oxygen-free environment. The coupling of this thermal energy as a laser source (pyrolyzer), an optical microscope, and the gas chromatographic tandem Mass Spectrometry (GC/MSMS) system compound a hyphenated analytical technique called Laser Micropyrolysis GC/MSMS (LmPy-GC/MSMS). This coupling contributes to improving the analytical processes during sample characterization, and it is considered a useful tool for the determination of the composition or structure of the organic compounds whether in individual components or samples of complex mixtures.

Molecular information on organic fossils in samples such as coals [1–4], source rocks [5–7], kerogen [8–14]; oil shales [4, 15, 16], and synthetic organic polymers [17, 18] had been successfully afforded by LmPy-GC/MSMS among existing pyrolysis technique, proving to be an excellent way to characterize the chemical composition of heterogeneous materials.

The analytical temperature is a crucial variable in pyrolysis techniques, being low temperatures are not analytically useful for efficient degradation while high temperatures are more efficient for breaking molecule bonds, however, caution needs to be taken not to destroy the molecule. In this sense selecting the adequate temperature at which macromolecules can be degraded in a wide array of products has great significance.

In the LmPy-GC/MSMS system is not possible to control the temperature. This laser provides a thermal flux as high as 1000°C/s which is enough to heat the sample in a very short time. The combination of high temperature and short time minimizes secondary pyrolysis reactions when laser pyrolysis is used and generates pyrolysis fragments characteristic of the original sample. This short-duration laser beam energy (1000°C/s) is collimated and coherent allowing the spread of very large amounts of thermal energy on a specific area of slim dimension, enabling the thermal cleavage of samples (e.g., isolated components or a macromolecule) into smaller components which in turn are analyzed by GC/MSMS.

The collimated and coherent laser irradiation allows focus on tiny or small areas of heterogeneous materials, which enables analysis of non-volatile, thermally labile components at the microscopic level. The focused and irradiated area can be less than 100 μm, using the appropriate lenses in the optical microscope. These microscopic areas permit isolated analysis of individual components into complex mixtures. In conclusion, the LmPy-GC/MSMS shows great potential to improve the understanding of organic composition in heterogeneous materials as well as isolated organic-walled microfossils.

Several authors [15, 16, 18–20] lay emphasis on their research on instrumental development and list diverse factors to be considered, such as (i) little knowledge about laser-material interactions; (ii) limitations in the sensibility of chromatographic techniques needed for products with low concentrations; (iii) need for interdisciplinary skill; (iv) possible interfacing-instruments difficulties; (v) the issue that all samples are not compatible with laser radiation to produce pyrolysis products and; (vi) the financial expense due to the acquisition of diverse instruments. On the other hand, the applicability of the LmPy-GC/MSMS as well as the use of other detection methods, such as NMR or FTIR, to characterize the molecular composition of distinctive organic samples, have been studied by many authors.

The kerogen of the Ordovician Estonian Kukersite sample was analyzed by Derenne et al. using spectroscopic NMR and FTIR, and off-line and flash pyrolytic methods [5]. Stout and Hall assayed two synthetic organic polymer samples by LmPy-GCMS [17] while Arouri et al. characterized Acritarch specimens [8],

Chitinozoan specimens from Silurian marine rocks were studied by Jacob et al. [11], and, Silva et al. compared the chemical composition of *Botryococcus braunii* and *Graphelmis prisca* microfossils [14]. Greenwood et al. demonstrated the credibility of the LmPy-GC/MS through the analysis of micro-sized quantities of various organic fossils [4, 6, 9, 21, 22]. Yoshioka and Takeda (2004) use IR laser micropyrolysis to analyze organic compounds in three macerals of an immature, sub-bituminous coal [23]. In addition, comparative studies between Curie-point pyrolyzer coupled to GC/MS (CP-Py-GC/MS) and LmPy-GC/MSMS were performed by Saundouk-Lincke et al. [12, 13] and Dutta et al. [10].

The favorable comparison of laser-derived molecular data with corresponding data from traditional methods, reported above, suggests that pyrolysis methods are suitable for molecular characterization of organic macromolecules. In addition, these researchers have also evidenced the possibility of a couple of combinations of LmPy-GC/MSMS with other micro or macroscale spectroscopic methods such as NMR, CP-Py, or FTIR. All of these results are useful to improve knowledge and understanding, allowing individual analysis of an isolated particle or smaller components inside complex mixtures in heterogeneous organic materials.

2. The LmPy-GC/MSMS hyphenated system

LmPy-GC/MSMS is a hyphenated analytical technique composed by: a neodymium-doped yttrium aluminum garnet (Nd-YAG; Nd: $Y_3Al_5O_{12}$) laser source, an optical device containing the sample chamber, a transfer line from the optical device to the gas chromatography equipment coupled to a triple quadrupole mass spectrometer (GC/MSMS) with a cryogenic entrapment (Liquid Nitrogen, −80°C) or Cold Trap in the PTV (programmed temperature vaporization) injector. **Figure 1** shows the schematic system of the LmPy-GC/MSMS technique and the Palynofacies and Organic Facies Laboratory (*LAFO*) hyphenated system located at the Federal University of Rio de Janeiro (*UFRJ*), Brazil. CSIRO (Commonwealth Scientific and Industrial Research Organization), Division of Petroleum Resources (Sydney, Australia) assembled this system exclusively for *LAFO/UFRJ* with financial support from *PETROBRAS*/Brazil.

The most common laser type used to emit light in the infrared spectrum (λ = 1064 nm) consists of an yttrium oxide and crystalline aluminum as a "host" doped with neodymium as a "guest." This device is known as the Nd-YAG laser and forms a variety of garnets. In the LmPy-GC/MSMS system, the Nd-YAG laser is the thermal energy source used to degrade large molecular-weight polymer carbon chains present in organic matter into smaller volatile species, under an oxygen-free environment. While the GC/MSMS technique separates and details the composition of molecular pyrolysis products (i.e., molecular fingerprinting). The association of these techniques has been useful to determine the composition or structure of original samples.

3. The LmPy-GC/MSMS system performance

An efficient and appropriate sample preparation method is necessary for good work involving the LmPy-GC/MSMS technique. Results interpretation quality is strongly influenced by factors such as standard sample collection, features of the

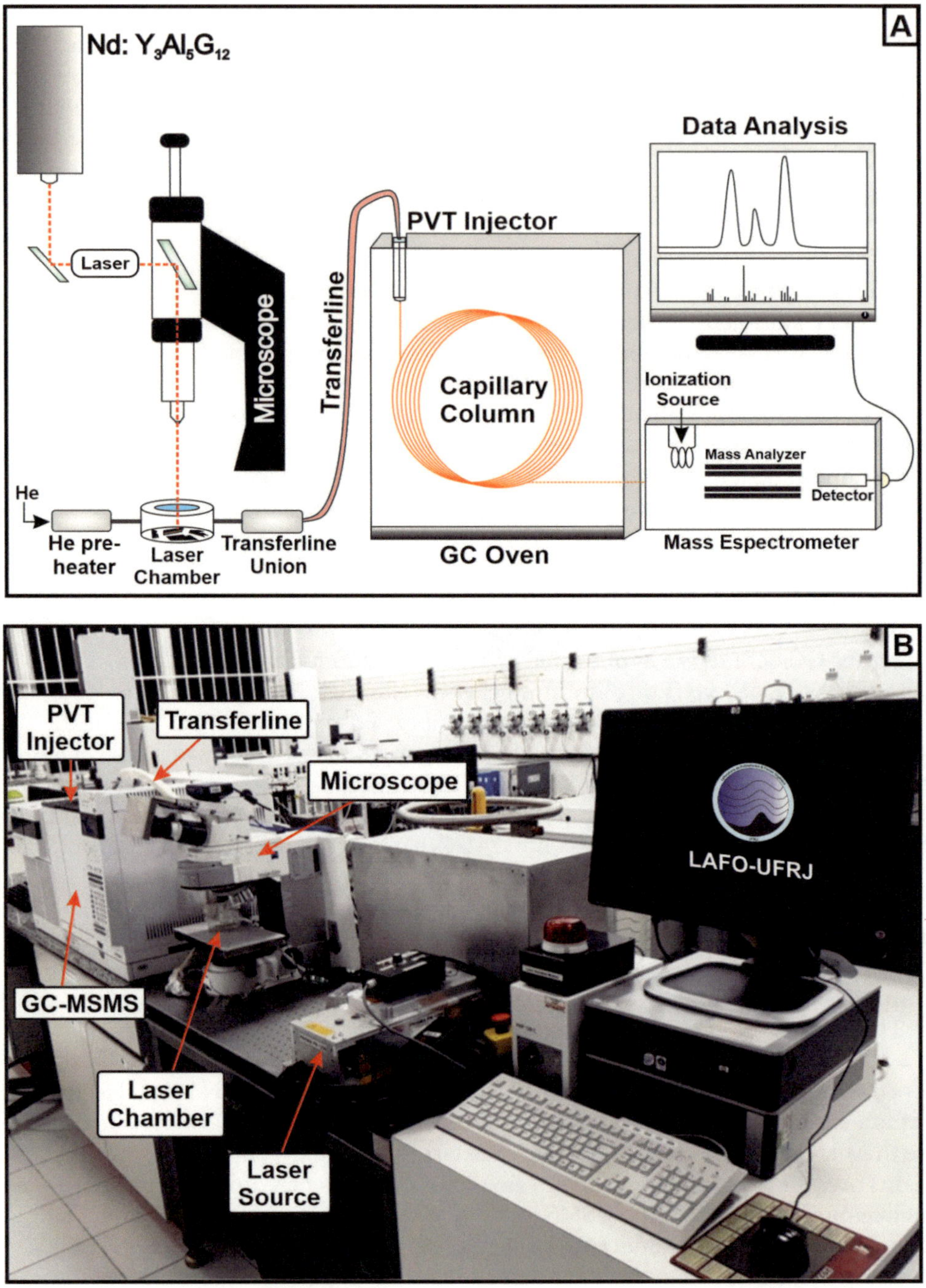

Figure 1.
Schematic LmPy-GC/MSMS system (a) and LmPy-GC/MSMS hyphenated system located at LAFO-UFRJ, Brazil (b).

samples, and preparation proceedings. Additionally, all these factors also have a significant effect on the utility of the LmPy-GC/MSMS analysis. The LmPy-GC/MSMS analysis can be carried out in samples such as whole rock (WR), kerogen concentrated (KC), individual organic components, or solid bitumen. However, analyses of isolated components are strongly recommended to avoid the influence of bordering material products.

3.1 Sample preparation for particulate organic components analysis

For analyzing individual organic components, a standard non-oxidative method is applied for the isolation of kerogen [24, 25]. The crushed (2–5 mm) sample is pre-extracted in the Soxhlet extraction apparatus using dichloromethane (CH_2Cl_2) for at least 24 hours. The crushed fraction must be treated with HCl and HF acids to remove carbonates and silicates. Heavy solution of $ZnCl_2$ (ρ = 1.95 at 20°C) is applied in order to concentrate the kerogen. The organic residue is sieved through a 10 mm nylon mesh using distilled water, isolated kerogen is the fraction retained in the sieve.

Particulate organic components are separated by size using a set of nylon membranes. Set of nylon membranes with pore diameters ranging from 5 to 200 μm. The sieving process follows the arrangement of the sieves (membranes) in a sequential nest: smallest mesh number (larger aperture, 120 to 200 μm) at the top and largest mesh number (smallest aperture, 20– 100 μm) at the bottom. The set of smallest aperture nylon membranes is used for the filtration process using a Kitasato flask and a vacuum pump. Organic residues separated by size after the filtration process are spread on slides using different solutions (water, detergent, and alcohol). The hand-picking technique is performed on the kerogen concentrate (KC) strewn slide. This slide is placed under the microscope and the organic particles are separated from each other, one by one, with the aid of a histological needle (**Figure 2a**). The components of the heterogeneous mixture are then identified and individually separated from each other (**Figure 2b–e**).

3.2 Assembling the LmPy-GC/MSMS system for analysis performing

3.2.1 Sample chamber assembly

The sample chamber is the module for the insertion of the sample. The chamber is connected to a gas injection system for capturing and transferring the generated

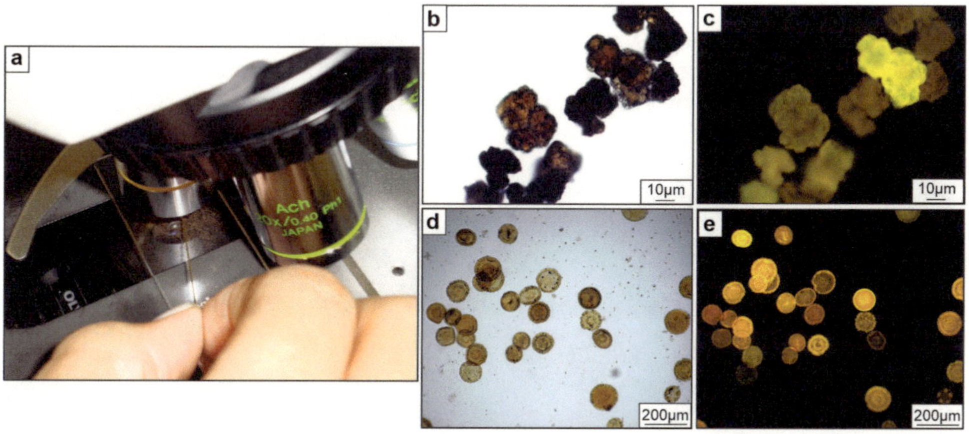

Figure 2.
Hand-picking technique (a). Hand-picked Botryococcus braunii (b, c) and Tasmanites (d, e) algae separates to LmPy-GC/MSMS analyses. Photomicrographs were taken under transmitted white light (TWL) (b, d) and fluorescence light (FM) (c, e).

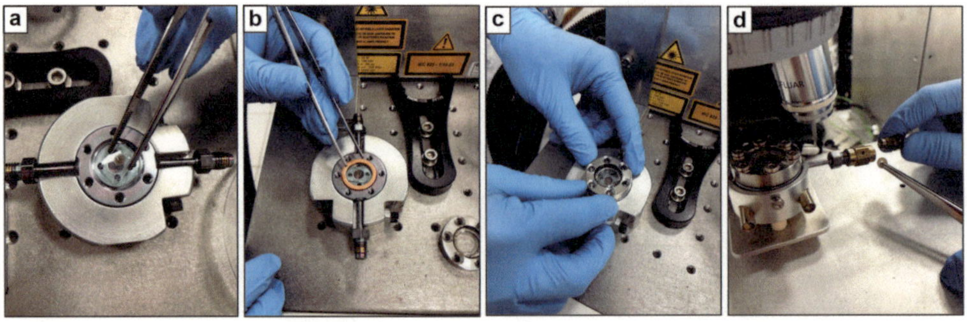

Figure 3.
Assembling the sample chamber: Sample in strewn slide (a), seal the chamber with copper gasket seal (b), close the chamber with a fused silica coating (c), and connecting the gas line (d).

products during pyrolysis. Samples (e.g., hand-picked organic material or solid bitumen) are placed in a glass slide and transferred inside the sample chamber compartment which has an inner diameter of 6 mm. The chamber is sealed with a copper gasket seal then, closed with a fused silica coating (VAR coating reduces reflectance from around 10–1%), and finally, connects to the gas line. The steps to assemble the sample chamber are shown in **Figure 3**.

3.2.2 Thermal extraction

Organic material is subjected to thermal extraction inside the sample chamber by heating at 100°C. This procedure is performed to remove adsorbed organic material on the surface of the samples. Finally, these extracted products are online transferred to the PTV injector via a transfer line heated at 300°C, using He as carrier gas.

3.2.3 Establishment of laser power conditions

In order to establish the most favorable parameters for the generation of primary pyrolysates, the laser energy, and irradiation time, among other laser conditions must be varied. These variations also provide a characterization of the performance and influence on the technique. **Figure 4** shows the spot size diameters obtained from a test performed in a Torbanite sample using different laser energy (13–20 A) at the same irradiation time (0.5 s).

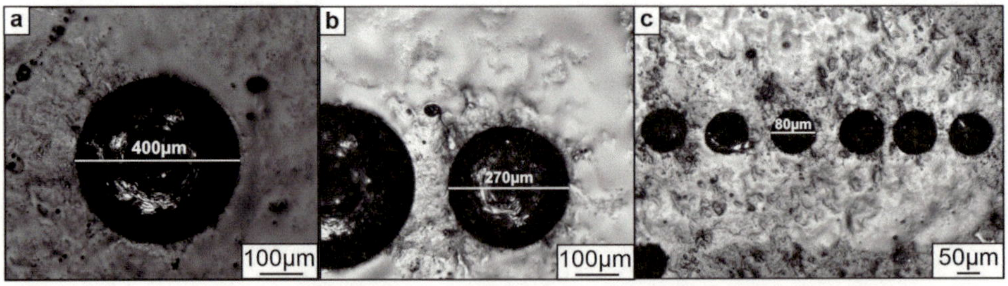

Figure 4.
Laser test in a Torbanite sample (Split mode) showing the spot size diameters obtained with different laser energy 20A (a), 17A (b), and 13A (c) using the irradiation time of 0.5 s.

3.2.4 Identification and selection of organic particles for pyrolysis

The organic particle is first identified through its optical properties (translucency, fluorescence, etc.), using a 20x objective and a 10x magnification front lens, and subsequently selected and marked with a spot size of range diameter of up to 100 μm for laser micropyrolysis analysis (**Figure 5**).

3.2.5 Analysis running

Due to the laser spot size depending on the sample size, this spot is controlled by adjusting the magnification in the microscope objective lens (10x, 20x, 40x, or 50x) and the laser power. Laser radiation is triggered, and the laser beam is directed throughout the prisms (lens) of the system until the entrance of the microscope as shown in **Figure 6**. The samples placed inside the sample chamber are pyrolyzed by focusing the laser radiation directly through the objective of the microscope and, generating the pyrolysis products. Multiple shots are necessary on the particles to obtain enough pyrolysis products (**Figure 7**) to be then transferred online to interface between the sample chamber and GC system. This transfer is carried out via a heated transfer line at 300°C, using He as carrier gas. Pyrolysis products are continuously

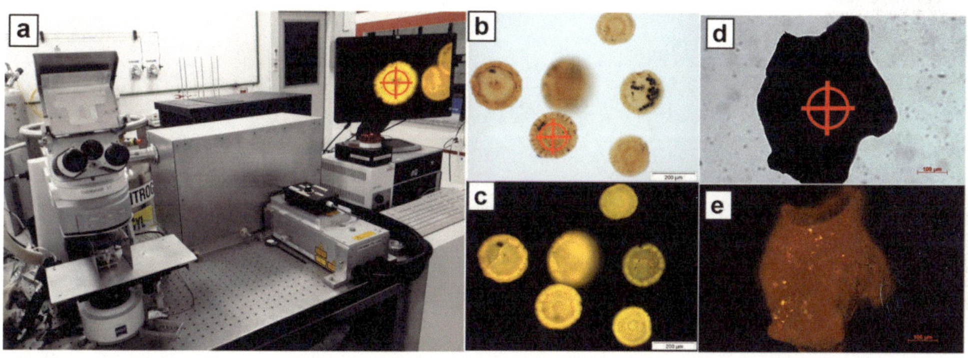

Figure 5.
Optical microscope equipment used to identify organic particles (a). Example of identification, selecting, and marking organic particles of Tasmanites algae (b, c) and solid bitumen (d, e) to LmPy-GC/MSMS analyses. Photomicrographs were taken under TWL (b, d) and FM (c, e).

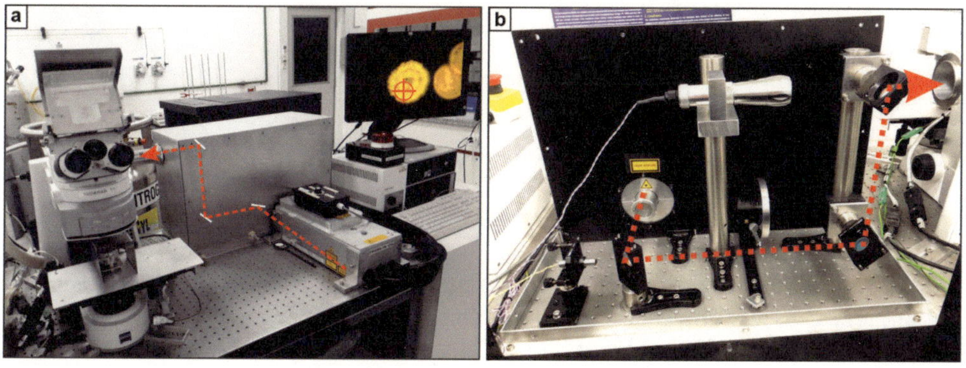

Figure 6.
Microscope receiving the laser beam (a) directed throughout lens (b) of the LmPy-GC/MSMS hyphenated system.

Figure 7.
Sample chamber at 300°C connected to an He flux interfacing (a) to transfer the pyrolysis products. Photomicrographs taken under TWL (b-d) and FM (e-g) showing the multiple shots on Botryococcus braunii particles.

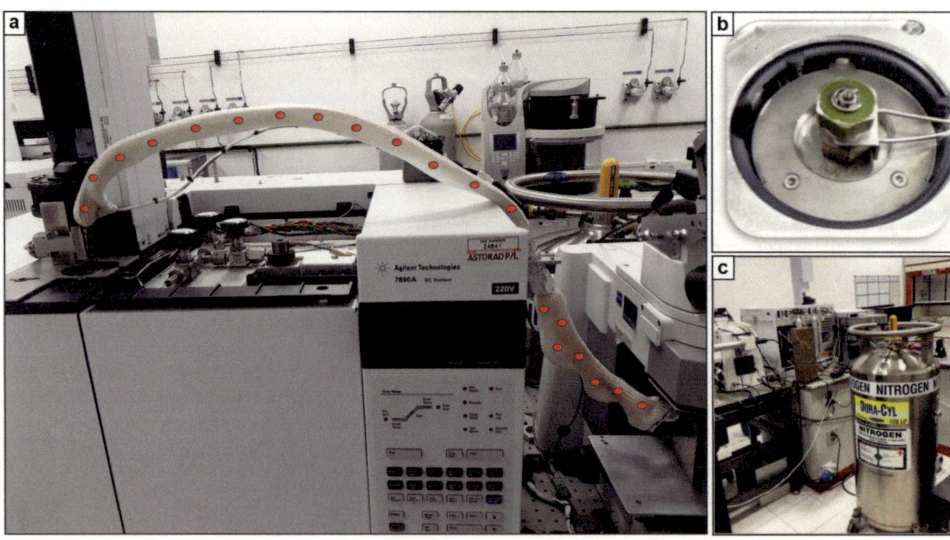

Figure 8.
Transfer line (300°C) of pyrolysis products from the microscope to the PTV injector in GC/MSMS equipment (a). PTV injector (b) cooled at −80°C with liquid nitrogen (c).

transferred to the PTV injector and then cryogenically entrapped in a cold trap (cryol collecting) using Liquid Nitrogen at −80°C.

Pyrolysis products are kept in the cold trapping long enough to ensure the complete trapping of the products from the sample chamber. After this time, the inlet temperature rapidly rises to vaporize the compounds and transfer the products to the Capillary GC Column. In this way, the pyrolysis products that were cryogenically entrapped at −80°C are heated up to 350°C and finally analyzed by GC/MSMS system (**Figure 8**).

4. Study cases

Applying examples of LmPy-GC/MSMS systems located at LAFO/UFRJ performed with samples of hand-picked *Botryococcus braunii algae* and solid bitumen are shown in this section.

4.1 LmPy-GC/MSMS of hand-picked *Botryococcus braunii* samples

The hand-picked *Botryococcus braunii* OWM (organic wall microfossil) was analyzed to check its molecular organic composition. **Figure 9** shows the isolated particles of *Botryococcus braunii* displaying craters generated by the laser radiations and drops of oil yielded. TICs from the LmPy-GC/MSMS analysis showing the distribution of n-alkanes and n-alkenes are shown in **Figure 10**.

To perform LmPy-GC/MSMS analyses on isolated particles (*Botryococcus braunii*), a thermal extraction is necessary (see Section 3.2.2). This extraction aims to release any molecules that may still be adsorbed in the isolated particles, even after performing Soxhlet extraction with DCM in the whole rock. TIC A in **Figure 10** represents the remaining molecules' profile. On the other hand, TICs B, C, and D (**Figure 10**) represent individual analyses of 6–8 particles of *Botryococcus braunii* performed with the goal of confirming the analyses' reproducibility. The n-alkanes and n-alkenes from C_8 to C_{25} were the main pyrolysis products. No cyclic or aromatic compounds were observed.

The experimental results indicated the presence of higher molecular weight n-alkanes, ranging from C_{25} to C_{31}. These compounds were also identified in the bitumen, as illustrated in **Figure 10A**, suggesting that they were not completely removed in the thermal extraction stage. It is important to note that the presence of aromatic compounds was not detected. This relevant information indicates that the temperature conditions were not severe, avoiding secondary reactions and cleavage of alkyl groups.

4.2 LmPy-GC/MSMS of solid bitumen samples

An isolated solid bitumen particle was analyzed to check its molecular organic composition. **Figure 11** shows the identification and selection of solid bitumen particles for pyrolysis and, of the craters and halo, and oil droplets produced in the

Figure 9.
Photomicrographs of hand-picked Botryococcus braunii particles with craters and oil drops generated by LmPy-GC/MSMS analysis. Photomicrographs were taken under TWL (a) and FM (b).

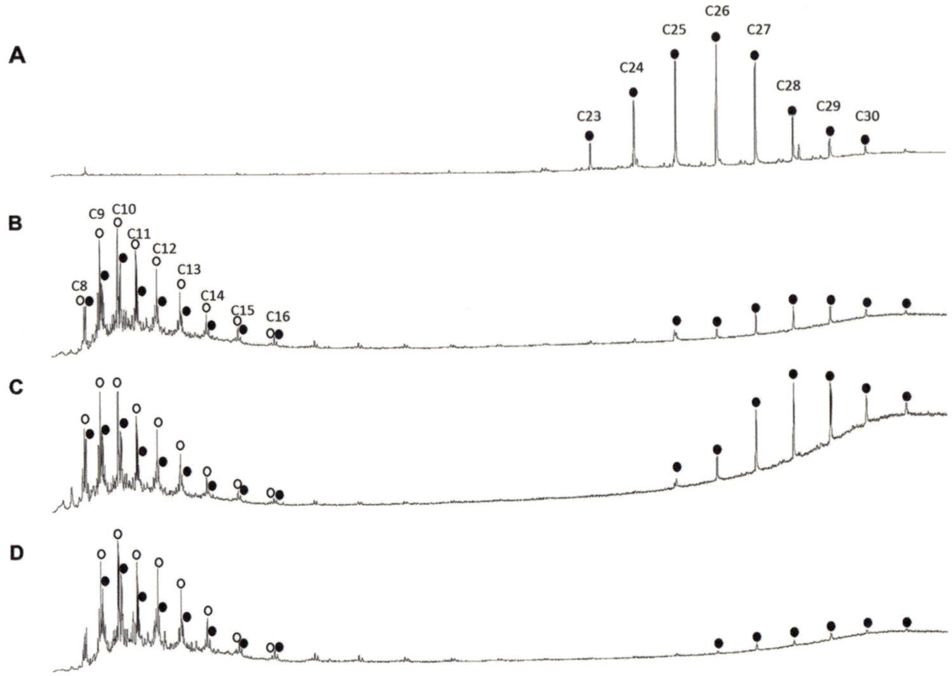

Figure 10.
Total ion chromatograms (TICs) from LmPy-GC/MSMS analysis of the Botryococcus braunii particles after thermal extraction (A) and after different analyses of the same sample but different particles (B, C, and D). n-alkanes (●) and n-alkenes (○) profile distributions.

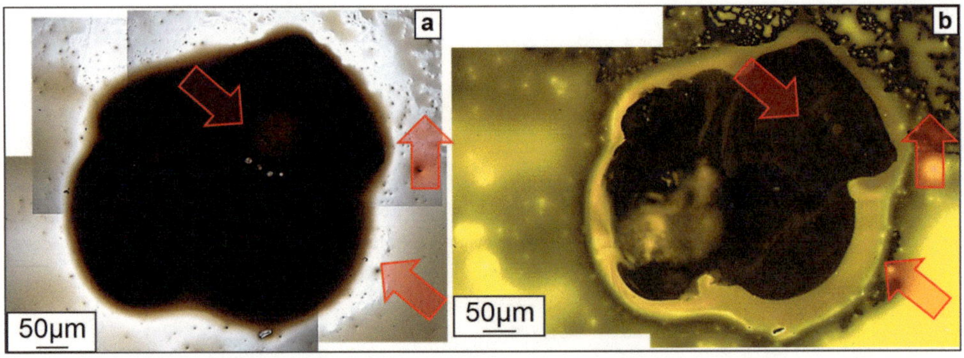

Figure 11.
Selected isolated solid bitumen particle after LmPy-GC/MSMS analysis showing craters and halo and oil droplets being produced by laser irradiation. Photomicrographs were taken under TWL (a) and FM (b).

sample after laser irradiation. The most favorable parameters, 25A, and 0.8 s, for primary pyrolysate generation, were obtained after testing different laser energy and irradiation time.

4.2.1 Geochemical characterization of the original bitumen sample

The total ion chromatogram (TIC) in **Figure 12**, shows the predominance of 8,14-Secohopane C_{29} hydrocarbon (m/z 123) and absence of other members of this

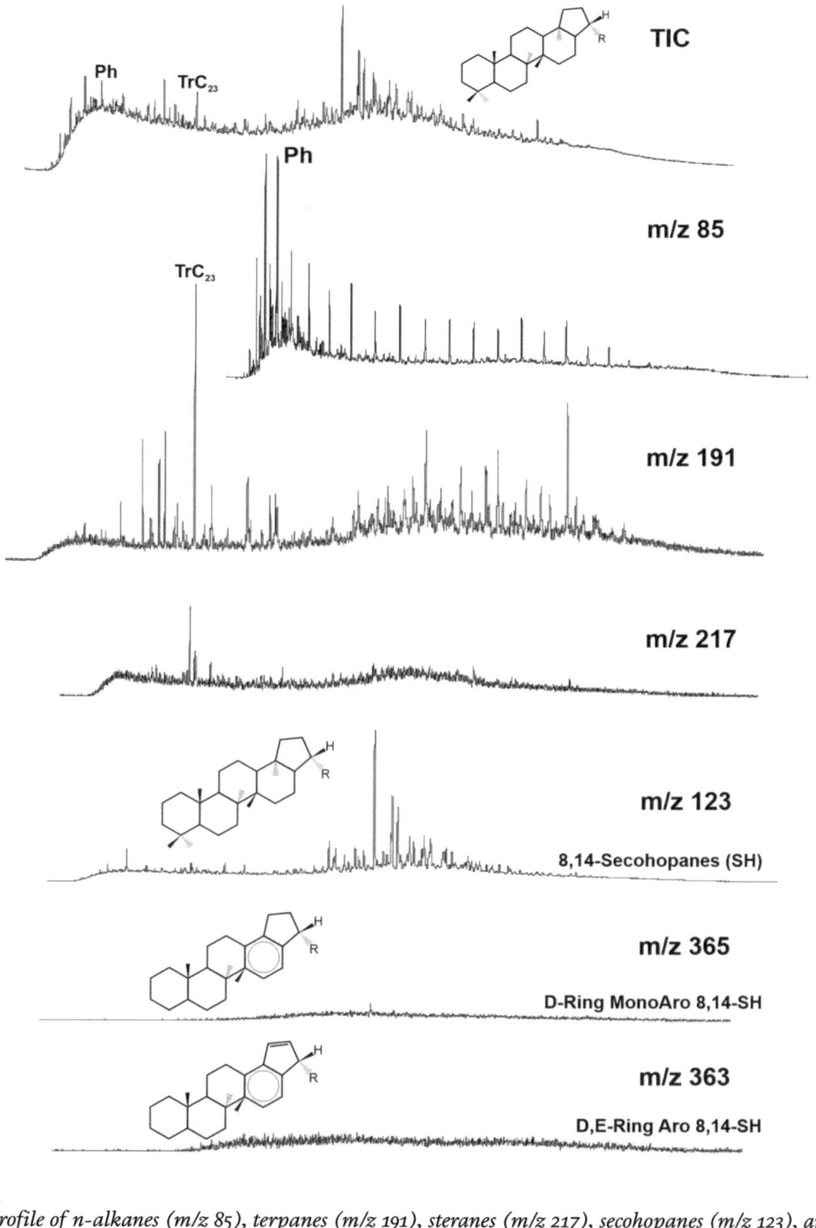

Figure 12.
TIC and profile of n-alkanes (m/z 85), terpanes (m/z 191), steranes (m/z 217), secohopanes (m/z 123), and aromatic secohopanes (m/z 363 and 365) compounds from the original solid bitumen sample.

major hydrocarbon series such as D-ring monoaromatic 8,14- secohopane (m/z 365) and D, E-ring aromatic 8,14-secohopane (m/z 363). Results showed low concentration of *n*-alkanes and isoprenoids (m/z 85), presence of tricyclic terpanes and unidentified pentacyclic terpanes (m/z 191), and absence of steranes (m/z 217).

4.2.2 Geochemical characterization of the compounds released during the LmPy-GC/MSMS analysis of bitumen sample

Results of chromatographic characterization (**Figure 13**) of compounds released during LmPy-GC/MSMS analysis show that the most abundant compound is the

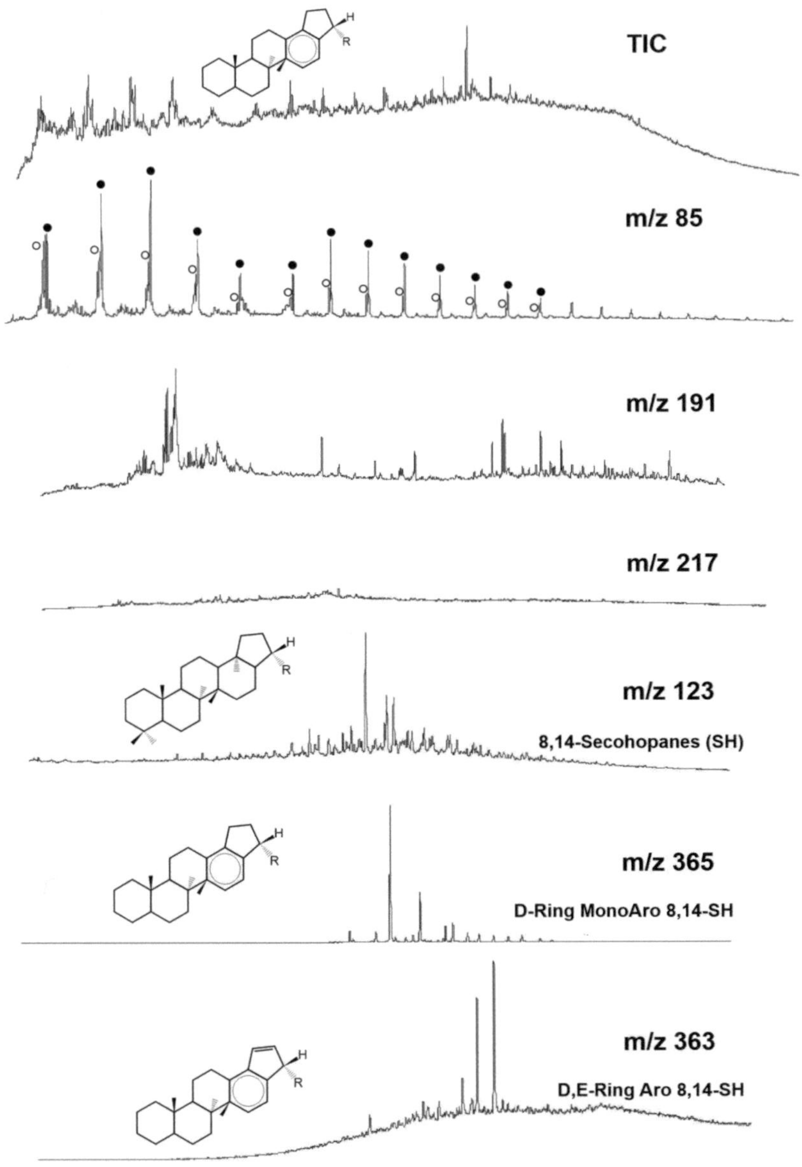

Figure 13.
TIC and profile of n-alkanes (●) and n-alkenes (○) (m/z 85), terpanes (m/z 191), steranes (m/z 217), secohopanes (m/z 123), and aromatic secohopanes (m/z 363 and 365) released during LmPy-GC/MSMS analyses at 25A for 0.8 s of bitumen sample.

monoaromatic 8,14 Secohopanoid D-ring (m/z 365), besides the secondary presence of the aromatic D, E-ring 8,14-secohopanoids (m/z 363) and 8,14, Secohopane C_{29} hydrocarbon (m/z 123). These aromatic compounds were absent in the original sample characterization. The laser experiments also produced *n*-alkanes and *n*-alkenes (m/z 85) and some terpanes (m/z 191). However, steranes (m/z 217) are absent.

4.2.3 Geochemical characterization of residual bitumen sample after LmPy-GC/MSMS analysis

In **Figure 14** can be observed, for the sample of residual bitumen after LmPy-GC/MSMS analysis, the predominance of the D-ring monoaromatic 8,14 Secohopanoid (m/z 365) and secondarily the presence of D, E-ring aromatic 8,14 secohopanoids

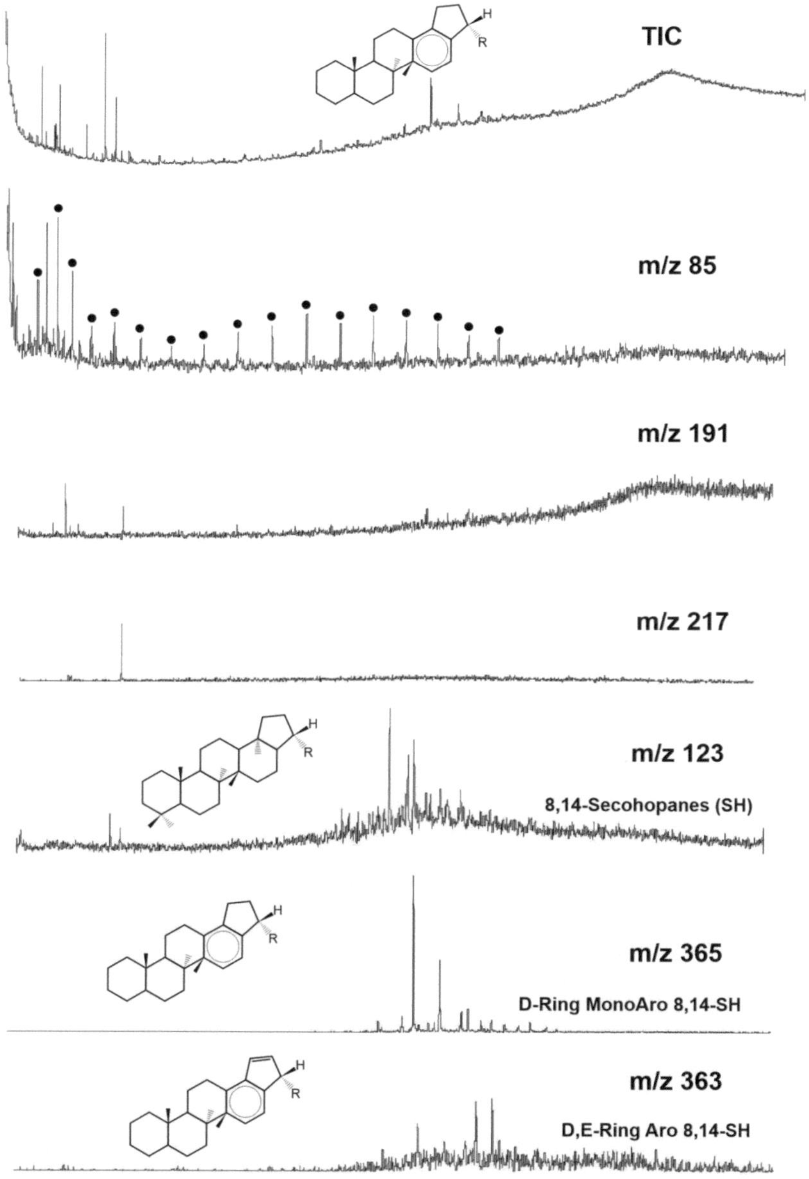

Figure 14.
TIC and profile of n-alkanes (●) (m/z 85), terpanes (m/z 191), steranes (m/z 217), secohopanes (m/z 123), and aromatic secohopanes (m/z 363 and 365) from residual bitumen sample after LmPy-GC/MSMS analyses at 25A for 0.8 s.

(m/z 363), and other secohopane hydrocarbons (m/z 123). The presence of *n*-alkanes (m/z 85) and the absence of terpanes (mz 191) and steranes (m/z 217).

5. LmPy-GC/MSMS system performance

The analyses of both types of samples showed in the chapter it was possibly observed that apart from the variation of laser parameters no significant difference, referent to the relative abundance of the hydrocarbon products, was observed between analyses of the same sample. Nevertheless, it was observed that parameters such as; the size of the formed craters and thus the intensity of the pyrolyzate distribution are degree focusing dependent as well as span time, and laser's energy settings (**Figure 15**).

6. Final remarks

The laser beam, in the LmPy system, is focused on smaller areas (than 100 μm) due to the use of focusing lenses. This makes it possible to analyze individual microscopic particles among a wide variety of small components in complex mixtures. The short duration of the laser beam enables the thermal cleavage of single particles or even macromolecules, using a huge amount of energy. This thermal interaction between the laser and matter triggers a sequence of pyrolysis reactions releasing compounds that, in turn, are examined by GC/MSMS.

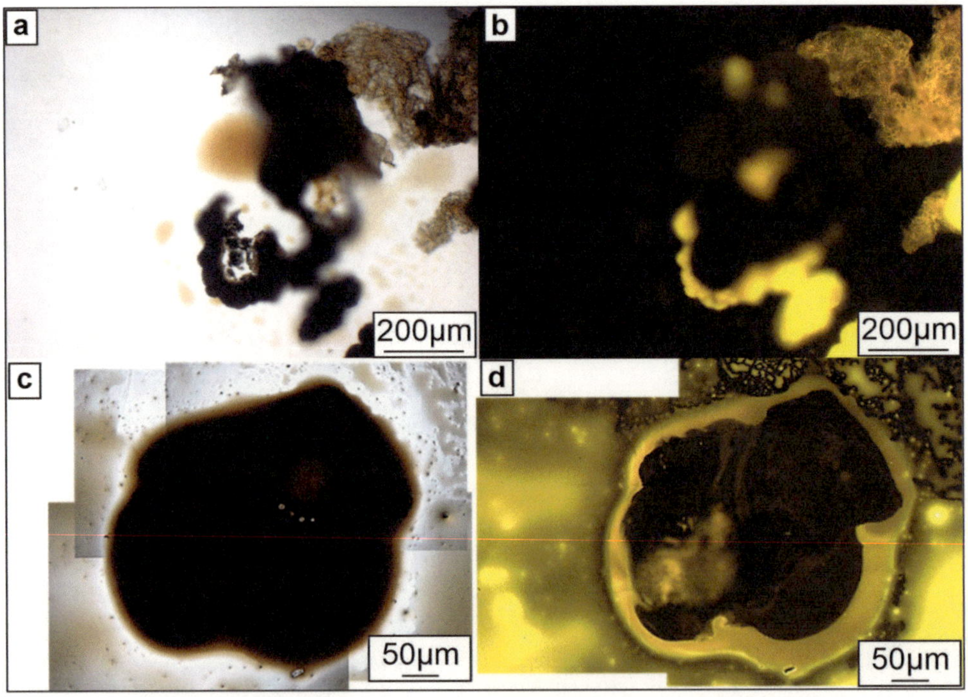

Figure 15.
Isolated Botryococcus braunii (a, b) and solid bitumen (c, d) particles after LaMPy analysis showing laser-generated craters and oil droplets produced (TWL: A, c and FM: b, d).

The level of temperature is critical to any pyrolysis analysis. The temperatures must be high enough to break molecular bonds without destroying the material. Hence, caution needs to be taken when pyrolysis temperature is selected. Overly high temperatures result in molecular degradation, which generates non-specific products while low temperatures are not analytically useful due to their low degradation potential. In this way, to obtain successful analytical pyrolysis one must select a temperature that allows the controlled degradation of macromolecules, producing a wide variety of products, quickly enough to be compatible with GC [26]. However, the control temperature in the LmPy system is not possible. On the other hand, the laser beam also provides a thermal flux (1000°C/s) enough to allow heating of polymeric materials in less than 2 seconds. This extremely fast laser incidence is a unique feature of micropyrolysis and has the fundamental purpose of reducing or annulling secondary reactions, allowing the production of original substances [15, 17].

Another characteristic of LmPy is the temperature applied. This temperature corresponds to the required energy to break covalent bonds, which can modify the molecular stereochemical features during the pyrolysis analysis. Consequently, the highly energetic, collimated, and coherent laser radiation is able to supply significant amounts of thermal energy in tiny and specific areas. These properties of directional coherent and monochromatic high-powered laser irradiation allow its use to thoroughly investigate non-volatile and thermally labile materials on a microscopic scale [3].

The use of the technique still has major hurdles to overcome, especially the high costs involved. Other difficulties are that the laser-material interactions are not fully understood, and not all samples couple efficiently to the laser to produce thermal degradation products. However, LmPy and other related methods have shown success in characterizing several organic components, significantly improving the possibility of studying the chemical composition of heterogeneous and small-sized materials, as well as identifying specific biomarkers in isolated OWM.

The destination of waste produced at the end of the consumption chain for industrialized products has attracted greater attention from various sectors of the economy, the scientific community, and public authorities. This concern with the reuse of raw materials is aimed at generating less environmental impact and making the consumption of products more sustainable [27].

In the field of pyrolysis, there are various initiatives to incorporate certain products into the circular economy. A good example is plastic, the mass production of which causes a number of environmental problems, and its final disposal is inefficient. Chemical recycling, a thermal process for breaking polymer bonds, has therefore been gaining ground in studies to increase the useful life of plastics [28]. One application widely used by researchers is pyrolysis as a technique for converting plastic waste into economically noble materials such as oils, gases, and ashes [29]. These measures do not close the loop, but they ensure more sustainable use of fossil energy to produce this waste and reduce greenhouse gas emissions.

LmPy-GC/MSMS is an experimental technique that can contribute to the development of circular models by enabling detailed analysis of individual elements in complex mixtures. It was initially developed to meet the specific demands of the oil exploration industry. However, its methodological proposal can be used in various economic or scientific segments with the aim of identifying substances and understanding the molecular structures of pyrolyzed products, offering alternatives for improving recycling processes, creating materials with greater added value, and eliminating undesirable waste. One of the advantages of LmPy-GC/MSMS is the low amount of sample required to pyrolyze these components in the order of micrograms.

Acknowledgements

The authors would like to thank CSIRO (Commonwealth Scientific and Industrial Research Organization), Division of Petroleum Resources (AUS) for assembling LmPy-GC/MSMS system, and PETROBRAS/Brazil for financial support. Special thanks to *CNPq (Conselho Nacional de Desenvolvimento Científico e Tecnológico)* for financial support for the development of research through the project "*Caracterização molecular de componentes orgânicos através do sistema Micropirólise a Laser acoplado a CG-EMEM* (401497/2016-0)".

Conflict of interest

The authors declare no conflict of interest.

Author details

João Graciano Mendonça Filho[1*], Noelia del Valle Franco Rondón[1], Luiz Guilherme Costa dos Santos[1], Ana Cristina de Oliveira Macedo[1], Milton Cézar da Silva[1], Thiago dos Santos Barbosa[1], Tais Freitas[2], Jaqueline Torres de Souza[1] and Antonio Donizeti de Oliveira[1]

1 Federal University of Rio de Janeiro, Palynofacies and Organic Facies Laboratory, Rio de Janeiro, Brazil

2 Federal University of Rio Grande do Sul, Rio Grande do Sul, Brazil

*Address all correspondence to: graciano@geologia.ufrj.br

References

[1] Vastola FJ, McGahan LJ. Development of laser micropyrolysis of coal macerals. Fuel. 1987;**66**(7):886-889. DOI: 10.1016/0016-2361(87)90325-5

[2] Stout SA. Lasers in organic petrology and organic geochemistry, II. In-situ laser micropyrolysis-GCMS of coal macerals. International Journal of Coal Geology. 1993;**24**(1-4):309-331. DOI: 10.1016/0166-5162(93)90017-5

[3] Greenwood PF, Zhang E, Vastola FJ, Hatcher PG. Laser micropyrolysis gas chromatography/mass spectrometry of coal. Analytical Chemistry. 1993;**65**(15):1937-1946. DOI: 10.1021/ac00063a004

[4] Greenwood PF, George SC, Hall K. Applications of laser micropyrolysis-gas chromatography-mass spectrometry. Organic Geochemistry. 1998;**29**(5-7):1075-1089. DOI: 10.1016/S0146-6380(98)00101-6

[5] Derene S, Largeau C, Casadevall E, Sinninghe Damsté JS, Tegelaar EW, de Leeuw JW. Characterization of Stonian Kukersite by spectroscopy and pyrolysis: Evidence for abundant alkyl phenolic moieties in an Ordovician, marine, type II/I kerogen. Organic Geochesmistry. 1990;**16**(4-6):873-888. DOI: 10.1016/0146-6380(90)90124-I

[6] Greenwood PF, Sherwood N, Willett GD. Chemical examination of some petroleum source rocks by laser pyrolysis-mass spectrometry and flash pyrolysis -gas chromatography/mass spectrometry. Journal of Analytical and Applied Pyrolysis. 1995;**31**:177-202. DOI: 10.1016/0165-2370(94)00823-J

[7] Vanderborgh NE, Jones CER. Laser microprobe mass analysis studies on coal and shale samples. Analytical Chemistry. 1983;**55**(3):527-532. DOI: 10.1021/ac00254a025

[8] Arouri K. Greenwood PF, Walter MR, a possible chlorophycean affinity of some Neoproterozoic acritarchs. Organic Geochemistry. 1999;**30**(10):1323-1337. DOI: 10.1016/S0146-6380(99)00105-9

[9] Greenwood PF, Arouri KR, George SC. Tricyclic terpenoid composition of tasmanites kerogen as determined by pyrolysis GC–MS. Geochimica et Cosmochimica Acta. 2000;**64**(7):1249-1263. DOI: 10.1016/S0016-7037(99)00326-9

[10] Dutta S, Greenwood PF, Brocke R, Schaefer RG, Mann U. New insights into the relationship between Tasmanites and tricyclic terpenoids. Organic Geochemistry. 2006;**37**(1):117-127. DOI: 10.1016/j.orggeochem.2005.08.010

[11] Jacob J, Paris F, Monod O, Miller MA, Tang P, George SC, et al. New insights into the chemical composition of chitinozoans. Organic Geochemistry. 2007;**38**(10):1782-1788. DOI: 10.1016/j.orggeochem.2007.06.005

[12] Al Saundouk-Lincke NA, Schwarzbauer J, Volk H, Hartkopf-Fröder C, Fuentes D, Young M, et al. Alteration of organic material during maturation: A pyrolytic and infrared spectroscopic study of isolated bisaccate pollen and total organic matter (lower Jurassic, Hils syncline, Germany). Organic Geochemistry. 2013;**59**:22-36. DOI: 10.1016/j.orggeochem.2013.03.006

[13] Al Saundouk-Lincke NA, Schwarzbauer J, Hartkopf-Fröder C, Volk H, Fuentes D, Young M, et al. The effect of different pyrolysis temperatures

on organic microfossils, vitrain and amber — A comparative study between laser assisted- and curie point-pyrolysis–gas chromatography/mass spectrometry. Journal of Analytical Applied. Pyrolysis. 2014;**107**:211-223. DOI: 10.1016/j.jaap.2014.03.004

[14] Freitas da Silva T, Mendonça Filho JG, da Silva MC, de Oliveira A, Torres de Souza J, Franco Rondón N. Botryococcus braunii versus Gloecapsomorpha prisca: Chemical composition correlation using laser micropyrolysis-gas chromatography/mass spectrometer (LmPy-GC/MSMS). International Journal of Coal Geology. 2016;**168**:71-79. DOI: 10.1016/j.coal.2016.08.010

[15] Stout A, Lin R. Lasers in organic petrology and organic geochemistry—I. Laser-induced fluorescence, thermal extraction, and pyrolysis. Organic Geochemistry. 1992;**18**(3):229-239. DOI: 10.1016/0146-6380(92)90064-5

[16] Greenwood PF, George SC, Wilson MA, Hall KJ. A new apparatus for laser micropyrolysis-gas chromatography/mass spectrometry. Journal of Analytical and Applied Pyrolysis. 1996;**38**(1-2):101-118. DOI: 10.1016/S0165-2370(96)00948-5

[17] Stout SA, Hall K. Laser pyrolysis-gas chromatography/mass spectrometry of two synthetic organic polymers. Journal of Analytical and Applied Pyrolysis. 1991;**21**(1-2):195-205. DOI: 10.1016/0165-2370(91)80025-4

[18] Meruva NK, Metz LA, Goode SR, Morgan SL. UV laser pyrolysis fast gas chromatography/time-of-flight mass spectrometry for rapid characterization of synthetic polymers: Instrument development. Journal of Analytical and Applied Pyrolysis. 2004;**71**(1):313-325. DOI: 10.1016/S0165-2370(03)00092-5

[19] Metz LA, Meruva NK, Morgan SL, Goode SR. UV laser pyrolysis fast gas chromatography/time-of-flight mass spectrometry for rapid characterization of synthetic polymers: Optimization of instrumental parameters. Journal of Analytical and Applied Pyrolysis. 2004;**71**(1):327-341. DOI: 10.1016/S0165-2370(03)00091-3

[20] Greenwood PF. Lasers used in analytical micropyrolysis. Journal of Analytical and Applied Pyrolysis. 2011;**92**(2):426-429. DOI: 10.1016/j.jaap.2011.08.001

[21] Greenwood PF, George S, Pickel W, Zhu Y, Zhong N. In situ analytical pyrolysis of coal macerals and solid bitumens by laser micropyrolisis GC-MS. Journal of Analytical and Applied Pyrolysis. 2001;**58-59**:237-253. DOI: 10.1016/S0165-2370(00)00208-4

[22] Greenwood PF, van Heemst JDH, Guthrie EA, Hatcher PG. Laser micropyrolysis GC–MS of lignin. Journal of Analytical and Applied Pyrolysis. 2002;**62**(2):365-373. DOI: 10.1016/S0165-2370(01)00135-8

[23] Yoshioka H, Takeda N. Analysis of organic compounds in coal macerals by infrared laser micropyrolysis. Journal of Analytical and Applied Pyrolysis. 2004;**71**(1):137-149. DOI: 10.1016/S0165-2370(03)00060-3

[24] Tyson RV. Sedimentary organic matter. In: Organic Facies and Palynofacies. Londres: Chapman & Hall; 1995. p. 615

[25] Mendonça, Filho JG, Menezes TR, Mendonça JO, Oliveira AD, Souza JT. Chapter 2: Kerogen: Composition and classification. In: 10th ICCP Course: Dispersed Organic Matter – Integrating Transmitted and Reflected Light

Microscopy. Potsdam, Germany. 2017. pp. 13-20

[26] Wampler TP. Introduction to pyrolysis-capillary gas chromatography. Journal of Chromatography A. 1999;**842**(1-2):207-220. DOI: 10.1016/S0021-9673(98)00921-2

[27] Geissdoerfer M, Savaget P, Bocken NMP, Hultink EJ. The circular economy - a new sustainability paradigm. Journal of Cleaner Production. 2017;**143**:757-768, ISSN 0959-6526,. DOI: https://doi.org/10.1016/j.jclepro.2016.12.048

[28] Duailibe A. Contribuição para reciclagem química de resíduos plásticos para o desenvolvimento sustentável [thesis]. Rio de Janeiro: Federal University of Rio de Janeiro; 2018

[29] Fiorelli FCC. Pirólise de Resíduos Plásticos: Revisão e Perspectivas Futuras [thesis]. Rio de Janeiro: Fluminense Federal University; 2020

Chapter 18

Biogas as a Sustainable Fuel and Feedstock: Properties, Purification, and Applications

Moses Jeremiah Barasa Kabeyi and Oludolapo Akanni Olanrewaju

Abstract

Biogas is a renewable energy resource derived from biomass mainly through anaerobic digestion. Methane (CH_4) and carbon dioxide (CO_2) and trace elements significantly affect the fuel. Hence, the undesirable components are undesirable and should be removed for various biogas applications. This makes it necessary to purify raw biogas to increase its utility making it suitable for many applications as a sustainable energy resource and process feedstock. Biogas can make substantial contribution to realization of sustainable development goals and the global energy transition. However, the wide-scale use of biogas is seriously constrained by the presence of harmful impurities as well as the low calorific value of biogas. The overall objective of this chapter is to investigate the properties of biogas, its cleaning, upgrading, and its sustainability as a transition energy resource for realization of sustainable development goals and the global energy transition. The chapter provides a comprehensive examination of the potential of biogas as a fuel and potential feedstock for various industrial processes. Furthermore, this study offers a comprehensive examination of the various technological alternatives that can be employed to harness biogas for the purpose of generating heat and electricity.

Keywords: raw biogas, biogas properties, biogas applications, energy potential of biogas, sustainability of biogas fuel, electricity from biogas, biogas reforming, liquid biofuels, Fischer-Tropsch

1. Introduction

Renewable energy sources like biogas and biomethane are important energy carriers when the society seeks to replace fossil fuels with sustainable energy sources [1]. Biogas is a renewable energy resource produced by anaerobic digestion of biomass mainly in the form of municipal waste, farm waste, food waste, and energy crops. Raw biogas typically consists of biomethane (30–80%), carbon dioxide (20–50%), and smaller amounts of nitrogen (1–10%) and varying composition of trace elements namely hydrogen sulfide, ammonia, hydrogen, and various volatile organic compounds depending on the feedstock of biogas. Life cycle assessment of biogas energy resource shows that biogas use effectively reduces greenhouse gas (GHG) emissions [2–4]. The production and use of biogas as an energy resource helps to diversify the energy systems while at the same time promoting sustainable biomass waste management recycling and disposal [2, 3, 5]. The growing energy demand and concerns over global climate change have increased demand for renewable and green energy sources. Various innovative approaches have been developed for bioenergy production and use as alternatives to fossil fuels as energy sources. Biogas is one of the viable energy options from organic wastes. The sustainability of biogas is enhanced further by production of organic manure as substitute for chemical fertilizers in crop production. Biogas can be used for heating, power generation, fuel, and processing. It could also be used to produce hydrogen, carbon dioxide, and biofuels sustainably. This vast range of applications makes biogas an attractive sustainable energy choice [2, 5, 6].

Over 59 billion m^3 biogas or about 35 billion m^3 methane equivalent of biogas by volume is produced globally on an annual basis with the European Union accounting for about half of this capacity. The energy generation capacity from biogas globally has expanded rapidly; for example it increased from 65 GW in 2010 to 120 GW in 2019, which is an increase of about 90% [2, 5]. Biogas is an important renewable energy resource whose production and use can aid in mitigation of greenhouse gas (GHG) emissions by substituting fossil fuels and promoting the utilization of biodegradable plant and animal waste feedstocks [7, 8]. As the modern society generates significant wastes, several waste disposal and treatment methods are employed to avoid environmental pollution with anaerobic digestion being one of the popular options as it additionally produces useful energy [9], which is a fundamental requirement for society [10, 11]. International treaties or agreements that commit countries to minimize their carbon footprint like the agenda 21 and Kyoto Protocol are motivating factors for production and use of biogas [12, 13], since they advocate for the transition to renewable and low carbon sources of energy to reduce the global greenhouse gas emissions particularly from the energy sector, which is dominated by fossil fuel sources of energy, with biogas being a very important option [14, 15].

Biogas can be produced from a variety of different types of biodegradable biomass like chicken manure, crop waste, and animal manure, through several processes but mainly by means of controlled anaerobic digestion. Raw biogas has feasible applications like heating and power generation as is, or it can be processed to biomethane and used to produce value-added chemicals. Purified or cleaned biogas has higher utility as a green fuel, and bio natural gas fed into natural gas pipes upon upgrading to remove carbon dioxide [8, 16], while reducing greenhouse gas emissions and contributing a huge potential for use as a renewable resource. In 2014, biogas accounted for almost 8% of Switzerland's renewable energy generation (excluding hydropower) and 0.29% of the country's overall energy consumption [17–19]. Biogas has a potential use as substitute for firewood and thus can protect the forest ecosystems. By 2040, it is projected that biogas will supply clean fuel for almost 200 million people, predominantly in Africa and Asia. Therefore, biogas is a significant player in realization of sustainable development goals [5, 20–22]. The SDGs and the energy transition to a green and low carbon energy and electricity mix will be much easier to accomplish with biogas as an alternative.

About 2.8 billion people globally rely on primary biomass fuel for their energy needs. Biogas production presents huge potential as a strategy to generate renewable energy products from biomass waste sources with significant economic and environmental benefits. The practical efficiencies indigestion is low, which calls for strategies available to overcome these barriers and create more efficient energy [19, 23]. It is the methane content of biogas that directly influences the energy value of biogas and hence the need to manage the process to ensure high methane composition and additionally control the polluting potential of organic residues having high contents of Biochemical Oxygen Demand (BOD) [24, 25]. With biogas proving to be a sustainable and renewable energy source [26, 27], the growing concerns over greenhouse gas emissions and climate change have generated significant interest in biogas as an energy resource for the global transition [28, 29], hence growing production of biogas from organic waste [30, 31] with common applications being power generation, heating applications like cooking, lighting, and biofuel production with biogas feedstock [6, 32, 33].

Providing modern energy services to their populations is a challenge for many developing nations; for instance, about 450 million people in India did not have access to modern energy for electricity and cooking [34, 35]. Technologies that conserve resources and revenue in agriculture are crucial to achieving sustainable food and energy production [34]. Coal, oil, and gas account for over 60% of global electricity generation, but renewables are steadily increasing their proportion, from 26% in 2015 to 28% in the first quarter of 2020, with variable renewables increasing from 8 to 9% during the same period [36]. Since 1990, the proportion of worldwide electricity output that comes from renewable sources has increased by an average of 2% per year, faster than the annual increase in electricity demand (1.8%). When compared to the annual capacity increases of solar photovoltaics (36.5%) and wind energy (23.0%), biogas's 11.5% ranked third fastest. The substantial role played by biomass in global energy generation is evidenced by the 9.7% annual growth rate of biofuels since 1990 [37]. Particularly, the relevance of biogas in agriculture and food production cannot be overemphasized. Beyond providing food for all of humanity, agriculture is the primary source of income for more than two-thirds of the world's population. In addition, nearly 82% of the global population is directly or indirectly involved in smallholder agriculture and other industries, making this economic activity the mainstay of many developing countries [34]. This shows that biogas is crucial to the energy transition and that agriculture and other carbon transformative processes should boost its production and consumption.

Biogas has a significant role to play in the impending energy transition as a renewable and sustainable energy resource because of its high energy conversion potential and widespread availability for power generation, for industrial applications as a renewable process feedstock, and for thermal energy applications [38–42]. In addition to producing biogas, the anaerobic digestion (AD) process leads to feedstock treatment during the treatment phase, and it can also produce digestate, which is a useful organic fertilizer that can replace chemical fertilizers in sustainable agriculture [43, 44]. An essential part of the global carbon cycle is the creation of biogas through microbial control, as natural anaerobic biodegradation releases 590 to 800 million tons of methane into the atmosphere annually [39, 45, 46]. Numerous different organic feedstocks are anaerobically digested to produce biogas, which can then be enhanced to produce natural gas such as biomethane [47–49]. The sustainable technique of turning biomass waste into biogas can lead to advantages including less carbon emissions, improved organic waste management, and increased resource use efficiency.

In terms of efficiency and cost, the Stirling engines and internal combustion engines theoretically offer the most viable options for converting biogas to electricity, particularly on a small scale. Comparatively, the electricity generated by internal combustion engines per kilowatt hour is cost competitive and efficient. The engines are available in a wider range of sizes, they are more flexible and efficient, and require less effort to operate and service [50–52]. For slightly larger plants, gas turbines are commonly utilized for power generation in the capacity range of 3 to 5 MW. Turbines and microturbines have less stringent fuel quality standard requirements compared to internal combustion engines making them ideal prime movers for raw biogas, which has a wide range of harmful impurities [53–55]. Both the steam turbine and the gas turbine power plants using biogas for power generation encounter higher total capital investment costs Biogas and biomethane provide a sustainable pathway to include rural businesses and communities in the energy transition. This can be accomplished by creating grid-connected electricity and reducing grid load by generating their own electricity and heat.

Biogas as an energy resource has the potential to contribute to sustainable energy transition if carefully produced through the process of anaerobic digestion of organic substrate under carefully regulated conditions and used in a manner that limits leakages of methane, which is a greenhouse gas [56, 57]. The energy content of biogas is contingent upon the methane composition, which is subject to influence from both the production process and the type of substrate employed. The composition of biogas may include elements like sulfur, which can make it inappropriate for use as a fuel in internal combustion engines and certain industrial chemical and thermal processes. The calorific value of biogas demonstrates variability because of its composition, primarily influenced by the relative abundance of methane. The heating value of biogas generally falls within the range of 21–23.5 MJ/m^3, suggesting that an approximate equivalence of 0.5–0.6 liters of diesel fuel or roughly 6 kWh of electricity can be attributed to 1 m^3 of biogas.

Biogas has significant potential as a sustainable and renewable energy resource, both currently and in the foreseeable future. Biogas energy can make a substantial contribution to the achievement of sustainable development goals and the global transition towards sustainable energy, because of its minimal greenhouse gas emissions, widespread availability, and access to raw materials. The widescale incorporation of biogas into the global energy mix has been constrained due to the presence of harmful impurities and low calorific value of raw biogas [11, 14, 31, 58]. The overall objective of this study is to investigate the properties of raw biogas as a fuel, its cleaning methods without necessarily upgrading to biomethane, and its sustainability as a transition energy resource for realization of sustainable development goals and the global energy transition. This chapter provides a comprehensive examination of the potential of biogas as a fuel and potential feedstock for various industrial processes. Furthermore, this study offers a comprehensive examination of the various technological alternatives that can be employed to harness biogas for the purpose of generating heat and electricity due to its significant energy potential and sustainability [59–61].

1.1 Problem statement

The heightening need for food and energy resulting from the growth of the global population, particularly in developing countries, has intensified the strain on energy production and consumption. The continued use of fossil fuels threatens humanity with unbearable consequences of climate change due to greenhouse gases, which are causing climate change [18, 34]. Although biogas has a great deal of promise to supply clean energy to both rural and urban populations, the prevalence of inoperable and abandoned biogas facilities has raised concerns about the sustainability of the technology [19, 62]. This poses questions concerning the dependable creation, distribution, and use of biogas fuel [35]. It is technically possible to convert this waste into valuable energy sources, thereby generating additional revenue [57, 63, 64]. The challenges facing biogas applications include lack of technical know-how, limited or lack of incentives and subsidies to make biogas competitive in several countries, and the low calorific value of and presence of harmful impurities in biogas. These challenges can be addressed by developing and accessing appropriate and affordable biogas technologies and development of a comprehensive policy on construction and operation of biodigesters, biogas utilization, and sustainable diversification in biogas energy products and services [2, 5, 6].

The implementation of biogas technology enables the harnessing of renewable energy through the utilization of crop and animal waste. The energy source exhibits significant potential for utilization in electricity generation and as a viable thermal energy solution for residential structures. The ultimate outcome of this process can be efficiently harnessed as a valuable fertilizer in rural agricultural environments and households, leading to a reduction in costs related to waste management and disposal [65]. Biogas energy can improve rural people's lives and economies. Many smallholder farmers in developing nations burn biomass for disposal, even though it can be used as a fertilizer, and with biogas technology, an impoverished organic fertilizer that is more affordable can be made by smallholder farmers [66]. Designing biogas systems so that they can reliably supply most or 100% of their energy demands requires defining a number of factors [67].

Sustainable farm-level biogas energy requires appropriate infrastructure design and selection [18, 68]. Another thing to consider in such system design is the natural decomposition of agricultural waste that results in the emission of substantial amounts of methane, a potent greenhouse gas, into the atmosphere. For instance, in the year 2015, the exclusive contribution of livestock manure to methane emissions in the United States of America amounted to approximately 10%. However, a mere 3% of the total livestock waste underwent recycling via anaerobic digestion [57]. This presents the main issue in utilizing as it lies in its unsteady production value and variations in quality. These factors can potentially disrupt the generation process or hinder the effectiveness of biogas applications, resulting in reduced reliability [69].

1.2 Rationale of biogas cleaning/purification

Both developed and developing countries face challenges with utilization of significant organic waste and the mounting pressure to replace fossil energy resources like natural gas, oil, and coal with renewable energy sources. Anaerobic digestion of organic biomass is now a mature technology while biogas in its raw form or cleaned and upgraded forms has multiple applications as a bioenergy substitute of fossil fuel sources. The production and utilization of biogas has also been growing with capacity growing by more than double between 2009 and 2022 [70]. Methane fermentation occurs naturally in the process of organic matter decay in oxygen deficient environments like swamps and in landfills leading to natural emission of methane in the atmosphere as a greenhouse gas. Controlled biogas production can be used to reduce these natural emissions [4, 5]. This research presents biogas as a fuel and feedstock for applications like production of biomethane, methanol, syngas, power generation through combustion, or fuel cells, among others [71–73].

Biogas has a very important role to play in the ongoing energy transition as a renewable and sustainable energy resource because of its high energy conversion potential and the widespread availability for power generation, for industrial applications as a renewable process feedstock, and for thermal energy applications [38–42]. Additionally, anaerobic digestion (AD) is an effective waste treatment and disposal method while the digestate from biodigesters is a rich organic fertilizer [43, 44]. Anaerobic digestion is a critical process in the global carbon cycle which releases 590 to 800 million tons of methane into the atmosphere annually through uncontrolled biodegradation. These harmful emissions can be mitigated through controlled anaerobic digestion for useful biogas production [39, 45, 46]. The benefits of biogas fuel and its production process include reduced less carbon emissions, better organic waste management, and increased resource use efficiency [39, 74].

Biogas contributes to the sustainable transition as a renewable fuel with multiple applications as a fuel characterized by a high methane content, which is generated through the process of anaerobic digestion of organic substrate under carefully regulated biochemical process conditions [56, 57]. Indigestible carbohydrates, proteins, and lipids are not present in the biomass substrate utilized for biogas production, which can complicate or slow down the digestion [75, 76]. Biogas can also be used as a feedstock for production of biomethane, carbon dioxide, hydrogen, and various biofuels for a wide range of applications including generation of heat and power and feedstock for various industrial processes [77, 78].

The whole world has got about 2.8 billion people who rely on primary biomass fuel like coal, firewood, crop waste, and dried animal waste for their energy needs in cooking and heating, which leads to high-level household pollution [7]. Biogas production presents huge potential as a strategy to generate renewable energy products from biomass waste sources with significant economic and environmental benefits. The practical efficiency in biodigestion is low, which calls for strategies available to overcome these barriers and create more efficient energy systems [23]. It is the methane content of biogas that determines the energetic potential of biogas since it directly influences the calorific value of the fuel hence the need to manage the process to ensure high methane composition. Organic bio digestion additionally reduces the polluting potential of organic residues having high contents of Biochemical Oxygen Demand (BOD), while the substrate can be used as a nutrient rich valuable fertilizer [24, 25].

The human population is steadily growing leading to increased demand for food and energy, which simultaneously aggravates the environmental challenges. Substituting fossil fuels with renewable energy alternatives is currently a major global issue of the twenty first century and is a key sustainable development objective. Implementation of biogas technologies will significantly transform costly, socially sensitive issues and environmentally damaging fossil fuel dependence, environmental pollution, Greenhouse Gas (GHG) emissions, and waste conversion into profitable options like electricity generation, heat production, and production of biofertilizers as well as provision of green vehicular fuel substitute of fossil gas. A national biogas production development program can significantly lead to development of new enterprises, boost income for rural communities, and create new jobs to improve people's socio-economic well-being [70].

1.2.1 Limitations/challenges facing biogas energy resources

The main challenges facing biogas production and use as an energy resource include high equipment cost and a lack of a governmental incentive programs in many countries, low reliability or lack of guarantee of long-term performance of biogas plants due to technological challenges, unpredictable investment environment, limited biogas distribution and storage capacity for some countries like Sweden, and low cost of commercial fertilizers as well as low cost of fossil fuels like gas and oil in some countries [24]. A study by [7] on household biogas digester use in rural China showed that there was rapid growth in the early part of the century in the use of household biogas digesters but the operations were never smooth because out of 1743 households interviewed, 42% adopted household biogas digesters, but on average they worked for just 6.66 months each year, which is an indicator of underutilization of the digestors.

1.2.2 Benefits of biogas as a renewable energy resource

Biogas has various applications that include production of biofuels as renewable energies; reduction in greenhouse gas emissions; removal of odors and flies, which are generally associated with uncontrolled biomass biodegrading; financial savings and revenue from biogas derived energy products like electricity and biomethane, waste treatment; and recycling of organic waste; reduced air and water pollution; alleviation of rural poverty; and better social conditions of rural or remote settlements, which include gender balance since the women who mainly bear the responsibility of looking for fuel will have an opportunity to engage in other tasks while forests and vegetation will be conserved due to reduced use of firewood and charcoal [2, 6, 79, 80].

The potential benefits associated with use of biogas as an energy resource including generating electrical energy from biogas may include the following:

i. Facilitates decentralized generation, which is done close to consumers with benefits like reduced transmission and distribution costs.

ii. Generates extra income and savings from the energy produced from biogas and sold to the grids.

iii. A decrease in utility-purchased electricity that must be offset.

iv. Creating opportunities for cogeneration, which is a more efficient form of energy exploitation.

v. Reducing greenhouse gas emissions from methane to the atmosphere, which would have been emitted freely to the atmosphere from uncontrolled biodegradation, etc.

The challenges limiting full use of biogas that should be overcome include lack of national generating technologies, the need to clean biogas before use, the challenge of economic feasibilities that requires incentives, and the lack of penalties for possible environmental damages from biogas schemes [24, 25].

2. Composition and properties of biogas

Biogas composition is a function of many factors like process design and the nature of the substrate with the main components of biogas being methane and carbon dioxide, and other minor components in varying proportions [1]. Methane and carbon dioxide are the primary components of the gas mixture known as biogas. Hydrogen sulfide, hydrogen gas, moisture, and siloxanes are a few others. Carbon monoxide, hydrogen, and methane are the combustible components, and they generate heat for a wide range of thermal and electrical uses [81]. The composition of biogas produced from any given feedstock and any given level of control over that process will be unique. However, the main components of biogas are methane (CH_4) and carbon dioxide (CO_2) [53]. With just one carbon and four hydrogen atoms, pure methane has no discernible odor. It is easier to ignite methane than it is to ignite air. A mixture of 5–15% of this substance with air is explosive. While methane alone is non-toxic, inhaling too much of it in a confined space can be fatal because it displaces oxygen. Since it may linger in the air for up to 15 years, methane is a potent greenhouse gas; on a 20-year time frame, it is nearly 20 times more effective at trapping heat than carbon dioxide [82, 83]. Because of the severe environmental concerns associated with the development and processing of biogas in the absence of effective regulation, this highlights the necessity for regulated bio-deration of organic wastes.

Biogas has multiple applications beyond just energy generation, including use as a feedstock in the petrochemical, hydrogen, and synthesis gas industries. Methane and carbon dioxide are the primary components of biogas, but other contaminants such as ammonia (NH_3), water vapor, hydrogen sulfide (H_2S), nitrogen (N_2), methyl siloxanes, oxygen, halogenated volatile organic compounds (VOCs), hydrocarbons, and carbon monoxide (CO) are often present as trace elements. For safety and efficiency reasons, it is necessary to filter out certain trace elements before using biogas as a fuel. Hydrogen sulfide and carbon dioxide are two examples. It is important to find ways to reduce the high operational expenses and energy consumption associated with biogas upgrading so that it can be used as a fuel [12, 13, 84].

2.1 Composition of biogas

Biogas is a mixture of gases consisting of methane and carbon dioxide as the main components, in addition to several trace elements like carbon monoxide, hydrogen sulfide, hydrogen gas, moisture, and siloxanes. Carbon monoxide, hydrogen, and methane are the combustible constituents with methane content being the main determinant of heating value of biogas. Combustion of biogas generates heat for various thermal and electricity applications with related emissions, mainly carbon dioxide, a greenhouse gas. Biogas composition of biogas varies based on the substrate used and the process control applied [2, 3].

Methane is the main constituent of biogas that generally accounts c 50–80% composition, while other constituents are carbon dioxide (20-45%), water vapor (2%), and trace gases like O_2 N_2, NH_3 H2 H_2S [56, 77]. Biogas may also contain siloxanes formed in anaerobic decomposition of materials commonly found in detergents and soaps. The composition varies with feedstock and process, for example, landfill has average methane composition of about 50%, while advanced waste treatment technologies yield 55%–75% methane content. Reactors with free liquids can generate biogas with 80%-90% methane content using in-situ gas purification techniques [2, 85, 86]. The average biogas composition is shown in **Table 1**.

Based on the data presented in **Table 1**, it can be observed that the predominant constituents of biogas consist of methane and carbon dioxide. The relative proportions of these components generally vary between 30 and 80%, with the specific range being contingent upon factors such as the quality of the feedstock utilized and the efficacy of process management.

	Element	Composition (%)
1	Methane	30–80%
2	Carbon dioxide	20–50%
3	Nitrogen	0.5–10%
4	Ammonia	0–0.05%
5	Moisture	2–8%
6	Hydrogen	0–1%
7	Hydrogen sulfide	0.1–3%
8	Oxygen	0.1–3%
9	Ammonia	0–0.5%
10	Carbon monoxide	0–0.1
11	R_2SiO	0–0.5 mg/m^3
12	C_xH_y	0–1%

Table 1.
Average composition of biogas [2, 16, 25, 49, 53, 56, 66, 87–89].

2.2 Thermodynamic properties of biogas

The H_2S concentration can be reduced through chemical, biological, and physical means, whereas the water content can be reduced through condensation in gas storage or along the gas stream [42, 77]. With rising methane content, biogas has a calorific value that varies between 5000 and 7000 kcal/m^3. A comparable volume of biogas (1 m^3) can be produced by: 0.7 m^3 of natural gas; 0.6 kilograms each of kerosene, gasoline, and butane; 3.5 kg each of wood and dung briquettes; 4 kWh each of electricity and carbon; and 0.43 kg each of butane [39, 90]. At 0.1013 Mpa and 273 K, biogas consists of 60% methane and 40% carbon dioxide. Molar mass 16.04, specific heat capacity 2.165 kJ/kg. K, and ignition temperature between 650 to 750°C are shown for biogas in **Table 2**.

From **Table 2**, it is noted that the thermodynamic properties of biogas are competitive with those of other fuels. Due to its advantages for the environment, biogas is a desirable fuel despite having somewhat worse thermal properties than fossil fuels [71, 92, 93]. **Table 3** lists the biogas equivalents of various fuels with remarks.

2.3 Energy and electricity potential of biomass

Biogas can be used for different purposes, including electricity and heat production (cogeneration), heat only, power only, biomethane, fuel for vehicles, high tech process energy, and in the chemical industry as feedstock material. With average caloric value of calorific value of biogas being 21–23.5 MJ/m^3, 1 m^3 of biogas is an equivalent of 0.5–0.6 liters of diesel fuel or 6 kWh in energy content. But due to inefficiencies and losses, 1 m^3 of biogas yields around 1.7 kWhe [5, 44, 59]. Biogas output and energy generation from various substrates are compared in **Table 4**.

From **Table 4**, it is noted that different feedstocks have different biogas potential with fats having the highest potential followed by maize silage and chicken droppings.

	Property	Symbol	Value
1	Specific heat capacity	Cp	2.165 kJ/kgK
2	Molar mass	M	16.04
3	Gas constant	R	0.518 kJ/kg
4	Normal density	g	1.2 g/l
5	Critical density	g_c	320 g/l
6	Relative (to air) density	g_r	0.83
7	Calorific value of biogas	LCV	22.6 Mj/m^3
8	Critical temperature	T	−2.5 °C
9	Critical pressure	p	7.3–8.9 Mpa
10	Flammability limit content in air	v	6–12%
11	Ignition temperature	T	650-750°C

Table 2.
Thermodynamic properties of biogas [77, 90, 91].

	Device/ technology	Application	Remarks
1	Fuel cell	Electricity generation	Though expensive, fuel cells are more efficient
2	Hydrogen production	Biomethane can be converted to hydrogen for use as combustion fuel or electricity and process chemicals.	Renewable hydrogen can be produced from biogas or biomethane by dry reforming, steam reforming, or hydrolysis
3	Biomethanation	Biomethane can be fed to natural gas supply as substitute for natural gas	Renewable replacement of fossil natural gas is feasible with use of biomethane
4	Diesel engine	Biomethane can be used as a diesel engine fuel in either dual fuel mode or pure gas engines	Diesel engines have more fuel flexibility and efficiency and can easily use biofuels as fossil fuel substitutes
5	Gas/petrol engine	Biomethane can be used as a fuel for petrol or gasoline engines with little or no modification	The gas/petrol engines are less efficient engines compared to diesel engines but are simpler and easier to convert to biogas engines
6	Stirling engine	Stirling engines are also called hot air engines.	The Stirling engines can operate on a wide variety of fuels due to high adaptability
7	Gas turbine	Based on size, gas turbines can be micro, small, or large gas turbines in open, closed, or combined cycle configuration	Turbines are simple in construction, are versatile, and can use raw biogas as well as biomethane and are easy to operate.
8	Cogeneration	In cogeneration, biomethane is burned to simultaneously produce useful heat and electricity.	Cogeneration with biomethane as a fuel can be applied on various conversion systems like Stirling engines, diesel engines, gas turbines, hydrogen, and fuel cells to increase system efficiency
9	Trigeneration	Trigeneration refers to generation of electricity and both heating and cooling from same fuel/energy resource simultaneously	Trigeneration is costly and complex but the most effective method of energy conversion

Table 3.
Summary of biogas to electricity conversion systems and technologies.

	Type	Electricity produced per ton fresh matter (kW·h)	Biogas yield per ton fresh matter (m^3)
1	Fat	1687.4	826–1200
2	Maize silage	409.6	200/220
3	Chicken litter/dung	257.3	126
4	Food waste (disinfected)	224.6	110
5	Municipal solid waste	207.2	101.5
6	Fruit wastes	151.6	74
7	Cattle dung	122.5	55–68
8	Horse manure	114.3	56
9	Sewage sludge	96.0	47
10	Pig slurry	23.5	11–25

Table 4.
Electricity produced from biogas yield per ton of fresh matter.

Pig slurry and sewage sludge have some lowest values of biogas potential. The conversion factors are 35% electrical efficiency, and 55% methane content [27]. When compared to maize silage, chicken manure, and sterilized food waste, fat has the highest biogas yield and largest biogas and electricity potential.

3. Energy content and potential of biogas

The energy content of biogas varies but a typical normal cubic meter of methane has a calorific value of around 10 kWh, but CO_2 being noncombustible has zero, making methane the ultimate determinant of the energy value of biogas. At 60% methane composition, then, the energy content of biogas would be less than 10 kWh per normal cubic meter, as CO_2 and other noncombustible impurities absorb instead of generating heat [1, 2, 94]. The primary factor influencing the thermal energy content of biogas is the methane composition within the gas mixture. The volumetric composition of methane in biogas widely varies based on factors like the feedstock used and process control [49, 95, 96]. This disparity suggests that biogas's LHV ranges between 16 and 28 MJ/m^3, rather than remaining stable at a single value. Combustion of biogas releases thermal energy that can be used for cooking or other uses like internal combustion engines [3]. Raw biogas has undesirable components like hydrogen sulfide, ammonia, and various volatile organic compounds whose presence and composition are determined by the feedstock used and process control applied [33, 97].

3.1 Biogas for engine applications

Biogas can be used as a fuel internal combustion engine although it requires some modifications on the engine design and ignition to handle gaseous biogas fuel instead of the traditional liquid fuels. Biogas has higher octane rating (MON = 130) enabling it to sustain higher compression ratios without engine knocking, hence better engine performance [51, 98–100]. Historically, the use of biogas as an engine fuel started during the Second World War (2WW), as Germany and its Nazi territories had challenges sourcing fuel vehicles and farm machinery [44, 62]. This forced them to resort to sewage gas and biogas from manure and digesters, which were bottled and used as fuel during the crises.

Biogas is an efficient, sustainable, and environmentally friendly fuel that can perform optimally in spark ignition engines modified to suit its specific properties. The performance and utility of biogas as engine fuel is increased by methane enrichment or the addition of hydrogen to improve its flame quality [1]. The main challenge with biogas as engine fuel is the gaseous nature, which means low volumetric density. Biogas as an engine fuel is improved by biogas upgrading, methane enrichment, prechamber combustion, alteration of ignition parameters, higher compression ratio, and addition of hydrogen for better performance and emissions management [98, 101].

The following are some guidelines to consider when selecting biogas engine fuel.

a. The amount of methane present must be high.

b. The amount of water and carbon dioxide (CO_2) should be minimized to maximize energy production based on high calorific value.

c. Acids produced by the condensation or combustion of sulfur corrode metals by eating away at their surface and should therefore be removed.

3.2 Energy content and equivalents of biogas

Biogas is an odorless and colorless gas that burns with a clear blue flame like natural gas but has a lower calorific value of 20–26 MJ/m^3 (537–700 Btu/ft^3) compared to natural gas whose caloric value is 39 MJ/m^3 (1028 Btu/ft^3) [6, 33, 94]. Energy equivalents of biogas are presented in **Table 5**.

	Alternative fuel	Biogas Equivalent	Remarks
1	1 kg of petrol	2.5 m^3 of biogas	Has less energy content than petrol
2	1 kg of carbon	2.33 m3 of biogas	Carbon has higher energy value than biogas
3	1 liter of kerosene	2.0 m^3	Biogas can limit fossil fuel dependence and use
4	1 kg of fuel oil	1.42 m^3 of biogas	Has lower energy content than fuel oil
5	1 m^3 of natural gas	1.43 m^3 of biogas	Has lower energy content, may need purification
6	1 kg charcoal	0.5 m^3 biogas	Biogas use limits forest destruction
7	1 kg of firewood	0.29 m^3	Biogas can be used to avoid deforestation
8	1 kWhr of electricity	0.25 m^3 of biogas	Biogas can substitute grid electricity use
9	1 kWhr of electricity	0.25 m^3 of biogas	Biogas can substitute grid electricity use
10	1 kg of dried cow dung	0.1 m^3	Biogas has more energy value than dried cow dung
11	1 kg of manure briquette	0.083 m^3 of biogas	Biogas has more energy value

Table 5.
Energy equivalents of biogas [77, 90, 91].

	Fuel/element	Calorific value (MJ/m^3)
1	Butane	117.23
2	Propane	92.11
3	Methane	35.56
4	Natural gas	31.82
5	Biogas	22.5 to 25
6	Biomethane	23.02

Table 6.
Calorific value of common gaseous fuels.

From **Table 5**, it is noted that biogas has a higher energy value than manure briquettes, but it has a lower energy content than fossil fuels. But it is a potential substitute for fossil fuels and cuts down on forest clearance at the same time.

The calorific value of the biogas generally varies between 22.5 and 25 MJ/m^3, or 6.25 and 10 kWh/m^3 assuming methane has heating value of 35.8 MJ/m^3. The energy content can be further enhanced if the raw biogas is upgraded to biomethane [24, 25]. Biogas with 55% CH_4 has an average calorific value of about 21.5 MJ/m^3, compared to pure methane (CH_4), which has a calorific value of about 35.56 MJ/m^3, which is one of the main reasons for upgrading by removal of carbon dioxide (CO_2) from raw biogas [4, 5, 70]. **Table 6** shows the energy value of biogas compared to other fuels like butane, natural gas, and propane.

From **Table 6**, it is noted that biogas has the lowest calorific value compared to methane, propane, butane, natural gas, and biomethane. Therefore, upgrading to biomethane can improve its calorific value making it more competitive thermodynamically.

4. Biogas purification/cleaning

The quality of biogas fuel for use in power generation and other applications can be improved using a variety of techniques. By guaranteeing constant fermentation through continuous supply of feedstock of the right quality and digestion at the ideal parameters of C/N ratio, PH, dilution, and digestor temperature, biogas output can be optimized. A little amount of oxygen or air can be injected into the headspace of a storage fermenter to help hydrogen sulfide (H_2S) be oxidized by microorganisms, leading to the removal of a significant portion of the sulfate from the gaseous phase. This technique is the most affordable and popular way to desulfurize biogas, removing up to 95% of the sulfur [27, 102].

The second method involves filtering the water with iron oxide and activated carbon as part of an external chemical treatment. Iron hydroxide is used in a reversible procedure that regenerates the filter by adding oxygen [99]. As an alternative, adsorption materials like iron-rich soils or leftovers from the steel or aluminum industries can be used. The governing equation is given by the relation.

$$Fe(OH)_2 + H_2S \rightarrow FeS + 2CH_2O \quad (1)$$

Activated carbon filters are used as standard components are provided for the system [77, 103].

Raw biogas is generally corrosive making it necessary to carefully select and maintain equipment material and to treat raw biogas. The corrosive nature of biogas is because of the presence of traces of hydrogen sulfide. The internal combustion engine and micro-turbine applications recommend allowable H_2S levels in the biogas, of 10 ppm or 0.0001% by volume for internal combustion engines and 5000 ppm. For microturbine. Different countries have also imposed different quality standards e.g., in Brazil the standard emission for sulfates according to CONAMA (the national environmental agency) is limited to 1.0 mg of S L^{-1}. Air pollutants containing sulfur include SO_2, SO_3, H_2S, and sulfates. Ammonia is also a corrosive element in raw biogas and can be released in the form of nitrogen oxides as a product of combustion. Nitrogen oxides are also harmful to human health and to the environment [24, 25].

5. Biogas applications

Biogas can be utilized in its raw, cleaned, or upgraded form for several applications as an energy resource and as a feedstock for several chemical processes. These applications include production of biomethane, hydrogen production, fuel cell energy source, production of biofuels, power generation, and thermal applications as a heat source [2, 5]. Biogas is used in combined heat and power (cogeneration), heating, feedstock for biomethane production used as a direct substitute of natural gas in applications like automotive fuel, industrial process heating, and feedstock in chemical industries [59, 96]. The various applications of biogas are discussed below.

5.1 Biomethane production

Biomethane is a product of biogas or syngas upgrading having superior properties to biogas, and a desirable substitute of natural gas. Compressed biomethane can be utilized as cooking gas in industrial settings, distributed by injecting it into natural gas mains, and packaged in containers or cylinders for home use. The main barrier is the processing cost, which is inversely proportionate to the type of technology used. The term "bio-CNG" refers to compressed fuel with a high methane content. To make bio-CNG, pure biogas that contains more than 97% methane is compressed to 20–25 MPa. Compressed bio-CNG has the same fuel characteristics, economics, engine performance, and emissions as regular CNG. Due to its high-octane number, bio CNG, like conventional CNG, offers outstanding thermal efficiency. As a result, it can be utilized as a direct substitute for regular compressed natural gas in gas pipes and other applications, such as a natural gas fuel source [104].

5.2 Hydrogen production

Hydrogen is an ideal raw material for a sustainable energy transformation, but with the challenge being where and how to get hydrogen from renewable sources. Renewable hydrogen can be produced using renewable energy sources and usually produced via water electrolysis [105]. Biogas has applications beyond electricity and biomethane production, as, through steam reforming, it can be used to manufacture green hydrogen, in a process where a catalyst refines and separates the hydrogen from the gas stream [41, 105]. The most common method used to manufacture hydrogen is by steam-reforming of natural gas, followed by pressure-swing adsorption to remove impurities. Small steam reformers in biogas plants are in commercial operation [105]. The biogas has low heating coefficient due to high composition of non-combustibles like carbon dioxide and water vapor which reduce the energy content [105]. Upon removal of carbon dioxide and water molecules, the methane (CH_4) can be used for hydrogen synthesis and bio-fuel production. Methane can be split to hydrogen molecules (H_2) in a process that can be done in a steam/methane reformer. In the process, high pressure and temperature steam is combined with the methane (CH_4) to produce flow of hydrogen molecules and CO molecules [48, 106].

It is through thermochemical processes of hydrocarbons that large-scale hydrogen production is manufactured through the reforming process. Biomethane has significant potential application in hydrogen manufacture as a substitute of fossil natural gas as a raw material for reforming processes. The demand for renewable hydrogen production is set to grow significantly due to concerns over fossil fuels depletion and greenhouse gas emissions, and associated concerns over global climate change. The selection of the reforming process is influenced by the availability of capital, hydrogen demand, purity hydrogen and the composition of the biogas feedstock used [8, 107].

Biogas can be used in fuel cells for power generation. This technology promises to play a key role in the hydrogen economy for sustainable energy transition. Hydrogen fuel cells can be used as prime movers in electric vehicles just like batteries do, in addition to application in power generation [21, 108]. Biomethane derived from biogas can be used as a source for renewable hydrogen, for stationary fuel cells in power generation and fuel cells as prime movers for electric vehicles (FCEVs). The hydrogen-powered FCEVs are environmentally attractive since they have no tailpipe emissions making them clean transport option and substitute for fossil fuel-powered vehicles [2, 107].

Use of biomethane for hydrogen production can increase energy sustainability for energy applications like fossil fuels. Hydrogen can be manufactured by autothermal reforming (ATR), electrolysis, or methane reforming (SMR) [109]. Biomethane can be used as a substitute for natural gas, which will provide a hedge against growing demand for natural gas [107].

Hydrogen fuel can be used to reduce emissions from engines that are widely used in transportation. Hydrogen fuel cells promise to provide an alternative to internal combustion (IC) engines particularly due to the clean exhaust emissions, renewal nature of the fuel, and higher efficiencies. Hydrogen fuel cell vehicles can achieve widespread acceptance except for existing challenges like waste heat removal in mobile applications [110].

5.3 Production of biofuels

The transport sector is important since it accounts for about 14% of the global greenhouse gas emissions [111]. Liquefied biomethane is not only an effective fuel for generators and other large machinery, but it may also be utilized as a building block in the production of other fuels and chemicals. As an alternative to fossil fuels and several other processed transport fuels, biomethane is already being used by several countries [50, 112].

Biofuels include the Bio-CNG, which is compressed biomethane like CNG in properties with industrial, automotive, and domestic applications. The process needs removal of impurities likes water, N_2, O_2, H_2S, NH_3, and CO_2 to achieve composition of >97% CH_4, <2% O_2 at 20–25 MPa. Bio-CNG occupies less than 1% of the volume at standard conditions [112, 113].

Biomethane can also be used in the industry as transport fuel by liquefying it at a high pressure ranging from 0.5 to 15 MP [4]. Through biological or chemical processes, biomethane made from biogas can be converted into methanol, diesel, LPG, and gasoline. As can be seen in the picture below, methane is partially oxidized to produce methanol [2].

$$CH_4 + 0.5\,O_2 \rightarrow CH_3OH \Delta H^0 = -128\,kJ/mol \tag{2}$$

In another method, methane is biologically converted from biomethane to methanol by using methanotrophic bacteria used in methanol production through the action of methane monooxygenase (MMO) enzyme [5].

Methanol can also be produced by reforming methane to syngas then followed by catalytic conversion of syngas to methanol as shown below [6].

$$2H_2 + CO \rightarrow CH_3OH + 2H_2 \qquad \Delta H^0 = -91\,kJ/mol \tag{3}$$

$$3H_2 + CO_2 \rightarrow CH_3OH + H_2O \qquad \Delta H^0 = -49\,kJ/mol \tag{4}$$

Then, the methanol-to-gasoline process can be utilized to convert the methanol into gasoline. Biogas or biomethane can be turned into methanol using the dry reforming, steam reforming, partial oxidation reforming, autothermal reforming (ATR), or Fischer-Tropsch (FT) process. Syngas, the main byproduct of the biomethane reforming procedure, can be used to produce a wide variety of long-chain hydrocarbons [114].

i. *Dry Reforming*

In dry reforming, CO and H_2 are produced by the reaction of methane (CH_4) and carbon dioxide (CO_2). The process uses CH_4 and CO_2 which are both greenhouse gases making it very attractive. However, the endothermic reaction reduces heat emitted in CO_2 production. Dry reforming is an effective method of creating synthesis gas with an H_2/CO ratio close to 1 is dry reforming [115]. The syngas ratio ($H_2/CO = 1$) produced by dry reforming is lower than that of steam reforming. In this reaction, the water gas shift reaction (WGS) influences H_2/CO ratio by decreasing it because of the reverse reaction that oxidises hydrogen to water. Through partial oxidation of methane with feeding water, the H_2/CO ratio is maintained between 1 and 2. This improves the responsiveness for shifting water and gas ahead. Due to the exothermic nature of partial oxidation, the energy requirement of the process is significantly reduced [114]. Dry reformation occurs within a temperature range of 700–1000°C [114, 116].

$$CH_4 + CO_2 \rightarrow 2CO + 2H_2 \qquad \Delta H^0 = 247\,kJ/mol \tag{5}$$

ii. *Steam Reforming and Water Shift Reaction*

This process combining methane in biomethane with water vapor generates CO and H_2 in the presence of a catalyst. The process is endothermic and takes place between 650 and 850°C, to produce hydrogen yield of 60–70% [115]. Steam reforming takes place between 700 and 900°C. The two-step chemical reaction is shown below.

$$CH_4 + H_2O \rightarrow CO + 3H_2 \qquad \Delta H^0 = 206\, kJ/mol \tag{6}$$

$$CO + H_2O \rightarrow CO_2 + H_2 \qquad \Delta H^0 = -41\, kJ/mol \tag{7}$$

The process of steam reforming is often followed by a water shift reaction to improve hydrogen generation.

iii. *Partial Oxidation Reforming (POR)*

Compared to steam reformation, which is very endothermic, this process produces hydrogen at a lower energy cost due to its mild exothermicity. H_2 and CO are produced by the partial oxidation at atmospheric pressure and between 700 and 900°C partial oxidation reforming. The H_2/CO ratio of 2 yield is achieved in full conversion with reduced soot formation. Methane reacts with oxygen to form carbon dioxide (CO_2) due to a decrease in CO selectivity. The high exothermicity of the combustion causes hotspots to emerge in the reactor bed and coke to deposit on the catalyst [115]. In this process, methane is oxidized to syngas as demonstrated below.

$$CH_4 + 0.5\, O_2 \rightarrow CO + 2H_2 \Delta H^0 = -25.2\, kJ/mol \tag{8}$$

iv. *Autothermal Reforming (ATR)*

Combining POR and SR in the presence of carbon dioxide results in autothermal reforming. Autothermal reforming (ATR) is a process wherein steam reforming occurs in a catalytic zone heated by heat generated by partial oxidation in the reactor. The process does not need external heating and the reactor is easy to stop and restart. Compared to partial oxidation reaction, the hydrogen yield is higher and consumes less oxygen [115].

5.3.1 Upgrading syngas

Dry reforming results in Syngas, which must be devoid of carbon dioxide before being fed into the Fischer-Tropsch reactor. Amines are highly selective in their ability to absorb carbon dioxide. Other applications of this technology are separation of CO_2 from flue gases, natural gas cleaning, and large-scale upgrading of biogas. Common solvents used in the process are alkanolamines like monoethanolamine (MEA), diethanolamine (DEA), and methyldiethanolamine (MDEA) [32, 117].

5.3.2 Fischer-Tropsch (FT) process

The Fischer-Tropsch (FT) synthesis, named after the German inventors Franz Fischer and Hans Tropsch, is a process used to manufacture liquid hydrocarbon fuels like coal-to-liquids (CTL) and/or gas-to-liquids (GTL) based on source of syngas [32, 117]. Fischer-Tropsch synthesis (FT synthesis) can be used to convert biomethane and natural gas into fuels at an industrial scale [114]. The Fischer-Tropsch (FT) process converts syngas to products like LPG, diesel, and jet fuels [117].

The Fischer-Tropsch synthesis (FT-synthesis) polymerizes the carbon and hydrogen atoms in syngas or biomethane to create long chain molecules. The process is run over iron or cobalt catalyst at 20–30 bars [14] in an overall exothermic process leading to polymerization of CH_2 to hydrocarbons with long chains called syncrude. The various reactions in Fischer-Tropsch process are summarized below.

$$(2n+1)*H_2+n*CO \rightarrow C_nH_{2n}+n*H_2O \tag{9}$$

$$2n*H_2+H_2+n*CO \rightarrow C_nH_{2n}+n*H_2O \tag{10}$$

Reactors used include multi-tubular fixed bed, circulating fluidized bed, fixed fluidized bed, and slurry reactor. The reactions for the slurry reactor conditions are 20–30 bar, and 200–300°C while the syngas H_2/CO ratio of 1–1.8 [114, 117]. For high temperature synthesis, fluidized-bed FT reactors are used to generate light hydrocarbons in the form of gaseous hydrocarbons and gasoline and generally have higher output. The catalysts used are Fe and Co, which are sensitive to sulfur compounds in syngas [117].

5.3.3 Biofuels from biomethane

Various biofuels can be made from biomethane for the transport sector e.g. methanol, compressed biogas (CBG), hydrogen, liquid biogas (LBG), dimethyl ether, and Fischer-Tropsch (FT) fuels [118]. Compressed biogas (CBG), liquefied biogas (LBG), syngas (used to make hydrogen, methanol, dimethyl ether (DME), and Fischer-Tropsch (FT) diesel), and biomethane (upgraded biogas) are all possible fuels that can be made through various processes [6, 33, 118].

5.4 Biomethane for gas and power grids

In many countries, governments have come up with national support schemes to promote the biomethane market. Support mechanisms include feed-in support schemes, green gas products, and quota obligations as market drivers in Europe. Biomethane production for many countries is based on organic waste as feedstock, but for Germany, which dominates Europe's feed-in market, it is based on energy crops. In Germany, the main driver is the feed-in tariff for renewable electricity through the Renewable Energy Sources Act (EEG). Biomethane support schemes mainly rely on mass balancing systems or 'book and claim' certificates. Existing mass balancing systems can contribute to international market development through the creation of common standards. As biomethane becomes popular and relevant in the energy systems, integration into power and gas grids and shift away from subsidies to markets and competition with natural gas will become major issues [119, 120].

Biomethane should meet some standard specifications with respect to storage and transport before it can be practically injected into existing natural gas networks. The presence of various components in different concentrations makes it difficult to inject biogas to the grid hence the need for upgrading. Pipeline designers should know what the exact thermodynamic properties of a gas mixture are, particularly in terms of density, and heating value, which may tend to vary greatly in biogas [16].

Biomethane has a very important role to play in the transition to renewable sources of energy. Demand-based production of biomethane for power generation directly links the gas grid and the electricity grids, which can help in balancing the power grid. The gas grid will shift from fossil fuel distribution to provide energy balancing service provider with short-term as well as seasonal storage options. There is increasing integration of decentralized biomethane feed-in into the gas grid for the gas grid infrastructure thus introducing new challenges. There are examples in Germany of facilities that feed in more biomethane to the local distribution network than the total discharge, which leads to the need to compress the excess gas and transfer it to a higher level [119].

Production of biomethane from energy crops has a negative impact on agriculture. On the other hand, use of digestion residues as fertilizers close to biogas production site improves local nutrient cycles. Production of energy-intensive nitrogen fertilizers and use of declining global phosphorous reserves can be avoided by use of bio-fertilizer from the digesters [51, 62].

Biomethane is the most efficient biofuel in terms of fuel production equivalent per area of crop land needed and is therefore expected to perform a larger role in the fuel/energy market because of government support, growing use in NGVs and reduction in GHG emissions. There is growing awareness of biomethane and a shift in perception from regarding biomethane as a sub-branch of biomass production to an independent renewable energy resource. And legislation and strategies are recognizing biomethane as an independent energy resource [119, 121].

The main sustainability challenge facing biomethane market is cost of subsidies and need for free market competition with fossil natural gas, which can be accelerated if the market price of natural gas rises. The European cap-and-trade for greenhouse gas emissions GHGs is another driving factor for the future. Since use of biomethane omits GHG emissions, there will not be compensation or penalties in the form of GHG certificates [119, 122].

The evolution of biomethane markets is expected to create its own demand and supply and enable and exchange between different countries since the green gas product market opened to international trade. Countries tend to create their own set of biomethane support schemes to address individual situations and therefore designed to address the priorities and challenges of specific countries. For the biomethane market to grow, countries should open their support schemes to biomethane imported from neighboring countries to encourage international trade in biomethane [6, 119].

5.5 Electricity from biogas and biomethane

Biogas can be used as fuel for power generation in various prime movers. They include internal combustion engines, gas turbines of varying sizes, and fuel cells, among others. The efficiency can be improved through combustion and conversion in set ups like cogeneration and tri-generation schemes [2, 50]. Diesel engines can run biogas as a direct substitute of natural gas. Biogas can also be upgraded to biomethane for applications like substitution of natural gas [123, 124]. Diesel engines can be run effectively either on pure diesel or in dual fuel mode with biogas and fossil fuel like diesel and petrol [123, 125].

Onsite electricity from biogas can be used directly to avoid or limit electricity power imports from the grid while excess generation can be used for a wide in fuel cells [5, 8, 126]. Various pathways for use of biomethane for power generation are summarized in **Table 3** below.

From **Table 3**, biomethane can be used through various conversion technologies with varying characteristics in thermal and electricity generation. The conversion can be done in cogeneration, trigeneration, and open conversion systems. Prime movers that can use biogas include internal combustion engines, gas turbines, fuel cells, and Stirling engines as well as production of fuels for application in transport, heat, and electricity generation [89].

In the transport sector, biomethane has a double role to play in emissions reduction i.e. as a direct fuel substitute of fossil fuels and as feedstock for production of biofuels/chemicals through the Fischer-Tropsch (FT) Process e.g. diesel, jet fuel, and gasoline, and through reforming processes to produce hydrogen and methanol [2, 100].

Biogas production, its upgrading, and use are associated with some limited greenhouse gas emissions like CO_2, CH_4, and N_2O whose quantities vary with the technology applied and the source of biogas or feedstock used. The use of biomethane reduces the negative environmental impact and pollution potential as a substitute for fossil fuel energy sources. Additionally, the anaerobic digestion and gasification used to produce biogas and syngas help keep the environment clean and healthy as a means of waste disposal [87, 127, 128]. **Figure 1** shows pathways for biogas use as a renewable source of energy.

From **Figure 1**, it is observed that upon cleaning and purification, biogas has got many applications like bio-CNG applications, CNG, heat and production, and reforming to produce syngas and fuels by the Fischer-Tropsch (FT) Process. Other energy and process products are methanol, ethanol, higher alcohols, and gasoline.

Diesel engines, petrol and oil-fired engines, turbines, microturbines, and Stirling engines are all viable options for converting biogas into mechanical power for use in power generation. Biogas can be used as a fuel in both spark ignition (petrol) and compression ignition (diesel) engines with varying degrees of modification. Internal combustion engines with a dual fuel mode can be used with minimal or no adaptation, in contrast to those that undergo a full engine conversion to petrol. Biogas is an important component in the development of sustainable energy since it may be used in fuel cells for direct conversion to electricity, and as a feedstock for the manufacture of hydrogen and transportation fuels. Biogas can also be used as a feedstock for manufacture of the Fischer-Tropsch (FT) fuels, in addition to being utilized for direct energy generation, cooking, and lighting. The biogas is cleaned and purified, then it is reformed into syngas and partially oxidized to create methanol, which can be used to make petrol. Alcohols, jet fuel, diesel, and petrol may all be made from syngas using the Fischer-Tropsch process [2, 5].

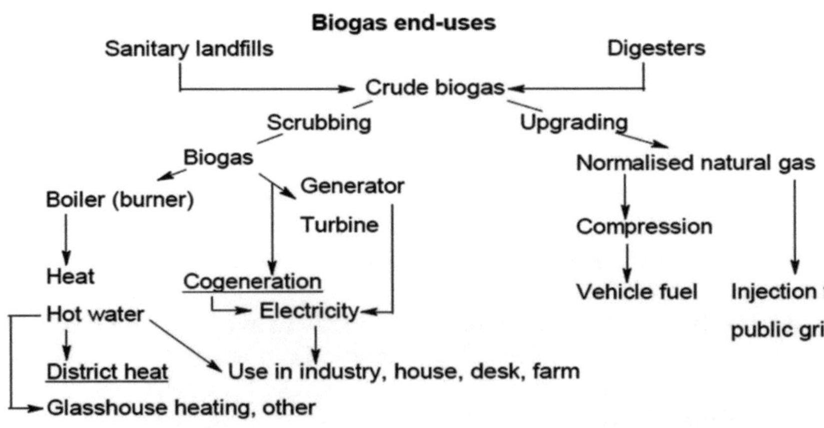

Figure 1.
Biogas applications.

6. Sustainability of biogas production and applications

Sustainable biogas systems incorporate systems for treatment of waste, environmental protection, conversion of low value material to higher value material, and power generation, heat, and advanced gaseous biofuels. Biogas and anaerobic digestion systems are dispatchable and hence can support intermittent renewable electricity [86]. Biogas adds value to the three pillars of sustainable development dimensions (SDDs) i.e., the first pillar of SDD is "Economic." SDG number 1 is related to this pillar, "No poverty," where biogas contributes by supplying affordable biofertilizer and eliminating the complex supply chain of fertilizer. Biogas contributes to SDG number 2, which is about "Zero hunger," by generating new jobs through new businesses creation and increasing the yield of crops due to affordable fertilizer, which enhances the soil fertility by replenishing lost nutrients and carbon. Biogas fuel contributes to SDG number 3, which is about "Decent work and economic growth," by increasing the gross domestic product through economic utilization of waste. Biogas also contributes to the SDG number 4, which is about "Industry, Innovation, and infrastructure," by building sustainable infrastructure and providing electricity to small-scale industries at local level [5, 8, 129].

6.1 Environmental and health impact of biogas energy resource

Biogas environmental benefits are valid and sustainable alternatives to fossil fuels that can reduce greenhouse gas (GHG) emissions and can enhance energy security. Biogas enables exploitation of agricultural and zootechnical byproducts and municipal wastes, due to their lower impact on the environment compared to other combustion-based energy options [130]. Biogas can reduce greenhouse gas emission by more than 100% by considering decreased use of fertilizer on account of substrate use and control of leakages of methane and nitrous oxides from biogas energy systems [1, 130, 131]. Biogas has environmental and health impacts to be considered before its selection and application as a fuel. Biogas has negative impacts on human health and on the air quality. Biogas has components that are potentially toxic to human health and the environment, by forming toxic substances in the process of combustion, or formation of toxic substances during photochemical aging in the atmosphere [39, 79]. Methanethane, which is odorless, consists of one atom of carbon and 4 atoms of hydrogen. It is lighter than air and is highly flammable and hence should be handled with care since it can form explosive mixtures with air at concentrations of 5 to 15%. Methane is not toxic but causes death due to asphyxiation through oxygen displacement in an enclosed environment [44, 126]. Methane is a very powerful greenhouse gas, which is about 20 times more potent than carbon dioxide and can remain in the atmosphere for up to 15 years. Therefore, uncontrolled biogas generation and handling is harmful to the environment and hence effort should be made to ensure controlled biodegradation of organic wastes to produce biogas energy resources [3, 122].

6.2 Life cycle assessment of biogas

The life cycle assessment of biogas shows that by deploying biogas technologies, energy related greenhouse gas emissions can effectively be reduced and hence reduce the climate impact of energy consumption. The use as well as production of biogas also help diversify energy systems while promoting sustainable waste management practice because anaerobic digestion also results in waste treatment [3, 132, 133]. Therefore there are significant environmental, economic, and social benefits of biogas production, using the circular agricultural waste utilization model [7, 39, 79]. Biogas is a renewable energy carrier used in electricity generation, in heat production, and as a transportation fuel. Most biogas is produced by anaerobic digestion of animal and plant materials [79, 99].

Health and environmental concerns over biogas use often hamper the social acceptance of biogas. Environmental factors considered in the use of biogas are direct emissions of gaseous pollutants e.g. nitrogen oxides (NO_x), impact of using waste biomass and digestate, nitrogen release and soil fertility, and methane leakages [130, 131, 134].

The biogas circular economy has multiple functions like waste treatment, reduction in greenhouse gas emissions, environmental production, and energy applications. The restorative and generative structure of circular economy seeks to maintain the products and materials always used and their maximum value [49, 64]. Biogas generation occurs at the last stage of anaerobic digestion with digestate, as the byproduct having valuable use as biofertilizer. Therefore biogas production is a value addition process for biowaste [30]. **Figure 2** shows the main concept of circular economy.

From **Figure 2**, it is noted that substitution of fossil fuels by biogas leads to significant environmental improvements. Generation of 1 MJ of electricity from biogas equivalently substitutes 0.4 and 0.9 MJ of fossil fuel and thus heat and power production reduces greenhouse gas emissions (GHG), by 75–90%. The emissions reduction in emissions is reduced further by substituting chemical fertilizers with digestate. Utilizing manure or food industry waste as a feedstock for biogas production is one of the feasible ways to reduce greenhouse gas emissions by values as high as 95%. Additional positive impacts include reduction in impacts like eutrophication and acidification dangers [19, 48, 62, 135]. Hence, biogas use is a strategy to control environmental problems that include eutrophication, acidification, air pollution, and greenhouse gas emissions with maximum benefits being realized through proper design and location of biogas systems [135].

Studies show that using biomethane as vehicle fuel can reduce greenhouse gas emission reductions of 27–62% as compared to emissions from conventional natural gas passenger vehicles or 41–70% compared to conventional petrol vehicles. The main challenge is limited suitable biomass to support widespread deployment of power-to-methane systems using biogenic carbon. Available biomethane

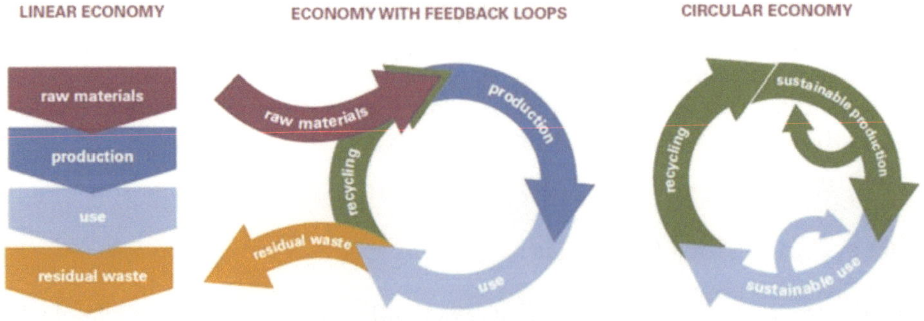

Figure 2.
Life cycle stages in the linear, recycling, and circular economy models.

production options offer attractive means to reduce greenhouse gas emissions in a future energy system with large amounts of intermittent renewable power generation [126, 136, 137].

Numerous studies have been undertaken to evaluate the environmental impacts of producing power from biogas from the anaerobic digestion of agricultural products and waste over their entire life cycles. Studies on the anaerobic digestion using tomato waste, maize silage, slurry, and tomato waste as feedstocks and for cogeneration applications suggested that systems using animal slurry offer best options, except for ecotoxicity to marine and terrestrial environment. The studies also indicate that it is more feasible to have smaller plants that use slurry and waste rather than big plants using maize silage to operate efficiently [12, 48, 138]. Electric power from biogas is environmentally more sustainable than grid electricity for many biogas schemes. However, compared to natural gas, biogas electricity is not always the best option in terms of impacts. Biogas also has higher impacts than other renewables, like solar photovoltaics, wind and solar. Overall, though, biogas electricity can help minimize the GHG emissions relative to a fossil-intensive electricity mix, although some other impacts may increase. Further analysis shows that, if the main objective is to mitigate climate change, then other renewables like wind, solar, and hydro offer greater potential to reduce GHG emissions than biogas. However, the environmental sustainability of biogas is much improved by cogeneration when heat is used, especially in terms of global warming potential, summer pollution, and the loss of ozone layer and abiotic resources.

6.3 Biogas in the circular economy

Circular economy refers to an economic concept that is the reverse of the linear system whose main objective is to minimize waste through reuse or recycling. Realizing a circular economy is a challenge for many countries because many biogas plants are generally small energy production facilities, having a smaller impact to waste management [86, 139]. Rural areas have an opportunity to diversify agriculture activities by adding value to the socio-economic systems in additional. Biogas plants should be implemented in a way that properly communicates with local communities to participate in the transformation. The operation of biogas plants is generally compatible with the goals of a circular economy that makes use of local energy resources. Locations for new biogas plants should minimize the need for substrate transport to increase their local character. Successful implementation of largescale biogas systems can help in achieving the Sustainable Development Goal 7 on achieving affordable and clean energy, and Goal 12 on responsible consumption and production [139].

Avoiding open digestate storage reduces methane emissions, and controlling how much of it is dispersed on land reduces ammonia emissions and other negative environmental effects [140]. **Figure 3** depicts the layout of a biogas plant in a circular design.

From **Figure 3**, the biogas role in the circular economy is presented. The first stage involves the substrate collection and preparation to feed the biodigesters where biogas is produced with digestate as a byproduct.

The traditional scenario model is such that biowastes from agricultural and livestock farming are disposed thus creating a cost for the household or the firm disposing of the waste [30]. In the circular economy scenario, manure and agricultural wastes are used to produce biogas and digestate as fertilizer, which help preserve the environment since biogas energy is an alternative to fossil energy and has socio-economic value to farmers and the biogas producer [30, 86].

Biogas production offers treatment and management of bio digestible animal and plant waste, which sterilizes waste and produces useful manure and sewage leading to significant volume and mass reduction, which effectively reduces the cost of waste treatment and disposal [44, 141, 142]. Use of biogas as a fuel significantly reduces greenhouse gas emissions while the use of animal manure as biodigester feedstock reduces methane emissions from the manure storage and use. Carbon dioxide (CO_2) emissions from biogas have a lower global warming potential than methane produced by fossil fuel. Therefore, biogas production and use reduce greenhouse gas emissions in accordance with the Kyoto protocol and Paris agreements. The negative effect is that reduced use of manure increases use of chemical fertilizers for fertility or nutrient replenishment to the soil [2, 30]. A few of the potential sustainability issues with employing agricultural wastes for biogas production are competition with feeding in animal husbandry and possible loss of nutrients in agricultural land [70].

To consider biogas applications like use as a vehicle fuel as sustainable requires a sustainability criterion to be met along the entire production chain, beginning with primary production to final use and disposal. Sustainability demands that they should not destroy areas of high biological value or generate excessive GHG emissions. For biofuels to decrease greenhouse gas emissions, the sustainability criteria require that the associated greenhouse gas emissions (GHG) emissions over the life cycle should be at least 35% lower compared to fossil fuels [1, 134].

Bio-based economy-enabled technologies lead to green electricity and heat generation and fossil fuel substitution in transport and power generation, and manufacture more value-added products and byproducts. Bio digestion industrial operations, agriculture, and other anthropogenic activities such as food waste (FW) can produce valuable energy sources, nutrient-rich manure, and specialty chemicals. Several anaerobic and microbial interventions sustain biomass valorization and related processes, leading to more efficient bio methanation [85].

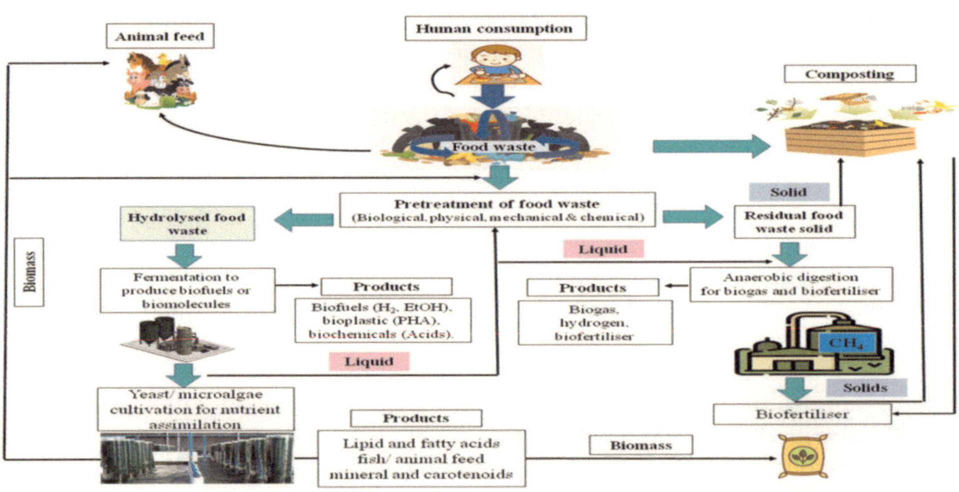

Figure 3.
Functions of biogas in the circular economy.

7. Conclusion

Resource overuse, climate change, and energy autonomy are challenges facing the volatile world. Biogas production and use represents a model example of the circular economy, which is an economy that *offers a path for sustainable growth amid growing energy demand and production requirements.* Anaerobic digestion is a biological process where biodegradable matter is degraded by the action of microorganisms in absence of oxygen to produce biogas, which is a clean fuel from renewable feedstocks with potential use in heat and electricity generation. Factors affecting bio digestion include type of raw materials, temperature, pH, mixing, level preparation and pretreatment, residence time, redox effect, and hydraulic pressure etc. The efficiency of anaerobic digestion can be evaluated in terms of bio-methane potential, gross energy yields, process energy demand, net energy yield, and energy efficiency. Energy crops are the most used substrate for biogas production.

Concerns about greenhouse gas emissions and depletion of fossil fuels have increased demand for sustainable renewable sources. Biogas production is one of the innovative and viable methods for production of energy from biowastes and biomaterials, with an extra benefit of producing biofertilizer to support sustainable agriculture. Huge quantities of biodegradable waste are generated by humanity globally, which can pose a serious threat to human and animal health and their environment. There are many wastes disposal and treatment methods employed to avoid environmental pollution. Among the many options, anaerobic digestion is a popular non-conventional energy source used to produce renewable energy from biodegradable biomass.

Biogas is a product of the anaerobic digestion process with many applications as in generation of renewable energy. The main component of biogas with energy value is methane, but it has impurities like moisture, carbon dioxide, siloxanes, hydrogen sulfide, siloxanes, hydrocarbons, oxygen, ammonia, oxygen, carbon monoxide, and nitrogen whose presence is undesirable as they reduce the calorific value of biogas and create operational problems in the energy systems. This necessitates biogas cleaning and application of multi-stage technologies to produce upgraded biogas called biomethane. Biomethane gas is a flexible and easy to store fuel with similar properties and applications to natural gas with no need to modify any equipment settings for natural gas devices and equipment.

The factors influencing the process and quality of biogas produces include the C/N (carbon nitrogen ratio, the pH, dilution of feedstock, composition and nutritive value of feedstock, residence period, mixing and stirring, temperature, presence of toxicants, loading time, and redox conditions. The main challenges facing biogas production and use include lack of technical training on biogas production and use and lack of subsidies and incentives. Therefore, besides easy access to appropriate and affordable biogas technologies, there is a need for developing a comprehensive policy on construction and operation of biodigesters as well as utilization of biogas and sale of biogas energy products and services. Biogas education and exposure, availability of simple and affordable biodigesters, and facilitating legal and policy environment are key tools needed to increase the adoption of biogas technology and its sustainable use.

Raw or cleaned biogas that is free from harmful contaminants has significant energy value and applications. The multifunctional uses of biogas produce include the production of biofuels and renewable energies, reduction in greenhouse gas emissions mainly methane and carbon dioxide, reduction in nuisances resulting from odors and flies associated with uncontrolled biodegrading of biomass, economic or financial savings for farmers, sustainable waste treatment and recycling of organic waste, minimization of water and air pollution, rural poverty alleviation, and better social conditions of rural or remote settlements like gender balance, as women who bear the responsibility of looking for firewood are relieved of the struggle, and conservation of forests and vegetation cover due to reduced use of firewood and charcoal.

Funding

There was no external funding provided for this research and publication charges were paid by the Durban University of Technology.

Conflict of interest

The authors declare that they have no conflict of interest.

Consent for publication

The authors have authority to publish research work.

Availability of data

The research has provided all data and information used and did not use any undeclared data and information. However, any datasets used and/or analyzed during the current study are available from the corresponding author on reasonable request.

Ethical approval and consent to participate

Not applicable.

Author details

Moses Jeremiah Barasa Kabeyi* and Oludolapo Akanni Olanrewaju
Industrial Engineering Department, Durban University of Technology, Durban, South Africa

*Address all correspondence to: moseskabeyi@yahoo.com

References

[1] Cook D, Davíðsdóttir B, Kristófersson DM. Energy projects in Iceland – Advancing the case for the use of economic valuation techniques to evaluate environmental impacts. Energy Policy. 2016;**94**:104-113. DOI: 10.1016/j.enpol.2016.03.044 [Accessed: Jul. 01, 2016]

[2] Kabeyi MJB, Olanrewaju OA. Biogas production and applications in the sustainable energy transition. Journal of Energy. 2022;**2022**(8750221):43. DOI: 10.1155/2022/8750221 [Accessed: Jul. 09, 2022]

[3] Li Y et al. Composition and toxicity of biogas produced from different feedstocks in California. Environmental Science & Technology. 2019;**53**(19): 11569-11579. DOI: 10.1021/acs.est.9b03003

[4] Kabeyi MJB, Olanrewaju OA. The potential of grid power generation from municipal solid waste for Nairobi city. In: Presented at the Second African International Conference on Industrial Engineering and Operations Management, Harare, Zimbabwe, 5-7 December 2020. 2020. p. 081 [Online]. Available from: http://ieomsociety.org/harare2020/papers/81.pdf

[5] Kabeyi MJB, Olanrewaju OA. Technologies for biogas to electricity conversion. Energy Reports. 2022;**8** (Supplement 16):774-786. DOI: 10.1016/j.egyr.2022.11.007 [Accessed: Dec. 01, 2022]

[6] Moses Jeremiah Barasa K, Oludolapo AO. Biomethane production and applications. In: Sevcan A, editor. Anaerobic Digestion - Biotechnology for Reactor Performance and Environmental Sustainability. Rijeka: IntechOpen; 2023 . Ch. 9. p. 45

[7] Huang X, Wang S, Shi Z, Fang L, Yin C. Challenges and strategies for biogas production in the circular agricultural waste utilization model: A case study in rural China. Energy. 2022;**241**:122889. DOI: 10.1016/j.energy.2021.122889 [Accessed: Feb. 15, 2022]

[8] Kabeyi MJB. Project and program evaluation consultancy with terms of reference, challenges, opportunities, and recommendations. International Journal of Project Management and Productivity Assessment (IJPMPA). 2020;**8**(2):47-68. DOI: 10.4018/IJPMPA.2020070103

[9] Singh B et al. Enhancing efficiency of anaerobic digestion by optimization of mixing regimes using helical ribbon impeller. Fermentation. 2021;**7**(4):251 [Online]. Available from: https://www.mdpi.com/2311-5637/7/4/251

[10] Kabeyi MJB, Olanrewaju O. Slaughterhouse waste to energy in the energy transition with performance analysis and design of slaughterhouse biodigestor. Journal of Energy Management and Technology. 2022; **6**(3):176-196. DOI: 10.22109/jemt.2021.292954.1309

[11] Kabeyi MJB. Sustainability in the energy transition to renewable and low carbon grid electricity generation and supply. Frontiers in Energy Research. 2022;**9**:1-45. DOI: 10.3389/fenrg.2021.743114

[12] Kabeyi MJB, Olanrewaju OA. Energy and environment: Opportunities and challenges for the energy transition. In: Presented at the Fifth European Conference on Industrial Engineering and Operations Management, Rome, Italy, July 26–28, 2022. Michigan: IEOM Society International; 2022. p. 427

[Online]. Available from: https://ieomsociety.org/proceedings/2022rome/427.pdf

[13] Kabeyi MJB, Olanrewaju OA. The role of electrification of transport in the energy transition. In: Presented at the Fifth European Conference on Industrial Engineering and Operations Management, Rome, Italy, July 26–28, 2022. Michigan: IEOM Society International; 2022. p. 426 [Online]. Available from: https://ieomsociety.org/proceedings/2022rome/426.pdf

[14] Sahota S et al. Review of trends in biogas upgradation technologies and future perspectives. Bioresource Technology Reports. 2018;**1**:79-88. DOI: 10.1016/j.biteb.2018.01.002 [Accessed: Jan 03, 2018]

[15] Abanades S, Abbaspour H, Ahmadi A, Das B, Ehyaei MA, Esmaeilion F, et al. A critical review of biogas production and usage with legislations framework across the globe. International Journal of Environmental Science and Technology. 2021;**19**(4):3377-3400. DOI: 10.1007/s13762-021-03301-6 [Accessed: May 16, 2021]

[16] Farzaneh-Gord M, Mohseni-Gharyehsafa B, Arabkoohsar A, Ahmadi MH, Sheremet MA. Precise prediction of biogas thermodynamic properties by using ANN algorithm. Renewable Energy. 2020;**147**:179-191. DOI: 10.1016/j.renene.2019.08.112 [Accessed: Jan 03, 2020]

[17] Holliger C, Fruteau de Laclos H, Hack G. Methane production of full-scale anaerobic digestion plants calculated from Substrate's biomethane potentials compares well with the one measured on-site. Frontiers in Energy Research, Original Research. 2017;**5**(12):1-9. DOI: 10.3389/fenrg.2017.00012 [Accessed: Jun. 09, 2017]

[18] Kabeyi M, Olanrewaju O. Performance analysis of a sugarcane bagasse cogeneration power plant in grid electricity generation. In: Presented at the 11th Annual International Conference on Industrial Engineering and Operations Management Singapore, March 7-11, 2021. Michigan, USA: IEOM Society International; 2021 [Online]. Available from: http://www.ieomsociety.org/singapore2021/papers/201.pdf

[19] Kabeyi M, Olanrewaju O. Preliminary design of a cogeneration plant for a 120 MW diesel engine power plant. In: Presented at the 12th Annual Istanbul International Conference on Industrial Engineering and Operations Management, Istanbul, Turkey, March 7–10, 2022. Michigan: IEOM Society International; 2022. p. 411 [Online]. Available from: https://ieomsociety.org/proceedings/2022istanbul/411.pdf

[20] IEA. Outlook for biogas and biomethane: Prospects for organic growth. In: International Energy Agency. Paris: IEA; 2020 [Online]. Available from: https://www.iea.org/reports/outlook-for-biogas-and-biomethane-prospects-for-organic-growth/an-introduction-to-biogas-and-biomethane

[21] Kabeyi MJB, Olanrewaju OA. Bagasse electricity potential of conventional sugarcane factories. Journal of Energy. 2023;**2023**:1-25. DOI: 10.1155/2023/5749122 [Accessed: May 19, 2023]

[22] Adnan AI, Ong MY, Nomanbhay S, Chew KW, Show PL. Technologies for biogas upgrading to biomethane: A review. Bioengineering. 2019;**6**(4):1-23. Available from: https://www.mdpi.com/2306-5354/6/4/92

[23] Beegle JR, Borole AP. Energy production from waste: Evaluation of anaerobic digestion and

bioelectrochemical systems based on energy efficiency and economic factors. Renewable and Sustainable Energy Reviews. 2018;**96**:343-351. DOI: 10.1016/j.rser.2018.07.057 [Accessed: Nov. 01, 2018]

[24] Salomon KR, Silva Lora EE. Estimate of the electric energy generating potential for different sources of biogas in Brazil. Biomass and Bioenergy. 2009; **33**(9):1101-1107. DOI: 10.1016/j.biombioe.2009.03.001 [Accessed: Sep. 01, 2009]

[25] Ogunjuyigbe ASO, Ayodele TR, Alao MA. Electricity generation from municipal solid waste in some selected cities of Nigeria: An assessment of feasibility, potential and technologies. Renewable and Sustainable Energy Reviews. 2017;**80**:149-162. DOI: 10.1016/j.rser.2017.05.177 [Accessed: Dec. 01, 2017]

[26] Kumar R, Jilte R, Ahmadi MH. Electricity alternative for e-rickshaws: An approach towards green city. International Journal of Intelligent Enterprise (IJIE). 2018;**5**(4): 333-344. DOI: 10.1504/IJIE.2018.10016762

[27] Achinas S, Achinas V, Euverink GJW. A technological overview of biogas production from biowaste. Engineering. 2017;**3**(3):299-307. DOI: 10.1016/J.ENG.2017.03.002 [Accessed: Jun. 01, 2017]

[28] Pasternak G. Chapter 9 - Electrochemical approach for biogas upgrading. In: Aryal N, Ottosen LDM, Kofoed MVW, Pant D, editors. Emerging Technologies and Biological Systems for Biogas Upgrading. Cambridge, Massachusetts: Academic Press; 2021. pp. 223-254

[29] Stevens DJ. Hot gas conditioning: Recent progress with larger-scale biomass gasification systems. In: National Renewable Energy Lab. (NREL), USA, August 2001. Colorado: National Renewable Energy Laboratory; 2001 [Online]. Available from: https://www.osti.gov/biblio/786288-hot-gas-conditioning-recent-progress-larger-scale-biomass-gasification-systems-update-summary-recent-progress

[30] Tagne RFT, Dong X, Anagho SG, Kaiser S, Ulgiati S. Technologies, challenges and perspectives of biogas production within an agricultural context. The case of China and Africa. Environment, Development and Sustainability. 2021;**23**(10):14799-14826. DOI: 10.1007/s10668-021-01272-9 [Accessed: Nov. 01, 2021]

[31] Kabeyi MJB, Olanrewaju OA. Relationship between electricity consumption and economic development. In: Presented at the International Conference on Electrical, Computer and Energy Technologies (ICECET), Cape Town-South Africa, 9–10 December 2021. New Jersey: Institute of Electrical and Electronics Engineers; 2022 [Online]. Available from: https://ieeexplore.ieee.org/stamp/stamp.jsp?tp=&arnumber=9698413

[32] Herz G, Reichelt E, Jahn M. Design and evaluation of a Fischer-Tropsch process for the production of waxes from biogas. Energy. 2017;**132**:370-381. DOI: 10.1016/j.energy.2017.05.102 [Accessed: Aug. 01, 2017]

[33] Kabeyi MJB, Olanrewaju OA, Akpan J. Biogas production and process control improvements. In: Eduardo J-L, Leila Queiroz Z, Rosangela Rodrigues D, editors. From Biomass to Biobased Products. Rijeka: IntechOpen; 2024. p. Ch. 9

[34] Sarkar S et al. Management of crop residues for improving input use efficiency and agricultural sustainability. Sustainability, Review. 2020;**12**(9808): 1-24. DOI: 10.3390/su12239808

[35] Palit D et al. The trials and tribulations of the village energy security programme (VESP) in India. Energy Policy. 2013;**57**(2013):407-417. DOI: 10.1016/j.enpol.2013.02.006

[36] I.E. Agency. Renewables Information: Overview. International Energy Agency. Available from: https://www.iea.org/reports/renewables-information-overview [Accessed: Sep 30, 2020]

[37] IEA. Global Energy Review 2020. Paris France: International Energy Agency; 2020. Available from: https://www.iea.org/reports/global-energy-review-2020/renewables [Accessed: Dec 16, 2020]

[38] Machado PG, Teixeira ACR, Collaço FMA, Mouette D. Review of life cycle greenhouse gases, air pollutant emissions and costs of road medium and heavy-duty trucks. WIREs Energy and Environment. 2021;**10**(4):1-16. DOI: 10.1002/wene.395

[39] Kabeyi MJB, Oludolapo AO. Preliminary Design of a Bagasse Based Firm Power Plant for a sugar factory. In: Presented at the South African Universities Power Engineering Conference (SAUPEC), Nortn West University, South Africa, 27–28 January 2021. New Jersey: Institute of Electrical and Electronics Engineers; 2021. p. 104 [Online]. Available from: https://ieeexplore.ieee.org/abstract/document/9377242

[40] Kabeyi MJB, Oludolapo AO. Central versus wellhead power plants in geothermal grid electricity generation. Energy, Sustainability and Society. 2021; **11**(7):1-23. DOI: 10.1186/s13705-021-00283-8. Art no. ESSO-D-20-00011R4

[41] Kabeyi MJB, Olanrewaju OA. Geothermal wellhead technology power plants in grid electricity generation: A review. Energy Strategy Reviews. 2022;**39**: 100735. DOI: 10.1016/j.esr.2021.100735 [Accessed: Jan. 01, 2022]

[42] Kabeyi MJB. Geothermal electricity generation, challenges, opportunities and recommendations. International Journal of Advances in Scientific Research and Engineering (IJASRE). 2019;**5**(8):53-95. DOI: 10.31695/IJASRE.2019.33408

[43] Lebuhn M, Munk B, Effenberger M. Agricultural biogas production in Germany - from practice to microbiology basics. Energy, Sustainability and Society. 2014;**4**(1):10. DOI: 10.1186/2192-0567-4-10 [Accessed: May 20, 2014]

[44] Kabeyi MJB, Oludolapo AO. Optimization of biogas production for optimal abattoir waste treatment with bio-methanation as solution to Nairobi slaughterhouses waste disposal. In: Presented at the Second African International Conference on Industrial Engineering and Operations Management, Harare, Zimbabwe, 5–7 December 2020. Michigan: IEOM Society International; 2020. p. 083 [Online]. Available from: http://ieomsociety.org/harare2020/papers/83.pdf

[45] Kabeyi MJB, Oludolapo AO. Development of a cereal grain drying system using internal combustion engine waste heat. In: Presented at the 11th Annual International Conference on Industrial Engineering and Operations Management, Singapore, March 7–11, 2021. Michigan: IEOM Society

International; 2021 [Online]. Available from: http://www.ieomsociety.org/singapore2021/papers/188.pdf

[46] Kabeyi MJB, Oludolapo AO. The potential of power generation from municipal solid waste. In: Presented at the Second African International Conference on Industrial Engineering and Operations Management, Harare, Zimbabwe, 5–7 December 2020. Michigan: IEOM Society International; 2020. p. 081 [Online]. Available from: http://ieomsociety.org/harare2020/papers/81.pdf

[47] Bond T, Templeton MR. History and future of domestic biogas plants in the developing world. Energy for Sustainable Development. 2011;**15**(4): 347-354. DOI: 10.1016/j.esd.2011.09.003 [Accessed: Dec. 01, 2011]

[48] Kabeyi M, Olanrewaju O. Diesel powerplants: Design and operation and performance enhancements. In: Presented at the Fifth European Conference on Industrial Engineering and Operations Management, Rome, Italy, July 26–28, 2022. Michigan: IEOM Society International; 2022. p. 425 [Online]. Available from: https://ieomsociety.org/proceedings/2022rome/425.pdf

[49] Kabeyi M, Olanrewaju O. Feasibility of conversion from diesel engine to natural gas power plants. In: Presented at the IECON 2022 – 48th Annual Conference of the IEEE Industrial Electronics Society, Brussels, Belgium, 17–20 Oct. 2022. New Jersey: Institute of Electrical and Electronics Engineers; 2022 [Online]. Available from: https://ieeexplore.ieee.org/document/9968428

[50] Kabeyi MJB, Olanrewaju OA. Cogeneration potential of an operating diesel engine power plant. Energy Reports. 2022;**8**(16):744-754. DOI: 10.1016/j.egyr.2022.10.447 [Accessed: Dec. 01, 2022]

[51] Kabeyi MJB, Oludolapo AO. Development of a biogas plant with electricity generation, heating and fertilizer recovery systems. In: Presented at the Second African International Conference on Industrial Engineering and Operations Management, Harare, Zimbabwe, 5–7 December 2020. Michigan: IEOM Society International; 2020. p. 082 [Online]. Available from: http://ieomsociety.org/harare2020/papers/82.pdf

[52] Kabeyi M, Olanrewaju O. Development of a cereal grain drying system using internal combustion engine waste heat. In: Presented at the 11th Annual International Conference on Industrial Engineering and Operations Management Singapore, March 7–11, 2021. Michigan: IEOM Society International; 2021 [Online]. Available from: http://www.ieomsociety.org/singapore2021/papers/188.pdf

[53] Barragán-Escandón A, Olmedo Ruiz JM, Curillo Tigre JD, Zalamea-León EF. Assessment of power generation using biogas from landfills in an equatorial tropical context. Sustainability. 2020;**12, 7**:2669 [Online]. Available from: https://www.mdpi.com/2071-1050/12/7/2669

[54] Kabeyi MJB. Sustainability in the energy transition to renewable and low carbon grid electricity generation and supply. Frontiers in Energy Research. 2022;**9**(743114):1-45. DOI: 10.3389/fenrg.2021.743114

[55] Kabeyi MJB, Olanrewaju OA. A techno-economic assessment of diesel to gas power plant conversion. In: Presented at the 12th Annual Istanbul International Conference on Industrial Engineering and Operations

Management, Istanbul, Turkey, March 7–10, 2022. Michigan: IEOM Society International; 2022. p. 406 [Online]. Available from: https://ieomsociety.org/proceedings/2022istanbul/406.pdf

[56] Arogo J, Ignosh J, Bendfeldt E. Biomethane production technology [Online]. Available from: https://www.pubs.ext.vt.edu/content/dam/pubs_ext_vt_edu/442/442-881/BSE-240.pdf

[57] Tanigawa S. Biogas: Converting waste to energy. 2017 [Online]. Available from: https://www.eesi.org/files/FactSheet_Biogas_2017.09.pdf

[58] Kabeyi MJB, Olenwaraju AO. Managing sustainability in electricity generation. In: Presented at the 2020 IEEE International Conference on Industrial Engineering and Engineering Management (IEEM), Singapore, 14–17 December 2020. New Jersey: Institute of Electrical and Electronics Engineers; 2020. IEEM20-P-0406

[59] Uddin W et al. Biogas potential for electric power generation in Pakistan: A survey. Renewable and Sustainable Energy Reviews. 2016;**54**:25-33. DOI: 10.1016/j.rser.2015.09.083 [Accessed: Feb. 01, 2016]

[60] Ugwu SN, Enweremadu CC. Enhancement of biogas production process from biomass wastes using iron-based additives: Types, impacts, and implications. Energy Sources, Part A: Recovery, Utilization, and Environmental Effects;**44**(2): 4458-4480. DOI: 10.1080/15567036.2020.1788675 [Accessed: Jun. 15, 2022]

[61] Ullah Khan I et al. Biogas as a renewable energy fuel – A review of biogas upgrading, utilisation and storage. Energy Conversion and Management. 2017;**150**:277-294. DOI: 10.1016/j.enconman.2017.08.035 [Accessed: Oct 15, 2017]

[62] Kabeyi MJB, Olanrewaju O. Optimum biodigestor design and operations. In: Presented at the Fifth European Conference on Industrial Engineering and Operations Management, Rome, Italy, July 26–28, 2022. Michigan: IEOM Society International; 2022. p. 424 [Online]. Available from: https://ieomsociety.org/proceedings/2022rome/424.pdf

[63] Kabeyi MJB, Olanrewaju OA. Fuel from plastic wastes for sustainable energy transition. In: Presented at the 11th Annual International Conference on Industrial Engineering and Operations Management, Singapore, March 7–11, 2021. Michigan, USA: IEOM Society International; 2021 [Online]. Available from: http://www.ieomsociety.org/singapore2021/papers/199.pdf

[64] Kabeyi MJB, Olanrewaju OA. Review and design overview of plastic waste-to-pyrolysis oil conversion with implications on the energy transition. Journal of Energy. 2023;**2023**:1821129. DOI: 10.1155/2023/1821129 [Accessed: May 22, 2023]

[65] Chen B, Hayat T, Alsaedi A.. Biogas systems in China. 2017

[66] Energypedia. Biogas basics. Energypedia. 2020. Available from: https://energypedia.info/wiki/Biogas_Basics [Accessed: Dec. 23, 2020]

[67] Kariuki LN. Indoor carbon Monoxide emission and particulates produced from combustion of carbon based Fuels and their health implications on rural households of Manyatta division, Embu Districts, Kenya [Msc

thesis]; Department of environmental Sciences, Kenyatta University. 2009

[68] Kabeyi MJBK, Olanrewaju OA. The use of smart grids in the energy transition. In: 2022 30th Southern African Universities Power Engineering Conference (SAUPEC), 25–27 Jan. 2022. New Jersey: Institute of Electrical and Electronics Engineers; 2022. pp. 1-8. DOI: 10.1109/SAUPEC55179.2022.9730635 [Online]. Available from: https://ieeexplore.ieee.org/abstract/document/9730635

[69] Farooque M, Leo A, Rauseo A, Wang J-Y. Efficient and ultra-clean use of biogas in the fuel cell - the DFC experience. Energy, Sustainability and Society. 2015;**5**(1):11. DOI: 10.1186/s13705-015-0041-0 [Accessed: Apr. 16, 2015]

[70] Kulichkova GI et al. Plant feedstocks and their biogas production potentials. The Open Agriculture Journal. 2020; **2020**(14):219-234. DOI: 10.2174/1874331502014010219

[71] Kabeyi MJB, Oludolapo AO. Performance analysis of an open cycle gas turbine power Plant in Grid Electricity Generation. In: Presented at the 2020 IEEE International Conference on Industrial Engineering and Engineering Management (IEEM), Singapore, Singapore, 14–17 December 2020. New Jersey: Institute of Electrical and Electronics Engineers; 2020. IEEM20-P-0438 [Online]. Available from: https://ieeexplore.ieee.org/stamp/stamp.jsp?tp=&arnumber=9309840

[72] Kabeyi MJB, Oludolapo AO. Performance analysis of diesel engine power plants for grid electricity supply. In: Presented at the 31ST Annual Southern African Institution for Industrial Engineering Conference, South Africa, 5th – 7th October 2020. Pretoria: Southern African Institution for Industrial Engineering Conference; 2020. p. 4423 [Online]. Available from: https://www.saiie.co.za/system/files/2021-11/SAIIE31%20Conference%20Proceedings.pdf

[73] Kabeyi MJB, Oludolapo AO. Fuel from plastic wastes for sustainable energy transition. In: Presented at the 11th Annual International Conference on Industrial Engineering and Operations Management, Singapore, March 7–11, 2021. Michigan: IEOM Society International; 2021 [Online]. Available from: http://www.ieomsociety.org/singapore2021/papers/199.pdf

[74] Kabeyi MJB, Oludolapo AO. The potential of grid power generation from municipal solid waste for Nairobi City. In: Presented at the 2nd African International Conference on Industrial Engineering and Operations Management, Harare, Zimbabwe, 5–7 December 2020. Michigan: IEOM Society International; 2020. p. 081 [Online]. Available from: http://ieomsociety.org/harare2020/papers/81.pdf

[75] Das S, Sherpa MT, Najar IN, Thakur N. Biomethane. Environmental Sustainability. 2020;**3**(4):453. DOI: 10.1007/978-981-13-8307-6_5

[76] Soeder DJ, Borglum SJ. 7 - energy economics. In: Soeder DJ, Borglum SJ, editors. The Fossil Fuel Revolution: Shale Gas and Tight Oil. Washington, D.C., United States: United States Environmental Protection Agency; 2019. pp. 213-247

[77] Energypedia. Biogas plant used for power generation.png. Energypedia. 2020. Available from: https://energyped

ia.info/wiki/Electricity_Generation_from_Biogas [Accessed: Dec. 16, 2020]

[78] Kabeyi MJB. Challenges of implementing thermal powerplant projects in Kenya, the case of Kipevu III 120MW power station, Mombasa Kenya, Masters, Department of Education Management, University of Nairobi, Nairobi, 5866. 2012. [Online]. Available from: http://erepository.uonbi.ac.ke:8080/xmlui/handle/123456789/11023

[79] Kabeyi MJB, Olanrewaju OA. The levelized cost of energy and modifications for use in electricity generation planning. Energy Reports. 2023;**9**:495-534. DOI: 10.1016/j.egyr.2023.06.036 [Accessed: Sep. 01, 2023]

[80] Kabeyi MJB, Olanrewaju OA. Smart grid technologies and application in the sustainable energy transition: A review. International Journal of Sustainable Energy. 2023;**42**(1):685-758. DOI: 10.1080/14786451.2023.2222298 [Accessed: Dec. 14, 2023]

[81] Ltd NNFCC. Biogas. NNFCC Ltd. Available from: https://www.nnfcc.co.uk/biogas; [Accessed: 2021]

[82] AgSTAR. Anaerobic digester/Biogas system operator guidebook. In: A Guidebook for Operating Anaerobic Digestion/Biogas Systems on Farms in the United States. Washington, D.C., United States: United States Environmental Protection Agency; 2020. EPA 430-B-20-003 [Online] Available from: https://www.epa.gov/sites/production/files/2020-11/documents/agstar-operator-guidebook.pdf [Accessed: May 17, 2020]

[83] Akpojaro J, Ofualagba G, Akpojaro MA. Electricity generation from cow dung biogas. Journal of Applied Sciences and Environmental Management. 2019;**23**(7):1301-1307. DOI: 10.4314/jasem.v23i7.17

[84] Awe OW, Zhao Y, Nzihou A, Minh DP, Lyczko N. A review of biogas utilisation, purification and upgrading technologies review. Waste and Biomass Valorization. 2017;**8**(2):267-283. DOI: 10.1007/s12649-016-9826-4 [Accessed: Mar. 2017]

[85] Singh PK, Mohanty P, Mishra S, Adhya TK. Food waste valorisation for biogas-based bioenergy production in circular bioeconomy: Opportunities, challenges, and future developments. Frontiers in Energy Research, Review. 2022;**10**:1-13. DOI: 10.3389/fenrg [Accessed: Aug 11, 2020]

[86] Burg V, Rolli C, Schnorf V, Scharfy D, Anspach V, Bowman G. Agricultural biogas plants as a hub to foster circular economy and bioenergy: An assessment using substance and energy flow analysis. Resources, Conservation and Recycling. 2023;**190**:106770. DOI: 10.1016/j.resconrec.2022.106770 [Accessed: Mar. 01, 2023]

[87] Bhardwaj S, Das P. A review: Advantages and disadvantages of biogas. International Research Journal of Engineering and Technology (IRJET). 2017;**4**(10):890-893

[88] Mambeli Barros R, Tiago Filho GL, da Silva TR. The electric energy potential of landfill biogas in Brazil. Energy Policy. 2014;**65**:150-164. DOI: 10.1016/j.enpol.2013.10.028 [Accessed: Feb. 01, 2014]

[89] Kabeyi M, Olanrewaju O. Slaughterhouse waste to energy in the energy transition with performance analysis and design of slaughterhouse biodigestor. Journal of Energy Management and Technology. 2022;

6(3):188-208. DOI: 10.22109/jemt.2021.292954.1309

[90] Asankulova A, Obozov AD. Biogas in Kyrgyzstan. Applied Solar Energy. 2007;**43**(4):262-265. DOI: 10.3103/S0003701X0704017

[91] Engines for biogas. 1988. 3-528-02032-6 [Online] Available from: https://biogas.ifas.ufl.edu/ad_development/documents/Engines%20for%20biogas.pdf

[92] Kabeyi MJB, Oludolapo AO. Design and modelling of a waste heat recovery system for a 250KW diesel engine for cereals drying. In: Presented at the Second African International Conference on Industrial Engineering and Operations Management Harare, Zimbabwe, December 7–10, 2020. Michigan: IEOM Society International; 78 [Online]. Available from: http://www.ieomsociety.org/harare2020/papers/78.pdf

[93] Kabeyi MJB. Feasibility of wellhead technology power plants for electricity generation. International Journal of Computer Engineering in Research Trends. 2020;**7**(2):1-16. DOI: 10.22362/ijcert/2020/v7/i02/v7i0201

[94] Kabeyi MJB, Olanrewaju AO. Biogas treatment and upgrading techniques. In: Presented at the Fourth Asia Pacific International Conference on Industrial Engineering and Operations Management. Michigan: IEOM Society International; 2023 [Online]. Available from: https://index.ieomsociety.org/index.cfm/article/view/ID/13606

[95] Cedigaz. Global biomethane market: green gas goes global. 20 March 2019. Retrieved 24 August 2023. Available from: https://www.cedigaz.org/global-biomethane-market-green-gas-goes-global/

[96] Kabeyi MJB, Olanrewaju O. Diesel to gas engine power plant conversion: A review and preliminary design for an operting power plant. Journal of Energy Management and Technology. 2023;**7**(2):103-115. DOI: 10.22109/jemt.2022.292982.1312

[97] Kabeyi MJB, Olanrewaju AO. Biogas as a sustainable fuel and feedstock: Properties, purification and applications. In: Sevcan A, editor. Anaerobic Digestion - Biotechnology for Reactor Performance and Environmental Sustainability. Rijeka: IntechOpen; 2023

[98] Bui VG et al. Flexible syngas-biogas-hydrogen fueling spark-ignition engine behaviors with optimized fuel compositions and control parameters. International Journal of Hydrogen Energy. 2023;**48**(18):6722-6737. DOI: 10.1016/j.ijhydene.2022.09.133 [Accessed: Feb. 28, 2023]

[99] Kabeyi MJB, Olanrewaju OA. Conversion from diesel to dual fuel power generation and implications on the transition. In: Presented at the Seventh North American International Conference on Industrial Engineering and Operations Management Orlando, Florida, USA June 12–14, 2022, Conference Paper. Michigan: IEOM Society International; 2022. p. 356 [Online]. Available from: https://ieomsociety.org/proceedings/2022orlando/356.pdf

[100] Kabeyi MJB, Olanrewaju OA. Conversion of diesel and petrol engines to biogas engines as an energy transition strategy. In: Presented at the Fourth African International Conference on Industrial Engineering and Operations Management, Nsukka, Nigeria, April 5–7, 2022. Michigan: IEOM Society International; 2022. p. 448 [Online]. Available from: https://ieomsociety.org/proceedings/2022nigeria/448.pdf

[101] Kukoyi TO, Muzenda E, Akinlabi ET, Mashamba A, Mbohwa C, Mahlatsi T. Biogas use as fuel in spark ignition engines. In: 2016 IEEE International Conference on Industrial Engineering and Engineering Management (IEEM), 4–7 Dec. 2016. Michigan: IEOM Society International; 2016. pp. 1064-1069. DOI: 10.1109/IEEM.2016.7798041

[102] Kabeyi MJB, Olanrewaju AO. Application of geothermal wellhead generators in sustainable power generation. Geothermal Resources Council Transactions. 2022;**46**(2022): 1692-1718. Art no. 1034703. Available from: https://www.geothermal-library.org/index.php?mode=pubs&action=view&record=1034703

[103] Kabeyi MJB, Oludolapo AO. Development of a biogas plant with electricity generation, heating and fertilizer recovery systems. In: Presented at the Second African International Conference on Industrial Engineering and Operations Management, IEOM Society International, 5–7 December 2020. Michigan: IEOM Society International; 2020. p. 082 [Online]. Available from: http://ieomsociety.org/harare2020/papers/82.pdf

[104] Kour D et al. Technologies for Biofuel Production: Current Development, Challenges, and Future Prospects. Cham: Springer; 2019

[105] Admin A. Green Hydrogen from Biogas. H2 International. Amsterdam, Netherlands: Elsevier; 2022. Available from: https://www.h2-international.com/2020/09/15/green-hydrogen-from-biogas/ [Accessed: Jul. 1, 2022]

[106] Karuppiah T, Azariah VE. Biomass Pretreatment for enhancement of biogas production. In: Anaerobic Digestion. London, UK: IntechOpen; 2019

[107] Saur G, Milbrandt A. Renewable hydrogen potential from biogas in the United States. In: NREL of the U.S. Department of Energy, Denver West Parkway, USA, July 2014. 2014 [Online]. Available from: https://www.nrel.gov/docs/fy14osti/60283.pdf

[108] NaturalGas.org. Natural gas and the environment. NaturalGas.org. Available from: http://naturalgas.org/environment/naturalgas/ [Accessed: 2021]

[109] Antonini C, Treyer K, Streb A, van der, Mijndert S, Bauer C, Mazzo M. Hydrogen production from natural gas and biomethane with carbon capture and storage – A techno-environmental analysis. Sustainable Energy & Fuels. 2020;**2020**(4):2967. DOI: 10.1039/D0SE00222D

[110] Fly A, Thring RH. System thermal and water balance in an evaporatively cooled PEM fuel cell vehicle. In: Vehicle Thermal Management Systems Conference Proceedings (VTMS11). Cambridge: Woodhead Publishing; 2013. pp. 267-277

[111] Boden TA, Marland G, Andres RJ. Global, Regional, and National Fossil-Fuel CO_2 Emissions. Oak Ridge, Tenn., U.S.A.: U.S. Department of Energy; 2017. [Online]. Available from: https://cdiac.ess-dive.lbl.gov/trends/emis/overview_2014.html [Accessed: Sep. 26, 2017]

[112] Lyng K-A, Brekke A. Environmental life cycle assessment of biogas as a fuel for transport compared with alternative fuels. Energies. 2019; **12**(3):532 [Online]. Available from: https://www.mdpi.com/1996-1073/12/3/532. [Accessed: Jul 30, 2020]

[113] Li Y. Converting Biogas to Transportation Fuels. Ohio, USA: Ohio State University; 2014 [Online].

Available from: https://ohioline.osu.edu/factsheet/AEX-653.2

[114] Ashraf MT, Schmidt JE, Bastidas Oyanedel JR. Conversion efficiency of biogas to liquids fuels through Fischer-Tropsch process. In: Presented at the 23rd European Biomass Conference and Exhibition, 2015. Cambridge, UK: Royal Society of Chemistry; 2015 [Online]. Available from: http://www.etaflorence.it/proceedings/?detail=11205

[115] Alves HJ, Bley Junior C, Niklevicz RR, Frigo EP, Frigo MS, Coimbra-Araújo CH. Overview of hydrogen production technologies from biogas and the applications in fuel cells. International Journal of Hydrogen Energy. 2013;**38**(13):5215-5225. DOI: 10.1016/j.ijhydene.2013.02.057 [Accessed: May 01, 2013]

[116] Kabeyi MJB, Olanrewaju OA. Sustainable energy transition for renewable and low carbon grid electricity generation and supply. Frontiers in Energy Research, Review. 2022;**9**(743114):1-45. DOI: 10.3389/fenrg.2021.743114 [Accessed: Mar. 24, 2022]

[117] National Energy Technology Laboratory. 10.2. Fischer-Tropsch synthesis. Pittsburg, US: National Energy Technology Laboratory; Available from: https://www.netl.doe.gov/research/coal/energy-systems/gasification/gasifipedia/ftsynthesis [Accessed: 2021]

[118] Dahlgren S. Biogas-based fuels as renewable energy in the transport sector: an overview of the potential of using CBG, LBG and other vehicle fuels produced from biogas. Biofuels, 2022;**13**(5):587-599. DOI: 10.1080/17597269.2020.1821571

[119] Stephan B. 19 - market development and certification schemes for biomethane. In: Wellinger A, Murphy J, Baxter D, editors. The Biogas Handbook. Cambridge: Woodhead Publishing; 2013. pp. 444-462

[120] Kabeyi MJB, Olanrewaju AO. Solar energy as a sustainable energy for power generation. In: Presented at the Sixth European Conference on Industrial Engineering and Operations Management, Lisbon, Portugal, July 18–20, 2023. Michigan: IEOM Society International; 2023, 244

[121] Kabeyi MJB, Olanrewaju OA. Performance analysis and evaluation of ethanol potential of Nzoia sugar company Ltd. Energy Reports. 2022;**8**(16):787-799. DOI: 10.1016/j.egyr.2022.11.006 [Accessed: Dec 01, 2022]

[122] Kabeyi MJB, Olanrewaju OA. Performance evaluation of Kipevu-III 120 MW power station and conversion to dual-fuel power plant. Energy Reports. 2022;**8**(16):800-814. DOI: 10.1016/j.egyr.2022.11.064 [Accessed: Dec. 01, 2022]

[123] Mitzlaff K. Engines for Biogas. Eschborn, Germany: Vieweg & Sohn Verlagsgesellschaft mbH; 1988. 3-528-02032-6 [Online]. Available from: https://biogas.ifas.ufl.edu/ad_development/documents/Engines%20for%20biogas.pdf [Accessed: Dec. 18, 2020]

[124] Kabeyi MJB, Olanrewaju OA. Performance analysis and electricity potential for Nzoia sugar factory. Energy Reports. 2022;**8**(16):755-764. DOI: 10.1016/j.egyr.2022.10.432 [Accessed: Dec. 01, 2022]

[125] Mungut M, Bore J, Kakinyi M. Housing characteristics and household population: Findings from the Kenya demographic and health survey 2014. Pan African Medical Journal. 2018;**30**:

603. DOI: 10.11604/pamj.2018.30.166.1477

[126] Kabeyi MJB, Olanrewaju OA. Electricity and gas potential of abattoir waste. In: Presented at the 12th Annual Istanbul International Conference on Industrial Engineering and Operations Management Istanbul, Turkey, March 7–10, 2022. Michigan: IEOM Society International; 2022. p. 403 [Online]. Available from: https://ieomsociety.org/proceedings/2022istanbul/403.pdf

[127] Mathur A, Verma L. Management and Utilization of Biogas Plant Slurry. Udaipur: Himanshu Publications; 1988 [Online]. Available from: https://www.econbiz.de/Record/management-and-utilization-of-biogas-plant-slurry-mathur/10000778141

[128] Li T, Tang D, Li Z, Du J, Zhou T, Jia Y. Development and test of a Stirling engine driven by waste gases for the micro-CHP system. Applied Thermal Engineering. 2012;**2012**:33-34. DOI: 10.1016/j.applthermaleng.2011.09.020

[129] Arshad M et al. Green electricity generation from biogas of cattle manure: An assessment of potential and feasibility in Pakistan. Frontiers in Energy Research, Original Research;**10**:1-10. DOI: 10.3389/fenrg.2022.911485 [Accessed: Aug. 25, 2022]

[130] Paolini V, Petracchini F, Segreto M, Tomassetti L, Naja N, Cecinato A. Environmental impact of biogas: A short review of current knowledge. Journal of Environmental Science and Health, Part A;**53**(10):899-906. DOI: 10.1080/10934529.2018.1459076 [Accessed: Aug. 24, 2018]

[131] Amato A, Tsigkou K, Becci A, Beolchini F, Ippolito NM, Ferella F. Life cycle assessment of biomethane vs. fossil methane production and supply. Energies;**16**(12). DOI: 10.3390/en16124555

[132] Kabeyi MJB, Olanrewaju AO. Environmental impact of energy resources. In: Presented at the International Conference on Industrial Engineering and Operations Management, Manila, Phillipines. Michigan: IEOM Society International; 2023. p. 612 [Online]. Available from: https://ieomsociety.org/proceedings/2023manila/612.pdf

[133] Kabeyi MJB, Olanrewaju AO. Conventional and exergetic life cycle assessment process and applications. In: Presented at the International Conference on Industrial Engineering and Operations Management, Manila, Phillipines, March 7–9, 2023. Michigan: IEOM Society International; 2023. p. 610 [Online]. Available from: https://ieomsociety.org/proceedings/2023manila/610.pdf

[134] Ola E. Environmental Technology Assessment of Natural Gas Compared to Biogas. In: Primoz P, editor. Natural Gas. IntechOpen; 2010. pp. 127-146. DOI: 10.5772/9837

[135] Liu X, Yan Z, Yue ZB. 3.10 - Biogas. In: Moo-Young M, editor. Comprehensive Biotechnology. 2nd ed. Burlington: Academic Press; 2011. pp. 99-114

[136] Zhang X, Witte J, Schildhauer T, Bauer C. Life cycle assessment of power-to-gas with biogas as the carbon source. Sustainable Energy & Fuels. 2020;**4**(3):1427-1436. DOI: 10.1039/C9SE00986H

[137] Kabeyi MJB, Olanrewaju OA. Performance analysis and modification of a slaughterhouse waste biogas plant for biogas and electricity generation. In:

Presented at the 11th Annual International Conference on Industrial Engineering and Operations Management Singapore, March 7–11, 2021. Michigan: IEOM Society International; 2021 [Online]. Available from: http://www.ieomsociety.org/singapore2021/papers/203.pdf

[138] Kabeyi MJB, Olanrewaju AO. Smart grid technologies, capabilities and applications in the sustainable energy transition: A review. International Journal of Sustainable Energy. 2023;**42** (1):685-758. DOI: 10.1080/14786451.2023.2222298

[139] Lubańska A, Kazak JK. The role of biogas production in circular economy approach from the perspective of locality. Energies;**16**(9). DOI: 10.3390/en16093801

[140] Fusi A, Bacenetti J, Fiala M, Azapagic A. Life cycle environmental impacts of electricity from biogas produced by anaerobic digestion. Frontiers in Bioengineering and Biotechnology, Original Research. 2016; **4**:1-17. DOI: 10.3389/fbioe.2016.00026 [Accessed: Mar. 11, 2016]

[141] Kabeyi MJB. Potential and challenges of bagasse cogeneration in the Kenyan sugar industry. International Journal of Creative Research Thoughts. 2022;**10**(4):379-526. DOI: 10.1729/Journal.30042. Art no. IJCRT_218740

[142] Kabeyi MJB. Investigating the challenges of bagasse cogeneration in the Kenyan sugar industry. International Journal of Engineering Sciences & Research Technology. 2020;**9**(5):7-64. DOI: 10.5281/zenodo.3828855 [Accessed: May, 2020]